308

Topics in Current Chemistry

Editorial Board:
K.N. Houk • C.A. Hunter • M.J. Krische • J.-M. Lehn
S.V. Ley • M. Olivucci • J. Thiem • M. Venturi • P. Vogel
C.-H. Wong • H. Wong • H. Yamamoto

Topics in Current Chemistry
Recently Published and Forthcoming Volumes

Fluorous Chemistry
Volume Editor: István T. Horváth
Vol. 308, 2012

Multiscale Molecular Methods in Applied Chemistry
Volume Editors: Barbara Kirchner, Jadran Vrabec
Vol. 307, 2012

Solid State NMR
Volume Editor: Jerry C. C. Chan
Vol. 306, 2012

Prion Proteins
Volume Editor: Jörg Tatzelt
Vol. 305, 2011

Microfluidics: Technologies and Applications
Volume Editor: Bingcheng Lin
Vol. 304, 2011

Photocatalysis
Volume Editor: Carlo Alberto Bignozzi
Vol. 303, 2011

Computational Mechanisms of Au and Pt Catalyzed Reactions
Volume Editors: Elena Soriano,
José Marco-Contelles
Vol. 302, 2011

Reactivity Tuning in Oligosaccharide Assembly
Volume Editors: Bert Fraser-Reid,
J. Cristóbal López
Vol. 301, 2011

Luminescence Applied in Sensor Science
Volume Editors: Luca Prodi, Marco Montalti, Nelsi Zaccheroni
Vol. 300, 2011

Chemistry of Opioids
Volume Editor: Hiroshi Nagase
Vol. 299, 2011

Electronic and Magnetic Properties of Chiral Molecules and Supramolecular Architectures
Volume Editors: Ron Naaman,
David N. Beratan, David H. Waldeck
Vol. 298, 2011

Natural Products via Enzymatic Reactions
Volume Editor: Jörn Piel
Vol. 297, 2010

Nucleic Acid Transfection
Volume Editors: Wolfgang Bielke,
Christoph Erbacher
Vol. 296, 2010

Carbohydrates in Sustainable Development II
Volume Editors: Amélia P. Rauter,
Pierre Vogel, Yves Queneau
Vol. 295, 2010

Carbohydrates in Sustainable Development I
Volume Editors: Amélia P. Rauter,
Pierre Vogel, Yves Queneau
Vol. 294, 2010

Functional Metal-Organic Frameworks: Gas Storage, Separation and Catalysis
Volume Editor: Martin Schröder
Vol. 293, 2010

C-H Activation
Volume Editors: Jin-Quan Yu, Zhangjie Shi
Vol. 292, 2010

Asymmetric Organocatalysis
Volume Editor: Benjamin List
Vol. 291, 2010

Ionic Liquids
Volume Editor: Barbara Kirchner
Vol. 290, 2010

Fluorous Chemistry

Volume Editor: István T. Horváth

With Contributions by

J.L. Aceña · C. Cai · M. Carreira · S. Catalán · M. Contel ·
R.H. Fish · S. Fustero · J.A. Gladysz · K. Hatanaka · D. He ·
I.T. Horváth · J. Hu · Y.-B. Huang · M. Jurisch · R.Y.-C. Kong ·
K.Y. Kwok · T.-B. Lai · P.K.S. Lam · E.I.H. Loi · H. Matsubara ·
L.T. Mika · B. Miriyala · M.B. Murphy · C. Ni · G. Pozzi · I. Ryu ·
J.-M. Vincent · S.G. Weber · W.-B. Yi · M.S. Yu · H. Zhang ·
W. Zhang · W. Zhang · X.-J. Zhang · X. Zhao

Editor
Prof. István T. Horváth
Department of Biology and Chemistry
City University of Hong Kong
Tat Chee Avenue
Kowloon
Hong Kong
istvan.t.horvath@cityu.edu.hk

ISSN 0340-1022　　　　　　e-ISSN 1436-5049
ISBN 978-3-642-25233-4　　e-ISBN 978-3-642-25234-1
DOI 10.1007/978-3-642-25234-1
Springer Heidelberg Dordrecht London New York

Library of Congress Control Number: 2011941773

© Springer-Verlag Berlin Heidelberg 2012
This work is subject to copyright. All rights are reserved, whether the whole or part of the material is concerned, specifically the rights of translation, reprinting, reuse of illustrations, recitation, broadcasting, reproduction on microfilm or in any other way, and storage in data banks. Duplication of this publication or parts thereof is permitted only under the provisions of the German Copyright Law of September 9, 1965, in its current version, and permission for use must always be obtained from Springer. Violations are liable to prosecution under the German Copyright Law.
The use of general descriptive names, registered names, trademarks, etc. in this publication does not imply, even in the absence of a specific statement, that such names are exempt from the relevant protective laws and regulations and therefore free for general use.

Printed on acid-free paper

Springer is part of Springer Science+Business Media (www.springer.com)

Volume Editor

Prof. István T. Horváth

Department of Biology and Chemistry
City University of Hong Kong
Tat Chee Avenue
Kowloon
Hong Kong
istvan.t.horvath@cityu.edu.hk

Editorial Board

Prof. Dr. Kendall N. Houk

University of California
Department of Chemistry and Biochemistry
405 Hilgard Avenue
Los Angeles, CA 90024-1589, USA
houk@chem.ucla.edu

Prof. Dr. Christopher A. Hunter

Department of Chemistry
University of Sheffield
Sheffield S3 7HF, United Kingdom
c.hunter@sheffield.ac.uk

Prof. Michael J. Krische

University of Texas at Austin
Chemistry & Biochemistry Department
1 University Station A5300
Austin TX, 78712-0165, USA
mkrische@mail.utexas.edu

Prof. Dr. Jean-Marie Lehn

ISIS
8, allée Gaspard Monge
BP 70028
67083 Strasbourg Cedex, France
lehn@isis.u-strasbg.fr

Prof. Dr. Steven V. Ley

University Chemical Laboratory
Lensfield Road
Cambridge CB2 1EW
Great Britain
Svl1000@cus.cam.ac.uk

Prof. Dr. Massimo Olivucci

Università di Siena
Dipartimento di Chimica
Via A De Gasperi 2
53100 Siena, Italy
olivucci@unisi.it

Prof. Dr. Joachim Thiem

Institut für Organische Chemie
Universität Hamburg
Martin-Luther-King-Platz 6
20146 Hamburg, Germany
thiem@chemie.uni-hamburg.de

Prof. Dr. Margherita Venturi

Dipartimento di Chimica
Università di Bologna
via Selmi 2
40126 Bologna, Italy
margherita.venturi@unibo.it

Prof. Dr. Pierre Vogel

Laboratory of Glycochemistry
and Asymmetric Synthesis
EPFL – Ecole polytechnique féderale
de Lausanne
EPFL SB ISIC LGSA
BCH 5307 (Bat.BCH)
1015 Lausanne, Switzerland
pierre.vogel@epfl.ch

Prof. Dr. Chi-Huey Wong

Professor of Chemistry, Scripps Research Institute
President of Academia Sinica
Academia Sinica
128 Academia Road
Section 2, Nankang
Taipei 115
Taiwan
chwong@gate.sinica.edu.tw

Prof. Dr. Henry Wong

The Chinese University of Hong Kong
University Science Centre
Department of Chemistry
Shatin, New Territories
hncwong@cuhk.edu.hk

Prof. Dr. Hisashi Yamamoto

Arthur Holly Compton Distinguished Professor
Department of Chemistry
The University of Chicago
5735 South Ellis Avenue
Chicago, IL 60637
773-702-5059
USA
yamamoto@uchicago.edu

Topics in Current Chemistry
Also Available Electronically

Topics in Current Chemistry is included in Springer's eBook package *Chemistry and Materials Science*. If a library does not opt for the whole package the book series may be bought on a subscription basis. Also, all back volumes are available electronically.

For all customers with a print standing order we offer free access to the electronic volumes of the series published in the current year.

If you do not have access, you can still view the table of contents of each volume and the abstract of each article by going to the SpringerLink homepage, clicking on "Chemistry and Materials Science," under Subject Collection, then "Book Series," under Content Type and finally by selecting *Topics in Current Chemistry*.

You will find information about the

– Editorial Board
– Aims and Scope
– Instructions for Authors
– Sample Contribution

at springer.com using the search function by typing in *Topics in Current Chemistry*.

Color figures are published in full color in the electronic version on SpringerLink.

Aims and Scope

The series *Topics in Current Chemistry* presents critical reviews of the present and future trends in modern chemical research. The scope includes all areas of chemical science, including the interfaces with related disciplines such as biology, medicine, and materials science.

The objective of each thematic volume is to give the non-specialist reader, whether at the university or in industry, a comprehensive overview of an area where new insights of interest to a larger scientific audience are emerging.

Thus each review within the volume critically surveys one aspect of that topic and places it within the context of the volume as a whole. The most significant developments of the last 5–10 years are presented, using selected examples to illustrate the principles discussed. A description of the laboratory procedures involved is often useful to the reader. The coverage is not exhaustive in data, but rather conceptual, concentrating on the methodological thinking that will allow the non-specialist reader to understand the information presented.

Discussion of possible future research directions in the area is welcome.

Review articles for the individual volumes are invited by the volume editors.

In references *Topics in Current Chemistry* is abbreviated *Top Curr Chem* and is cited as a journal.

Impact Factor 2010: 2.067; Section "Chemistry, Multidisciplinary": Rank 44 of 144

Preface

The term fluorous was introduced, as the analogue of the term aqueous, to emphasize the fact that a chemical transformation is primarily controlled by a reagent or a catalyst designed to dissolve preferentially in the fluorous phase in 1994 [1]. The strikingly similar appearance of the oil-vinegar and the methanol-perfluoromethylcyclohexane biphasic systems is obvious, though the visualization and use of fluorous systems required the synthesis of a fluorous soluble dye [2], such as a perfluoroalkylated iron phathalocyanine, or reagents or catalysts [1].

The fluorous phase was defined as the fluorocarbon (mostly perfluorinated alkanes, dialkyl ethers and trialkyl amines) rich phase of a biphasic system. It was also emphasized that perfluoroaryl groups do offer dipole-dipole interactions, making them less compatible with the fluorous biphasic concept than perfluoroalkyl groups or fluorous ponytails. The temperature dependent phase behavior of the fluorous biphasic system was not the first, but its use to control reactivity in a single liquid phase was probably the first thermoregulated homogeneous catalytic system providing reaction in one phase at higher temperature and separation of the product from the fluorous catalyst at low temperature [1].

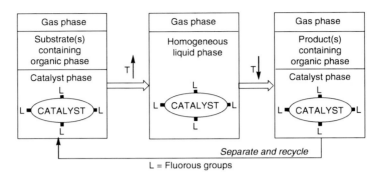

L = Fluorous groups

While the original definition was useful at the birth of fluorous chemistry to catch the imagination of the scientific and engineering communities, its meaning and scope have significantly changed due to novel discoveries and applications. The fluorous liquid-liquid biphasic concept was soon expanded to fluorous solid phase extraction [3] and fluorous chromatography using fluorous silica for the separation of molecules with fluorous tags [4].

The temperature regulated solubility of the fluorous compounds themselves [5, 6] has resulted in another paradigm shift by bringing the solid fluorous reagents or catalysts and the reactants into a single phase at higher temperature and offering facile separation of the product(s) at lower temperature.

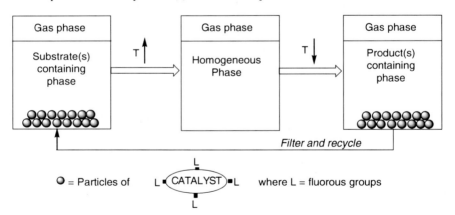

The most recent advance was the introduction of the fluorous release and catch concept [7]. A fluorous catalyst, which has limited or no solubility in the reaction mixture at room temperature and entrapped in a Teflon tape, is released to the reaction mixture at higher temperature, where it acts as a homogeneous catalyst. When the reaction is completed, the reaction mixture is cooled back to room temperature during which the fluorous catalyst returns to the Teflon tape.

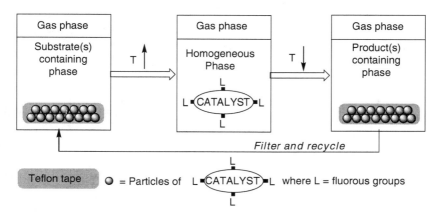

The publication of the Handbook of Fluorous Chemistry in 2004 [8] was followed by the first International Symposium of Fluorous Technologies (ISoFT'05) in Bordeaux, France in 2005. After two additional meetings, ISoFT'07 in Yokohama-Kamakura, Japan in 2007 and ISoFT'09 (as part of the 19th International Symposium on Fluorine Chemistry), in Jackson Hole, Wyoming, USA in 2009, ISoFT'11 will be held in Hong Kong in 2011. The current volume of Topics in Current Chemistry on Fluorous Chemistry is dedicated to ISoFT'11 and contains a broad range of articles addressing the synthesis, characterization, and applications of fluorous compounds in chemistry, material science, and biology.

November 2011

István T. Horváth
Department of Biology and Chemistry
City University of Hong Kong

References

1. Horváth IT, Rábai J (1994) Facile catalyst separation without water. Fluorous biphase hydroformylation of olefines. Science 266:72–75
2. British Patent 840,725 (1960) to Minnesota Mining and Manufacturing Company, Process of Perfluoroalkylating Aromatic Compounds
3. Curran DP, Hadida S, He M (1997) Thermal allylations of aldehydes with a fluorous allylstannane. Separation of organic and fluorous products by solid phase extraction with fluorous reverse phase silica gel. J Org Chem 275:6714–6715
4. Curran DP, Luo ZY (1999) Fluorous synthesis with fewer fluorines (light fluorous synthesis): separation of tagged from untagged products by solid-phase extraction with fluorous reverse-phase silica gel. J Am Chem Soc 121:9069–9072
5. Wende M, Meier R, Gladysz JA (2001) Fluorous catalysis without fluorous solvents: A friendlier catalyst recovery/recycling protocol based upon thermomorphic properties and liquid/solid phase separation. J Am Chem Soc 123:11490–11491

6. Ishihara K, Kondo S, Yamamoto H (2001) 3,5-Bis(perfluorodecyl)phenylboronic acid as an easily recyclable direct amide condensation catalyst. Synlett 1371–74
7. Dinh LV, Gladysz JA (2005) "Catalyst-on-a-Tape" Teflon: A New Delivery and Recovery Method for Homogeneous Fluorous Catalysts. Angew Chem Int Ed 44:4095–4097
8. Gladysz JA, Curran DP, Horváth IT, (2004) Handbook of Fluorous Chemistry. Wiley-VCH: Weinheim

Contents

Structural, Physical, and Chemical Properties of Fluorous Compounds 1
John A. Gladysz and Markus Jurisch

Selective Fluoroalkylation of Organic Compounds by Tackling the "Negative Fluorine Effect" 25
Wei Zhang, Chuanfa Ni, and Jinbo Hu

Synthetic and Biological Applications of Fluorous Reagents as Phase Tags 45
Santos Fustero, José Luis Aceña, and Silvia Catalán

Chemical Applications of Fluorous Reagents and Scavengers 69
Marvin S. Yu

Fluorous Methods for the Synthesis of Peptides and Oligonucleotides 105
Bruhaspathy Miriyala

Fluorous Organic Hybrid Solvents for Non-Fluorous Organic Synthesis 135
Hiroshi Matsubara and Ilhyong Ryu

Fluorous Catalysis: From the Origin to Recent Advances 153
Jean-Marc Vincent

Fluorous Organocatalysis 175
Wei Zhang

Thiourea Based Fluorous Organocatalyst 191
Yi-Bo Huang, Wen-Bin Yi, and Chun Cai

Fluoroponytailed Crown Ethers and Quaternary Ammonium Salts as Solid–Liquid Phase Transfer Catalysts in Organic Synthesis 213
Gianluca Pozzi and Richard H. Fish

Fluorous Hydrogenation 233
Xi Zhao, Dongmei He, László T. Mika, and István T. Horváth

Fluorous Hydrosilylation 247
Monica Carreira and Maria Contel

Fluorous Hydroformylation 275
Xi Zhao, Dongmei He, László T. Mika, and István T. Horváth

Incorporation of Fluorous Glycosides to Cell Membrane and Saccharide Chain Elongation by Cellular Enzymes 291
Kenichi Hatanaka

Teflon AF Materials 307
Hong Zhang and Stephen G. Weber

Ecotoxicology of Organofluorous Compounds 339
Margaret B. Murphy, Eva I.H. Loi, Karen Y. Kwok, and Paul K.S. Lam

Biology of Fluoro-Organic Compounds 365
Xiao-Jian Zhang, Ting-Bong Lai, and Richard Yuen-Chong Kong

Index 405

Structural, Physical, and Chemical Properties of Fluorous Compounds

John A. Gladysz and Markus Jurisch

Abstract The sizes and structures of fluorous molecules are analyzed, particularly with respect to the helical conformations of perfluoroalkyl segments and their phase separation in crystal lattices. Basic molecular properties, bond energies, and special bonding motifs are reviewed. Solubility, adsorption, and related phenomena are treated. Miscibilities of fluorous solvents, and partition coefficients of solutes in fluorous/organic biphase mixtures, are analyzed. Electronic effects and NMR properties are discussed, and some reactions involving the fluorinated parts of fluorous substances are presented.

Keywords Bond energies · Conformations · Electronic effects · Fluorous · Miscibilities · NMR · Partition coefficient · Reactivity · Solubilities

Contents

1 Introduction ... 2
2 Structural Properties of Fluorous Compounds ... 3
3 Physical Properties of Fluorous Compounds ... 6
   3.1 Basic Molecular Properties ... 6
   3.2 Bond Energies and Special Bonding Motifs ... 6
   3.3 Solubility, Adsorption, and Related Phenomena ... 7
   3.4 Miscibilities of Fluorous Solvents ... 9
   3.5 Partition Coefficients ... 10
   3.6 Electronic Effects ... 15
   3.7 NMR Properties ... 16
4 Chemical Properties of Fluorous Compounds ... 16
References ... 19

J.A. Gladysz (✉) and M. Jurisch
Department of Chemistry, Texas A&M University, P.O. Box 30012, College Station, TX 77842-3012, USA
e-mail: gladysz@mail.chem.tamu.edu

1 Introduction

The objective of this chapter is to summarize the most salient aspects of the structural, physical, and chemical properties of fluorous compounds from the perspective of a practitioner of fluorous chemistry. Although an attempt has been made to cite recent publications, some of the topics covered are grounded in an older literature, and here readers are referred to authoritative treatises that provide additional details.

It is assumed that the reader is familiar with the central concepts and definitions associated with fluorous chemistry [1]. Towards this end, it should be kept in mind that there are three types of orthogonal liquid phases – fluorous, organic, and aqueous – as diagrammed in Fig. 1. These also have solid phase counterparts. A few representative fluorous solvents are also depicted in Fig. 1. These include two hydrofluoro ethers, which may offer environmental advantages [2–5]. It is also important to emphasize that perfluoroarenes and similar unsaturated compounds are *not* fluorous. They exhibit significantly greater polarities and polarizabilities.

Many fluorous molecules are comprised of nonfluorous and fluorous domains. In these cases, the fluorous domain can be viewed as a phase label, which is often a "ponytail" [6]. The most common and extensively investigated ponytails have the formula $(CH_2)_m(CF_2)_{n-1}CF_3$, which will be abbreviated $(CH_2)_mR_{fn}$. Their properties will be extensively treated in this chapter. However, there is currently intense interest in the development of ponytails that are functionalized and/or based upon smaller perfluorinated units, in part to promote biodegradability [7–12]. These share many of the properties of the $(CH_2)_mR_{fn}$ ponytails described below, but can be somewhat less fluorophilic.

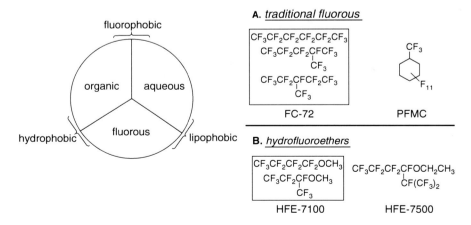

Fig. 1 Three orthogonal liquid phases (*left*) and representative fluorous solvents (*right*)

2 Structural Properties of Fluorous Compounds

Historically, fluorous chemistry has been centered around perfluoroalkyl groups or segments, for which there is a rich structural literature. Two articles, both of which include J. D. Dunitz as an author, do a particularly thorough job of framing the issues most of interest to fluorous chemists [13, 14].

First, consider the obvious issue of size. By the three most widely employed sets of van der Waals radii – Pauling, Bondi [15], and Williams and Houpt [16] – that of fluorine is considerably larger than hydrogen (1.35 vs 1.20 Å; 1.47 vs 1.20 Å; 1.44 vs 1.15 Å). Accordingly, fluorination always increases the sizes of alkyl groups, and the molecular volumes of fluorocarbons are greater than those of the corresponding hydrocarbons. Sizes can be parameterized in a number of ways and, according to one scale, that of a trifluoromethyl group is most comparable to that of an isopropyl group [17, 18].

Despite these differences, it is sometimes claimed in studies involving fluorinated molecules that the steric effect of fluorine or a fluorine/hydrogen substitution is small. As emphasized by Smart [18], this is not necessarily inconsistent. It can well be that the fluorinated molecule is accommodated equally well in a biological receptor, or that a transition state is essentially unaffected. A steric effect is a function of a specific chemical or physical process.

There is only one crystal structure of an n-perfluoroalkane [13, 14, 19], namely n-perfluorohexane. It exhibits, in contrast to n-alkanes, a helically twisted chiral backbone with CF_2–CF_2–CF_2–CF_2 torsion angles of 163–165° as opposed to 180° (anti; the energy minimum of n-butane). Computational studies predict analogous conformations, which relieve certain electrostatically repulsive interactions while at the same time introducing new attractive interactions [20–22]. It has been noted that, for a perfluoroalkane with a planar all-anti carbon backbone (CF_2–CF_2–CF_2–CF_2 torsion angles of 180° and C–C–C bond angles of 110°), the fluorine atoms would be ca. 2.52 Å apart, or somewhat less than the sum of the van der Waals radii [13, 14, 23]. With the experimentally observed helical backbone and 116° C–C–C bond angles, the fluorine/fluorine separations increase to ca. 2.75 Å.

Unlike some areas of synthetic chemistry, crystals of fluorous molecules suitable for diffraction studies are only obtained in a small fraction of cases. The standing policy in the authors' laboratory has been to determine an X-ray crystal structure at every opportunity. A high quality crystal structure of a relatively simple derivative of n-perfluorohexane, the branched carboxylic acid $CF_3(CF_2)_5CH(CH_3)CO_2H$, has recently been reported [24]. The features closely correspond with those noted above. A high quality crystal structure of the more complicated cationic ruthenium fluorous phosphine complex **1** shown in Fig. 2 lent itself to detailed analysis [25]. The average CF_2–CF_2–CF_2–CF_2 torsion angle was 167.2°, with the F–CF_2–CF_2–F torsion angles generally falling into two regimes, ca. 40–50° and ca. 70–80° (see **I**). Importantly, vibrational circular dichroism (VCD) studies have established that the helical conformations of n-perfluoroalkyl segments persist in solution [26].

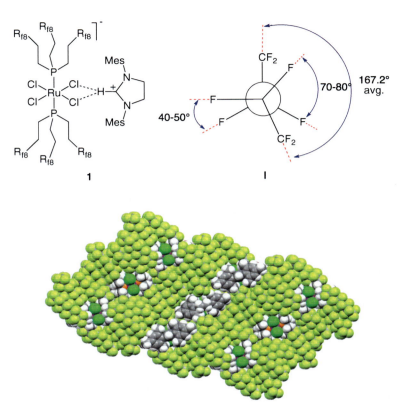

Fig. 2 Structural data for the fluorous ruthenium salt **1** (*upper left*): packing diagram with fluorine atoms in *yellow-green*, chlorine atoms in *forest green*, and nitrogen atoms in *orange* (*bottom*); average torsion angles about CF_2–CF_2–CF_2–CF_2 linkages (*upper right*, **I**)

Dunitz has also compared the packing motif of *n*-perfluorohexane vis-à-vis those of hydrocarbons [13, 14]. With molecules that have fluorous and nonfluorous subunits, one commonly observes a "phase separation" in the crystal lattice. This is vividly illustrated in Fig. 2 (bottom) for the ruthenium salt **1** (yellow green atoms = fluorous domain). The onset of domain formation has been systematically mapped in a series of gold phosphine complexes by increasing the size of the fluorinated substituent [27]. It should also be noted that substantial numbers of liquid crystals with fluorous substituents have been characterized [28, 29].

The helical chiralities of $(CF_2)_n$ segments can also manifest themselves in supramolecular phenomena. Ternary mixtures of suitably functionalized calixarenes, alkali or alkaline metal iodides, and α,ω-diiodoperfluoroalkanes can yield complex assemblies that feature infinite halogen-bonded chains, $[I^- \cdots I (CF_2)_n I]_{n'} \cdots$ (see also below) [30, 31]. In the case of the barium complex shown in Fig. 3, the complex crystallizes in a chiral space group with all of the $(CF_2)_8$ segments of *identical* chirality and entwined in double helices. Interestingly, in

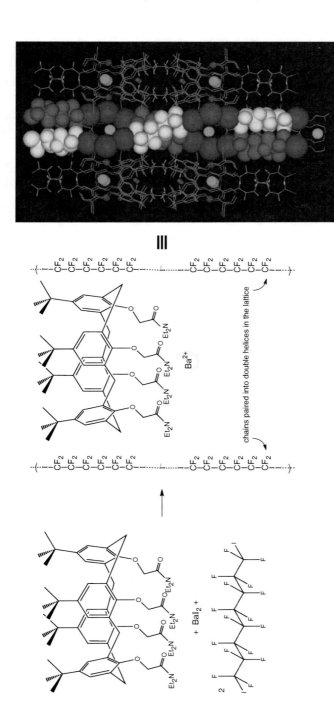

Fig. 3 A supramolecular assembly derived from a calixarene, BaI_2, and $I(CF_2)_8I$ (1:1:2) illustrating homochiral helical $(CF_2)_8$ segments and halogen bonding

contrast to the systems analyzed above, the $(CF_2)_8$ segments feature two gauche torsion angles (62(2)°). In any case, chemists have synthesized many compounds with stunning helical structures, but those of the *n*-perfluoroalkanes can be regarded as the most basic.

3 Physical Properties of Fluorous Compounds

3.1 Basic Molecular Properties

It has often been noted that the enthalpies of vaporization, boiling points, and molecular polarizabilities of *n*-perfluoroalkanes are very close to those of the corresponding *n*-alkanes [13, 14]. In view of the much larger atomic number and weight of fluorine, these similarities may seem surprising. However, the additional electrons are tightly bound by the electronegative fluorine nucleus, with the result that the atomic polarizability is close to that of hydrogen [13, 32]. Nonetheless, consistent with the size effects noted above, the molecular volumes of perfluoroalkanes are significantly larger. The densities are also much greater, which is an important factor in fluorous liquid/liquid biphase chemistry.

3.2 Bond Energies and Special Bonding Motifs

Fluorine forms the strongest single bonds of any element to boron, carbon, silicon, and hydrogen. With carbon, the bond strengths increase with the degree of fluorination, with D°(C–F) (kcal/mol) = 108.3 for CH_3F, 119.5 for CH_2F_2, 127.5 for $CHCF_3$, and 130.5 for CF_4 [33]. Although the carbon hydrogen bond strengths in this series are less affected (D°(C–H) (kcal/mol) = 104.3 in CH_4 vs 106.7 in $CHCF_3$), these trends are clearly reflected in the enhanced robustness of highly fluorinated and perfluorinated alkanes. Many other bond strength trends involving fluorine substituents have been analyzed [33].

Unsurprisingly in view of the above data, R_{fn} groups do not show any tendency to engage in hydrogen bonding. Suitably functionalized fluorous molecules can form hydrogen bonds, but the "ground rules" are more or less the same as for nonfluorous molecules [34, 35].

In contrast, perfluoroalkyl iodides and related species are ideally suited to engage in halogen bonding, as exemplified in Fig. 3. These interactions involve a Lewis base donor and a molecule with a halogen atom with an appreciable partial positive charge [36, 37]. Obviously, a highly electron-withdrawing perfluoroalkyl group renders the halogen atom a potent acceptor. This phenomenon has recently been used as a design element in recoverable fluorous catalysts [38].

3.3 Solubility, Adsorption, and Related Phenomena

The solubilities of both fluorous and nonfluorous solutes in fluorous solvents are of interest. These are largely determined by two parameters: solute polarity and size. The first represents the familiar "like dissolves like" paradigm. The second is unique to perfluorinated solvents, which because of low intermolecular forces can form substantial cavities (free volumes) that can accommodate small molecules. In this context, there is much literature involving gas solubilities in fluorocarbons. These data correlate with the isothermal compressibility of the solvent [39], which supports the cavity-based solubility model.

With regard to gas solubilities, there are some frequent misconceptions that stem from the well known use of certain highly fluorinated fluids as blood substitutes. It is true that oxygen is quite soluble in perfluoroalkanes, especially relative to water, where a strong hydrogen bonding network must be disrupted. However, data are often presented as mole fractions in the literature, whereas most molecular chemists would compare molarities.

For example, the solubility of oxygen in perfluoro(methylcyclohexane) or PFMC is about five times greater than that in THF on a mole fraction basis [40]. In converting to molar units, the much higher molecular weights of perfluoroalkanes serve to diminish differences, whereas their higher densities amplify them. The net result is that is that oxygen is only about twice as soluble in PFMC as in THF on a molar basis. The result with hydrogen is similar. Hence, it is accurate to state that gas solubilities are somewhat higher in perfluoroalkanes than common organic solvents. However, the differences are more modest than often thought.

Unfortunately, quantitative solubility data for the types of solutes that would be of greatest interest to fluorous chemists are scarce [41–43]. However, some generalizations are possible. First, it is usually possible to fine-tune the solubility of a fluorous solute by varying the lengths and numbers of the R_{fn} segments. Compounds with R_{f6} segments commonly display good solubilities in fluorous media. However, analogs with R_{f10} segments are distinctly less soluble, and in some cases insoluble. R_{f8} segments are intermediate, although in the authors' experience they can be relied upon to be "soluble enough" for most purposes.

One way to conceptualize these trends is to view the ponytails as short pieces of Teflon®, which does not dissolve in any common fluorous or nonfluorous solvent. As the ponytails become longer, many physical properties of the molecule approach those of the fluoropolymer.

In a related vein, numerous researchers have now observed that fluorous solutes often exhibit highly temperature dependent solubilities. Of course, this phenomenon is not restricted to fluorous molecules. However, it is possible to fine tune the perfluoroalkyl segments such that the solute has essentially no solubility in a fluorous or *organic* solvent at room temperature (i.e., is engineered to be below a certain limit or tolerance) but appreciable solubility at 60–120 °C. This phenomenon can be used to conduct homogeneous reactions at elevated temperatures, with catalyst or reagent recovery by solid/liquid phase separation at lower temperatures [44–46]. This topic has been reviewed [47–49], and some of the many catalysts thus employed include the ketone $(R_{f8})_2C=O$, the stannane $(R_{f10}(CH_2)_2)_3SnH$, the phosphines $(R_{f8}(CH_2)_m)_3P$ ($m = 2, 3$), the phosphonium salt $(R_{f8}(CH_2)_2)_3(R_{f6}(CH_2)_2)P]^+\ I^-$, the boronic acid $(3,5\text{-}C_6H_3(R_{f10})_2)B(OH)_2$, the Brønsted acid $(4\text{-}R_{f10}CH_2OC_6H_4)CH(SO_2CF_3)_2$, the Lewis acid $Yb(N(SO_2R_{f8})_2)_3$, fluorous IBX oxidants, and the rhodium complexes $((R_{fn}(CH_2)_2)_3P)_3RhCl$ ($n = 6, 8$).

Fluorous molecules normally show good solubilities in supercritical CO_2 [50–52]. Only modest fluorine content is normally required, and hence many lightly fluorinated systems have been employed as catalysts. Furthermore, CO_2 pressure can also increase the solubilities of fluorous solutes in organic solvents [53]. This nonthermal "solubility switch" can be exploited as a means of catalyst recovery by liquid/solid phase separation [54].

Fluorous molecules can be adsorbed onto a variety of fluorous supports, such as fluorous silica gel and fluoropolymers, including Teflon® and Gore-Tex®, as illustrated by their use in various catalyst recovery protocols [47, 55]. It is important to emphasize that this does not imply a significant enthalpic attraction, although a very small amount would be expected. Rather, these phenomena reflect more that the fluorous solute has "nowhere else to go" – i.e., dispersal elsewhere in the system would come at the expense of more favorable interactions between more polar species. Recently, the permeability of Teflon® tape to certain nonfluorous solutes has been used to effect the controlled delivery of certain reagents [56]. Physical studies of solute transport through Teflon® films have also been reported [57].

Finally, it should be noted that n-perfluoroalkanes can be scavenged from mesitylene solutions into cylindrical guest molecules that have been developed by Rebek [58]. As sketched in Fig. 4, the greatest association constants are found

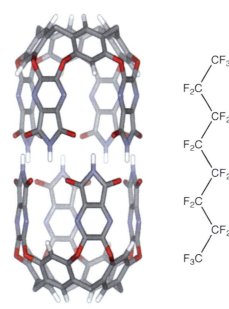

Fig. 4 A container molecule that binds n-perfluorooctane and n-perfluorononane

for the chain lengths that are best accommodated within the container. The driving force is mainly connected to the filling of space, as opposed to any fluorophilic interactions.

3.4 Miscibilities of Fluorous Solvents

Although fluorous and organic solvents are regarded as orthogonal, they frequently become miscible at elevated temperature, a process that is favored entropically. This is exploited in many protocols for fluorous/organic liquid/liquid biphase catalysis [59]. With binary solvent systems, it is customary to specify a "consolute" or "upper critical solution" temperature [40], above which phase separation cannot occur, whatever the composition. However, plots as a function of mole or volume fraction are more informative, as exemplified for toluene and the fluorous ionic liquid 2 in Fig. 5 [60].

Miscibilities can also be strongly affected by solutes or dissolved species. It is well known that homogeneous mixtures of aqueous and certain organic solvents can often be induced to phase separate or "salt out" by adding a suitable material, and fluorous biphase systems can behave similarly.

Another common misconception regarding liquid/liquid biphase systems involves the composition of each layer. Just because two phases do not mix does not mean that

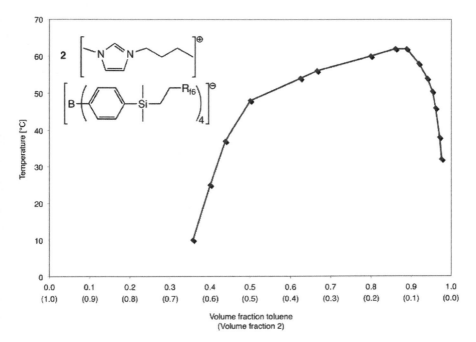

Fig. 5 Temperatures at which the fluorous ionic liquid 2 and toluene become miscible

each phase consists of a single species. For example, the ether phase of an ether/water biphase mixture contains considerable water, which is the reason that, after phase separation, it is common to dry the ether layer over Na_2SO_4 or another agent. In the case of a 50:50 v/v toluene/PFMC mixture at 25 °C, the authors' coworkers have measured ratios of 98.4:1.6 (molar), 94.2:5.8 (mass), and 97.1:2.9 (volume) in the upper organic layer, and 3.8:96.2, 1.0:99.0, and 2.0:98.0 in the lower fluorous layer [61]. Thus, some leaching of the fluorous solvent into the nonfluorous solvent (and vice versa) occurs under the conditions of fluorous/organic biphase catalysis.

In parallel to the effect on fluorous solute solubility described above, CO_2 pressure can function as a "miscibility switch" for fluorous and organic solvents [62]. For some applications, this may have advantages over temperature, such as with thermally labile substrates or catalysts. The pressures necessary to mix 1:1 volumes of perfluorohexane and organic solvents at room temperature vary from 16.3–19.4 bar for ethyl acetate, THF, and chloroform to 44.4–45.6 bar for the strongly associating solvents DMF, nitromethane, ethanol, and methanol. Acetic and propionic acid, which form dimers in solution, have lower miscibility pressures (27.5 bar).

3.5 Partition Coefficients

Partition coefficients quantify the equilibrium distribution of a solute between two immiscible phases, which are most often but not necessarily liquids. They see extensive use throughout chemistry, and their thermodynamic nuances have been analyzed in detail [63]. In order to rationally separate fluorous and nonfluorous substances from fluorous/nonfluorous liquid/liquid biphase systems, or design and optimize fluorous catalysts and reagents, libraries of partition coefficients are necessary. Partition coefficients constitute a direct measure of fluorophilicity, a term that is used interchangeably with fluorous phase affinity.

Some investigators prefer to express partition coefficients as ratios that have been normalized to 100 (e.g., 98.3:1.7), others as ratios with either the less populated phase or the nonfluorous phase set to 1 (e.g., 57.8:1), and still others as logarithmic values. The abbreviation P indicates a concentration ratio with the nonfluorous phase in the denominator. The natural logarithm of the PFMC/toluene ($CF_3C_6F_{11}/CH_3C_6H_5$) concentration ratio, $\ln\{[c(\text{PFMC})]/[c(\text{toluene})]\}$, has been given the abbreviation f, for fluorophilicity [64]. Hundreds of partition coefficients were tabulated in 2003 [64] and several trends are illustrated by the data for PFMC/ toluene mixtures in Table 1.

The n-alkanes, despite being very nonpolar, show high affinities for the toluene phase over PFMC (entries 1–6). The partition coefficients increase monotonically with alkane size (5.4:94.6 for decane to 1.1:98.9 for hexadecane). This is in accord with the size affect discussed above. The n-alkenes (entries 7–12) have slightly higher toluene phase affinities, consistent with their slightly greater polarities (4.8:95.2 for 1-decene to 0.9:99.1 for 1-hexadecene). When the side-chain of

Table 1 Selected PFMC/toluene (CF$_3$C$_6$F$_{11}$/CH$_3$C$_6$H$_5$) partition coefficients (room temperature) [64]

Entry	Solute	Partitioning fluorous:organic (P)
1	CH$_3$(CH$_2$)$_8$CH$_3$	5.4:94.6 ($P = 0.057$)
2	CH$_3$(CH$_2$)$_9$CH$_3$	4.2:95.8 ($P = 0.044$)
3	CH$_3$(CH$_2$)$_{10}$CH$_3$	3.4:96.6 ($P = 0.035$)
4	CH$_3$(CH$_2$)$_{11}$CH$_3$	2.4:97.6 ($P = 0.025$)
5	CH$_3$(CH$_2$)$_{12}$CH$_3$	1.9:98.1 ($P = 0.019$)
6	CH$_3$(CH$_2$)$_{14}$CH$_3$	1.1:98.9 ($P = 0.011$)
7	CH$_3$(CH$_2$)$_7$CH=CH$_2$	4.8:95.2 ($P = 0.050$)
8	CH$_3$(CH$_2$)$_8$CH=CH$_2$	3.7:96.3 ($P = 0.038$)
9	CH$_3$(CH$_2$)$_9$CH=CH$_2$	2.5:97.5 ($P = 0.026$)
10	CH$_3$(CH$_2$)$_{10}$CH=CH$_2$	1.9:98.1 ($P = 0.019$)
11	CH$_3$(CH$_2$)$_{11}$CH=CH$_2$	1.6:98.4 ($P = 0.016$)
12	CH$_3$(CH$_2$)$_{13}$CH=CH$_2$	0.9:99.1 ($P = 0.009$)
13	R$_{f8}$CH=CH$_2$	93.5:6.5 ($P = 14.4$)
14	Cyclohexanone	2.2:97.8 ($P = 0.022$)
15	Cyclohexanol	1.6:98.4 ($P = 0.016$)
16	Cyclohexyl–OSi(CH$_3$)$_2$C$_6$H$_5$	0.8:99.2 ($P = 0.008$)
17	CF$_3$CH$_2$OH	14.5:85.5 ($P = 0.170$)
18	(CF$_3$)$_2$CHOH	26.7:73.3 ($P = 0.364$)
19	R$_{f6}$(CH$_2$)$_2$OH	52:48 ($P = 1.1$)
20	R$_{f6}$(CH$_2$)$_3$OH	44:56 ($P = 0.79$)
21	R$_{f8}$(CH$_2$)$_2$OH	73.5:26.5 ($P = 2.77$)
22	R$_{f8}$(CH$_2$)$_3$OH	64:36 ($P = 1.8$)
23	R$_{f10}$(CH$_2$)$_3$OH	80.5:19.5 ($P = 4.14$)
24	R$_{f8}$(CH$_2$)$_3$NH$_2$	70.0:30.0 ($P = 2.33$)
25	R$_{f8}$(CH$_2$)$_4$NH$_2$	63.2:36.8 ($P = 1.72$)

(continued)

Table 1 (continued)

Entry	Solute	Partitioning fluorous:organic (P)
26	$R_{f8}(CH_2)_5NH_2$	56.9:43.1 ($P = 1.32$)
27	$R_{f7}CH_2NH(CH_3)$	74.5:25.5 ($P = 2.92$)
28	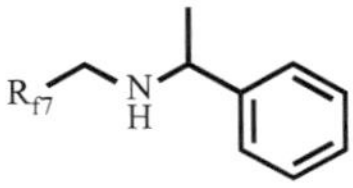 R_{f7}–CH$_2$–NH–CH(CH$_3$)–C$_6$H$_5$	30:70 ($P = 0.42$)
29	$R_{f8}(CH_2)_3NH(CH_3)$	71:29 ($P = 2.4$)
30	$[R_{f8}(CH_2)_3]_2NH$	96.5:3.5 ($P = 27.6$)
31	$[R_{f8}(CH_2)_4]_2NH$	95.1:4.9 ($P = 19.4$)
32	$[R_{f8}(CH_2)_5]_2NH$	93.0:7.0 ($P = 13.3$)
33	$R_{f7}CH_2N(CH_3)_2$	82.2:17.8 ($P = 4.62$)
34	$R_{f8}(CH_2)_3N(CH_3)_2$	79.8:20.2 ($P = 3.94$)
35	$[R_{f8}(CH_2)_3]_2N(CH_3)$	97.4:2.6 ($P = 37.7$)
36	$[R_{f8}(CH_2)_3]_3N$	>99.7:<0.3 ($P > 332$)
37	$[R_{f8}(CH_2)_4]_3N$	>99.7:<0.3 ($P > 332$)
38	$[R_{f8}(CH_2)_5]_3N$	99.5:0.5 ($P = 199$)
39	C_6H_6	6:94 ($P = 0.063$)
40	C_6HF_5	22.4:77.6 ($P = 0.289$)
41	C_6F_6	28.0:72.0 ($P = 0.389$)
42	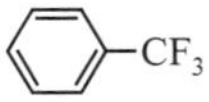 C$_6$H$_5$–CF$_3$	12.4:87.6 ($P = 0.142$)
43	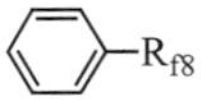 C$_6$H$_5$–R$_{f8}$	77.5:22.5 ($P = 3.46$)
44	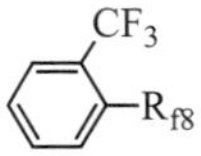 2-CF$_3$-C$_6$H$_4$-R$_{f8}$	81.8:18.2 ($P = 4.48$)
45	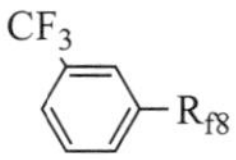 3-CF$_3$-C$_6$H$_4$-R$_{f8}$	91.5:8.5 ($P = 10.7$)
46	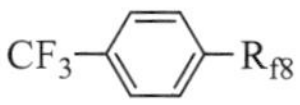 4-CF$_3$-C$_6$H$_4$-R$_{f8}$	89.4:10.6 ($P = 8.41$)

(continued)

Table 1 (continued)

Entry	Solute	Partitioning fluorous:organic (P)
47	R_{f8}—C$_6$H$_4$—R_{f8}	99.3:0.7 ($P = 145$)
48	C$_6$H$_5$—(CH$_2$)$_3$R$_{f8}$	49.5:50.5 ($P = 0.980$)
49	HO—C$_6$H$_4$—(CH$_2$)$_2$R$_{f8}$	20:80 ($P = 0.25$)
50	1,2-C$_6$H$_4$[(CH$_2$)$_3$R$_{f6}$]$_2$	73.7:26.3 ($P = 2.80$)
51	1,2-C$_6$H$_4$[(CH$_2$)$_3$R$_{f8}$]$_2$	91.2:8.8 ($P = 10.4$)
52	1,2-C$_6$H$_4$[(CH$_2$)$_3$R$_{f10}$]$_2$	97.4:2.6 ($P = 37.5$)
53	1,3-C$_6$H$_4$[(CH$_2$)$_3$R$_{f8}$]$_2$	90.7:9.3 ($P = 9.75$)
54	1,4-C$_6$H$_4$[(CH$_2$)$_3$R$_{f8}$]$_2$	91.1:8.9 ($P = 10.2$)
55	3,5-(CF$_3$)$_2$C$_6$H$_3$—R$_{f8}$	98.3:1.7 (P = 58.6)
56	1,3,5-C$_6$H$_3$[(CH$_2$)$_3$R$_{f8}$]$_3$	>99.7:<0.3 ($P > 332$)
57	[(R$_{f6}$(CH$_2$)$_2$)$_3$P]$_3$RhCl	99.86:0.14 ($P = 713$)
58	[(R$_{f8}$(CH$_2$)$_2$)$_3$P]$_3$RhCl	99.88:0.12 ($P = 832$)
59[a]	(Ar$_3$P)$_3$RhCl Ar = R$_{f6}$(CH$_2$)$_2$(CH$_3$)$_2$Si–4-C$_6$H$_4$	99.7:0.3 ($P = 293$)
60[a]	[R$_{f6}$(CH$_2$)$_2$(CH$_3$)$_2$Si—C$_6$H$_4$—P]$_3$	52:48 ($P = 1.1$)

[a]PFMC/n-octane or CF$_3$C$_6$F$_{11}$/n-C$_8$H$_{18}$ at 0 °C

1-decene is perfluorinated to give $R_{f8}CH=CH_2$ (entry 13), the partition coefficient nearly reverses, to 93.5:6.5.

The toluene phase affinity of a typical alcohol, cyclohexanol (entry 15, 1.6:98.4), is higher than that of the less polar ketone cyclohexanone (entry 14, 2.2:97.8). That of the corresponding dimethylphenyl silyl ether is higher still (entry 16, 0.8:99.2).

A number of fluorous alcohols have been examined. The short-chain, relatively polar species CF_3CH_2OH and $(CF_3)_2CHOH$ exhibit poor fluorophilicities (14.5:85.5 and 26.7:73.3; entries 17 and 18). As the perfluoroalkyl segment lengthens in the series $R_{f6}(CH_2)_3OH$, $R_{f8}(CH_2)_3OH$, and $R_{f10}(CH_2)_3OH$ (entries 20, 22, 23), the fluorous phase affinities increase from 44:56 to 64:36 to 80.5:19.5. As would be expected, when a methylene spacer is removed from the first two compounds, the fluorophilicities also increase (52:48 and 73.5:26.5; entries 19 and 21). Similar trends are found with respect to spacers for all other functional groups.

Fluorous amines also exhibit a variety of representative trends. The effect of the number of ponytails can be seen in systems of the type $[R_{f8}(CH_2)_3]_xNH_{3-x}$ (entries 24, 30, and 36). As x increases from one (primary amine) to three (tertiary amine), the fluorous phase affinities increase monotonically from 70.0:30.0 to 96.5:3.5 to the point where no GLC-detectable concentration in toluene remains (>99.7:<0.3). When the number of the methylene groups in each ponytail of the tertiary amine is increased to five, a small amount of the amine can again be detected in the toluene phase (99.5:0.5; entry 38). The effect of the spacer length can also been seen in the secondary amines (entries 30–32). Similar trends are observed with fluorous trialkylphosphines [65].

Turning to arenes, both pentafluorobenzene and hexafluorobenzene preferentially partition into toluene (22.4:77.6 and 28.0:72.0; entries 40 and 41), apropos to their nonfluorous nature noted above. Benzene exhibits an even greater toluene phase affinity (6:94; entry 39). However, the introduction of a single $R_{f8}(CH_2)_3$ ponytail levels the playing field, giving a partition coefficient of 49.5:50.5 (entry 48). This value is similar to those obtained when an $R_{f8}(CH_2)_3$ moiety is capped with an iodide or thiol. With $R_{f8}C_6H_5$ (entry 43), which lacks methylene spacers, the fluorous phase affinity increases (77.5:22.5) but the electronic properties of the arene ring are strongly perturbed.

As shown in entries 51, 53, and 54, benzenes with two $R_{f8}(CH_2)_3$ ponytails exhibit appreciable fluorophilicities, with partition coefficients of 91.2:8.8 to 90.7:9.3. The substitution pattern has little influence. As seen with the alcohols, when the perfluoroalkyl segment of the ponytail is shortened, the fluorous phase affinity decreases (73.7:26.3 for $R_{f6}(CH_2)_3$; entry 50), and when it is lengthened the fluorous phase affinity increases (97.4:2.6 for $R_{f10}(CH_2)_3$; entry 52). Benzenes with three $R_{f8}(CH_2)_3$ ponytails partition, within detection limits, completely into PFMC, at least when arrayed in a 1,3,5-pattern (entry 56). With more polar monofunctional benzenes, at least three $R_{f8}(CH_2)_3$ ponytails are required for high fluorous phase affinities.

Compounds that are catalyst precursors are of particular interest. The rhodium complexes in entries 57 and 58, which feature three phosphine ligands of the formula $(R_{fn}(CH_2)_2)_3P$, exhibit very high fluorophilicities (99.86:0.14 for $n = 6$

and 99.88:0.12 for $n = 8$). Entry 59 illustrates an interesting effect. The central rhodium is surrounded by three fluorous triarylphosphines that have only one ponytail per ring and exhibit a PFMC/n-octane partition coefficient of 52:48 (entry 60). Nonetheless, the complex is highly fluorophilic, with a partition coefficient of 99.7:0.3. Similar phenomena, in which the "sum is greater than the parts," have been observed with other compounds that are aggregates of fluorous building blocks. In such systems, the ponytails are thought to be deployed in a maximally efficient way around the periphery of the molecule. Other effects may also be in play [65].

An early rule of thumb stated that, for a molecule to be preferentially soluble in a fluorous liquid phase, 60% of the molecular weight should be fluorine-derived [59]. However, exceptions in both directions are now well known. For example, in compounds that already contain a long perfluoroalkyl group, the introduction of a CF_3 moiety or "pigtail" sometimes imparts a fluorophilicity significantly greater than might be expected. Thus, the R_{f8}-monosubstituted benzene in entry 43 can be compared with the R_{f8}/CF_3- and R_{f8}/R_{f8}-disubstituted benzenes in entries 44 through 47 and the $R_{f8}/CF_3/CF_3$-trisubstituted benzene in entry 55. Although the partition coefficients for the R_{f8}/CF_3 compounds (91.5:8.5 (*meta*), 89.4:10.6 (*para*), 81.8:18.2 (*ortho*)) indicate fluorophilicities less than that of the R_{f8}/R_{f8} compound (99.3:0.7), they are distinctly greater than that of the R_{f8}-monosubstituted compound (77.5:22.5). The $R_{f8}/CF_3/CF_3$ compound (98.3:1.7) is nearly as fluorophilic as the R_{f8}/R_{f8} compound. Thus, although trifluoromethylbenzene itself has a very poor fluorous phase affinity (entry 42, 12.4:87.6), CF_3 groups represent legitimate design elements for enhancing fluorophilicities once a longer R_{fn} segment is in place.

There have been a number of efforts to parameterize partition coefficient data such that fluorophilicities can be predicted [63, 66–69]. These have involved 3D QSAR descriptors, neural networks, and Mobile Order and Disorder (MOD) theory. The most rigorous treatments require estimations of the Hildebrand solubility parameters of the solute and fluorous and nonfluorous phases, and their respective molar volumes. The reader is referred to the original papers for further details.

3.6 Electronic Effects

The electron-withdrawing properties of the fluorous ponytails are felt far into the nonfluorous domains of fluorous molecules [6]. These have been evidenced by shifts in electron-density-sensitive IR bands, gas phase ionization potentials, X-ray photoelectron spectroscopy, calorimetric experiments, acid–base equilibrium constants, cyclic voltammetry data, and reactivity trends.

With the insight of computational data, it is clear that it is very challenging to "completely" insulate a reactive site from a perfluoroalkyl group in a fluorous molecule. With ponytails of the formula $(CH_2)_m R_{fn}$ there are still readily detectable effects upon lengthening the spacer from four to five methylene groups. The magnitudes are such that solution equilibria can be significantly affected. Computationally, the asymptotic limit is reached with seven to eight methylene groups. As a bottom line, a fluorous compound should be "reactive enough" for the purpose at hand, and a small residual electronic effect is for many applications of no significant consequence.

3.7 NMR Properties

A recent monograph provides an excellent resource for ^{19}F NMR (as well as some ^{1}H and ^{13}C NMR) spectroscopic properties of fluorinated molecules [70]. This merits consultation before reporting data for new compounds, as there are a number of counterintuitive trends. For example, triplets are commonly observed for $CF_2CF_2CF_3$ and $XCF_2CF_2CF_2$ signals, with the former typically ca. 10 Hz. However, these actually represent four-bond couplings ($^4J_{FF}$, F–C–C–C–F) and *not* three-bond vicinal couplings ($^3J_{FF}$), which for some reason are much smaller [71, 72].

Overall, the ^{19}F NMR spectra of fluorous compounds are (like the ^{1}H spectra of *n*-alkanes) usually not very informative, and often not reported. The ^{13}C data for the fluorous segments are only reported in rare cases, as these are strongly coupled to multiple ^{19}F nuclei, and the signals and coupling constants are difficult to deconvolute. Figure 6 summarizes ^{19}F NMR and ^{13}C NMR assignments that were carefully made for the fluorous alcohol $R_{f6}CH_2CH_2OH$, the carboxylic acids $R_{f6}CH_2CO_2H$ and $R_{f5}CH_2CH_2CO_2H$, the phosphine $P(p-C_6H_4SiMe_2(CH_2)_2R_{f6})_3$, and the tertiary amine $N((CH_2)_3R_{f8})_3$ [70, 73, 74]. Analogous data have been reported for the closely related carboxylic acids $R_{f6}CO_2H$ [75] and $R_{f6}CH_2CH_2CO_2H$ [24]. In each case, extensive series of 2D NMR experiments were necessary.

Another type of ponytail that sees occasional use, $CF_3(CF_2)_2OCF(CF_3)CF_2OCF(CF_3)CH_2$–, is derived from the hexafluoropropene-1,2-oxide trimer. The ^{19}F NMR signals and coupling constants of a number of derivatives have been carefully assigned [76].

4 Chemical Properties of Fluorous Compounds

Textbooks of organofluorine chemistry cover some reactions of relevance to fluorous chemistry. However, given the inertness of aliphatic perfluorocarbons and perfluoroalkyl groups, it is not surprising that there is little literature involving reactions. Nonetheless, some topics of interest can be identified.

First, there are several reports on the chemical modification of tetrafluoroethylene polymers (PTFE), of which Teflon® is an example [77]. These require harsh conditions and highly reactive species, such as sodium/liquid ammonia, the radical anion sodium naphthalenide, and liquid alkali metal amalgams. When

$^3J_{FH} = 18.7\ ^3J_{HH} = 6.0$

-85.6 -130.4 -126.8 -125.7 -127.7 -117.7
CF$_3$—CF$_2$—CF$_2$—CF$_2$—CF$_2$—CF$_2$—CH$_2$—CH$_2$OH
118.1 108.4 111.2 112.0 111.7 118.8 34.4 55.0
$^1J_{FC}$=287 $^1J_{FC}$=255 $^2J_{FC}$=21.2
$^2J_{FC}$=33.1 $^2J_{FC}$=31.7
$^4J_{FC}$=10

$^3J_{FH} = 18.7$

-80.8 -125.8 -122.4 -122.5 -121.4 -111.5
CF$_3$—CF$_2$—CF$_2$—CF$_2$—CF$_2$—CF$_2$—CH$_2$—CO$_2$H
118.24 109.8 111.4 111.9 112.2 117.3 37 165.4
$^1J_{FC}$=287 $^1J_{FC}$=257 $^2J_{FC}$=22 $^3J_{FC}$=2
$^2J_{FC}$=33 $^2J_{FC}$=30
                                            $^3J_{FH}$=18
$^4J_{FC}$=10 $^3J_{HH}$=8

-80.9 -126.0 -122.3 -123.4 -114.2
CF$_3$—CF$_2$—CF$_2$—CF$_2$—CF$_2$—CH$_2$—CH$_2$—CO$_2$H
118.4 109.6 111.7 112.1 119.4 27.2 25.6 172.7
$^1J_{FC}$=288 $^1J_{FC}$=253
$^2J_{FC}$=33 $^2J_{FC}$=33

-118.1 -126.5 -123.2 -122.2 -123.3 -115.8
(CF$_3$—CF$_2$—CF$_2$—CF$_2$—CF$_2$—CF$_2$—CH$_2$—CH$_2$—Si(CH$_3$)$_2$-p-C$_6$H$_4$)$_3$P
118.1 109.4 111.3 112.1 112.3 119.3 26.6 5.62
$^1J_{FC}$=287 $^1J_{FC}$=272 $^1J_{FC}$=272 $^1J_{FC}$=271 $^1J_{FC}$=254
$^2J_{FC}$=33.3 $^2J_{FC}$ NR $^2J_{FC}$=31.7 $^2J_{FC}$=38.8 $^2J_{FC}$=30.5
$^3J_{FC}$=2.2 $^4J_{FF}$=14 $^4J_{FF}$ NR $^4J_{FF}$=14 $^4J_{FF}$ NR $^4J_{FC}$=15
$^4J_{FF}$=9.8

-80.8 -125.8 -122.3 -122.5 -122.5 -121.2 -123.5 -113.9
(CF$_3$—CF$_2$—CF$_2$—CF$_2$—CF$_2$—CF$_2$—CF$_2$—CF$_2$—CH$_2$—CH$_2$—CH$_2$)$_3$N
118.1 109.5 111.3 111.9 111.8 112.3 112.2 120.3 29.4 19.0 53.5

Fig. 6 ^{19}F (*green* type) and ^{13}C (*black* type) NMR data for representative fluorous molecules with R$_{f5}$–R$_{f8}$ segments (chemical shifts (δ) in ppm; J values in Hz with NR = not resolved)

atmospheres of silicon tetrahalides are introduced (SiF$_4$, SiCl$_4$, SiBr$_4$), subsequent hydrolyses lead to silicic acid functionalized PFTE surfaces. The ozonolysis of partially fluorinated polymers introduces alkylperoxide and hydroperoxide functionalities. Physical methods such as plasma treatment, or X-ray, γ-ray, laser, electron beam, and ion beam irradiation, have also been extensively employed to derivatize fluoropolymers.

Second, there is a side reaction that can crop up unexpectedly with fluorous compounds that contain ponytails of the type $(CH_2)_m CF_2 R_{fn-1}$. Namely, moderately strong bases can promote HF elimination at the alkane/perfluoroalkane juncture, leading to unsaturated fragments $(CH_2)_{m-1} CH = CFR_{fn-1}$ [78]. This becomes more pronounced when the CH_2CF_2 protons are activated. The authors' group encountered this in connection with Wittig reactions that afforded systems with $ArCH_2CH = CH_2R_{fn}$ linkages [78]. As summarized in Fig. 7, it proved advantageous to generate the necessary ylides from fluorous triphenylphosphonium salts with the weaker base K_2CO_3 at higher temperatures as opposed to the stronger base n-BuLi at lower temperatures.

Third, a reaction that is well known in the organofluorine literature, and potentially very useful for the introduction of branched ponytails, deserves emphasis. Trisubstituted perfluoroalkenes are easily obtained, and fluoride ion readily adds to the less substituted terminus to give the corresponding tertiary carbanions [79–81]. These can be derivatized by activated electrophiles, as illustrated in Fig. 8. Such sequences have in the authors' opinion been underutilized.

Finally, there would be great interest in any method for the well defined functionalization of R_{fn} moieties, as opposed to the "shotgun" protocols described

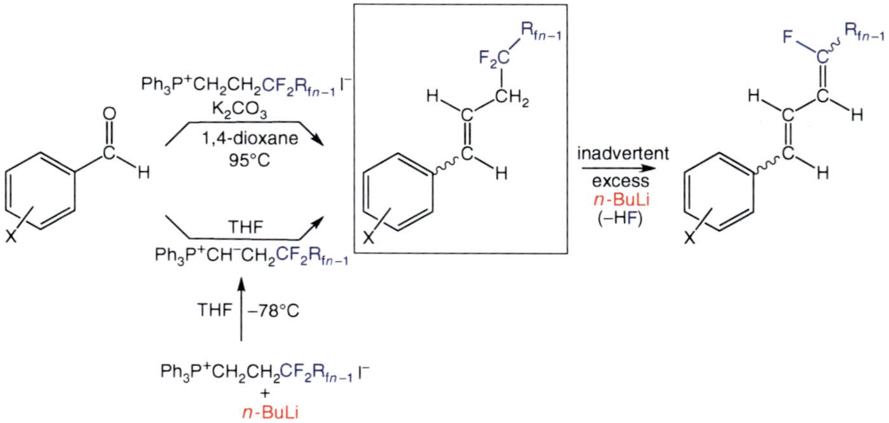

Fig. 7 HF elimination across a CH_2CF_2 juncture

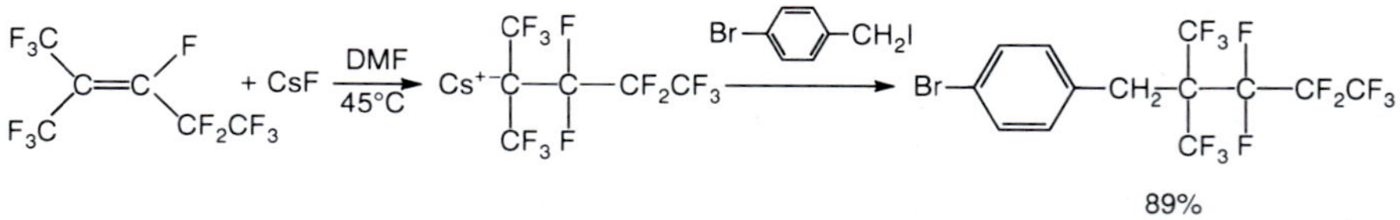

Fig. 8 Elaboration of trisubstituted perfluoroalkenes into compounds with branched ponytails

for PTFE above. This derives in part from increasing concern regarding the environmental persistence of many organofluorine compounds [82], as well as possible adverse physiological properties [83]. The authors' group has been heavily involved with transition metal adducts of fluorous ligands, and have been particularly attentive for any indication of carbon–fluorine bond activation, but to no avail. Recently, a number of creative new approaches to carbon–fluorine bond activation have been reported [84, 85], giving some hope that solutions to this long-standing problem may someday be achieved.

Acknowledgment The authors thank the Welch Foundation (A-1656) for support.

References

1. Horváth IT, Curran DP, Gladysz JA (2004) Fluorous chemistry: scope and definition. In: Gladysz JA, Curran DP, Horváth IT (eds) Handbook of fluorous chemistry. Weinheim, Wiley/VCH, pp 1–4
2. Yu MS, Curran DP, Nagashima T (2005) Increasing fluorous partition coefficients by solvent tuning. Org Lett 7:3677
3. Curran DP, Bajpai R, Sanger E (2006) Purification of fluorous Mitsunobu reactions by liquid–liquid extraction. Adv Synth Catal 348:1621
4. Chu Q, Yu MS, Curran DP (2007) New fluorous/organic biphasic systems achieved by solvent tuning. Tetrahedron 63:9890
5. Chu Q, Yu MS, Curran DP (2008) CBS reductions with a fluorous prolinol immobilized in a hydrofluoroether solvent. Org Lett 10:749
6. Gladysz JA (2004) Ponytails: structural and electronic considerations. In: Gladysz JA, Curran DP, Horváth IT (eds) Handbook of fluorous chemistry. Wiley/VCH, Weinheim, pp 41–55
7. Rábai J, Szabó D, Borbás EK, Kövesi I, Kövesdi I, Csámpai A, Gömöry Á, Pashinnik VE, Shermolovich YG (2002) Practice of fluorous biphase chemistry: convenient synthesis of novel fluorophilic ethers via a Mitsunobu reaction. J Fluor Chem 114:199
8. Jiang Z-X, Yu YB (2007) The design and synthesis of highly branched and spherically symmetric fluorinated oils and amphiles. Tetrahedron 63:3982
9. Szabó D, Bonto A-M, Kövesdi I, Gömöry A, Rábai J (2005) Synthesis of novel lipophilic and/or fluorophilic ethers of perfluoro-*tert*-butyl alcohol, perfluoropinacol and hexafluoroacetone hydrate via a Mitsunobu reaction: typical cases of ideal product separation. J Fluor Chem 126:641
10. Szabó D, Mohl J, Bálint A-M, Bodor A, Rábai J (2006) Novel generation ponytails in fluorous chemistry: syntheses of primary, secondary, and tertiary (nonafluoro-*tert*-butyloxy) ethyl amines. J Fluor Chem 127:1496
11. Jiang Z-X, Yu YB (2007) The synthesis of a geminally perfluoro-*tert*-butylated β-amino acid and its protected forms as potential pharmacokinetic modulator and reporter for peptide-based pharmaceuticals. J Org Chem 72:1464
12. Jiang Z-X, Yu YB (2008) The design and synthesis of highly branched and spherically symmetric fluorinated macrocyclic chelators. Synthesis 215
13. Dunitz JD, Gavezzotti A, Schweizer WB (2003) Molecular shape and intermolecular liaison: hydrocarbons and fluorocarbons. Helv Chim Acta 86:4073
14. Dunitz JD (2004) Organic fluorine: odd man out. ChemBioChem 5:614
15. Bondi A (1964) Van der Waals volumes and radii. J Phys Chem 68:441
16. Williams DE, Houpt DJ (1986) Fluorine nonbonded potential parameters derived from crystalline perfluorocarbons. Acta Crystallogr B42:286

17. Hansch C, Leo A, Hoekman D (1995) Exploring QSAR. American Chemical Society, Washington DC
18. Smart BE (2001) Fluorine substituent effects (on bioactivity). J Fluor Chem 109:3
19. Kuduva SS, Boese R (2003) Cambridge Crystallographic Data Centre, Deposition 220154. Refcode OLAWUT
20. Albinsson B, Michl J (1996) Anti, ortho, and gauche conformers of perfluoro-n-butane: matrix-isolation IR spectra and calculations. J Phys Chem 100:3418
21. Watkins EK, Jorgensen WL (2001) Perfluoroalkanes: conformational analysis and liquid-state properties from ab initio and Monte Carlo calculations. J Phys Chem A 105:4118
22. Jang SS, Blanco M, Goddard WA III, Caldwell G, Ross RB (2003) The source of helicity in perfluorinated N-alkanes. Macromolecules 36:5331
23. Bunn CW, Howells ER (1954) Structures of molecules and crystals of fluorocarbons. Nature (London) 174:549
24. Baker RJ, McCabe T, O'Brien JE, Ogilvie HV (2010) Thermomorphic metal scavengers: a synthetic and multinuclear NMR study of highly fluorinated ketones and their application in heavy metal removal. J Fluor Chem 131:621
25. da Costa RC, Hampel F, Gladysz J (2007) Crystal structure of an unusual bis(fluorous phosphine) ruthenium(III) complex derived from a fluorous Grubbs' catalyst. Polyhedron 26:581
26. Monde K, Miura N, Hashimoto M, Taniguchi T, Inabe T (2006) Conformational analysis of chiral helical perfluoroalkyl chains by VCD. J Am Chem Soc 128:6000
27. Barnes NA, Brisdon AK, Brown FRW, Cross WI, Crossley IR, Fish C, Herbert CJ, Pritchard RG, Warren JE (2011) Synthesis of gold(I) fluoroalkyl and fluoroalkenyl-substituted phosphine complexes and factors affecting their crystal packing. Dalton Trans 40:1743
28. Kirsch P, Bremer M (2010) Understanding fluorine effects in liquid crystals. ChemPhysChem 11:357 and references therein
29. Rocaboy C, Hampel F, Gladysz JA (2002) Syntheses and reactivities of disubstituted and trisubstituted fluorous pyridines with high fluorous phase affinities: solid state, liquid crystal, and ionic liquid-phase properties. J Org Chem 67:6863 and references therein
30. Casnati A, Liantonio R, Metrangolo P, Resnati G, Ungaro R, Ugozzoli F (2006) Molecular and supramolecular homochirality: enantiopure perfluorocarbon rotamers and halogen-bonded fluorous double helices. Angew Chem Int Ed 45:1915; Angew Chem 118:1949
31. Casnati A, Cavallo G, Metrangolo P, Resnati G, Ugozzoli F, Ungaro R (2009) The role of building-block metrics in the halogen-bonding-driven self-assembly of calixarenes, inorganic salts and diiodoperfluoroalkanes. Chem Eur J 15:7903
32. van Duijnen PT, Swart M (1998) Molecular and atomic polarizabilities: Thole's model revisited. J Phys Chem A 102:2399
33. Smart BE (1995) Physical and physicochemical properties. In: Hudlický M, Pavlath AE (eds) Chemistry of organic fluorine compounds, ACS Monograph 187. ACS, Washington DC, p 979
34. For some lead references, see [34] and [35]. Palomo C, Aizpurua JM, Loinaz I, Fernandez-Berridi MJ, Irusta L (2001) Scavenging of fluorinated N,N'-dialkylureas by hydrogen binding: a novel separation method for fluorous synthesis. Org Lett 3:2361
35. O'Neal KL, Weber SG (2009) Molecular and ionic hydrogen bond formation in fluorous solvents. J Phys Chem B 113:149
36. Legon AC (2008) The interaction of dihalogens and hydrogen halides with Lewis bases in the gas phase: an experimental comparison of the halogen bond and the hydrogen bond. Struct Bond 126:17
37. Metrangolo P, Meyer F, Pilati T, Resnati G, Terraneo G (2008) Halogen bonding in supramolecular chemistry. Angew Chem Int Ed 47:6114; Angew Chem 120:6206
38. Dordonne S, Crousse B, Bonnet-Delpon D, Legros J (2011) Fluorous tagging of DABCO through halogen bonding: recyclable catalyst for the Morita–Baylis–Hillman reaction. Chem Commun 47:5855

39. Serratrice G, Delpuech J-J, Diguet R (1982) Isothermal compressibilities of fluorocarbons. Relationship to gas solubility. Nouv J Chem 6:489
40. Gladysz JA, Emnet C (2004) Fluorous solvents and related media. In: Gladysz JA, Curran DP, Horváth IT (eds) Handbook of fluorous chemistry. Wiley/VCH, Weinheim, pp 11–23
41. For some recent measurements of oxygen, water and halogen solubilities, see [41]–[43]. Costa Gomes MF, Deschamps J, Menz J-H (2004) Solubility of dioxygen in seven fluorinated liquids. J Fluor Chem 125:1325
42. Freire MG, Gomes L, Santos LMNBF, Marrucho IM, Coutinho JAP (2006) Water solubility in linear fluoroalkanes used in blood substitute formulations. J Phys Chem B 110:22923
43. Podgorsek A, Stavber S, Zupan M, Iskra J, Padua AAH, Costa Gomes MF (2008) Solvation of halogens in fluorous phases. Experimental and simulation data for F_2, Cl_2, and Br_2 in several fluorinated liquids. J Phys Chem B 112:6653
44. van Vliet MCA, Arends IWCE, Sheldon RA (1999) Perfluoroheptadecan-9-one: a selective and reusable catalyst for epoxidations with hydrogen peroxide. Chem Commun 263
45. Wende M, Meier R, Gladysz JA (2001) Fluorous catalysis without fluorous solvents: a friendlier catalyst recovery/recycling protocol based upon thermomorphic properties and liquid/solid phase separation. J Am Chem Soc 123:11490
46. Ishihara K, Kondo S, Yamamoto H (2001) 3,5-Bis(perfluorodecyl)phenylboronic acid as an easily recyclable direct amide condensation catalyst. Synlett 1371
47. Gladysz JA (2009) Catalysis involving fluorous phases: fundamentals and directions for greener methodologies. In: Anastas P (ed), Crabtree RH (vol ed) Handbook of green chemistry, vol 1: Homogeneous catalysis. Wiley/VCH, Weinheim, p 17
48. Additional recent literature Vuluga D, Legros J, Crousse B, Bonnet-Delpon D (2010) Fluorous 4-N,N-dimethylaminopyridine (DMAP) salts as simple recyclable acylation catalysts. Chem Eur J 16:1776
49. Miura T, Nakashima K, Tada N, Itoh A (2011) An effective and catalytic oxidation using recyclable fluorous IBX. Chem Commun 47:1875
50. Jessop PG, Ikariya T, Noyori R (1999) Homogeneous catalysis in supercritical fluids. Chem Rev 99:475
51. For recent references see [51] and [52]. Berven BM, Koutsantonis GA, Skelton BW, Trengove RD, White AH (2009) Highly fluorous complexes of ruthenium and osmium and their solubility in supercritical carbon dioxide. Inorg Chem 48:11832
52. Harwardt T, Franciò G, Leitner W (2010) Continuous-flow homogeneous catalysis using the temperature-controlled solvent properties of supercritical carbon dioxide. Chem Commun 46:6669
53. Jessop PG, Olmstead MM, Ablan CD, Grabenauer M, Sheppard D, Eckert CA, Liotta CL (2002) Carbon dioxide as a solubility 'switch' for the reversible dissolution of highly fluorinated complexes and reagents in organic solvents: application to crystallization. Inorg Chem 41:3463
54. Ablan CD, Hallett JP, West KN, Jones RS, Eckert CA, Liotta CL, Jessop PG (2003) Use and recovery of a homogeneous catalyst with carbon dioxide as a solubility switch. Chem Commun 2972
55. For a recent reference, see Motreff A, Belin C, da Costa RC, El Bakkari M, Vincent J-M (2010) Self-adaptive hydrophilic and coordinating Teflon surfaces through a straightforward physisorption process. Chem Commun 46:6261
56. van Zee NJ, Dragojlovic V (2009) Phase-vanishing reactions with PTFE (Teflon) as a phase screen. Org Lett 11:3190
57. Zhao H, Zhang J, Wu N, Zhang X, Crowley K, Weber SG (2005) Transport of organic solutes through amorphous teflon AF films. J Am Chem Soc 127:15112
58. Purse BW, Rebek J Jr (2005) Encapsulation of oligoethylene glycols and perfluoro-n-alkanes in a cylindrical host molecule. Chem Commun 722
59. Horváth IT (1998) Fluorous biphase chemistry. Acc Chem Res 31:641

60. van den Broeke J, Winter F, Deelman B-J, van Koten G (2002) A highly fluorous roomtemperature ionic liquid exhibiting fluorous biphasic behavior and its use in catalyst recycling. Org Lett 4:3851
61. Juliette JJJ, Rutherford D, Horváth IT, Gladysz JA (1999) Transition metal catalysis in fluorous media: practical application of a new immobilization principle to rhodium-catalyzed hydroborations of alkenes and alkynes. J Am Chem Soc 121:2696
62. West KN, Hallett JP, Jones RS, Bush D, Liotta CL, Eckert CA (2004) CO_2-induced miscibility of fluorous and organic solvents for recycling homogeneous catalysts. Ind Eng Chem Res 43:4827
63. Leo A, Hansch C, Elkins D (1971) Partition coefficients and their uses. Chem Rev 71:525
64. Kiss LE, Kövesdi I, Rábai J (2001) An improved design of fluorophilic molecules: prediction of the ln P fluorous partition coefficient, fluorophilicity, using 3D QSAR descriptors and neural networks. J Fluor Chem 108:95
65. Gladysz JA, Emnet C, Rábai J (2004) Partition coefficients involving fluorous solvents. In: Gladysz JA, Curran DP, Horváth IT (eds) Handbook of fluorous chemistry. Weinheim, Wiley/VCH, pp 56–100
66. Huque FTT, Jones K, Saunders RA, Platts JA (2002) Statistical and theoretical studies of fluorophilicity. J Fluor Chem 115:119
67. de Wolf E, Ruelle P, van den Broeke J, Deelman B-J, van Koten G (2004) Prediction of partition coefficients of fluorous and nonfluorous solutes in fluorous biphasic solvent systems by mobile order and disorder theory. J Phys Chem B 108:1458
68. Mercader AG, Duchowicz PR, Sanservino MA, Fernández FM, Castro EA (2007) QSPR analysis of fluorophilicity for organic compounds. J Fluor Chem 128:484
69. de Wolf ACA (2002) Fluorous phosphines as green ligands for homogeneous catalysis; solving problems in fluorous catalysis, Chap 7. Doctoral Thesis, University of Utrecht
70. Dolbier WR (2009) Guide to fluorine NMR for organic chemists. Wiley, Hoboken, NJ
71. White HF (1966) Fluorine resonance spectra-structure correlations for perhalogenated propanes. Anal Chem 38:625
72. Foris A (2004) ^{19}F and ^{1}H NMR spectra of halocarbons. Magn Reson Chem 42:534
73. Richter B, de Wolf E, van Koten G, Deelman B-J (2000) Synthesis and properties of a novel family of fluorous triphenylphosphine derivatives. J Org Chem 65:3885
74. Szlávik Z, Tárkányi G, Gömöry Á, Tarczay G, Rábai J (2001) Convenient syntheses and characterization of fluorophilic perfluorooctyl-propyl amines and ab initio calculations of proton affinities of related model compounds. J Fluor Chem 108:7
75. Ribeiro AA (1997) ^{19}F, ^{13}C single- and two-bond 2D NMR correlations in perfluoroheptanoic acid. J Fluor Chem 83:61
76. Kysilka O, Rybáčková M, Skalický M, Kvíčalová M, Cvačka J, Kvíčala J (2008) HFPO trimer-based alkyl triflate, a novel building block for fluorous chemistry. Preparation, reactions and ^{19}F gCOSY analysis. Coll Czech Chem Commun 73:1799
77. Kang ET, Zhang Y (2000) Surface modification of fluoropolymers via molecular design. Adv Mater 12:1481
78. Rocaboy C, Rutherford D, Bennett BL, Gladysz JA (2000) Strategy and design in fluorous phase immobilization: a systematic study of the effect of 'pony tails' $(CH_2)_3(CF_2)_{n-1}CF_3$ on the partition coefficients of benzenoid compounds. J Phys Org Chem 13:596
79. Bayliff AE, Bryce MR, Chambers RD, Matthews RS (1985) Direct observation of simple fluorinated carbanions. J Chem Soc Chem Commun 1018
80. Zhang Q, Luo Z, Curran DP (2000) Separation of "light fluorous" reagents and catalysts by fluorous solid-phase extraction: synthesis and study of a family of triarylphosphines bearing linear and branched fluorous tags. J Org Chem 65:8866
81. Chambers RD, Magron C, Sandford G (1999) Reactions involving fluoride ion. Part 44.1 Synthesis and chemistry of aromatics with bulky perfluoroalkyl substituents. J Chem Soc Perkin Trans 1:283
82. Ravishankara AR, Solomon S, Turnipseed AA, Warren RF (1993) Atmospheric lifetimes of long-lived halogenated species. Science 259:194

83. Lopez-Espinosa M-J, Fletcher T, Armstrong B, Genser B, Dhatariya K, Mondal D, Ducatman A, Leonardi G (2011) Association of perfluorooctanoic acid (PFOA) and perfluorooctane sulfonate (PFOS) with age of puberty among children living near a chemical plant. Environ Sci Technol doi: 10.1021/es1038694, and references therein
84. Amii H, Uneyama K (2009) C-F bond activation in organic synthesis. Chem Rev 109:2119
85. Douvris C, Ozerov OV (2008) Hydrodefluorination of perfluoroalkyl groups using silylium-carborane catalysts. Science 321:1188

Selective Fluoroalkylation of Organic Compounds by Tackling the "Negative Fluorine Effect"

Wei Zhang, Chuanfa Ni, and Jinbo Hu

Abstract The presence of fluorine on a carbanion center will dramatically influence the nucleophilic alkylation reactions. Based on our own experience, we noticed that the fluorine substitution on the carbanionic carbon poses a negative effect in many nucleophilic fluoroalkylation reactions [we propose this effect as "negative fluorine effect (NFE)"]. Two factors were believed to contribute to the NFE: (1) thermal instability of fluorinated carbanions caused by α-elimination (self-decomposition) and (2) the intrinsic nucleophilicity of fluorinated carbanion influenced by the fluorine atoms (such as hard/soft nature of the fluorinated carbanions). By tackling the NFE, our research group has attempted to design nucleophilic fluoroalkylation reactions with fluorinated sulfones and related reagents. These results were summarized as four methods to modulate the fluoroalkylation reactions: (1) changing the number of fluorine atoms, (2) slightly changing the neighboring groups, (3) changing the metal counterion, including using carbon-metal covalent bond to tune the reactivity, and (4) enhancing the generation of carbene species.

Keywords Difluorocarbene · Fluorinated sulfones · Fluorinated sulfoximines · Fluoroalkylations · Negative fluorine effect

Contents

1 Introduction 26
2 The "Negative Fluorine Effect" in Fluoroalkylation Reactions 28
   2.1 The Unique Effect of Fluorine Substitution on the Carbanions 28
   2.2 The "Negative Fluorine Effect" 29

W. Zhang, C. Ni, and J. Hu (✉)
Key Laboratory of Organofluorine Chemistry, Shanghai Institute of Organic Chemistry, Chinese Academy of Sciences, 345 Lingling Road, Shanghai 200032, China
e-mail: jinbohu@sioc.ac.cn

3 Modulating the Fluoroalkylation Reactions by Tackling the "Negative Fluorine Effect" . 30
  3.1 Modulating the Fluoroalkylation Reactions by Changing the Number of Fluorine Atoms 31
  3.2 Modulating the Fluoroalkylation Reactions by Slightly Changing the Neighboring Groups 33
  3.3 Modulating the Fluoroalkylation Reactions by Changing the Metal Counterion 37
  3.4 Modulating the Fluoroalkylation Reactions by Enhancing the Generation of Carbene Species 39
4 Conclusion 41
References 42

Abbreviations

Bn	Benzyl
DFT	Density functional theory
dr	Diastereomeric ratio
ee	Enantiomeric excess
i-Pr	*iso*-Propyl
KHMDS	Potassium *bis*(trimethylsilyl)amide
LiHMDS	Lithium *bis*(trimethylsilyl)amide
NaHMDS	Sodium *bis*(trimethylsilyl)amide
Ph	Phenyl
Py	Pyridine
TBAT	Tetrabutylammonium triphenyldifluorosilicate
t-Bu	*tert*-Butyl
TMS	Trimethylsilyl
Ts	Tosyl (*p*-toluenesulfonyl)

1 Introduction

Due to the high electronegativity and small size of fluorine, the replacement of hydrogen atoms by fluorine in organic compounds often results in a profound change in their physical and chemical properties, including the stability, high lipophilicity, and bioavailability [1–4]. In the past decades, organofluorine compounds have found broad applications in both life sciences and material sciences, and around 20% of all pharmaceuticals and 30–40% of all agrochemicals on the market contain fluorine [5, 6]. Among nearly 3,200 known naturally occurring organohalogen compounds, only very few (around 13) are organofluorine compounds (all of them are monofluorinated) [7]. However, despite their rarity in nature, till 2010 about four million organic compounds which contain at least one C–F bond have been artificially synthesized by chemists, and those occupied about 55% of the approximate 7.2 million man-made organohalogen compounds [a SciFinder search (October, 2010) revealed >4,000,000 structures containing at least one C–F bond and >7,200,000 structures containing at least one C–X (X = F, Cl, Br, I) bond].

Many fluorinated compounds can be constructed by using simple fluorine-containing building blocks. However, some desired fluorinated molecules are complex and, therefore, selective fluorination and fluoroalkylation synthetic methodologies are preferred for the synthesis of these molecules [8, 9]. In the field of fluoroalkylation chemistry, although both free radical fluoroalkylation and electrophilic fluoroalkylation are well known, nucleophilic fluoroalkylation often bring about many advantages and thus it has become one of the most important methods for incorporation of fluorinated moieties into organic molecules [10]. Nucleophilic fluoroalkylation typically features the transfer of a fluoroalkyl group (such as perfluoroalkyl, CF_3, CF_2H, and CH_2F) to an electrophile, in which either a free α-fluoro carbanion, an equivalent of α-fluoro carbanion (i.e., a species that has similar reactivity character to an α-fluoro carbanion, such as pentacoordinate silicon species) or a fluoroalkyl metal species (R_fM) is involved.

During the past decades, nucleophilic perfluoroalkylation using organometallic reagents of lithium, magnesium, and zinc, among others, has been extensively studied, but these reagents do not have attractive profiles of reactivity, selectivity, stability, and convenience, and do not generally apply to the trifluoromethylation case [11, 12]. In 1989 Prakash and Olah reported the first nucleophilic trifluoromethylation reactions of carbonyls by using the stable organosilicon reagent (trifluoromethyl)trimethylsilane ($TMSCF_3$) under the initiation of a fluoride [13], then nucleophilic trifluoromethylation was creatively solved and perfluoroalkyl silanes have become the most widely used reagents in nucleophilic perfluoroalkylation. Compared to trifluoromethylation chemistry, the analogous difluoromethylation and monofluoromethylation (selective introduction of a CF_2H or CH_2F group into organic molecules) are less studied. It is found that difluoromethyl- and monofluoromethyl-containing compounds often exhibit unique biological properties, and there is a growing interest in developing new synthetic methods for nucleophilic di- and monofluoromethylation. Due to the lower polarization of the C–Si bond, initial results showed that nucleophilic difluoromethylation with compound R_3Si-CF_2H had to be conducted under harsh conditions, which made it less likely to become a widely used difluoromethylation reagent [14]. Similarly, the analogous R_3Si-CH_2F is even more stable and less likely to be a viable monofluoromethylation reagent.

Selective di- and monofluoromethylation are generally accomplished by two strategies: one is the direct transfer of a "CF_2H" or "CH_2F" moiety into organic molecules [15]; the other is the transfer of a functionalized moiety (such as "CF_2R" or "CFR_2"), followed by removal of the functional or auxiliary groups [16]. The introduction of functional groups to fluorinated carbanions can facilitate the nucleophilic fluoroalkylation reactions. In this review we will discuss the negative effect of fluorine substitution on the reactivity of carbanions, and summarize our research results as four methods to modulate the fluoroalkylation reactions by tackling the "negative fluorine effect."

2 The "Negative Fluorine Effect" in Fluoroalkylation Reactions

On the basis of the nature of the substituents on the anionic or metal-bearing fluorinated carbon atom, the fluoroalkyl anions can be classified as activated and unactivated. When an electron-withdrawing group (such as sulfonyl, sulfinyl, sulfanyl, carbonyl, phosphinyl, perfluoroalkyl, ester, cyano, nitro, etc.) is present on the anionic carbon, the fluoroalkyl anion is activated; otherwise, in the absence of such groups, it is unactivated. In the case of an activated fluoroalkyl anion, it is usually stabilized by one or two neighboring functional groups, and the negative charge on the anionic carbon is usually delocalized to the neighboring groups. However, for the unactivated fluoroalkyl anions (such as CF_3^-, CHF_2^-, CH_2F^-, etc.), the negative charge has to be more localized on the carbanionic carbon.

2.1 The Unique Effect of Fluorine Substitution on the Carbanions

In order to understand better the fluorine effect in nucleophilic fluoroalkylation reactions, the unique effect of fluorine substitution on the carbanions (R_f^-) can be discussed from three aspects: the acidity (for the generation of the anion), thermodynamic stability, and kinetic stability (lifetime of the anion).

First, for the formation of a fluorinated carbanion (R_f^-) by deprotonation of its conjugate acid (R_fH), the inductive effect of fluorine substitution can increase the acidity of a hydrofluorocarbon, thus favoring the production of the carbanion. The calculated enthalpy values of C–H ionization (ΔH_{calcd}, kcal/mol) in the gas phase are as follows: 368.9 (CF_3–H) < 391.3 (CHF_2–H) < 406.3 (CH_2F–H) < 416.8 (CH_3–H), which reveals that the production of CF_3^- by deprotonation will be much easier than CH_3^- [2].

Second, for the thermodynamic stability of a fluorinated carbanion itself, fluorine substitution is more effective in α-stabilization of a methyl anion than other halogen atoms and hydrogen substitution itself. The calculated enthalpy values by Bickelhaupt and coworkers for homolytic cleavage of the C–F bonds in CH_2F^- (117.0 kcal/mol) is higher than the cleavage of C–H bonds in CH_3^- (103.2 kcal/mol), which indicates that CH_2F^- is thermodynamically more stable than CH_3^- [17].

Third, for the decomposition of a fluorinated carbanion via α-elimination of a fluoride (kinetic aspect) in the presence of a hard metal cation, despite the fact that α-fluorine substitution on the carbanion has a certain stabilization influence through inductive effect, the strong repulsion between the electron lone pair on the anionic carbon and those on fluorine atom can decrease the stability of the carbanion [18]. These fluoromethyl metal species display carbenoid reactivity [19, 20]. The formation of carbene and metal fluoride species will act as a substantial driving force for its decomposition. In this regard, the kinetic stability (lifetime) of the carbanions decreases in the following order: $CH_3^- > CH_2F^- > CHF_2^- > CF_3^-$.

2.2 The "Negative Fluorine Effect"

Due to the lack of systematic study of per-, poly-, tri-, di-, and mono-fluoroalkylation reactions, the chemical reactivities of various fluorinated methyl anions in nucleophilic fluoroalkylation reactions were not well summarized. As the "chemical chameleon" in organic synthesis, the sulfone functionality (such as $PhSO_2R$) possesses strong electron-withdrawing ability and is ideal for various types of reactions, and the versatile transformations of the sulfone functionality make the subsequent intermediates suitable for the generation of a range of important products which are otherwise difficult to obtain. For instance, difluoromethyl phenyl sulfone ($PhSO_2CF_2H$, **1**) has been extensively employed as a difluoromethylation reagent [21, 22]. The (phenylsulfonyl)difluoromethyl group has been introduced to a series of electrophiles with reagent **1**, and the (phenylsulfonyl) difluoromethyl substituted intermediates so obtained could undergo many further transformations [10, 23].

In recent years, our research group studied the nucleophilic ring-opening fluoroalkylation of epoxides with di- and monofluoro(phenylsulfonyl)-methyllithium (Scheme 1) [24]. It was found that α-fluorine substitution on the carbanion dramatically decreases the carbanion's reactivity towards epoxides. According to the product yields, we can see that the nucleophilicity of halogenated sulfone reagents toward propylene oxide decreases in the following order: (1) for fluorinated carbanions $(PhSO_2)_2CF^- > PhSO_2CHF^- > PhSO_2CF_2^-$ and (2) for different halogen-substituted carbanions $PhSO_2CCl_2^- > PhSO_2CF_2^-$. The difficulty of the ring-opening reaction between a fluorinated carbanion and an epoxide can be attributed to (1) the low thermal stability of fluorinated carbanion (R_f^-) caused by its high tendency to undergo α-elimination of a fluoride ion (or another leaving group) and (2) its intrinsic low nucleophilicity towards epoxides.

This negative (unfavorable) influence of fluorine substitution on the reactivity of the carbanions was proposed as "Negative Fluorine Effect (NFE)" (Scheme 2) [24, 25]. The word "negative" implies both the *negative* (unfavorable) influence and the *negatively* charged species (carbanion). For the desired nucleophilic reaction between fluorinated carbanion **3** (R^1, R^2 = H, halogen, alkyl, aryl, etc.) and electrophile E^+, two factors are involved. Factor 1 is the α-elimination reaction

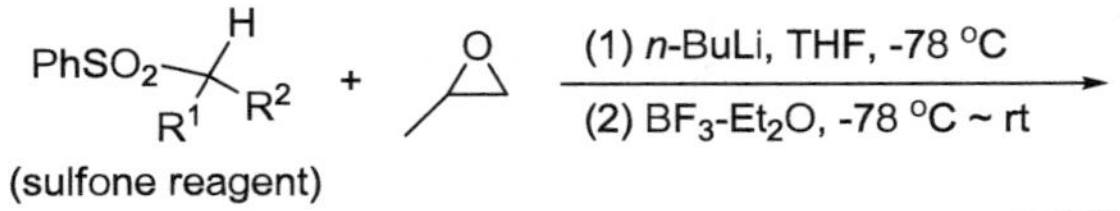

$PhSO_2\overset{H}{\underset{R^1\ R^2}{\diagdown\diagup}}$ + (epoxide) → $\overset{OH}{\underset{2}{\diagdown\diagup}}R$

(1) *n*-BuLi, THF, −78 °C
(2) $BF_3\text{-}Et_2O$, −78 °C ~ rt

(sulfone reagent)

2a R = $PhSO_2CF_2$	0%
2b R = $PhSO_2CCl_2$	72%
2c R = $PhSO_2CHF$	67%
2d R = $PhSO_2CHCl$	71%
2e R = $(PhSO_2)CF$	91%

Scheme 1 Nucleophilic ring-opening fluoroalkylation of epoxides

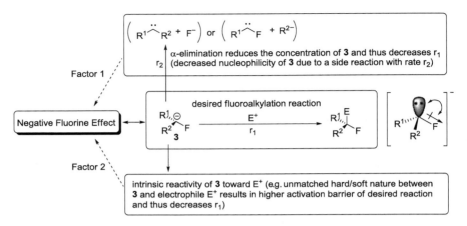

Scheme 2 The "negative fluorine effect" in nucleophilic fluoroalkylation reaction

(with reaction rate r_2) which acts as a competitive side reaction of the desired fluoroalkylation reaction (with reaction rate r_1), and the former reaction reduces the concentration of carbanion **3** and thus decreases r_1. Factor 2 is the intrinsic reactivity of fluorinated carbanion **3** toward electrophile E^+. This can be influenced by many factors, such as the match of the hard/soft nature between **3** and electrophile E^+. For instance, the increased hardness of **3** arising from more fluorine substitution can result in a higher activation barrier of the desired reaction with soft electrophiles (such as alkyl iodides) [26, 27], and thus decreases the desired reaction rate r_1.

It should be noted that Factor 1 (self-decomposition of carbanion **3**) should always have a negative (unfavorable) influence on the desired nucleophilic fluoroalkylation reaction, no matter what kind of electrophile is applied. However, the influence of Factor 2 is independent from Factor 1, which depends on the intrinsic reactivity of **3** toward an electrophile. Combining Factors 1 and 2 together, we can understand that there often exists a unique and quite subtle fluorine effect in many nucleophilic fluoroalkylation reactions involving a fluorinated carbanion.

3 Modulating the Fluoroalkylation Reactions by Tackling the "Negative Fluorine Effect"

As mentioned above, the presence of fluorine on the carbanion center will dramatically influence the nucleophilic alkylation reactions. Although the "NFE" could not be used to summarize all reactivity aspects of fluoroalkyl carbanion in nucleophilic fluoroalkylation reactions, it is at least helpful to understand many nucleophilic fluoroalkylation reactions. During the past several years, our group has attempted to design nucleophilic fluoroalkylation reactions by tackling the NFE with fluorinated

Selective Fluoroalkylation of Organic Compounds

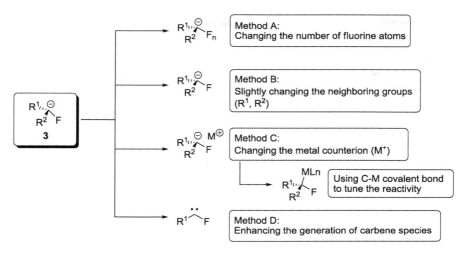

Scheme 3 Methods to modulate the fluoroalkylation reactions

sulfones and related reagents. Our research results were summarized as four methods to modulate the fluoroalkylation reactions: (1) changing the number of fluorine atoms, (2) slightly changing the neighboring groups (R^1, R^2), (3) changing the metal counterion, including using carbon–metal (C–M) covalent bond to tune the reactivity, and (4) enhancing the generation of carbene species (Scheme 3).

3.1 Modulating the Fluoroalkylation Reactions by Changing the Number of Fluorine Atoms

As shown in Scheme 2, the unmatched hard/soft nature between fluorinated carbanion **3** and electrophile E^+ will decrease the desired reaction rate. Recently, we studied the nucleophilic fluoroalkylation reaction of α,β-enones with different fluorinated carbanion (Scheme 4) [28]. It was found that the two fluorines substituted sulfone (**1**) gave the 1,2-addition product solely, while the one fluorine substituted sulfone (**4**) gave the mixture of 1,2- and 1,4-addition products. The number of α-fluorine atoms in carbanions affects its hard/soft nature, which made the 1,2- and 1,4-addition product ratio significantly different. This disclosed the order of softness of halogenated carbanions, which can be given as follows: $(PhSO_2)_2CF^- \approx PhSO_2CCl_2^- > PhSO_2CHF^- > PhSO_2CF_2^- (\geq CF_3^-)$.

The introduction of an electron-withdrawing sulfonyl group to the fluoroalkyl anions can stabilize the fluorinated carbanion via electron-delocalization, and reduce the electron repulsion between the electron pairs on the carbanionic carbon and those on the fluorine atoms. The sulfonyl group also acts as a good auxiliary group that can be readily removed by reductive desulfonylation after desired transformations.

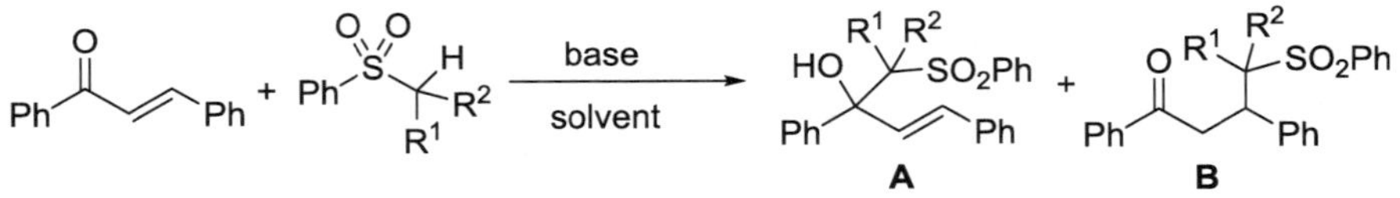

Sulfone	R¹	R²	Yield (A+B)	A:B
1	F	F	97	100:0
4	F	H	98	42:58
5	F	SO$_2$Ph	98	0:100
6	Cl	Cl	95	0:100

Scheme 4 1,2- and 1,4-additions to α,β-enones with carbanions

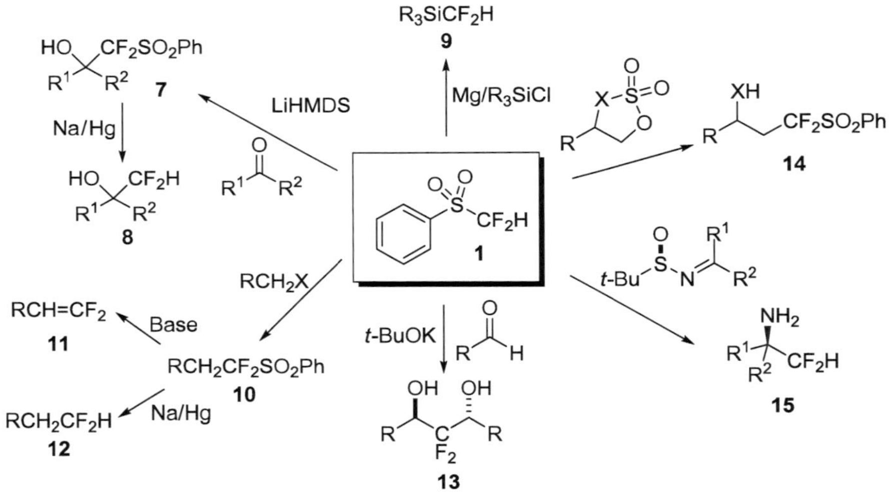

Scheme 5 Nucleophilic difluoromethylation reactions with PhSO$_2$CF$_2$H (**1**)

Difluoromethyl phenyl sulfone (**1**) was first prepared by Hine in 1960 [22], and in 1989 Stahly reported the preparation of difluoromethylated alcohol from aldehyde by using the nucleophilic reaction with PhSO$_2$CF$_2$H, followed by reductive desulfonylation (Scheme 5) [21]. In 2003 Prakash and coworkers reported that PhSO$_2$CF$_2$H could be used as a nucleophilic difluoromethylation reagent in the presence of a reducing metal (such as magnesium). By using this protocol, chlorotrialkylsilanes was successfully difluoromethylated to give difluoromethylsilanes **9** in practically useful yields (Scheme 5) [29]. Thereafter, PhSO$_2$CF$_2$H reagent was extensively used as a robust "CF$_2$H⁻" synthetic equivalent, and the reagent has been efficiently applied to react with a wide range of substrates, such as primary alkyl halides [30, 31], aldehydes, ketones and esters [28, 32–35], imines [36, 37], cyclic sulfates and sulfamidates [25] (Scheme 5).

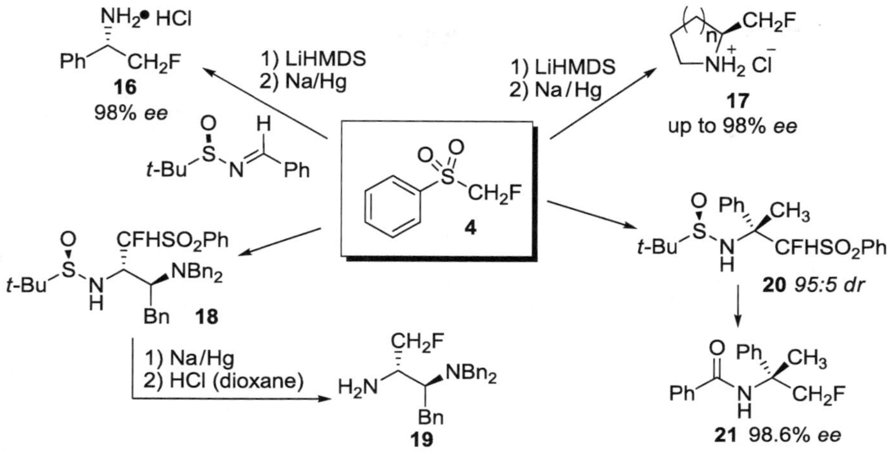

Scheme 6 Nucleophilic monofluoromethylation reactions with PhSO$_2$CH$_2$F (4)

By decreasing the number of fluorine atoms substituted on the fluorinated carbanion center, the nucleophilic monofluoromethylation is relatively easier than difluoromethylation. According to our study on the reactivity difference of mono- and difluoromethyl anions in nucleophilic fluoroalkylation reactions, the PhSO$_2$CHF$^-$ anion (generated from PhSO$_2$CH$_2$F, 4) possesses higher thermal stability than PhSO$_2$CF$_2^-$ anion and is suitable for pregeneration. In 2006 we reported the first stereoselective nucleophilic monofluoromethylation with the reagent PhSO$_2$CH$_2$F (Scheme 6) [38]. It turned out that the reaction between homochiral N-*tert*-butylsulfinyl aldimines and 4 in the presence of LiHMDS readily gave (phenylsulfonyl)fluoromethylated products, which could be converted to monofluoromethylated chiral amines 16 and 17 (via removal of both phenylsulfonyl and *tert*-butylsulfinyl groups) in good yields with excellent optical purity (Scheme 6). The reagent 4 was also successfully used by us in the stereoselective synthesis of monofluoromethylated vicinal ethylenediamine 19 (Scheme 6) [37], as well as in the nucleophilic monofluoromethylation of α,β-unsaturated ketones [28]. Recently we achieved the highly stereoselective monofluoromethylation of N-*tert*-butylsulfinyl ketimines by using the pregenerated PhSO$_2$CHF$^-$ anion [39], and this stereocontrol mode of the diastereoselective monofluoromethylation of ketimines is opposite to the other known fluoroalkylation of N-*tert*-butylsulfinyl aldimines, which suggests that a cyclic six-membered transition state is involved in the reaction.

3.2 Modulating the Fluoroalkylation Reactions by Slightly Changing the Neighboring Groups

For a long time, the phenylsulfonyl group at the α-position of fluorinated carbanions is the most frequently used neighboring group due to its versatility. Recently

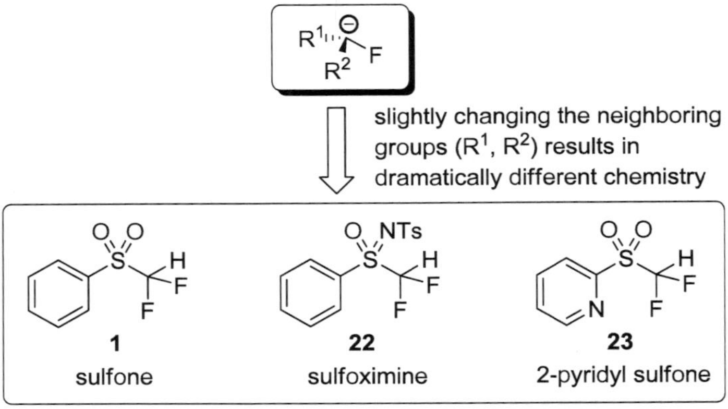

Scheme 7 Slightly changing the neighboring groups of fluorinated carbanions

we found that when one oxygen atom in compound **1** was replaced by the NTs group, the chemical reactivity of the resulting compound *N*-tolyl-*S*-difluoromethyl-*S*-phenylsulfoximine (**22**) is significantly different from **1** (Scheme 7). Also, by careful modification of the aryl group of reagent **1**, difluoromethyl 2-pyridyl sulfone (**23**) was developed (Scheme 7). These slight changes of neighboring groups of fluorinated carbanions result in dramatically different and very interesting chemistry.

Sulfoximines have been widely used in organic synthesis [40], but the fluorinated sulfoximines are poorly studied. In 1988 Finch and coworkers reported that monofluorinated sulfoximines could react with carbonyl compounds to yield hydroxyl adducts, which can be converted into fluoroolefins (but with poor Z/E-selectivity) by reduction with aluminum amalgam [41]. In 2009 we successfully prepared the first α-difluoromethyl sulfoximine compound **22** by using the copper-catalyzed nitrene transfer reaction, and the compound **22** was found to be a novel and efficient electrophilic difluoromethylation reagent for transferring CF_2H group to S-, N-, and C-nucleophiles (Scheme 8) [42]. According to the deuterium-labeling experiments, a difluorocarbene mechanism was proposed in this difluoromethylation reaction with reagent **22**. Under the same reaction conditions, the reagent $PhSO_2CF_2H$ (**1**) does not react with the S-nucleophiles to give electrophilic fluoroalkylation products. In addition, by using the sulfoximine **22** and aldehyde, the nucleophilic fluoroalkylation reaction does not occur, while the reagent **1** readily gives adduct products. This remarkably different reactivity pattern between reagent **22** and the $PhSO_2CF_2H$ (**1**) provides important insights into the unique chemical reactivities of fluorinated sulfones and sulfoximines.

The trifluoromethylated sulfoximine derivative (fluorinated Johnson reagent) was developed by Shibata and coworkers for the electrophilic trifluoromethylation of carbon nucleophiles [43]. Recently, by using monofluoromethylsulfoxinium salts, the electrophilic monofluoromethylation reaction was developed, and the unique inherent preference of the CH_2F cation for the oxygen atom in the alkylation of enolates was observed [44].

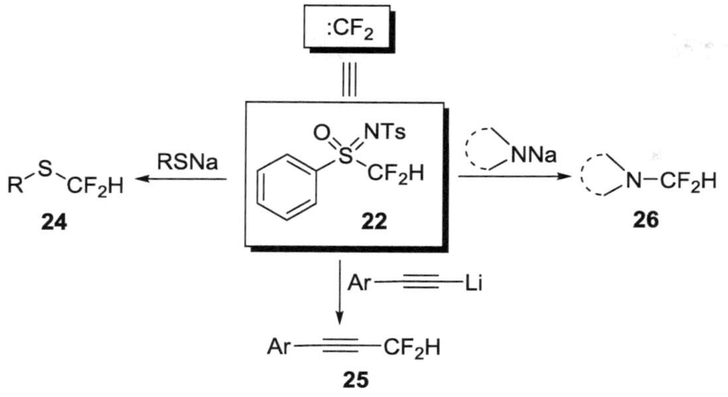

Selective Fluoroalkylation of Organic Compounds

Scheme 8 Difluoromethylation of S-, N-, and C-nucleophiles with reagent **22**

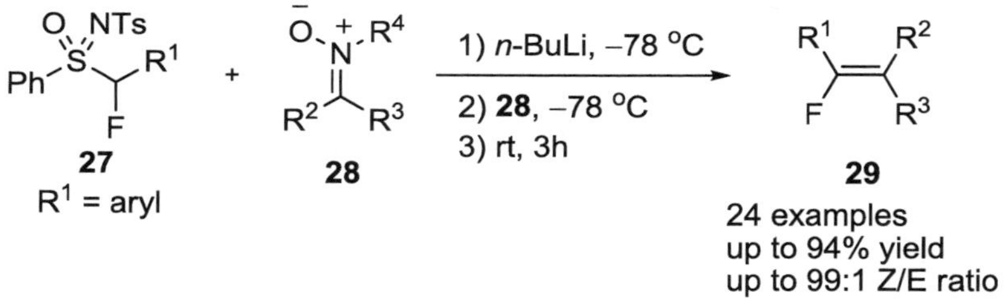

Scheme 9 Synthesis of fluorinated alkenes by using fluorinated sulfoximines **27** and nitrones

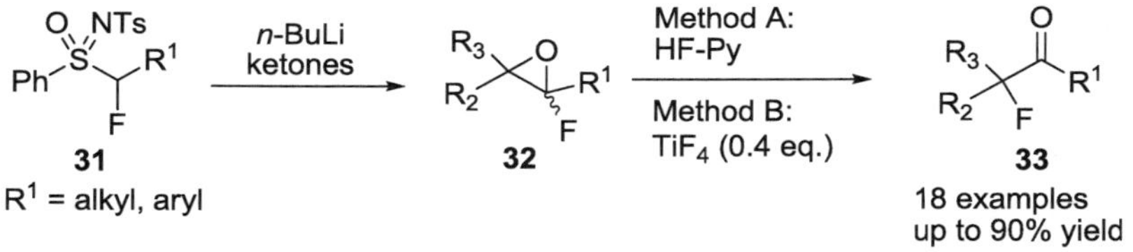

Scheme 10 O-Cyclization reaction with α-fluorosulfoximines **31** and relative ring-opening reactions

In 2009 we successfully prepared α-monofluorinated sulfoximines **27** (Scheme 9) [45]. By using the reagent **27** and nitrones **28**, monofluoroalkenes **29** were synthesized with excellent Z/E stereoselectivity (Scheme 9) [45]. This novel fluoroolefination reaction between a sulfoximine and a nitrone proceeds via an addition– elimination pathway. When fluorinated sulfone compound PhSO$_2$CF$_2$TMS (**30**, in place of a sulfoximine) was used to react with a nitrone, a fluoroalkylated hydroxylamine product was obtained, and no alkene product was formed, which demonstrates a remarkable reactivity difference between fluorinated sulfoximines and sulfones.

In 2010 we reported an O-cyclization reaction between α-fluorosulfoximines **31** and ketones to synthesize monofluorinated epoxides **32** (Scheme 10) [46]. The obtained fluoroepoxides **32** were found to undergo readily an interesting

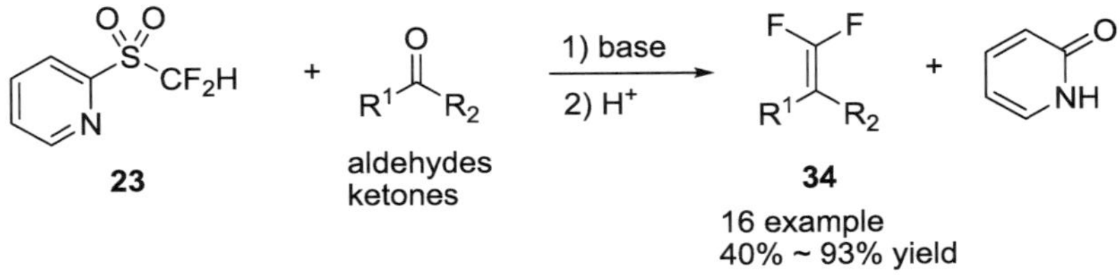

Scheme 11 *gem*-Difluoroolefination reactions with difluoromethyl 2-pyridyl sulfone

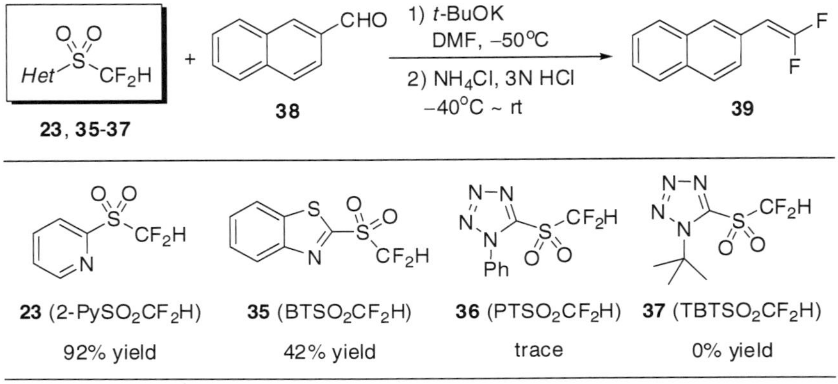

Scheme 12 *gem*-Difluoroolefination with different sulfones

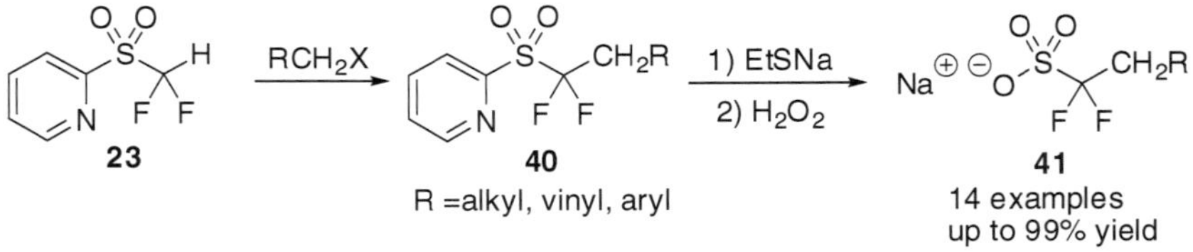

Scheme 13 Synthesis of various α,α-difluorosulfonates with reagent **23**

ring-opening process (involving both a C−F bond cleavage and another C−F bond formation) in the presence of titanium tetrafluoride or pyridinium poly(hydrogen fluoride) to afford α-fluorinated ketones **33**. The sulfoximinyl group of compound **31** not only retains the strong electron-withdrawing ability to stabilize the carbanion, but also displays better leaving group ability for further transformations.

By slightly changing the phenyl group, difluoromethyl 2-pyridyl sulfone (**23**) was developed by us, which was found to act as a novel and efficient *gem*-difluoroolefination reagent for both aldehydes and ketones (Scheme 11) [47]. In this Julia-Kocienski olefination reaction, the fluorinated sulfinate salt intermediate in the reaction is relatively stable under basic conditions, and it decomposes after protonolysis.

Difluoromethyl heteroaryl sulfones **35–37** were also prepared to investigate their reactivities in difluoroolefination reactions [47]. It was found that heteroaryl sulfones **35–37** show lower reactivity in the difluoroolefination reaction with 2-naphthaldehyde **38**. The high reactivity of 2-pyridyl sulfone **23** (compared with other sulfones **35–37**) may be due to the fact that 2-PySO$_2$CF$_2^-$ anion possesses the best nucleophilicity (among four HetSO$_2$CF$_2^-$ anions, Het = 2-Py, BT, PT, and TBT) toward carbonyl compounds (Scheme 12).

Another important application of difluoromethyl 2-pyridyl sulfone (**23**) was reported by Prakash and coworkers (Scheme 13) [48]. Alkyl α,α-difluorosulfonates **41** were efficiently synthesized from the reagent **23** and halides. This result not only extends the synthetic application of fluorinated sulfones but also provides a unique solution for the long-standing challenge in the nucleophilic difluoro(sulfonato) methylation reaction.

3.3 Modulating the Fluoroalkylation Reactions by Changing the Metal Counterion

Based on our understanding of the proposed "NFE," the nucleophilic fluoroalkylation reactions can be improved by attaching an electron-withdrawing group, which makes the fluorinated carbanions thermally more stable and "softer." Moreover, during the past few years, our research results have showed that the metal counterion (M$^+$) of the fluorinated carbanions may play an important role in C$_F$–C bond formation reaction.

In 2010 we reported that the in situ generated PhSO$_2$CF$_2^-$ anion can react with a variety of structurally diverse N-tert-butylsulfinyl ketimines, which will give a variety of structurally diverse homochiral α-difluoromethyl tertiary carbinamines, including α-difluoromethyl allylic amines and α-difluoromethyl propargylamines in excellent yields and with very high diastereoselectivity (Scheme 14) [49].

The reaction between the PhSO$_2$CF$_2^-$ anion and α-unsaturated N-tert-butylsulfinyl ketimines is highly dependent on the metal counterions, which is different from aldimines and β-acetylenic N-tert-butylsulfinyl ketimines. The former preferred potassium base to other metal bases, while the later showed little base preference. It is believed that the kinetically preferred generation of the PhSO$_2$CF$_2^-$ anion by KHMDS and nucleophilic addition of the PhSO$_2$CF$_2^-$ anion to ketimines over the undesired aza-enolization of ketimines are the key factors for the success of difluoromethylation of simple ketimines.

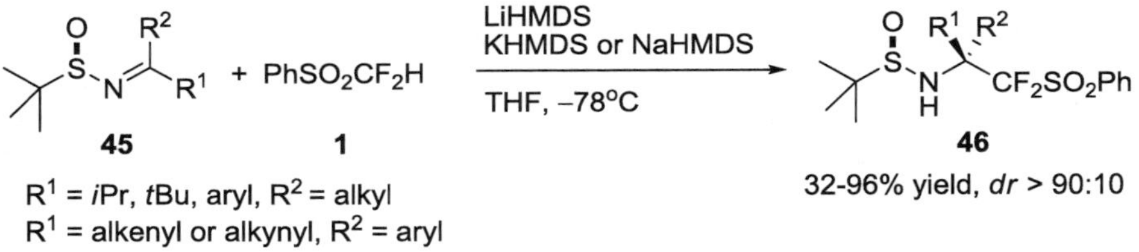

R^1 = iPr, tBu, aryl, R^2 = alkyl
R^1 = alkenyl or alkynyl, R^2 = aryl

32–96% yield, dr > 90:10

Scheme 14 Nucleophilic difluoromethylation of ketimines

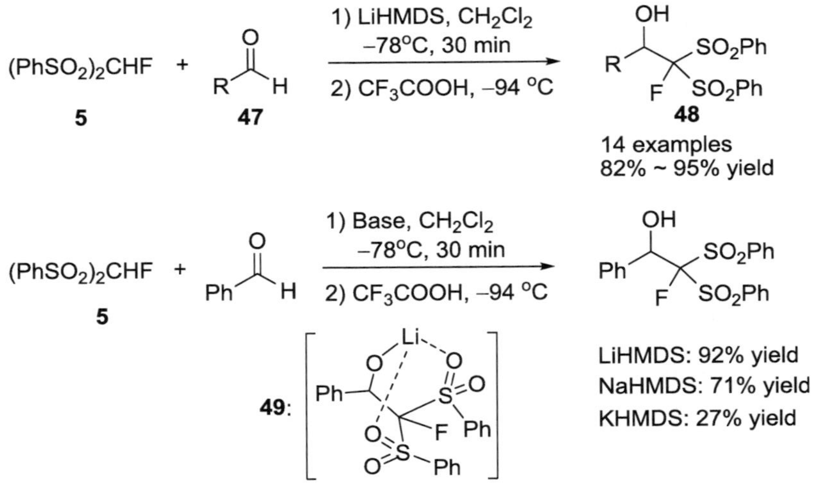

Scheme 15 Nucleophilic additions to aldehyde with (PhSO$_2$)$_2$CFH

The nucleophilic addition reaction between (PhSO$_2$)$_2$CFH (**5**) and an aldehyde, a reaction that was previously believed to be unattainable, was recently accomplished by us (Scheme 15) [50]. Among the lithium, sodium, and potassium hexamethyldisilazides, LiHMDS serves as the best base for this addition reaction, which indicates that the strong Li–O coordination in the carbinolate intermediate **49** plays a very important role in the success of the reaction (Scheme 15). Furthermore, under the same reaction conditions, the similar nucleophilic addition reactions between (PhSO$_2$)$_2$CHX (X = Cl or H) and PhCHO were unsuccessful. That means the fluorine substitution is also very important for the nucleophilic fluoroalkylation reaction, which was further supported by DFT calculations.

The copper-mediated perfluoroalkyl (or trifluoromethyl) cross-coupling reaction, involving an "R$_f$-Cu" species, are well documented; however, corresponding difluoromethylation involving a "RCF$_2$-Cu" intermediate was less studied. It is believed that the carbon–metal covalent bond in fluorinated carbanion will tune its reactivity. Recently, we systematically studied the Cu-mediated fluoroalkylation reactions with iododifluoroacetamides **50** (Scheme 16) [51]. It was found that three types of reactions (cross-coupling, intramolecular cyclization, and homocoupling reactions) may coexist in this Cu-mediated reaction between the reagent **50** and aryl/alkenyl iodides. The selectivity among these three types of reactions could be controlled by tuning the substituents on the nitrogen atom of iododifluoroacetamides, and/or by removing the cross-coupling reaction partner (aryl/alkenyl halides).

By using PhSO$_2$CF$_2$TMS/CuI/CsF in DMF, PhSO$_2$CF$_2$Cu (**54**) was successfully prepared, which was found to undergo efficiently cross-coupling reactions with propargyl chlorides and alkynyl halides to give relative allenes (**55**) and alkynes (**56**) (Scheme 17) [52].

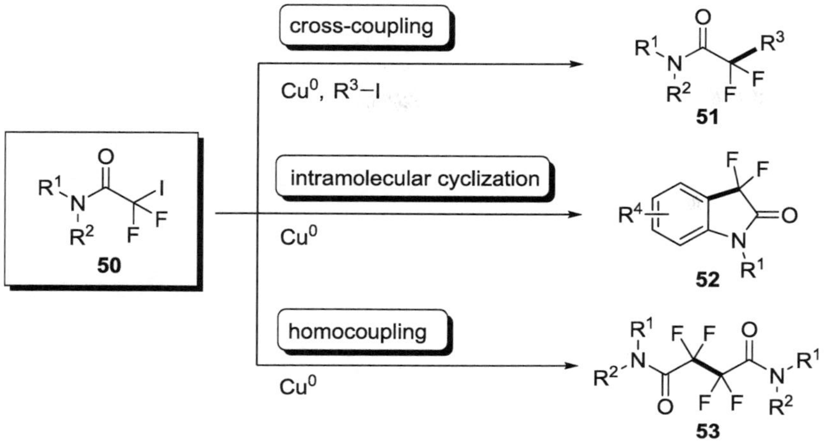

Scheme 16 Cu-mediated fluoroalkylation reactions with iododifluoroacetamides **50**

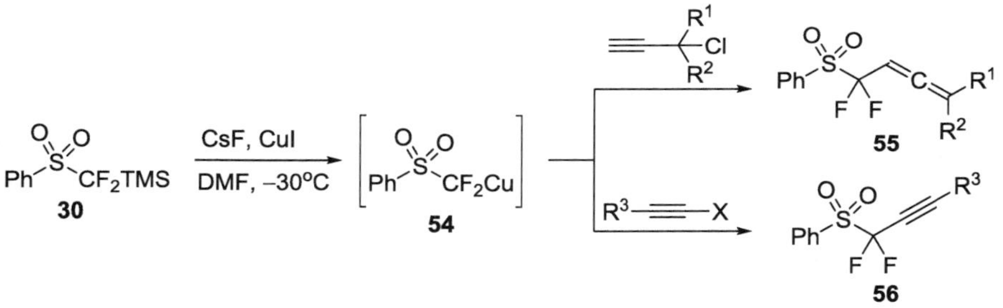

Scheme 17 Cross-coupling reactions with PhSO$_2$CF$_2$Cu

3.4 Modulating the Fluoroalkylation Reactions by Enhancing the Generation of Carbene Species

As mentioned in the discussion of "NFE," the Factor 1 (self-decomposition of carbanion **3**) should always have a negative effect for nucleophilic fluoroalkylation reactions; however, by enhancing this decomposition, novel and efficient difluorocarbene reagents were successfully developed by our research group (Scheme 18) [42, 53–56]. Difluorocarbene is the most stable dihalocarbene due to the interaction of the carbene center with the lone pairs of fluorine [57]. Typical synthetic applications of difluorocarbene include (1) homocoupling at very high temperature to produce tetrafluoroethylene [58], (2) reacting with O-, S-, N-, P-, and C-nucleophiles to give difluoromethylated products, and (3) undergoing [2 + 1] cycloaddition with alkynes or alkenes [57].

In 2006 we reported a novel and non-ODS-based (ODS = ozone-depleting substance) difluorocarbene reagent **57** (Scheme 19) [53]. 2-Chloro-2,2-

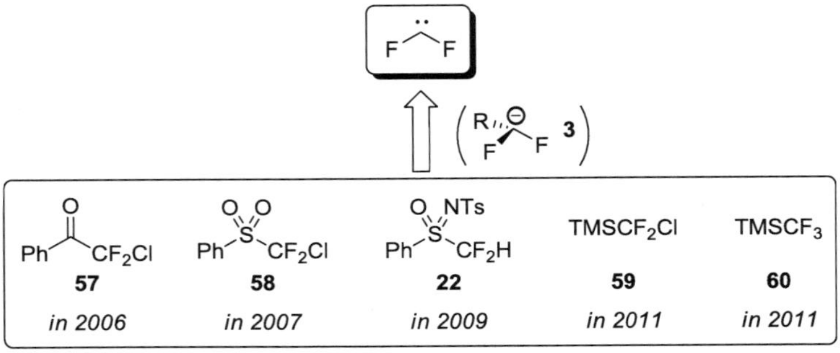

Scheme 18 New difluorocarbene reagents

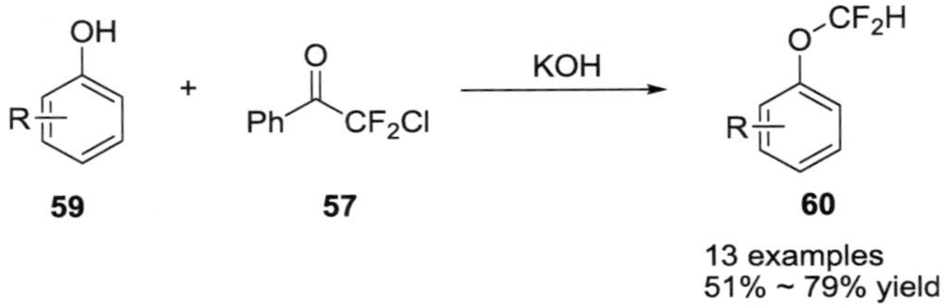

Ph-C(O)-CF$_2$Cl	Ph-S(O)$_2$-CF$_2$Cl	Ph-S(O)(NTs)-CF$_2$H	TMSCF$_2$Cl	TMSCF$_3$
57	**58**	**22**	**59**	**60**
in 2006	in 2007	in 2009	in 2011	in 2011

Scheme 19 Difluoromethylation with carbene reagent **57**

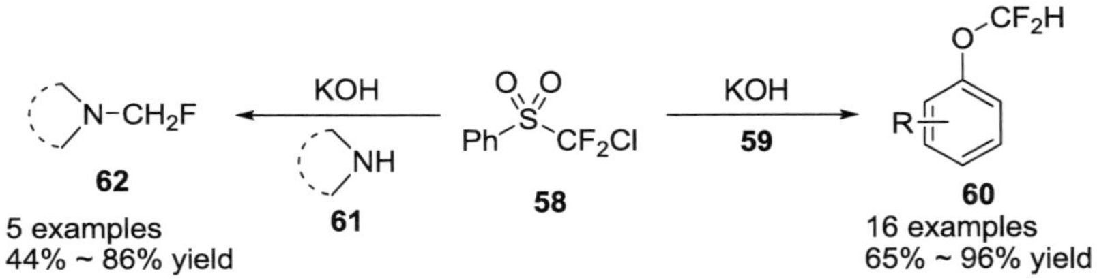

Scheme 20 Difluoromethylation with carbene reagent **58**

difluoroacetophenone (**57**) was found to react readily with a variety of structurally diverse phenol derivatives (**59**) to produce aryl difluoromethyl ethers (**60**) in good yields.

In 2007 another environmentally friendly difluorocarbene reagent was developed by us (Scheme 20) [54]. Chlorodifluoromethyl phenyl sulfone (**58**) was prepared with non-Freon- or Halon-based procedures, and the compound was found to be an efficient difluorocarbene reagent for O- and N-difluoromethylation of phenols and N-heterocycles.

As mentioned above, the reagent **22** (PhSO(NTs)CF$_2$H) can be used to transfer the CF$_2$H group to S-, N-, and C-nucleophiles with a difluorocarbene mechanism [42]. However, besides the reagent **22**, PhCOCF$_2$Cl (**57**) and PhSO$_2$CF$_2$Cl (**58**) were found to be incapable of undergoing [2 + 1] cycloaddition reactions with alkynes or alkenes.

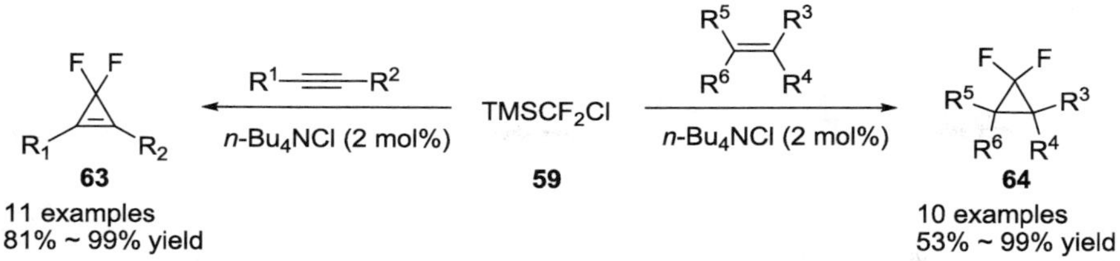

Scheme 21 Difluoromethylation with carbene reagent **59**

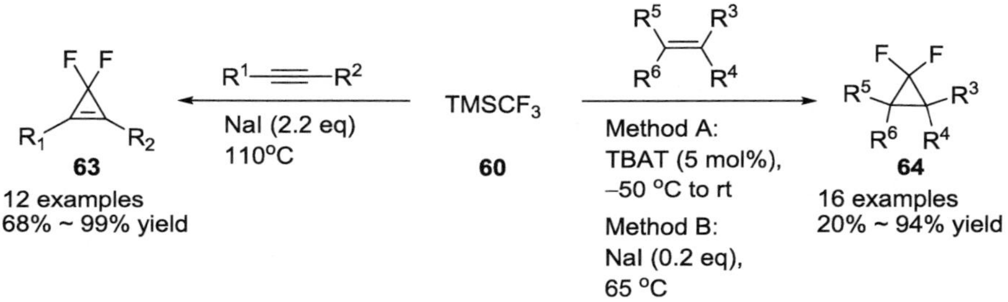

Scheme 22 [2 + 1] Cycloaddition by using TMSCF$_3$ as a difluorocarbene reagent

In 2011 we reported that TMSCF$_2$Cl (**59**) can be used as a relatively nontoxic difluorocarbene precursor (Scheme 21) [55]. Under mild and neutral conditions, by using chloride ion as a catalyst, difluorocarbene can be efficiently generated from TMSCF$_2$Cl (**59**), which will undergo [2 + 1] cycloaddition reactions with alkynes and alkenes to give *gem*-difluorocyclopropenes (**63**) and difluorocyclopropanes (**64**).

Trifluoromethyltrimethylsilane (TMSCF$_3$ **60**), known as Ruppert–Prakash reagent, is the most widely used nucleophilic trifluoromethylating agent for a variety of applications. Recently, an efficient method for the generation of difluorocarbene from TMSCF$_3$ (**60**) was successfully developed (Scheme 22) [56]. It was found that TBAT was able to initiate decomposition of **60** to generate difluorocarbene at low temperatures to give corresponding *gem*-difluorocyclopropane (**64**) in good yields (Method A) and, promoted by NaI, the [2 + 1] cycloaddition reactions of alkenes and alkynes at higher temperatures were also successful (Method B).

4 Conclusion

The incorporation of fluorine atom(s) into organic compounds has become a useful strategy in drug design and new functional material development. The presence of fluorine on a carbanion center will dramatically influence the nucleophilic alkylation reactions. Based on our research, we noticed that, in many cases, the fluorine substitution on the carbanionic carbon poses a negative effect in nucleophilic fluoroalkylation reactions. We proposed this effect as "NFE." At least two factors were believed to contribute to the NFE: (1) thermal instability of fluorinated carbanions caused by α-elimination (self-decomposition) and (2) the intrinsic nucleophilicity of fluorinated carbanion influenced by the fluorine atoms (such as hard/soft nature of the fluorinated carbanions). Although a comprehensive understanding of the unusual reactivity of fluorinated carbanions needs more investigations and insights, our discussion in terms of NFE is hoped to be helpful to understand many nucleophilic fluoroalkylation reactions. During the past several years, our group has attempted to design nucleophilic fluoroalkylation reactions by tackling the NFE with fluorinated sulfones and related reagents. Our research results were summarized as four methods to modulate the fluoroalkylation reactions: (1) changing the number of fluorine atoms, (2) slightly changing the neighboring groups, (3) changing the metal counterion, including using carbon–metal covalent bond to tune the reactivity, and (4) enhancing the generation of carbene species.

Acknowledgments We are grateful to the following agencies for financial support: the National Natural Science Foundation of China (20825209, 20772144, 20832008) and the Chinese Academy of Sciences.

References

1. Müller K, Faeh C, Diederich F (2007) Fluorine in pharmaceuticals: looking beyond intuition. Science 317:1881
2. Bégué J-P, Bonnet-Delpon D (2008) Bioorganic and medicinal chemistry of fluorine. Wiley, Hoboke
3. Hagmann WK (2008) The many roles for fluorine in medicinal chemistry. J Med Chem 51:4359
4. Prakash GKS, Chacko S (2008) Novel nucleophilic and electrophilic fluoroalkylation methods. Curr Opin Drug Discov Dev 11:793
5. Kirsch P (2004) Modern fluoroorganic chemistry: synthesis, reactivity, applications. Wiley-VCH, Weinheim
6. Banks RE, Smart BE, Tatlow JC (1994) Organofluorine chemistry: principles and commercial applications. Plenum, New York
7. O'Hagan D, Schaffrath C, Cobb SL, Hamilton JTG, Murphy CD (2002) Biochemistry: biosynthesis of an organofluorine molecule. Nature 416:279
8. Uneyama K (2006) Organofluorine chemistry. Blackwell, Oxford
9. Chambers RD (2004) Fluorine in organic chemistry. Blackwell, Oxford
10. Prakash GKS, Hu J (2007) Selective fluoroalkylations with fluorinated sulfones, sulfoxides, and sulfides. Acc Chem Res 40:921
11. Farnham WB (1996) Fluorinated carbanions. Chem Rev 96:1633
12. Prakash GKS, Yudin AK (1997) Perfluoroalkylation with organosilicon reagents. Chem Rev 97:757
13. Prakash GKS, Krishnamurti R, Olah GA (1989) Synthetic methods and reactions. 141. Fluoride-induced trifluoromethylation of carbonyl compounds with trifluoromethyltrimethylsilane (TMS-CF$_3$). A trifluoromethide equivalent. J Am Chem Soc 111:393
14. Hagiwara T, Fuchikami T (1995) Difluoroalkylation of carbonyl compounds with (1,1-difluoroalkyl)silane derivatives. Synlett 7:717

15. Hu J, Zhang W, Wang F (2009) Selective difluoromethylation and monofluoromethylation reactions. Chem Commun 7465
16. Ni C, Hu J (2011) Selective nucleophilic fluoroalkylations facilitated by removable activation groups. Synlett 6:770
17. Bickelhaupt FM, Hermann HL, Boche G (2006) α-Stabilization of carbanions: fluorine is more effective than the heavier halogens. Angew Chem Int Ed 45:823
18. Faustov VI, D'yachenko AI, Nefedov OM (1979) Unstable organolithium halide compounds. Communication 7. Thermodynamic calculations of the mechanism of decomposition of organolithium halide compounds. Izvestiya Akademii Nauk SSSR, Seriya Khimicheskaya 2183
19. Boche G, Lohrenz JCW (2001) The electrophilic nature of carbenoids, nitrenoids, and oxenoids. Chem Rev 101:697
20. Capriati V, Florio S (2010) Anatomy of long-lasting love affairs with lithium carbenoids: past and present status and future prospects. Chem Eur J 16:4152
21. Stahly GP (1989) Nucleophilic addition of difluoromethyl phenyl sulfone to aldehydes and various transformations of the resulting alcohols. J Fluor Chem 43:53
22. Hine J, Porter JJ (1960) Formation of difluoromethylene from difluoromethyl phenyl sulfone and sodium methoxide. J Am Chem Soc 82:6178
23. Hu J (2009) Nucleophilic, radical, and electrophilic (phenylsulfonyl) difluoromethylations. J Fluor Chem 130:1130
24. Ni C, Li Y, Hu J (2006) Nucleophilic fluoroalkylation of epoxides with fluorinated sulfones. J Org Chem 71:6829
25. Ni C, Liu J, Zhang L, Hu J (2007) A remarkably efficient fluoroalkylation of cyclic sulfates and sulfamidates with PhSO$_2$CF$_2$H: facile entry into β-difluoromethylated or β-difluoromethylenated alcohols and amines. Angew Chem Int Ed 46:786
26. Fleming I (1976) Frontier orbitals and organic chemical reactions. Wiley, New York
27. Ho T-L (1977) Hard and soft acids and bases principle in organic chemistry. Academic, New York
28. Ni C, Zhang L, Hu J (2008) Nucleophilic fluoroalkylation of α, β-enones, arynes, and activated alkynes with fluorinated sulfones: probing the hard/soft nature of fluorinated carbanions. J Org Chem 73:5699
29. Prakash GKS, Hu J, Olah GA (2003) Preparation of tri- and difluoromethylsilanes via an unusual magnesium metal-mediated reductive tri- and difluoromethylation of chlorosilanes using tri- and difluoromethyl sulfides, sulfoxides, and sulfones. J Org Chem 68:4457
30. Prakash GKS, Hu J, Wang Y, Olah GA (2004) Nucleophilic difluoromethylation of primary alkyl halides using difluoromethyl phenyl sulfone as a difluoromethyl anion equivalent. Org Lett 6:4315
31. Prakash GKS, Hu J, Wang Y, Olah GA (2004) Difluoromethyl phenyl sulfone, a difluoromethylidene equivalent: use in the synthesis of 1,1-difluoro-1-alkenes. Angew Chem Int Ed 43:5203
32. Prakash GKS, Hu J, Mathew T, Olah GA (2003) Difluoromethyl phenyl sulfone as a selective difluoromethylene dianion equivalent: one-pot stereoselective synthesis of anti-2,2-difluoropropane-1,3-diols. Angew Chem Int Ed 42:5216
33. Prakash GKS, Hu J, Wang Y, Olah GA (2005) Convenient synthesis of difluoromethyl alcohols from both enolizable and non-enolizable carbonyl compounds with difluoromethyl phenyl sulfone. Eur J Org Chem 2218
34. Ye J-D, Liao X, Piccirilli JA (2005) Synthesis of 2′-C-difluoromethylribonucleosides and their enzymatic incorporation into oligonucleotides. J Org Chem 70:7902
35. Liu J, Ni C, Wang F, Hu J (2008) Stereoselective synthesis of α-difluoromethyl-β-amino alcohols via nucleophilic difluoromethylation with Me$_3$SiCF$_2$SO$_2$Ph. Tetrahedron Lett 49:1605
36. Li Y, Hu J (2005) Facile synthesis of chiral α-difluoromethyl amines from N-(tert-butylsulfinyl)aldimines. Angew Chem Int Ed 44:5882

37. Liu J, Li Y, Hu J (2007) Stereoselective synthesis of di- and monofluoromethylated vicinal ethylenediamines with di- and monofluoromethyl sulfones. J Org Chem 72:3119
38. Li Y, Ni C, Liu J, Zhang L, Zheng J, Zhu L, Hu J (2006) Stereoselective nucleophilic monofluoromethylation of N-(tert-butanesulfinyl)imines with fluoromethyl phenyl sulfone. Org Lett 8:1693
39. Liu J, Zhang L, Hu J (2008) Stereoselective monofluoromethylation of N-tert-butylsulfinyl ketimines using pregenerated fluoro(phenylsulfonyl)methyl anion. Org Lett 10:5377
40. Reggelin M, Zur C (2000) Sulfoximines. Structures, properties, and synthetic applications. Synthesis 1:1
41. Boys ML, Collington EW, Finch H, Swanson S, Whitehead JF (1988) α-Fluoromethyl-n-methyl-phenylsulfoximine: a new fluoromethylenation reagent. Tetrahedron Lett 29:3365
42. Zhang W, Wang F, Hu J (2009) N-Tosyl-S-difluoromethyl-S-phenylsulfoximine: A new difluoromethylation reagent for S-, N-, and C-nucleophiles. Org Lett 11:2109
43. Noritake S, Shibata N, Makamura S, Toru T, Shiro M (2008) Fluorinated Johnson reagent for transfer-trifluoromethylation to carbon nucleophiles. Eur J Org Chem 3465
44. Nomura Y, Tokunaga E, Shibata N (2011) Inherent oxygen preference in enolate monofluoromethylation and a synthetic entry to monofluoromethyl ethers. Angew Chem Int Ed 50:1885
45. Zhang W, Huang W, Hu J (2009) Highly stereoselective synthesis of monofluoroalkenes from α-fluorosulfoximines and nitrones. Angew Chem Int Ed 48:9858
46. Zhang W, Hu J (2010) Efficient synthesis and ring-opening reactions of monofluorinated epoxides derived from α-fluorosulfoximines. Adv Synth Catal 352:2799
47. Zhao Y, Huang W, Zhu L, Hu J (2010) Difluoromethyl 2-pyridyl sulfone: a new gem-difluoroolefination reagent for aldehydes and ketones. Org Lett 12:1444
48. Prakash GKS, Ni C, Wang F, Hu J, Olah GA (2011) From difluoromethyl 2-pyridyl sulfone to difluorinated sulfonates: a protocol for nucleophilic difluoro(sulfonato)methylation. Angew Chem Int Ed 50:2559
49. Liu J, Hu J (2010) Highly diastereoselective synthesis of α-difluoromethyl amines from N-tert-butylsulfinyl ketimines and difluoromethyl phenyl sulfone. Chem Eur J 16:11443
50. Shen X, Zhang L, Zhao Y, Zhu L, Li G, Hu J (2011) Nucleophilic fluoromethylation of aldehydes with fluorobis(phenylsulfonyl)methane: the importance of strong Li-O coordination and fluorine substitution for C-C bond formation. Angew Chem Int Ed 50:2588
51. Zhu J, Zhang W, Zhang L, Liu J, Zheng J, Hu J (2010) Copper-mediated fluoroalkylation reactions with iododifluoroacetamides: controlling the selectivity among cross-coupling, intramolecular cyclization, and homocoupling reactions. J Org Chem 75:5505
52. Zhu J, Wang F, Huang W, Zhao Y, Ye W, Hu J (2011) Copper-mediated fluoroalkylation reactions with [(phenylsulfonyl)-difluoromethyl] trimethylsilane: synthesis of PhSO$_2$CF$_2$-containing allenes and alkynes. Synlett 7:899
53. Zhang L, Zheng J, Hu J (2006) 2-Chloro-2,2-difluoroacetophenone: a non-ODS-based difluorocarbene precursor and its use in the difluoromethylation of phenol derivatives. J Org Chem 71:9845
54. Zheng J, Li Y, Zhang L, Hu J, Meuzelaar GJ, Federsel H-J (2007) Chlorodifluoromethyl phenyl sulfone: a novel non-ozone-depleting substance-based difluorocarbene reagent for O- and N-difluoromethylations. Chem Commun 48:5149
55. Wang F, Zhang W, Zhu J, Li H, Huang K-W, Hu J (2011) Chloride ion-catalyzed generation of difluorocarbene for efficient preparation of gem-difluorinated cyclopropenes and cyclopropanes. Chem Commun 47:2411
56. Wang F, Luo T, Hu J, Wang Y, Krishnan HS, Jog PV, Ganesh SK, Prakash GKS, Olah GA (2011) Synthesis of gem-difluorinated cyclopropanes and cyclopropenes using TMSCF$_3$ as a difluorocarbene source. Angew Chem Int Ed. 50:7153
57. Brahms LS, Dailey WP (1996) Fluorinated carbenes. Chem Rev 96:1585
58. Hudlicky M, Pavlath AE (1995) Chemistry of organic fluorine compounds II. American Chemical Society, Washington DC

Synthetic and Biological Applications of Fluorous Reagents as Phase Tags

Santos Fustero, José Luis Aceña, and Silvia Catalán

Abstract The search for new and better techniques for the purification of organic compounds has made the recent emergence of fluorous chemistry in the field of organic synthesis possible. Using fluorous reagents as phase tags allows for the access of different synthetic routes, and this has been translated into a time reduction and higher simplicity compared to standard, nonfluorous procedures. The synthesis in fluorous phase of target molecules can be pursued in a parallel or combinatorial manner in order to access chemical libraries with structural and/or stereochemical diversity. The preparation of radiolabeled molecules also benefits from the advantages of fluorous synthesis. Finally, biochemical tools such as microarrays or proteomic techniques are improved by means of fluorous-tagged compounds.

Keywords Diversity-oriented synthesis · Fluorous synthesis · Library scaffolds · Microarrays · Radiochemistry

Contents

1 Introduction 46
2 Synthesis in Fluorous Phase 47
    2.1 Fluorous Target-Oriented Synthesis 47
    2.2 Fluorous Diversity-Oriented Synthesis 50

S. Fustero (✉)
Departamento de Química Orgánica, Universidad de Valencia, 46100 Burjassot, Spain

Laboratorio de Moléculas Orgánicas, Centro de Investigación Príncipe Felipe, 46012 Valencia, Spain
e-mail: santos.fustero@uv.es

J.L. Aceña and S. Catalán
Laboratorio de Moléculas Orgánicas, Centro de Investigación Príncipe Felipe, 46012 Valencia, Spain

    2.3 Fluorous Mixture Synthesis 56
3   Fluorous Radiochemistry 59
4   Biological Applications of Fluorous Reagents 62
    4.1 Fluorous Microarrays 62
    4.2 Fluorous Proteomics 64
5   Conclusions 64
References 65

1 Introduction

The efficiency of a synthetic process is judged primarily by the overall yield of the chemical reactions involved, although the ease with which the resulting products are purified must also be taken into account. Therefore, the design of new and better techniques to facilitate the purification of organic compounds is as important as the development of new synthetic methods. In this context, the use of fluorous reagents and materials offers the opportunity for time reduction and higher simplicity in a similar way as solid-phase synthesis but incorporates several advantages derived from conducting reactions in homogeneous media [1].

In theory, any organic compound can be transformed into its corresponding fluorous analog by attaching one or more fluorous groups ("tags"), composed of a perfluoroalkyl chain and a suitable spacer in order to maintain the reactivity of other functional groups present in the original molecule. In contrast, the distinct physical properties of fluorous compounds allow for their easy separation from nonfluorous species using appropriate purification techniques. Early work in fluorous chemistry was based mostly on the use of heavy fluorous tags (fluorine atoms contribute to more than 60% of molecular weight), which implied that the resulting compounds are soluble only in perfluorinated solvents [2]. Nowadays, techniques such as fluorous solid-phase extractions (F-SPE) are preferred for the purification of light fluorous molecules (with a low overall content of fluorine) with the aid of fluorous silica and conventional nonfluorous solvents, which circumvent the expense of relatively costly fluorous materials [3–5].

The potential of fluorous tags has inspired the development of many fluorous reagents, scavengers, and catalysts, which are able to mediate synthetic operations involving nonfluorous species. The reversed possibility is the use of fluorous reagents as phase tags in order to prepare target compounds in fluorous phase. This approach involves the attachment of the fluorous tag(s) to a starting substrate(s), and in most cases it can be conveniently removed at the end of the synthesis [6]. A number of these reagents act as fluorous protecting groups and many of them are now commercially available.

The aim of this chapter is to highlight some of the most recent applications of fluorous reagents when employed as phase tags. As mentioned before, their main purpose is the fluorous synthesis of biomolecules, natural products, etc., which may, on the one hand, be conducted either in a standard manner or, on the other hand, by using parallel or combinatorial techniques. An emerging field is the application of fluorous chemistry in the synthesis of radiolabeled molecules. Finally, biological applications of fluorous-tagged compounds are also discussed [7].

2 Synthesis in Fluorous Phase

2.1 Fluorous Target-Oriented Synthesis

The significance of a synthetic route in fluorous phase relies on several aspects, including the accessibility of the fluorous starting material, the ease with which it is introduced into an organic substrate, and most importantly the efficiency of the detagging process. In fact, a convenient, orthogonal method for the removal of a fluorous tag is always needed. Furthermore, the possibility of creating a new functionality during the detagging process represents an added value.

Currently, peptidomimetics are popular targets in the field of medicinal chemistry since they have led to a large number of drug candidates. Nevertheless, their synthesis in fluorous phase may be hampered by the difficult elimination of the fluorous tag without affecting their structural integrity. For instance, the fluorous synthesis of fluorinated retropeptides developed by Fustero et al. using the commercially available alcohol $C_8F_{17}(CH_2)_3OH$ turned out to be impractical because its ultimate removal caused epimerizations in the final products. Accordingly, a fluorous analog of 2-(trimethylsilyl)ethanol 2 was designed and synthesized in two steps from iodide 1 (Scheme 1). Compound 2 reacted with Fmoc-protected valine and subsequently with 2-(trifluoromethyl)acryloyl chloride to give Michael acceptor 4. Addition of a second valine unit produced compound 5 with high diastereoselectivity, and final transesterification reaction with TBAF and BnBr in the presence of molecular sieves afforded retropeptide 6 without epimerization of the sensitive CF_3 group. All synthetic intermediates were purified by means of F-SPE [8].

A further example of the advantageous use of the ^{F}TMSE group allowed for the asymmetric synthesis of β,β-difluorinated cyclic quaternary α-amino acid derivatives. The anchoring of 2 was carried out through the alkoxycarbonylation of chiral imidoyl iodide 7 to yield fluorous imino ester 8 (Scheme 2). Next, an allylation process took place with total diastereoselectivity, and was followed by cyclization of diene 9 through a ring-closing metathesis (RCM) reaction to afford fluorous amino ester 10. Finally, removal of the fluorous tag as above (TBAF/BnBr) furnished the target compound 11 [9, 10].

Gouverneur et al. easily removed a related silicon-derived fluorous tag by adapting their previously reported nonfluorous method based on an electrophilic fluorodesilylation reaction [11]. Thus, addition of allylmagnesium bromide to fluorous chlorosilane 12 afforded allylsilane 13 in good yield (Scheme 3). Subsequent cross metathesis (CM) reaction with several olefins produced

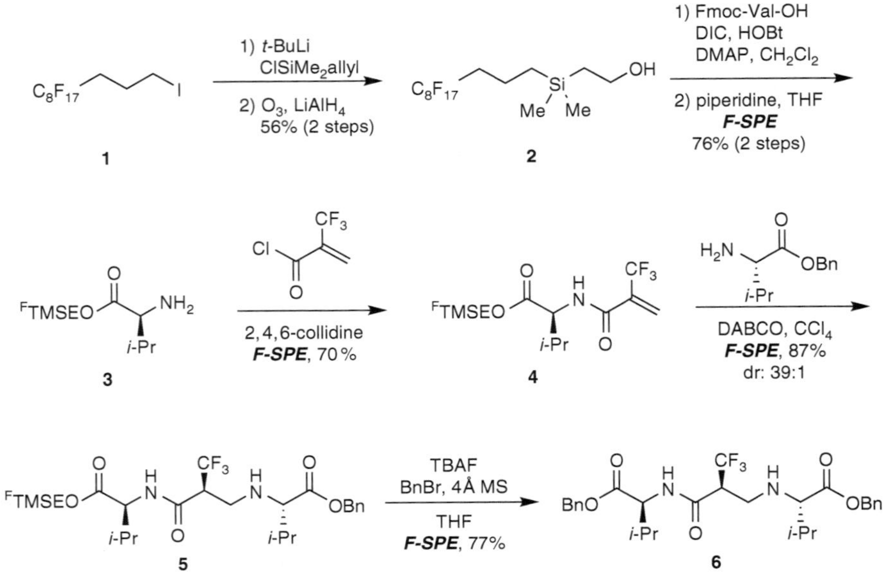

Scheme 1 Fluorous synthesis of fluorinated retropeptide **6**. ^{F}*TMSE* $C_8F_{17}(CH_2)_3SiMe_2(CH_2)_2$, *DIC* N,N'-diisopropylcarbodiimide, *HOBt* 1-hydroxybenzotriazole

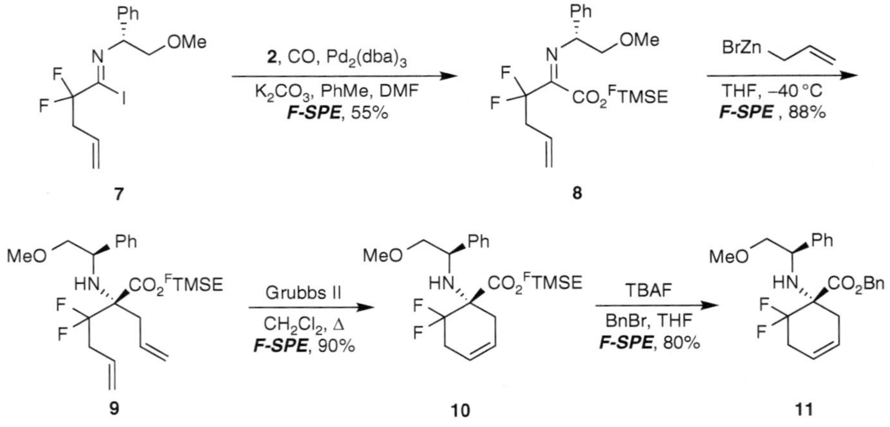

Scheme 2 Fluorous synthesis of 2,2-difluoro-1-aminocyclohexanecarboxylic acid derivative **11**. ^{F}*TMSE* $C_8F_{17}(CH_2)_3SiMe_2(CH_2)_2$, *dba* dibenzylideneacetone

compounds **14** as mixtures of geometrical isomers. Final detagging of **14** by reaction with Selectfluor led to allylic fluorides **15** after purification by F-SPE [12].

A similar procedure for the detagging process involved the oxidation of the fluorous silyl group. In this case, the starting fluorous reagent was bromobenzene **16**, which was metallated with *n*-BuLi and reacted with allylchlorodiisopropylsilane

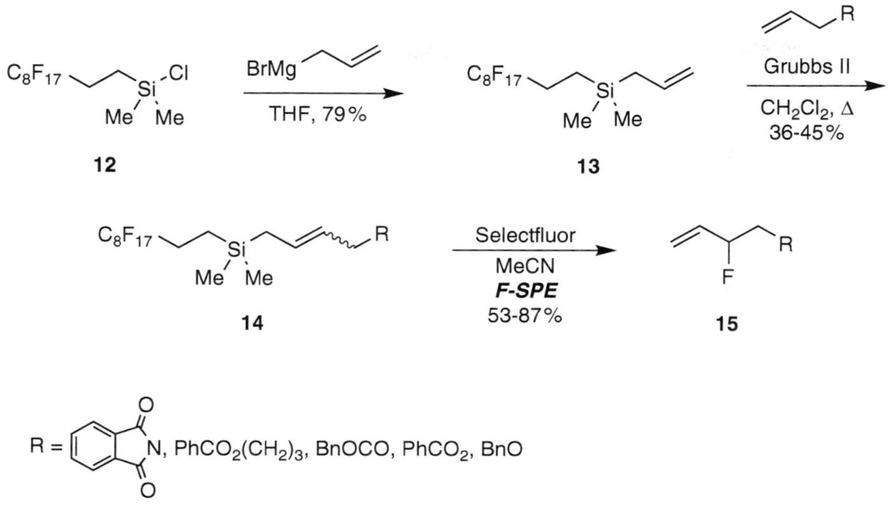

Scheme 3 Fluorous synthesis of allylic fluorides **15**

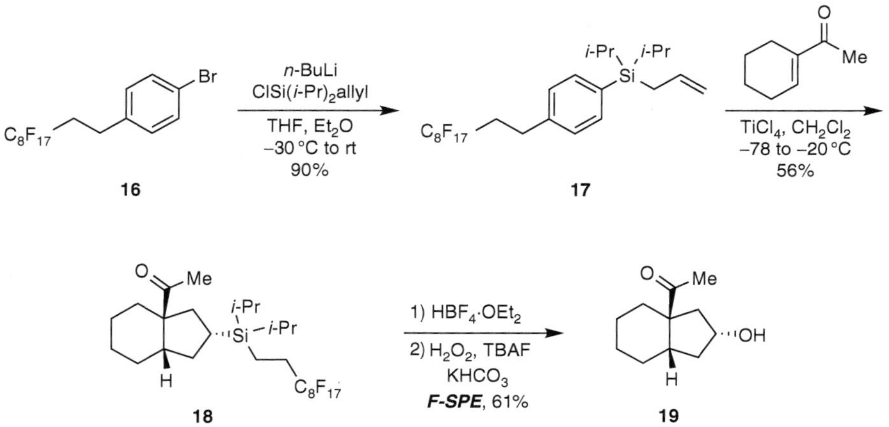

Scheme 4 Oxidative detagging of a fluorous silane

to yield **17** (Scheme 4). Further cyclization with 1-acetylcyclohexene in the presence of TiCl$_4$ gave silane **18**, and removal of the fluorous moiety by sequential treatment with tetrafluoroboric acid and H$_2$O$_2$/TBAF/KHCO$_3$ afforded alcohol **19** [13].

A fluorous synthetic equivalent of ammonia containing an easily cleavable N–O bond has recently been described by Kristensen et al. for the convenient preparation in fluorous phase of nitrogen-containing compounds such as ureas, amides, sulfonamides, or carbamates. This method is illustrated by the synthesis of the drug itopride (**23**) (Scheme 5). The fluorous ammonia derivative **20** was obtained in

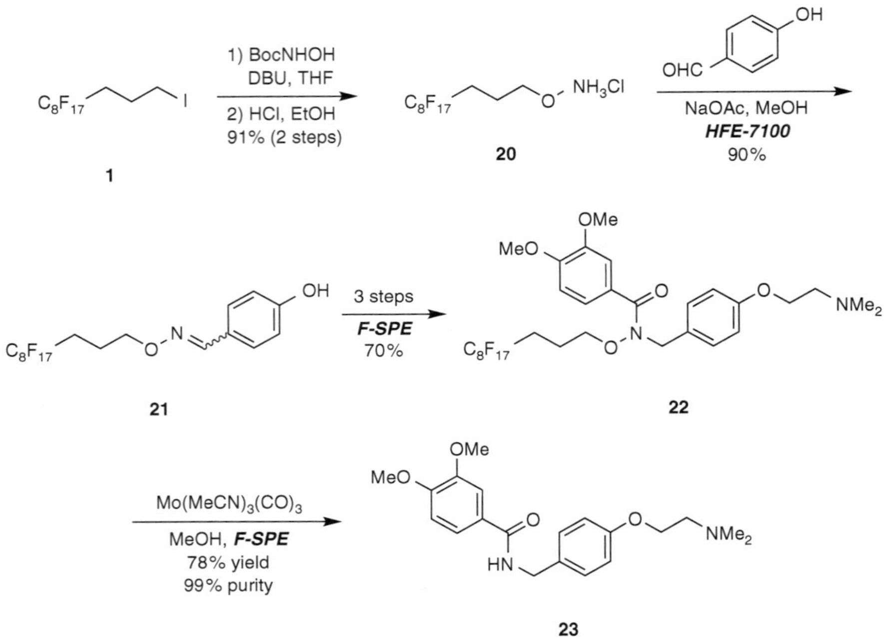

Scheme 5 Fluorous synthesis of itopride (**23**) using a fluorous ammonia surrogate

two steps from iodide **1**, and next reacted with 4-hydroxybenzaldehyde to produce the corresponding oxime **21**, purified by liquid–liquid extraction with HFE-7100 (C$_4$F$_9$OMe). The rest of synthetic intermediates were purified by F-SPE, and the selective removal of the fluorous tag in **22** was accomplished by reaction with a molybdenum complex [14]. A further application of this fluorous reagent was its Cu-catalyzed coupling with iodoaromatic compounds in order to produce anilines in high yields and purities [15].

2.2 Fluorous Diversity-Oriented Synthesis

Nowadays, drug discovery strategies increasingly demand the production of new libraries of potentially bioactive compounds in a rapid and efficient high-throughput manner. An emerging approach is diversity-oriented synthesis (DOS), which relies on the preparation of libraries of structurally diverse, drug-like compounds from a common intermediate in order to cover as much chemical space as possible [16, 17]. Given the possibilities of fluorous methods, their current relevance for the parallel synthesis of compound libraries is not surprising.

The fluorous-phase synthesis of structurally diverse nitrogen-containing heterocycles reported by Procter et al. was based on a Pummerer cyclative capture

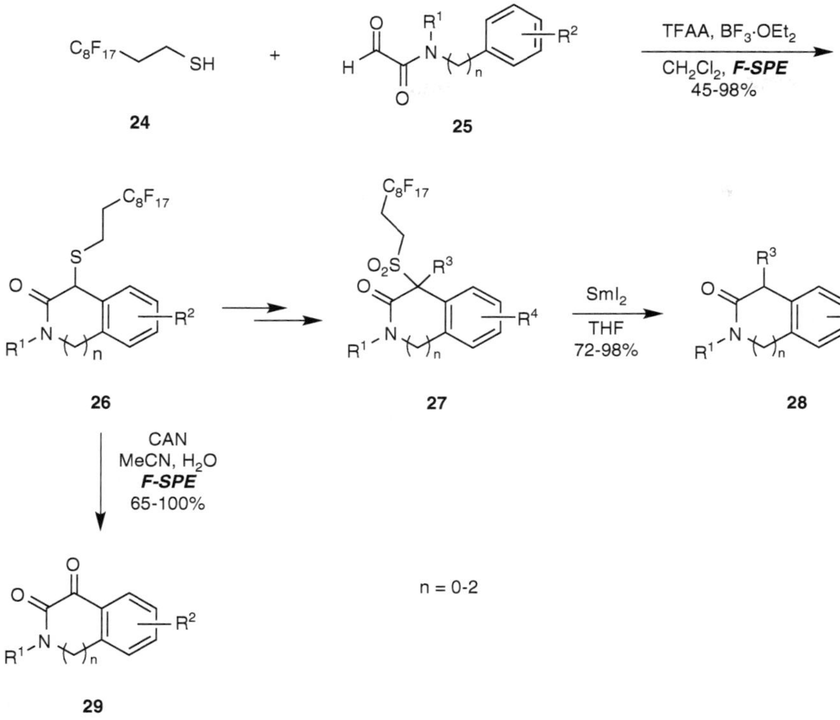

Scheme 6 Fluorous synthesis of nitrogen-containing heterocycles **28** and **29** via a Pummerer cyclative reaction

reaction, starting from fluorous thiol **24** and several substituted glyoxamides **25** to form oxindoles ($n = 0$), tetrahydroisoquinolinones ($n = 1$), or tetrahydrobenzazepinones ($n = 2$) **26** (Scheme 6). The introduction of further structural diversity also led to the attainment of tagged molecules **27**, and the sulfur atom provided different oxidation states as well as a variety of cleavage strategies. For instance, reductive elimination of the fluorous tag was carried out in sulfone **27** by reaction with SmI$_2$ to reveal deprotected products **28** with no need of purification, whereas an oxidative cleavage of the sulfide moiety in **26** with ceric ammonium nitrate (CAN) afforded compounds **29** [18, 19].

The same group also developed an SmI$_2$-mediated cleavage-cyclization sequence that removed the fluorous tag and introduced more diversity through the formation of new heterocyclic rings. Thus, fluorous sulfinyl oxindole **30** ($n = 0$) reacted with divinylsulfone to afford adduct **31**, and subsequent detagging with SmI$_2$ produced a samarium enolate which ultimately led to spirocyclic sulfone **32** (Scheme 7). Alternatively, reaction of sulfone **30** ($n = 2$) with 2-nitrobenzyl bromide to give compound **33** was followed by sequential treatment with SmI$_2$ and AcOH to furnish indoloquinoline **34** [20].

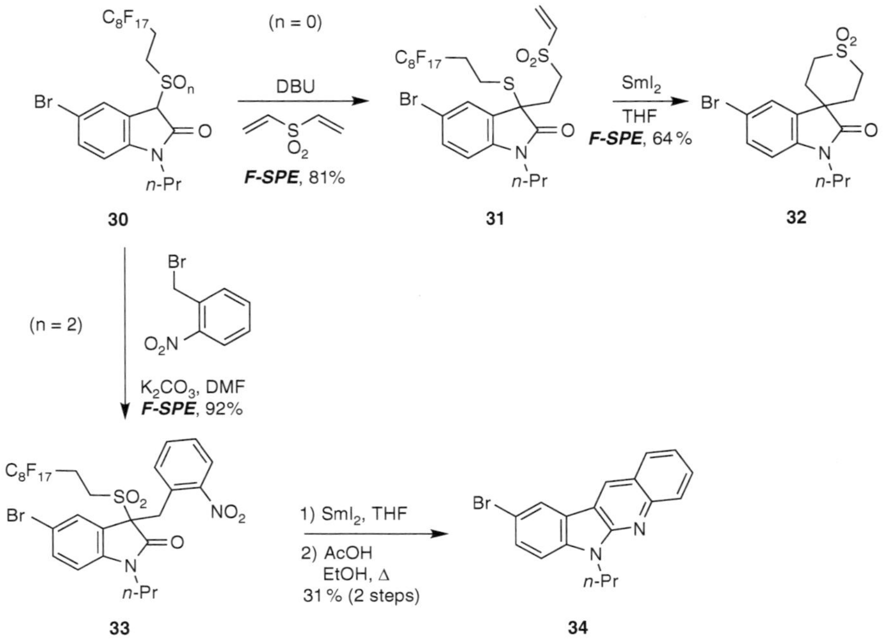

Scheme 7 SmI$_2$-mediated detagging and cyclization

As previously developed in solid-phase methodology, cyclization-assisted cleavage of the phase tag has been applied successfully to the fluorous synthesis of heterocyclic compounds. Thus, this simple detagging process may contribute to prepare valuable chemical libraries of potential biologically active molecules within a reasonable amount of time. For instance, the synthesis of fluorinated uracils and thiouracils **38** described by Fustero et al. involved first the reaction of fluorous-tagged acetates **35** with α,α-difluorinated nitriles to yield β-enamino esters **36** (Scheme 8). Next, addition of isocyanates or isothiocyanates produced ureas or thioureas **37**, which underwent cyclization and simultaneous cleavage to afford the final compounds **38** in a one-pot transformation. It should be mentioned that two different fluorous tags were employed in this synthesis with similar results [21].

A simple but also efficient protocol for the preparation of a library of 1,3-oxazoles and thiazoles **43** used a fluorous tag which was removed by basic hydrolysis with LiOH (Scheme 9). In this case, fluorous glycine **39** was the starting substrate and, after transformation into benzophenone imine **40**, consecutive functionalization with two acid chlorides afforded α-amido-β-ketoesters **41**. Cyclization of **41** was pursued by reaction with I$_2$ and PPh$_3$ to give 1,3-oxazoles, whereas treatment with Lawesson's reagent led to the corresponding 1,3-thiazoles. This synthesis was also performed using standard, non-fluorous or solid-phase methods, but the fluorous strategy proved to be superior in terms of speed and practicability [22].

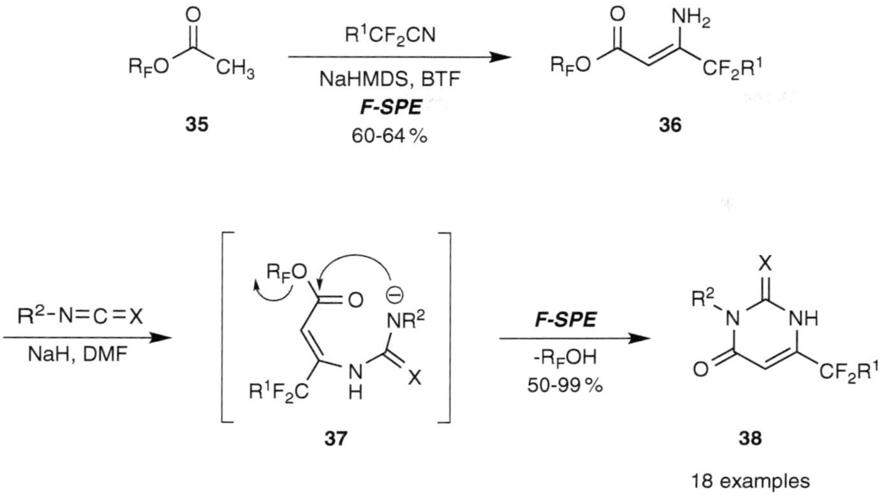

Scheme 8 Fluorous parallel synthesis of uracils and thiouracils **38**. *BTF* benzotrifluoride

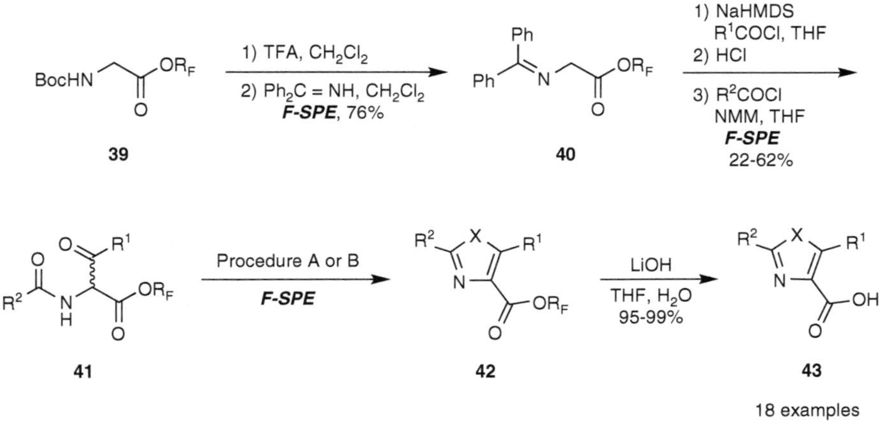

$R_F = C_8F_{17}(CH_2)_3$, $C_8F_{17}(CH_2)_2$-p-$C_6H_4CH_2$
X = O, S

Scheme 9 Fluorous parallel synthesis of 1,3-oxazoles and thiazoles **43**

Procedure A (X = O): Ph$_3$P, I$_2$, Et$_3$N (63-98%)
Procedure B (X = S): Lawesson's reagent (54-82%)

$R_F = C_8F_{17}(CH_2)_3$

Nelson et al. reported the parallel synthesis of a variety of heterocyclic compounds in fluorous phase with the aid of a new fluorous tag designed for its later removal by means of an RCM reaction. The fluorous reagent **41** was prepared using a Fukuyama–Mitsunobu reaction for the attachment of the fluorous moiety to

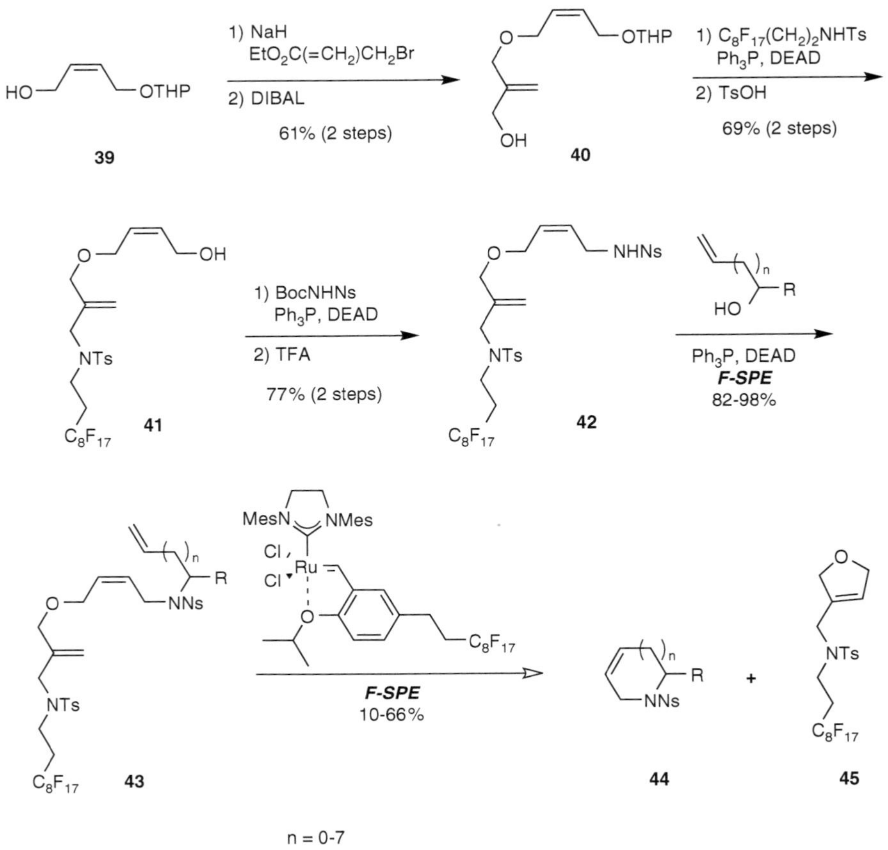

Scheme 10 Fluorous parallel synthesis of nitrogen-containing heterocycles **44**. *Ns* 2-nitrobenzenesulfonyl

allylic alcohol **40**, available in turn from monoprotected diol **39** (Scheme 10). The derived fluorous sulfonamide **42** was next coupled with different unsaturated alcohols, and the resulting compounds **43** were purified by F-SPE. Subsequent RCM afforded the corresponding small- or medium-ring or macrocyclic nitrogen-containing heterocycles **44**, which were separated from all the fluorous species (including unreacted starting materials and byproducts such as tetrahydrofuran **45**) by another F-SPE, since only metathesized products were released from the fluorous tag. However, it should be noted that a fluorous analog of the Hoveyda–Grubbs second generation catalyst had to be employed in the metathesis process in order to facilitate the purification of heterocycles **44** [23].

In a similar fashion, fluorous silanes **46** were also accessed from allylic alcohol **41**, and again an RCM protocol promoted the removal of the fluorous tag to afford silicon-tethered diols **47** (Scheme 11). In this case, a nonfluorous metathesis catalyst was suitable, which was eliminated during work-up [P(CH$_2$O)$_3$/Et$_3$N] [24].

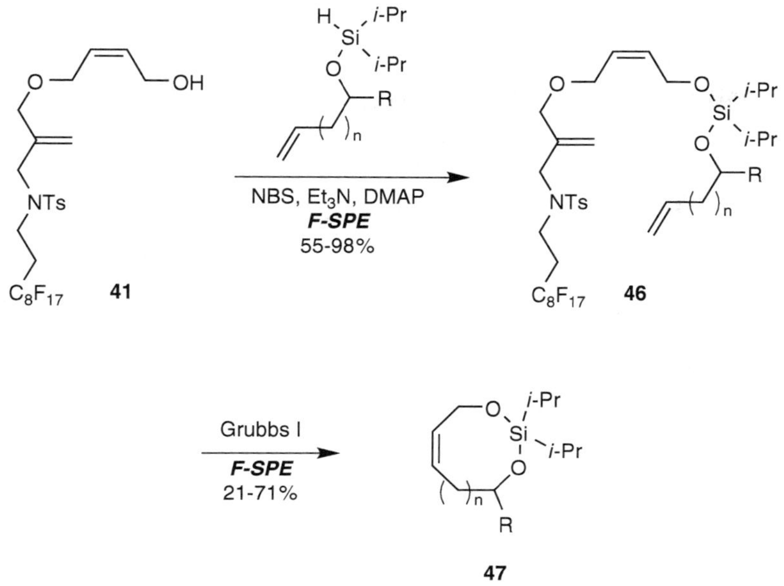

Scheme 11 Fluorous synthesis of silicon-tethered diols 47

Later on, both strategies based on fluorous reagents **41** and **42** were combined in order to produce a chemical library of 96 compounds displaying a high structural diversity (more than 80 different scaffolds) in a very efficient manner in terms of yield and purity [25].

Nelson et al. also developed a more elaborated fluorous tag that remained appended after the metathesis process. In this manner, assembly of precursor **49** was followed by RCM reaction to produce tricyclic compound **50**, which was still fluorous-tagged but now through a newly created acetal moiety (Scheme 12). This molecule contained a diene motif that could be employed in a Diels–Alder cycloaddition to furnish adduct **51**. The choice of starting substrates and the combination of inter- and intramolecular Diels–Alder reactions as well as further derivatizations produced a large array of molecular scaffolds. Final release of the fluorous tag by means of acetal hydrolysis only affected those products coming from successful metathesis reactions to furnish compounds such as **52** [26, 27].

Multicomponent reactions are widely employed for the straightforward assembly of compound libraries displaying many points of structural diversity. Zhang et al. have incessantly exploited this strategy in fluorous phase, also with the aid of microwave-assisted reactions [28]. An illustrative example is the preparation of a 31-membered library of biaryl-substituted 1,4-benzodiazepine-2,5-diones **59** (Scheme 13). The synthesis started with a multicomponent Ugi reaction between carboxylic acids **53**, fluorous-tagged benzaldehydes **54**, amines **55**, and isocyanides **56** to afford compounds **57** in good yields. Next, acid-mediated Boc deprotection and in situ cyclization formed the heterocycle framework **58**, microwave heating

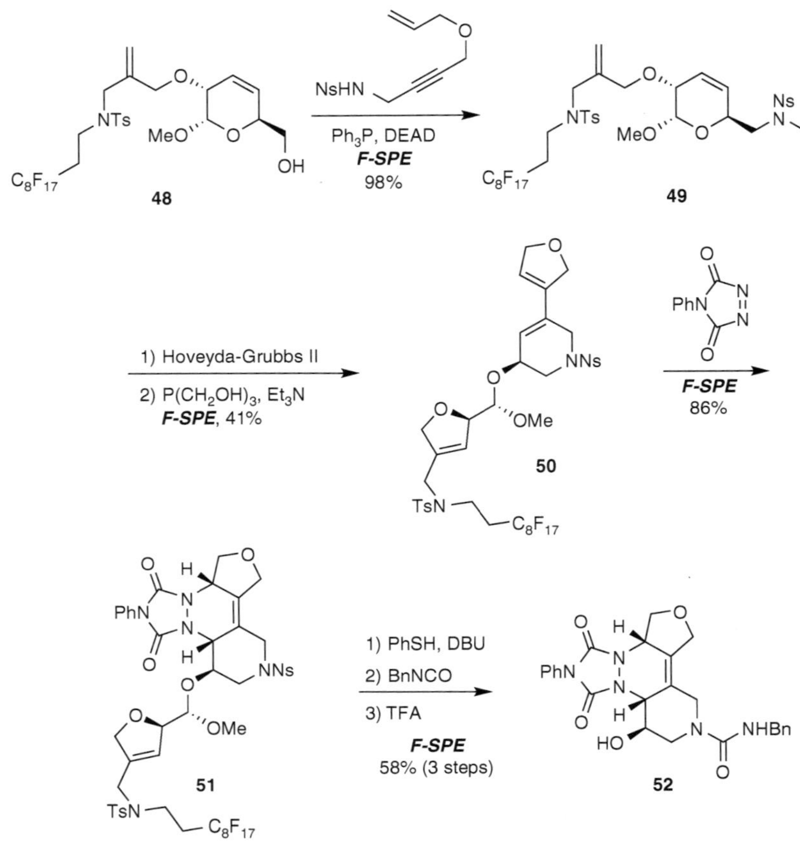

Scheme 12 Fluorous synthesis of heterocyclic compound **52** via metathesis and Diels–Alder reactions

being essential for driving the reaction to completion. Finally, Suzuki coupling of **58** with a series of aryl boronic acids also promoted by microwave irradiation led to the final compounds **59** in moderate yields. All products and intermediates were easily purified by means of F-SPE [29].

2.3 Fluorous Mixture Synthesis

Fluorous molecules with different fluorine contents can be separated by chromatography when using a fluorous stationary phase, most commonly by means of fluorous HPLC. This feature made the development of fluorous mixture synthesis (FMS) possible, a combinatorial technique pioneered by Curran's group with the aim of performing synthetic sequences using mixtures of fluorous compounds.

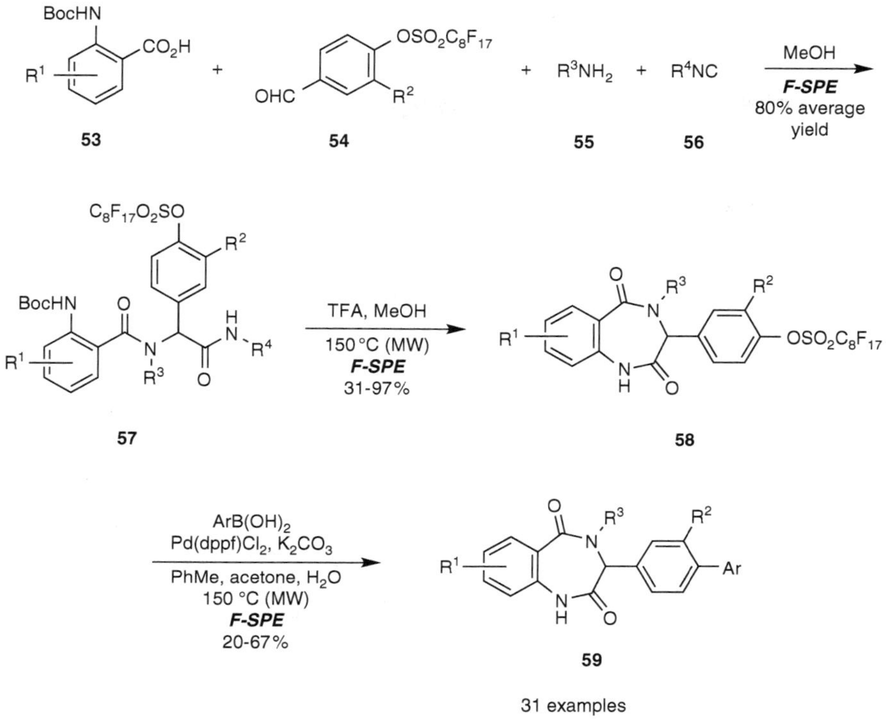

Scheme 13 Fluorous multicomponent synthesis of biaryl-substituted benzo-1,4-diazepine-2,5-diones **59**. *dppf* 1,1′-bis(diphenylphosphino)ferrocene

Each molecule in the mixture is unambiguously encoded by a different fluorous tag, and a final demixing is carried out at the end of the synthesis. The result is a dramatic reduction in the overall number of reactions required for the preparation of each individual molecule [30].

FMS has been particularly useful for the stereochemical elucidation of natural products to prepare all possible diastereoisomers. Many examples have been reported by Curran et al. and the most recent deals with the synthesis of the four possible isomers of petrocortyne A (**63**) (Scheme 14). Achiral ketone **60** was reduced with both enantiomers of the Corey–Bakshi–Shibata (CBS) reagent and each enantiomeric alcohol was further protected with either fluorous or nonfluorous TIPSOTf. The resulting mixture of compounds **61** worked as a quasiracemate and, after splitting in two, the corresponding primary iodides reacted with both enantiomeric forms of the lithium-derived anion of a chiral propargyl alcohol. This time, hydroxyl protection of each diastereoisomer was carried out with a different fluorous TIPS group, and they were then combined to afford a mixture of the four diastereoisomers of **62**, each having a different number of fluorine atoms ranging from 7 to 18. At the end of the synthesis, demixing by ^{F}HPLC and removing all silyl protecting groups furnished petrocortyne A and its three diastereoisomers [31, 32].

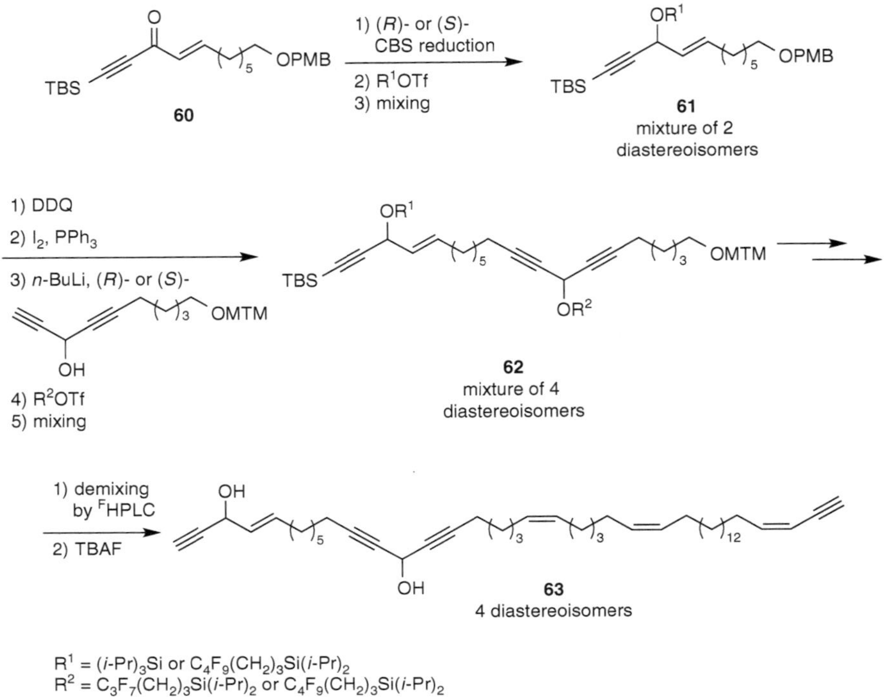

Scheme 14 Fluorous mixture synthesis of petrocortyne A (**63**). *MTM* methylthiomethyl

The preparation of compound libraries can also be achieved by FMS as demonstrated by Winssinger et al. in the production of a family of resorcylic acid lactones **67** (Scheme 15). The key step was the coupling of several mixtures of iodides **65** with a series of functionalized aromatic esters **64**. Each member of the mixture contained a different fluorous PMB group and demixing of compounds **66** was accomplished by chromatography on fluorous silica, with no need of ^{F}HPLC. Removal of the fluorous group in each individual compound and final cyclization gave a collection of 36 macrocyclic lactones **67** with a variety of substitutions for further biological assays [33, 34].

Fluorous diastereomeric mixture synthesis (FDMS) is a new method recently introduced by Curran, Zhang et al., allowing for the preparation of several fluorous-tagged diastereoisomers that are treated as a single compound and finally detagged and separated by nonfluorous chromatography. As proof of concept, the synthesis of hydantoin-fused hexahydrochromeno[4,3-*b*]pyrroles **72** was accomplished from benzaldehyde **68** (Scheme 16). Microwave-assisted condensation of **68** with fluorous-tagged alanine **69** and in situ intramolecular 1,3-dipolar cycloaddition afforded a mixture of four diastereoisomers **70**, purified by F-SPE. The lack of selectivity in the cycloaddition reaction is important in order to produce a large amount of each diastereoisomer. Next, derivatization with PMP-NCO yielded ureas **71**,

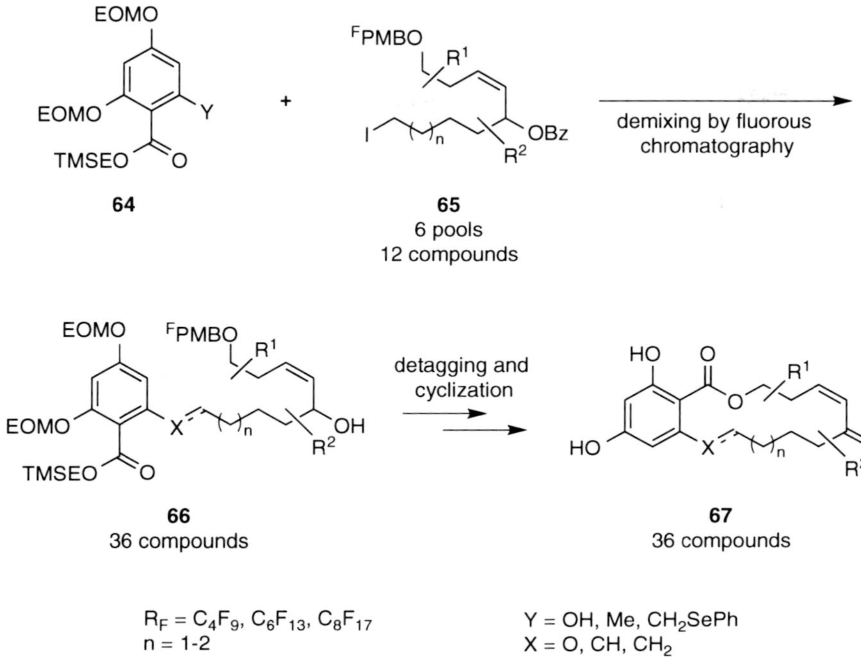

Scheme 15 Fluorous mixture synthesis of resorcylic acid lactones **67**. ^{F}*PMB* p-[R$_F$(CH$_2$)$_3$O]C$_6$H$_4$CH$_2$, *EOM* ethoxymethyl, *TMSE* 2-(trimethylsilyl)ethyl

which were transformed into the target compounds **72** in the detagging process. This cyclization step also caused epimerization of the ester group and hence eight diastereoisomers were produced although only six were finally isolated by reverse-phase chromatography [35].

3 Fluorous Radiochemistry

Fluorous chemistry has proved to be a useful choice for the production of radiolabeled compounds, especially when fast and efficient purification techniques are required due to the short half-lives of some radioactive isotopes as well as the need for samples with very high radiochemical purity. Thus, the introduction of labeled atoms during the removal of a fluorous tag would enable its easy separation from the nonfluorous, radiolabeled compound by F-SPE. This strategy has been demonstrated by Valliant et al. by using heavy fluorous stannanes **74**, in turn prepared from organozinc reagents **73** and a fluorous tin bromide (Scheme 17). After hydrolysis towards acids **75** and further coupling with several amines to give amides **76**, iodination with Na[^{125}I] and iodogen as iodide activator produced a

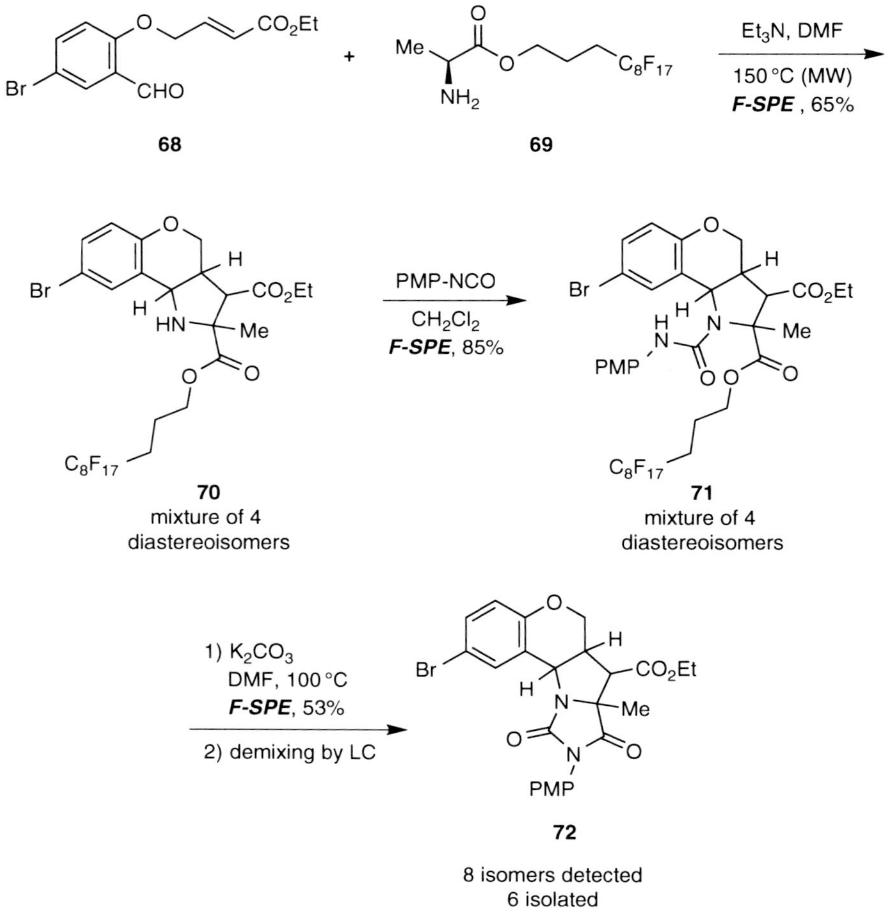

Scheme 16 Fluorous diastereomeric mixture synthesis of hydantoin-fused hexahydrochromeno [4,3-*b*]pyrroles **72**. *PMP* 4-(methoxy)phenyl

small family of radiolabeled benzamides **77** in high radiochemical yields and purities after F-SPE purification. In this manner, time-consuming HPLC separations were avoided. In contrast, compounds **74–76** had to be purified by liquid–liquid extractions with FC-72 (perfluorohexane) because of their heavy fluorous character [36]. This method has been successfully applied to the synthesis of radiopharmaceuticals such as [^{125}I]-meta-iodobenzylguanidine (**78**) and [^{125}I]-5-iodo-2′-deoxyuridine (**79**) in highly effective specific activities (Fig. 1) [37, 38].

Compounds labeled with ^{18}F are widely employed in positron emission tomography (PET) experiments despite their somewhat short half-life (109.7 min) and hence their production by means of fluorous synthesis procedures is particularly attractive. In this context, Gouverneur et al. used sulfonyl chloride **82**, readily

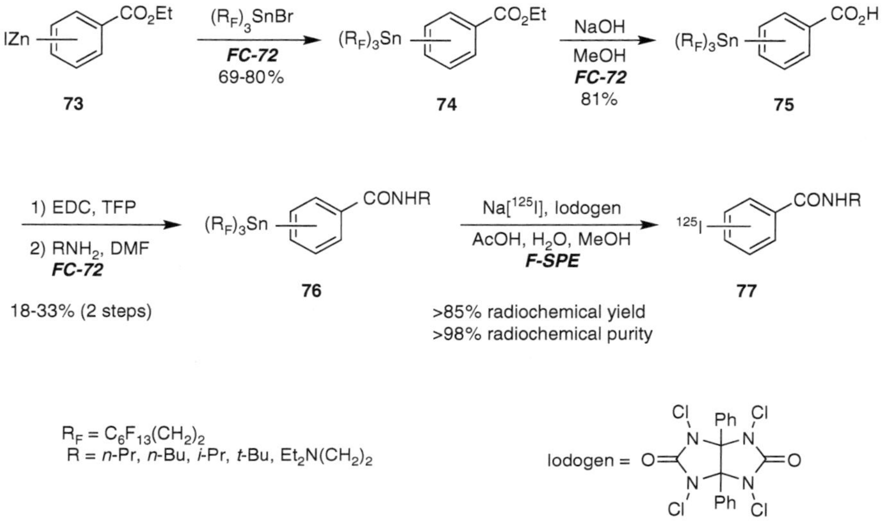

Scheme 17 Fluorous synthesis of radiolabeled benzamides **77**. *EDC* 1-ethyl-3-(3-dimethylaminopropyl)carbodiimide, *TFP* 2,3,5,6-tetrafluorophenol

Fig. 1 Structures of [^{125}I]-meta-iodobenzylguanidine (**78**) and [^{125}I]-5-iodo-2'-deoxyuridine (**79**)

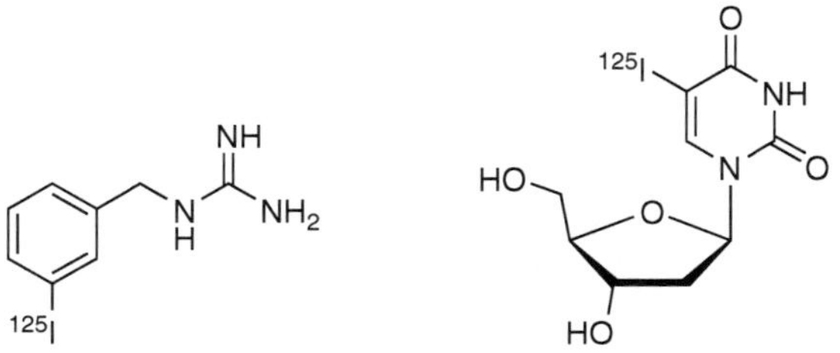

available from commercial fluorous iodide **80**, for the attachment of conveniently functionalized alcohols (Scheme 18). For example, coupling of **82** with 2-azidoethanol provided fluorous azide **83**, and subsequent reaction with [^{18}F]$^-$/K$^+$-Kryptofix 222, a nucleophilic source of ^{18}F obtained in a cyclotron, promoted the detagging, leading to the formation of [^{18}F]-2-fluoroethylazide (**84**), in better radiochemical yield than previous approaches relying on distillation of the final product [39].

Radiolabeled azide **84** was next employed in a Staudinger reaction with fluorous phosphine **88**, in turn prepared from fluorous phosphonium borate **85** (Scheme 19). This fluorous molecule contained two C$_4$F$_9$ groups rather than a single C$_8$F$_{17}$ tag and also included a thioester moiety, which served to react with azide **84** to produce amide **89** in high radiochemical yield [40].

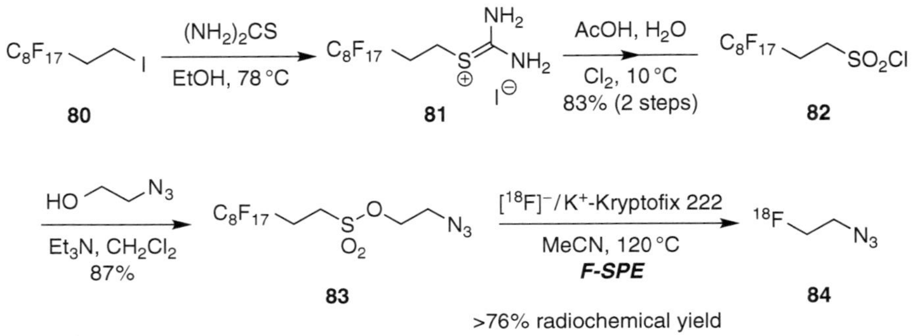

Scheme 18 Fluorous synthesis of [^{18}F]-2-fluoroethylazide (**84**)

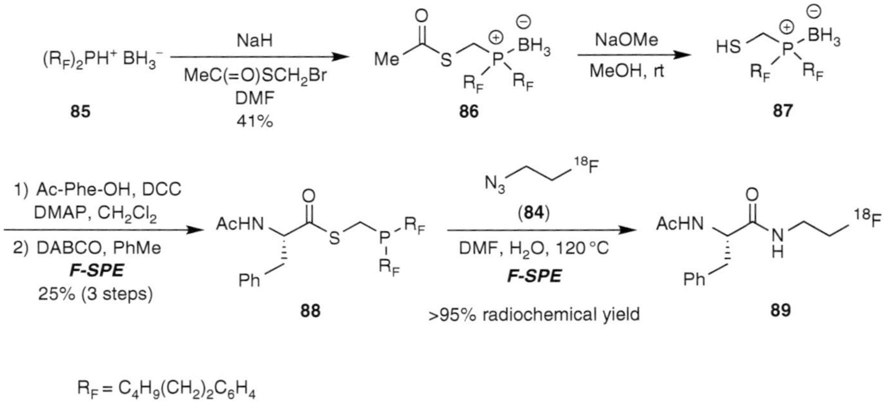

Scheme 19 Staudinger reaction for the production of radiolabeled compound **89**

4 Biological Applications of Fluorous Reagents

4.1 Fluorous Microarrays

Since microarray-based technologies were used for the first time as analytical tools for DNA experiments, this screening method has revolutionized the field of molecular biology tremendously. Besides gene analysis, microarray techniques have recently been extended to the analysis of biologically significant molecules such us proteins, carbohydrates, or even natural and synthetic small molecules with the aim of identifying potential therapeutic leads [41].

In standard microarrays, molecules must be attached to an appropriate solid surface, typically a glass or silicon slide, via a covalent bond. This fact implies certain chemical modifications in the original molecule to assist the anchoring onto

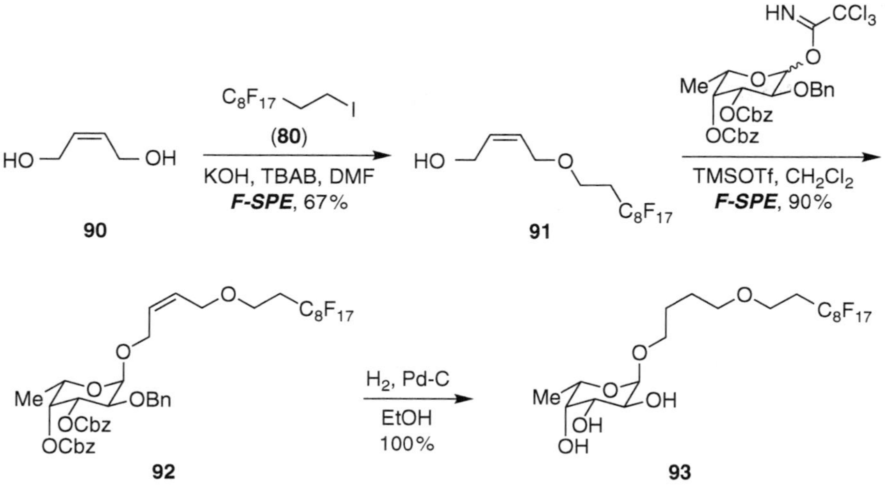

Scheme 20 Synthesis of carbohydrate **93** for the formation of fluorous microarrays

the surface, which can hamper the overall synthesis. To circumvent this, noncovalent interaction-based methods have been extensively developed over the last few years. In this context, fluorous tags have found application in carbohydrate microarrays, wherein fluorous-tagged carbohydrates are efficiently attached to a fluorinated solid surface by means of affinity fluorous interaction [42]. In this sense, fluorous carbohydrates were designed by Pohl et al. with the aim of preparing microarrays for further biological assays. Thus, monoprotection of diol **90** with substoichiometric amounts of fluorous iodide **80** afforded compound **91**, which reacted with trichloroacetimidates to produce fluorous carbohydrate **92** (Scheme 20). F-SPE separations were used for the purification of synthetic intermediates, and final deprotection by hydrogenation afforded **93**. A microarray of compound **93** was then formed by immobilization onto a glass slide coated with a Teflon/epoxy mixture [43]. In subsequent work, a related tag derived from iodide **1** containing a propylene linker proved to be more stable towards undesired β-elimination reactions [44, 45].

The use of fluorous phase techniques in small-molecule microarrays has also meant a notable advance with regard to the discovery of new protein–ligand interactions. In this context, Spring et al. carried out affinity studies based on fluorous-tagged biotin **94** incubated with Cy5-labeled avidin (Fig. 2). Amongst different fluorous reagents tested, a C_8F_{17} group with a polyethylene glycol linkage was found to be the best tag used to screen a small collection of amide compounds [46].

With the same purpose, Schreiber et al. developed a fluorous-tagged small-molecule microarray to identify new histone deacetylase inhibitors from different chemical libraries. In this way, systematic modifications of positive tagged molecules may lead to the discovery of an even stronger protein ligand [47].

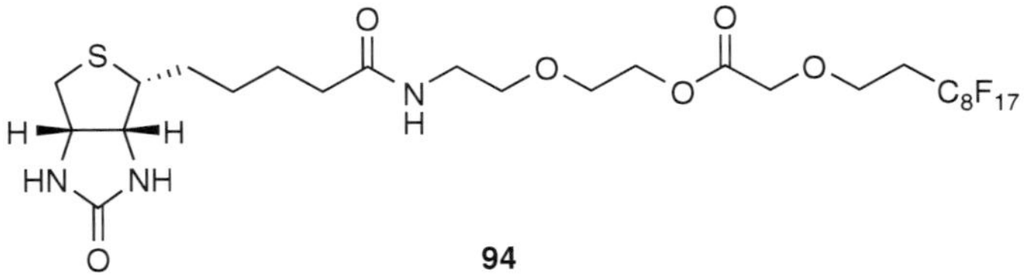

Fig. 2 Structure of fluorous-tagged biotin **94** for the preparation of fluorous microarrays

4.2 Fluorous Proteomics

As shown in the previous section, perfluoroalkyl groups have been successfully used as tags not only in organic synthesis for easy, rapid separations with fluorous silica gel, but also in biochemistry. Although it is still the same concept, that is, fluorous phase tags as a tool for separation, it was introduced to proteomics techniques with the aim of facilitating the structural analysis of proteins.

Mass spectrometry (MS) has become a leading experimental method for identifying proteins, in which a set of fragments from a specific peptide ion is generated and characterized. In this way, the three-dimensional structure can be rebuilt from the outer layer of the protein to the interior. In practice, a protein sample first undergoes a tagging process with the corresponding affinity reagent, which recognizes specific points of union in the protein structure. The resulting mixture is then loaded onto a solid support, and unlabeled peptides are removed from the initial mixture, washed off with appropriate mixtures of organic solvents. Finally, tagged peptides are eluted from the affinity resin and characterized by MS. Although many affinity reagents such as biotin–streptavidin, lectins, or immobilized chelated metals are known in this field, fluorous tags have recently been applied successfully to this purpose.

Fluorous tags make it possible to overcome some limitations linked to other affinity reagents. For instance, fluorous affinity tags are not susceptible to fragmentation during MS analysis and, hence, spectral interpretation is easier. Furthermore, F-SPE allows for a selective immobilization of the fluorous-tagged peptides onto the fluorous silica gel cartridge and a subsequent highly efficient recovery of those labeled fragments is possible when they are washed off with a fluorophilic solvent. From an economic point of view, fluorous phase tags are also more attractive than other strategies employed in current proteomic techniques [48–50].

5 Conclusions

Fluorous phase synthesis has undergone incredible growth, especially over the last decade. For this purpose, fluorous tags are conveniently attached to the starting substrates of the synthetic routes so that fluorous purification techniques may be used after each reaction step to facilitate and accelerate the overall synthetic process. An obvious expansion of this method has been the application of fluorous synthesis in the parallel or combinatorial production of compound collections. Most importantly, practical detagging protocols must also be developed in order to release the final products at the end of the synthesis.

Although the use of perfluoroalkyl groups for labeling small molecules emerged to circumvent specific limitations of solid phase synthesis, fluorous reagents have become a powerful tool not only in organic synthesis but also in chemical biology. For instance, the design of fluorous microarrays based on noncovalent interactions has broadened the usefulness of this interesting field. It is likely that new contributions will be reported in the near future, having high impact on both organic synthesis and biochemistry fields.

Acknowledgments We would like to thank MICINN (CTQ2010-19774) of Spain and *Generalitat Valenciana* (PROMETEO/2010/061) for their financial support. J.L.A. expresses his gratitude to MICINN for a *Ramón y Cajal* contract and S. C. expresses her thanks to MICINN for a *Juan de la Cierva* contract.

References

1. Gladysz JA, Curran DP, Horváth IT (eds) (2004) Handbook of fluorous chemistry. Wiley-VCH, Weinheim
2. Horváth IT, Rábai J (1994) Facile catalyst separation without water: Fluorous biphase hydroformylation of olefins. Science 266:72–75
3. Curran DP (2001) Fluorous reverse phase silica gel. A new tool for preparative separations in synthetic organic and organofluorine chemistry. Synlett 1488–1496
4. Curran DP (2006) Organic synthesis with light-fluorous reagents, reactants, catalysts, and scavengers. Aldrichim Acta 39:3–9
5. Zhang W, Curran DP (2006) Synthetic applications of fluorous solid-phase extraction (F-SPE). Tetrahedron 62:11837–11865
6. Zhang W (2009) Fluorous linker-facilitated chemical synthesis. Chem Rev 109:749–795
7. Curran DP (2008) Fluorous tags unstick messy chemical biology problems. Science 321:1645–1646
8. Fustero S, García Sancho A, Chiva G, Sanz-Cervera JF, del Pozo C, Aceña JL (2006) Fluorous (trimethylsilyl)ethanol: A new reagent for carboxylic acid tagging and protection in peptide synthesis. J Org Chem 71:3299–3302
9. Fustero S, Sánchez-Roselló M, Rodrigo V, Sanz-Cervera JF, Piera J, Simón-Fuentes A, del Pozo C (2008) Solution-, solid-phase, and fluorous synthesis of β,β-difluorinated cyclic quaternary α-amino acid derivatives: A comparative study. Chem Eur J 14:7019–7029
10. Fustero S, Rodrigo V, Sánchez-Roselló M, Mojarrad F, Vicedo A, Moscardó T, del Pozo C (2008) Solution and fluorous phase synthesis of β,β-difluorinated 1-amino-1-cyclopentane carboxylic acid derivatives. J Fluorine Chem 129:943–950
11. Thibaudeau S, Gouverneur V (2003) Sequential cross-metathesis/electrophilic fluorodesilylation: A novel entry to functionalized allylic fluorides. Org Lett 5:4891–4893
12. Boldon S, Moore JE, Gouverneur V (2008) Fluorous synthesis of allylic fluorides: C–F bond formation as the detagging process. Chem Commun 3622–3624
13. Boldon S, Moore JE, Gouverneur V (2009) Oxidative detagging of fluorous organosilanes. J Fluorine Chem 130:1151–1156

14. Nielsen SD, Smith G, Begtrup M, Kristensen JL (2010) Synthesis and application of a new fluorous-tagged ammonia equivalent. Chem Eur J 16:4557–4566
15. Nielsen SD, Smith G, Begtrup M, Kristensen JL (2010) Amination of aryl iodides using a fluorous-tagged ammonia equivalent. Eur J Org Chem 3704–3710
16. Schreiber SL (2000) Target-oriented and diversity-oriented organic synthesis in drug discovery. Science 287:1964–1969
17. Spandl RJ, Bender A, Spring DR (2008) Diversity-oriented synthesis; a spectrum of approaches and results. Org Biomol Chem 6:1149–1158
18. McAllister LA, McCormick RA, Brand S, Procter DJ (2005) A fluorous-phase Pummerer cyclative-capture strategy for the synthesis of nitrogen heterocycles. Angew Chem Int Ed 44:452–455
19. McAllister LA, McCormick RA, James KM, Brand S, Willetts N, Procter DJ (2007) A fluorous, Pummerer cyclative-capture strategy for the synthesis of N-heterocycles. Chem Eur J 13:1032–1046
20. James KM, Willetts N, Procter DJ (2008) Samarium(II)-mediated linker cleavage-cyclization in fluorous synthesis: Reactions of samarium enolates. Org Lett 10:1203–1206
21. Fustero S, Catalán S, Flores S, Jiménez D, del Pozo C, Aceña JL, Sanz-Cervera JF, Mérida S (2006) First fluorous synthesis of fluorinated uracils. QSAR Comb Sci 25:753–760
22. Sanz-Cervera JF, Blasco R, Piera J, Cynamon M, Ibáñez I, Murguía M, Fustero S (2009) Solution versus fluorous versus solid-phase synthesis of 2,5-disubstituted 1,3-azoles. Preliminary antibacterial activity studies. J Org Chem 74:8988–8996
23. Leach SG, Cordier CJ, Morton D, McKiernan GJ, Warriner S, Nelson A (2008) A fluorous-tagged linker from which small molecules are released by ring-closing metathesis. J Org Chem 73:2753–2759
24. Cordier C, Morton D, Leach S, Woodhall T, O'Leary-Steele C, Warriner S, Nelson A (2008) An efficient method for synthesising unsymmetrical silaketals: Substrates for ring-closing, including macrocycle-closing, metathesis. Org Biomol Chem 6:1734–1737
25. Morton D, Leach S, Cordier C, Warriner S, Nelson A (2009) Synthesis of natural-product-like molecules with over eighty distinct scaffolds. Angew Chem Int Ed 48:104–109
26. O'Leary-Steele C, Cordier C, Hayes J, Warriner S, Nelson A (2009) A fluorous-tagged "safety catch" linker for preparing heterocycles by ring-closing metathesis. Org Lett 11:915–918
27. O'Leary-Steele C, Pedersen PJ, James T, Lanyon-Hogg T, Leach S, Hayes J, Nelson A (2010) Synthesis of small molecules with high scaffold diversity: Exploitation of metathesis cascades in combination with inter- and intramolecular Diels-Alder reactions. Chem Eur J 16:9563–9571
28. Kadam A, Zhang Z, Zhang W (2011) Microwave-assisted fluorous multicomponent reactions - A combinatorial chemistry approach for green organic synthesis. Curr Org Synth 8:295–309
29. Zhou H, Zhang W, Yan B (2010) Use of cyclohexylisocyanide and methyl 2-isocyanoacetate as convertible isocyanides for microwave-assisted fluorous synthesis of 1,4-benzodiazepine-2,5-dione library. J Comb Chem 12:206–214
30. Luo Z, Zhang Q, Oderaotoshi Y, Curran DP (2001) Fluorous mixture synthesis: A fluorous-tagging strategy for the synthesis and separation of mixtures of organic compounds. Science 291:1766–1769
31. Curran DP, Sui B (2009) A "shortcut" mosher ester method to assign configurations of stereocenters in nearly symmetric environments. Fluorous mixture synthesis and structure assignment of petrocortyne A. J Am Chem Soc 131:5411–5413
32. Sui B, Yeh EA-H, Curran DP (2010) Assignment of the structure of petrocortyne A by mixture syntheses of four candidate stereoisomers. J Org Chem 75:2942–2954
33. Dakas P-Y, Barluenga S, Totzke F, Zirrgiebel U, Winssinger N (2007) Modular synthesis of radicicol A and related resorcylic acid lactones, potent kinase inhibitors. Angew Chem Int Ed 46:6899–6902
34. Jogireddy R, Dakas P-Y, Valot G, Barluenga S, Zirrgiebel U, Winssinger N (2009) Synthesis of a resorcylic acid lactone (RAL) library using fluorous-mixture synthesis and profile of its selectivity against a panel of kinases. Chem Eur J 15:11498–11506

35. Lu Y, Geib SJ, Damodaran K, Sui B, Zhang Z, Curran DP, Zhang W (2010) Fluorous diastereomeric mixture synthesis (FDMS) of hydantoin-fused hexahydrochromeno[4,3-b] pyrroles. Chem Commun 46:7578–7580
36. Donovan A, Forbes J, Dorff P, Schaffer P, Babich J, Valliant JF (2006) A new strategy for preparing molecular imaging and therapy agents using fluorine-rich (fluorous) soluble supports. J Am Chem Soc 128:3536–3537
37. Donovan AC, Valliant JF (2008) A convenient solution-phase method for the preparation of meta-iodobenzylguanidine (mibg) in high effective specific activity. Nuclear Med Biol 35:741–746
38. McIntee JW, Sundararajan C, Donovan AC, Kovacs MS, Capretta A, Valliant JF (2008) A convenient method for the preparation of fluorous tin derivatives for the fluorous labeling strategy. J Org Chem 73:8236–8243
39. Bejot R, Fowler T, Carroll L, Boldon S, Moore JE, Declerck J, Gouverneur V (2009) Fluorous synthesis of ^{18}F-radiotracers with the [^{18}F] fluoride ion: Nucleophilic fluorination as the detagging process. Angew Chem Int Ed 48:586–589
40. Carroll L, Boldon S, Bejot R, Moore JE, Declerck J, Gouverneur V (2011) The traceless Staudinger ligation for indirect ^{18}F-radiolabelling. Org Biomol Chem 9:136–140
41. Duffner JL, Clemons PA, Koehler AN (2007) A pipeline for ligand discovery using small-molecule microarrays. Curr Opin Chem Biol 11:74–82
42. Pohl NL (2008) Fluorous tags catching on microarrays. Angew Chem Int Ed 47:3868–3870
43. Ko K-S, Jaipuri FA, Pohl NL (2005) Fluorous-based carbohydrate microarrays. J Am Chem Soc 127:13162–13163
44. Mamidyala SK, Ko K-S, Jaipuri FA, Park G, Pohl NL (2006) Noncovalent fluorous interactions for the synthesis of carbohydrate microarrays. J Fluorine Chem 127:571–579
45. Jaipuri FA, Collet BYM, Pohl NL (2008) Synthesis and quantitative evaluation of glycero-D-manno-heptose binding to concanavalin A by fluorous-tag assistance. Angew Chem Int Ed 47:1707–1710
46. Nicholson RL, Ladlow ML, Spring DR (2007) Fluorous tagged small molecule microarrays. Chem Commun 3906–3908
47. Vegas AJ, Bradner JE, Tang W, McPherson OM, Greenberg EF, Koehler AN, Schreiber SL (2007) Fluorous-based small-molecule microarrays for the discovery of histone deacetylase inhibitors. Angew Chem Int Ed 46:7960–7964
48. Brittain SM, Ficarro SB, Brock A, Peters EC (2005) Enrichment and analysis of peptides subsets using fluorous affinity tags and mass spectrometry. Nat Biotechnol 23:463–468
49. Go EP, Uritboonthai W, Apon JV, Trauger SA, Nordstrom A, O'Maille G, Brittain SM, Peters EC, Siuzdak G (2007) Selective metabolite and peptide capture/mass detection using fluorous affinity tags. J Proteome Res 6:1492–1499
50. Qian J, Cole RB, Cai Y (2011) Synthesis and characterization of a 'fluorous' (fluorinated alkyl) affinity reagent that labels primary amine groups in proteins/peptides. J Mass Spectrom 46:1–11

Chemical Applications of Fluorous Reagents and Scavengers

Marvin S. Yu

Abstract Fluorous modified reagents and scavengers have been widely used in the synthesis of small molecules and small molecule libraries. This chapter highlights some of those applications based on type of transformation and reagent or scavenger.

Keywords FLLE · Fluorous reagents · Fluorous scavengers · FSPE

Contents

1 Introduction 70
2 Fluorous Reagents 71
    2.1 Acylations with Fluorous Reagents 72
    2.2 Aminations with Fluorous Reagents 76
    2.3 Fluorous Mitsunobu Reaction 78
    2.4 Reductions with Fluorous Reagents 81
    2.5 Oxidations with Fluorous Reagents 83
    2.6 Miscellaneous Transformations 89
3 Fluorous Scavengers 91
    3.1 Nucleophilic Fluorous Scavengers 92
    3.2 Electrophilic Fluorous Scavengers 95
    3.3 Miscellaneous Fluorous Scavengers 97
4 Conclusions 99
References 100

M.S. Yu (✉)
Fluorous Technologies, Inc, 970 William Pitt Way, Pittsburgh 15238, PA, USA
e-mail: M.yu@fluorous.com

1 Introduction

The separation of a desired product from other components in a reaction mixture is a problem that chemists have faced since the earliest days of the science. Methods which have been developed to facilitate purification include physical methods (recrystallization, distillation, filtration, etc.) that take advantage of innate physical properties of the new molecule itself or partitioning based methods (chromatography, extraction, etc.) that exploit differing levels of interaction between molecules within a mixture and separation media. In either case the goal is to obtain the product in high yield and purity with a minimal amount of effort. Molecular modification through tagging is an oft used strategy to change either physical characteristics or partitioning behavior in order to facilitate purifications.

Molecular tagging strategies can generally be placed in one of three categories: substrate-tagging, reagent-tagging, or scavenger strategies (Fig. 1). Substrate-tagging

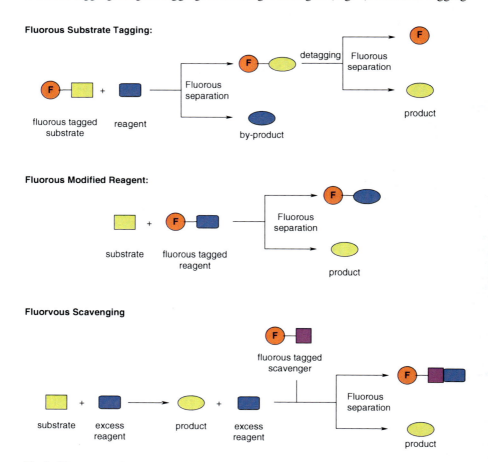

Fig. 1 Fluorous tagging, reagent, and enrichment strategies

strategies are those that temporarily tag the product of interest and use that tag for purification. After one or more reactions, the tag is then removed to provide the desired product. Reagent tagging, on the other hand, places a permanent tag on a reagent and leaves the substrate of interest untagged. Excess reagents and by-products can then be removed using the tag once the reaction is complete. The tag is therefore generally used for only one reaction. Scavenger strategies are similar to reagent-tagged strategies in that the tag is permanently affixed to the scavenger. However, in this approach, the scavenger's role is to effect tagging of a substrate or reagent used in excess in the actual reaction of interest while leaving the desired product unchanged. The tagged excess reagent can then easily be removed by a phase separation. All three strategies have been employed using fluorous tags.

Fluorous modified reagents and scavengers have been used extensively to simplify purification by both physical separations and partition based separations [1]. Physical fluorous separations include thermomorphic or other precipitation based methods, while partitioning based methods include fluorous solid-phase extraction (FSPE) or fluorous liquid-liquid extraction (FLLE). Applications include targeted small molecule synthesis, library synthesis, biomolecule synthesis, and analytical sample enrichment. In general the fluorous modified reagents are designed to have similar reactivity to their traditional analogs, thereby minimizing reaction method development while providing a handle for facile separation.

This chapter will highlight applications using fluorous reagents and scavengers for the synthesis of small molecules. The subject is organized primarily by transformation type so as to highlight the versatility and range of chemistries accessible using fluorous methods. Synthesis of the reagents and experimental details will not be discussed at length. In most cases, the fluorous reagents described will be stoichiometric in nature, although some catalytic processes are included. Fluorous organocatalysts and metal-catalyzed processes are described in other chapters.

2 Fluorous Reagents

Fluorous reagents are compounds that have permanent fluorous domains and are used to conduct a chemical transformation on a substrate. The by-products of the fluorous reagent can then be removed from the reaction mixture by a fluorous separation technique. Typically, the fluorous domain is located some distance away from the reactive center of the reagent in order to minimize changes in reactivity due to the electron-withdrawing effects of the fluorous domain. A wide range of transformations have been conducted using fluorous reagents and they have found extensive use in library synthesis [2, 3].

2.1 Acylations with Fluorous Reagents

The reaction of carboxylic acids with amines and alcohols to produce amides and esters is one of the most common transformations in organic chemistry. The net condensation is usually mediated by an acyl activating agent which transforms the carboxyl functionality to a more electrophilic species which is attacked by a nucleophilic amine or alcohol. There are a number of different types of fluorous acyl activating reagents, all of which are designed to facilitate post-reaction purification.

2.1.1 Fluorous Carbodiimides

Fluorous carbodiimide **1** was introduced by Palomo who demonstrated its utility in the synthesis of dipeptides [4]. The fluorous urea formed as the reagent by-product was removed from the reaction mixture by a novel hydrogen-bonding assisted FLLE (Fig. 2).

Curran later introduced a number of carbodiimides **2–5** (Fig. 3) which were developed in order to use either fluorous solid phase extraction (FSPE) or reverse fluorous solid phase extraction (RFSPE) as the separation method [5]. It was found that carbodiimides **2–4** behaved well from a reaction standpoint, but were problematic due to either gelling or solubility problems. Carbodiimide **5**, however, not only was effective in amide formations, but the resultant fluorous urea was also easily removed using either FSPE or RFSPE.

2.1.2 Fluorous Triazines

Fluorous mono- and di-chloro-triazines have also been prepared and applied to acylation reactions by Dembinski and Zhang. The first example, reported by Dembinski, was fluorous monochloro-triazine **6** (Fig. 4) which was used to form both dipeptides and simple amides [6]. Triazine **6** is the fluorous analog to chloro-dimethoxy-triazine

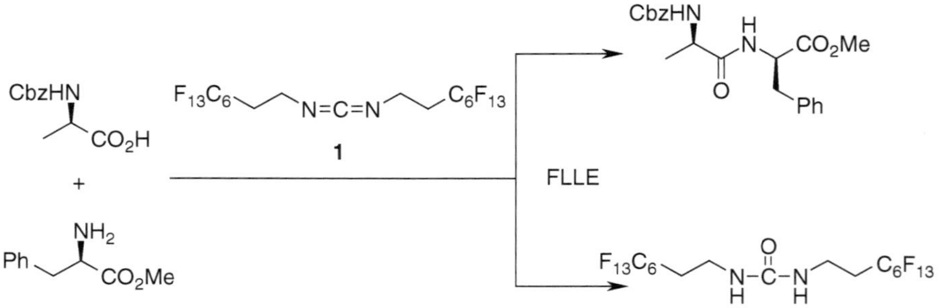

Fig. 2 Acylation with fluorous carbodiimide

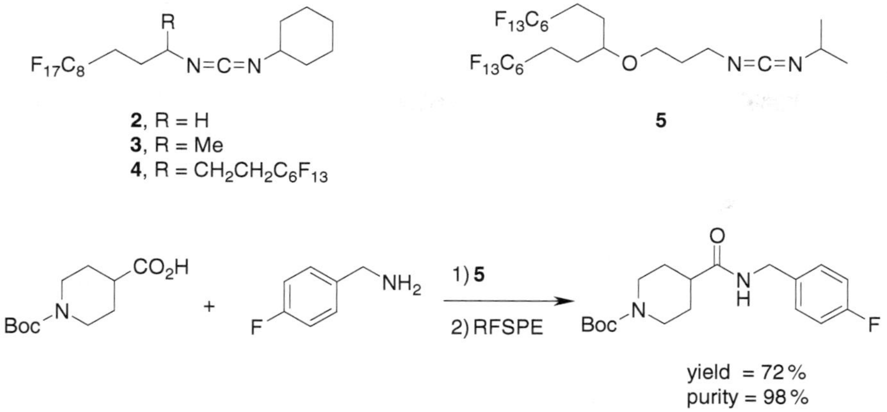

Fig. 3 Fluorous carbodiimides reported by Curran et al.

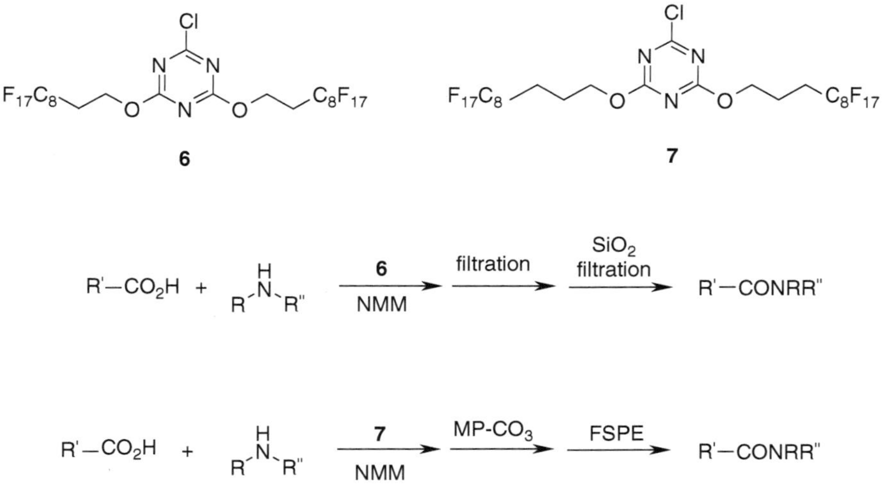

Fig. 4 Fluorous triazine based acylating agents

(CDMT) and, similarly to CDMT, a tertiary amine, N-methyl-morpholine (NMM) is used as a co-reagent. The doubly tagged hydroxy-triazine by-product was found to be insoluble in the reaction solvent and was removed by filtration.

Zhang used a similar triazine **7** in the parallel synthesis of a small 24-member amide library [7]. Reaction of a carboxylic acid and a primary or secondary amine in the presence of **7** and NMM was followed by treatment with solid supported macroporous carbonate resin (MP-CO$_3$) to form free NMM. The entire library was then purified simultaneously using a 24-well plate-to-plate FSPE providing amides in good yield and purity.

2.1.3 Fluorous Mukaiyama's Salt

2-Halo-pyridimium salts, commonly known as Mukaiyama salts, have been used as carboxyl activating agents for acylations and several fluorous versions have been prepared and used in amidations and esterifications. Zhang et al. reported the use of pyridinium salt **8a** (Fig. 5) in conjunction with HOBt for the preparation of the same library produced using triazine **7** [7]. The researchers also added a fluorous amine to scavenge excess activated carboxylic acid and once again used MP-CO$_3$. Purification was conducted using plate-to-plate FSPE. In comparison to triazine **7**, the same library was produced in greater yield and purity although it was operationally more complex than using **7**.

Since that report other fluorous Mukaiyama salts have been reported by Matsugi [8–10]. The triflate salt **8b** was found to be significantly more reactive than **7**, thereby obviating the need for HOBt. The pyridone by-product **9** was removed by FSPE. The same researchers later prepared fluorous Mukaiyama's salt **8c** using a C$_{10}$F$_{21}$ fluorous chain. With **8c** the resultant pyridone was precipitated from the reaction mixture by addition of water by taking advantage of the greater hydrophobicity of the Rf$_{10}$ chain compared to the Rf$_8$. Filtration provided a solution

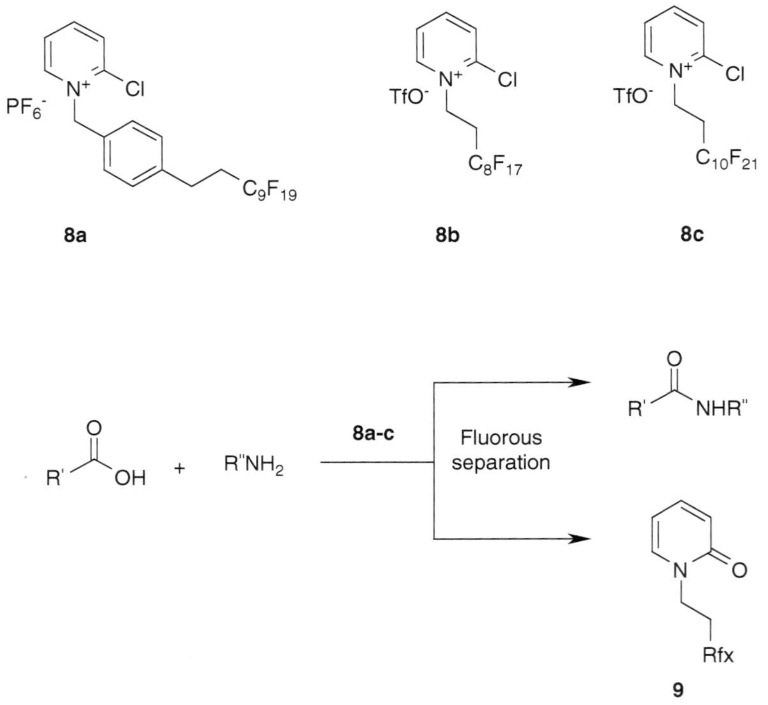

Fig. 5 Fluorous Mukaiyama's salts in acylations

of the product which was isolated through standard extraction. This results in a process which is operationally simpler than that reported for **7**.

2.1.4 Fluorous Distannoxanes

Tin oxides and distannoxanes have been demonstrated to catalyze transesterification reactions; however, the removal of the tin residues can be problematic. In response, fluorous versions of these organotins have been reported [11]. Otera has prepared a number of different distannoxanes **10** (Fig. 6) that have been used in transesterifications under fluorous biphasic conditions [12, 13]. Curran, meanwhile, prepared a fluorous version of dibutyltin oxide **11**. Compound **11** is not monomeric but rather a mixture of polymeric forms [14]. Porco used **11** as a catalyst to form macrolactones from α,ω-hydroxyesters [15] and in the formation of polypropionate libraries [16]. The fluorous tin residues were then easily separated from the desired products by either FSPE or FLLE.

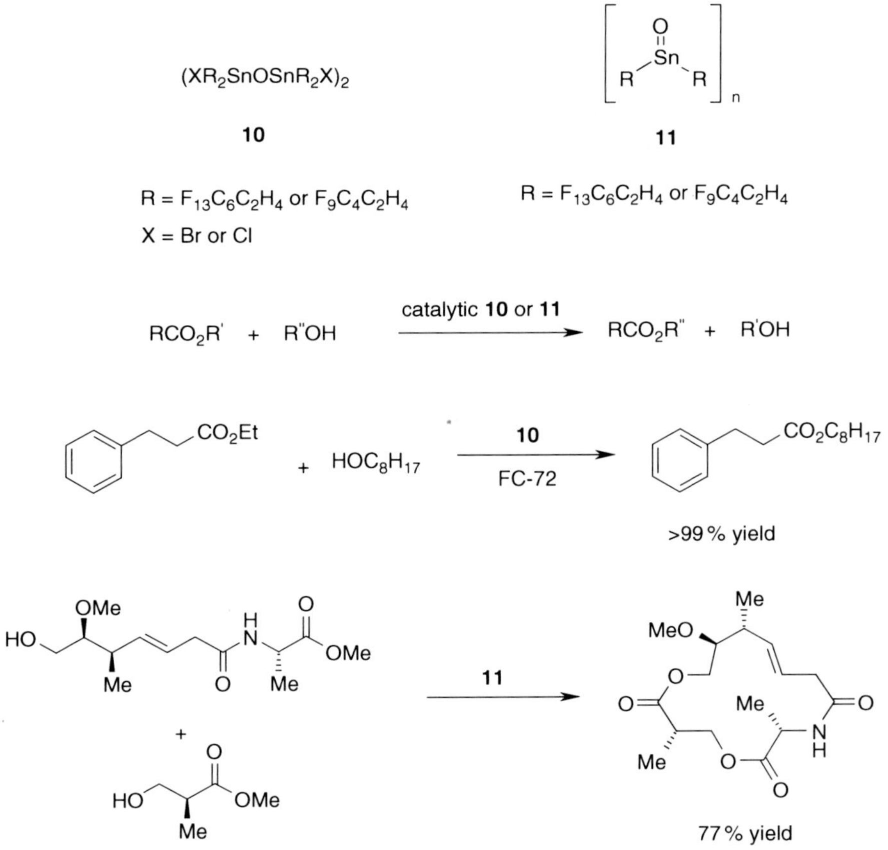

Fig. 6 Transesterification using fluorous tin oxides and distannoxanes

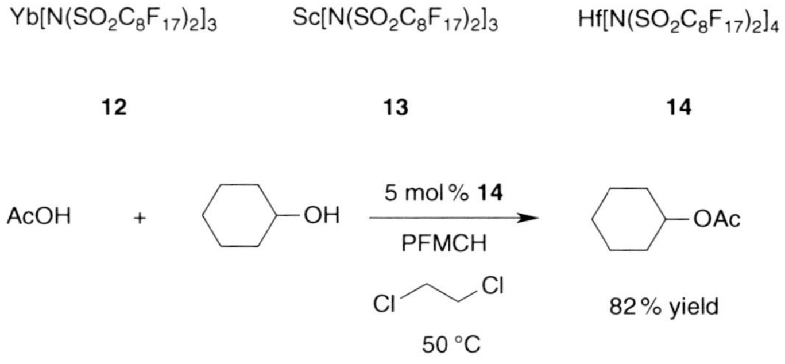

Yb[N(SO$_2$C$_8$F$_{17}$)$_2$]$_3$ Sc[N(SO$_2$C$_8$F$_{17}$)$_2$]$_3$ Hf[N(SO$_2$C$_8$F$_{17}$)$_2$]$_4$

12   13   14

AcOH + C$_6$H$_{11}$–OH →(5 mol% 14, PFMCH, ClCH$_2$CH$_2$Cl, 50 °C)→ C$_6$H$_{11}$–OAc   82% yield

Fig. 7 Fluorous metal sulfonamide Lewis acids

2.1.5 Fluorous Metal Sulfonamides

Fluorous Lewis acid catalysts produced by reaction of fluorous sulfonamide with metal salts have proven to be strongly Lewis acidic and highly effective at catalyzing esterification and transesterification reactions between carboxylic acids and esters and alcohols [17, 18]. In general, these catalysts, as exemplified by compounds **12–14** (Fig. 7), have been used in fluorous biphasic applications which use FLLE as the separation method. Perfluoromethylcyclohexane (PFMCH) has most often been the fluorous solvent with a number of different organic solvents used. Nishikido has also described a continuous process which could lead to highly scalable industrial processes [19]. Besides esterifications a number of other Lewis acid catalyzed transformations have also been reported including Friedel–Crafts acylations, Baeyer–Villiger oxidations, and Mannich reactions [20].

2.2 Aminations with Fluorous Reagents

The introduction of a nitrogen atom into a molecule, particularly an aromatic ring, has gained a high degree of use and interest within the synthetic community over the last decade, as exemplified by the Buchwald–Hartwig amination [21, 22]. The nitrogen synthons are varied, but are often protected versions of amines which are then later deprotected to provide the desired amine product. This allows for the introduction of fluorous domains which can be used as fluorous aminating reagents.

2.2.1 Fluorous Benzophenone Imine

The first reported fluorous aminating reagent was fluorous benzophenone imine (F-BPI) **15** (Fig. 8) described by Herr [23]. The authors found that typical Buchwald–Hartwig conditions could be used to form the substituted benzophenone imine **16** from aryl bromides, iodides, or triflates. Intermediate **16** was easily

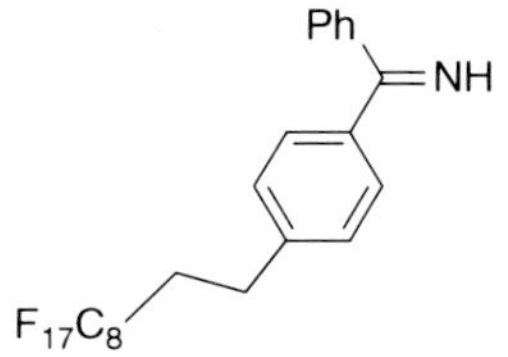

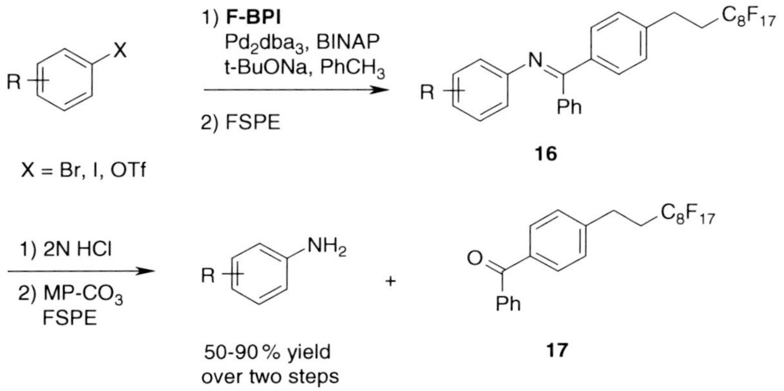

Fig. 8 Amination using F-BPI

purified by FSPE and the corresponding aniline **17** isolated by FSPE following hydrolysis of imine **16**. A number of different anilines were produced in this fashion in good yield and high purity.

2.2.2 Fluorous Carbamate

In 2007 Trabanco et al. published their results using F-Boc carbamate **18** (Fig. 9) in amidation reactions [24]. They found that yields were comparable to that reported by Herr using **15** and to those reported by Buchwald using standard carbamates. Unlike imine **15**, which necessarily limits products to primary anilines, secondary anilines could also be formed using substituted carbamates. The researchers also demonstrated that either FSPE or FLLE could be used as the fluorous separation method. FLLE separation using solvent tuning techniques on a "light" fluorous intermediate **19** is noteworthy, since it provides a method by which the process can be scaled without solid phase extraction, which can be more difficult to conduct on larger scales. In addition to aryl halides, it was demonstrated that heteroaryl halides such quinolines are also suitable substrates for the reaction.

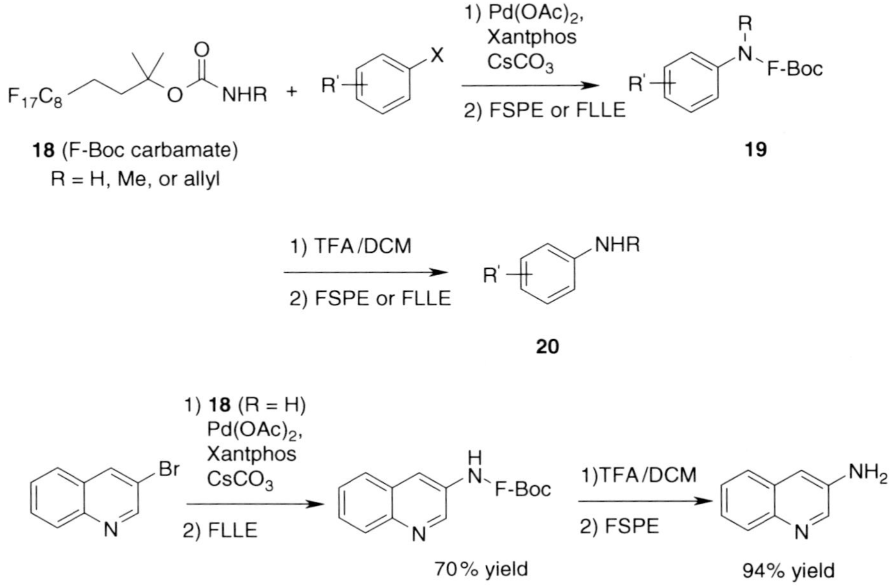

Fig. 9 F-Boc carbamates in aminations

2.3 Fluorous Mitsunobu Reaction

The Mitsunobu reaction between an acidic nucleophile and an alcohol has been used extensively in organic synthesis. The reaction's generality and mild conditions have proven valuable in many library and total synthesis efforts. One problematic aspect of the reaction, however, is that purification can be burdensome. The utility of the reaction, along with the associated problems in separation, are aptly demonstrated by the large number of variations that have been reported [25, 26]. Many of these were designed purely with post-reaction separation in mind and include solid phase reagents, polymerizable reagents, acid degradable reagents, and water-soluble reagents. Not surprisingly, fluorous based diazodicarboxylates and phosphines have also been employed.

Initial reports of a fluorous Mitsunobu reaction used fluorous diazodicarboxylate **21** (Fig. 10) and fluorous phosphine **22** [27]. Since all reagents and by-products were fluorous, the product could easily be purified using FSPE. Unlike other phase tag methods, such as solid-phase where only one of the reagents can be solid supported, all the by-products from a fluorous Mitsunobu reaction could be removed in a single operation.

The reactivity of **21**, however, was found to be less than ideal with less acidic nucleophiles such as phenols which were essentially unreactive. It was presumed that the electron-withdrawing effects of the fluorous domains hampered reactivity, and thus diazodicarboxylate **23** with an additional methylene spacer inserted was

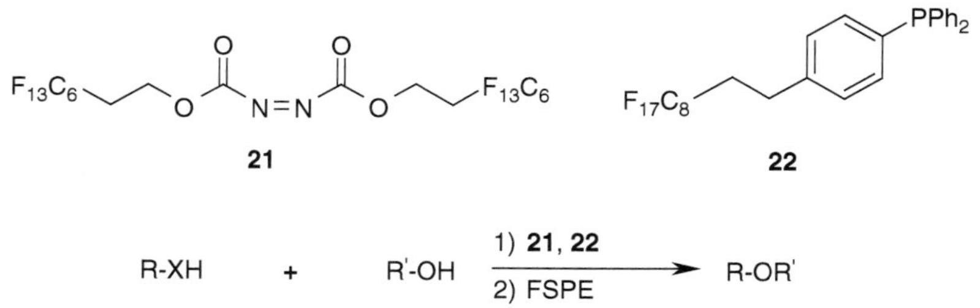

Fig. 10 Fluorous Mitsunobu reagents and general reaction

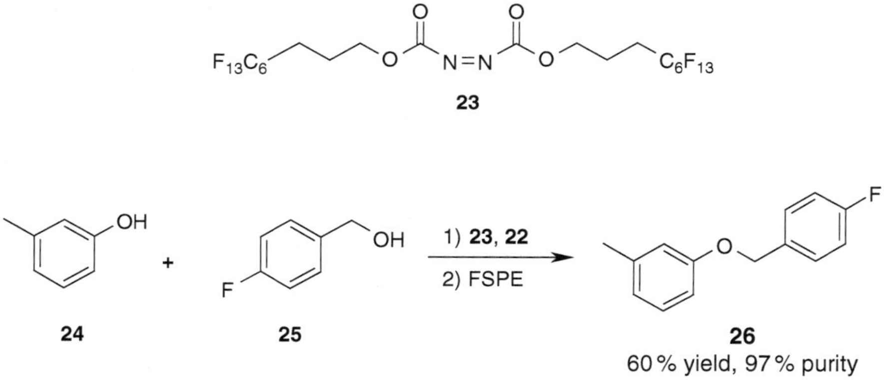

Fig. 11 Fluorous Mitsunobu reaction of electron-rich phenol using 23

prepared [28]. As expected, **23** (Fig. 11) in conjunction with **22** was found to retain much of the reactivity found in DEAD or DIAD. Phenols, including those with electron rich substituents, were found to be suitable Mitsunobu substrates with minimal loss in yield. The formation of ether **26** from the Mitsunobu reaction of phenol **24** and benzyl alcohol **25** proceeded in 60% yield compared to 0% yield using **21** and 76% using traditional triphenylphosphine and DIAD. The lower yield, however, is offset by the high purity of the products obtained after simple FSPE. In this instance **25** was obtained in 97% purity. An excellent example of the utility of the fluorous Mitsunobu was demonstrated by Winssinger et al. in their synthesis of radicicol A analogs where the key macrolactonization used dizaodicarboxylate **23** and phosphine **22** [29].

In an effort to lower the total fluorine count to produce a more environmentally friendly reagent, Chu et al. reported the use of alternate fluorous diazodicarboxylate **27** [30]. In this instance the two C_6F_{13} chains had been replaced by O-*t*-C_4F_9 domains which had previously been suggested to be more fluorophilic than the *n*-C_4F_9 isomer [31]. Shorter fluoroalkyls are known to be less persistent and bioaccumulative than longer perfluoroalkyls. Fluorous reagent **27** (Fig. 12) was indeed found to be an effective reagent for Mitsunobu reactions, but was not quite

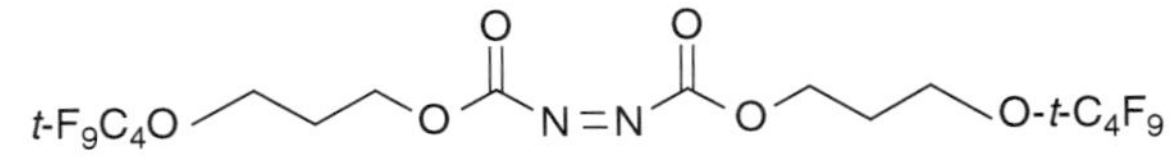

27

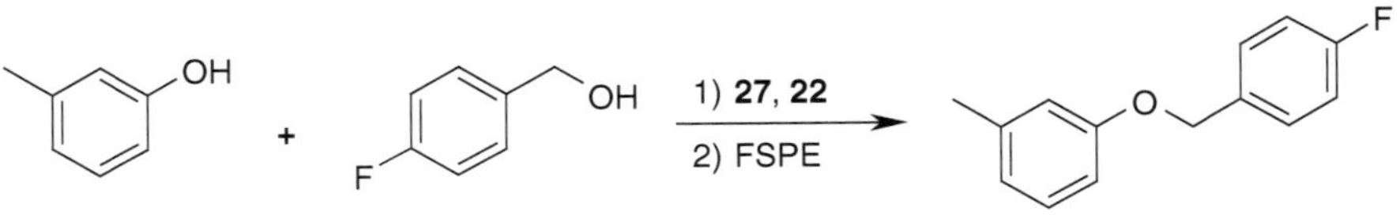

24 **25** **26**

64 % yield, 97 % purity

Fig. 12 Fluorous Mitsunobu using diazodicarboxylate 27

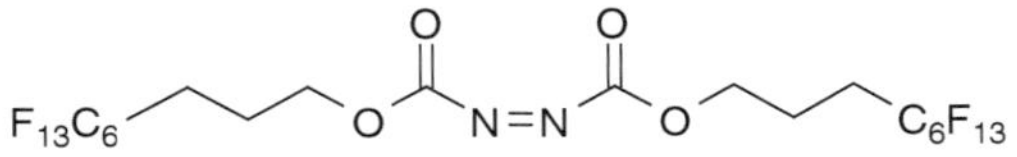

23

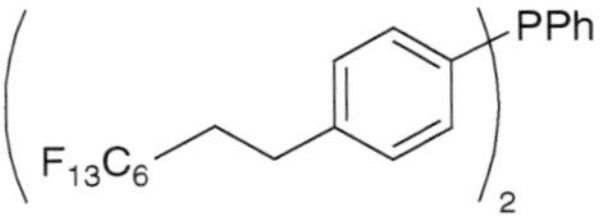

28

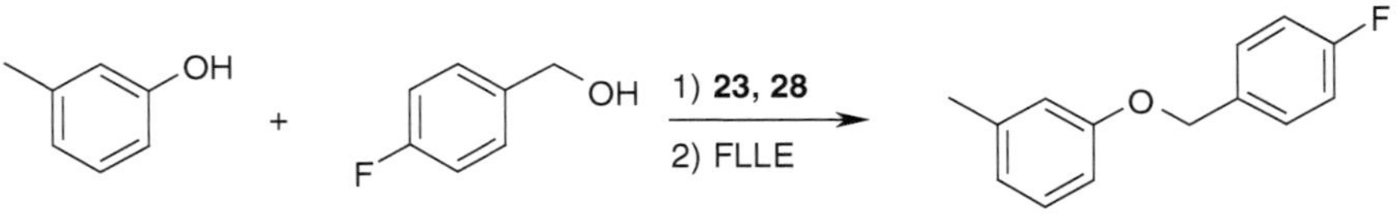

24 **25** **26**

81 % yield, 98 % purity

Fig. 13 Fluorous Mitsunobu using FLLE separation

as well retained upon FSPE although this issue was easily overcome by lowering the overall loading on the fluorous silica gel.

Fluorous Mitsunobu reactions using FLLE purification have also been developed within the Curran labs [32]. In these instances **23** (Fig. 13) was used with either phosphine **21** or bis-fluorous tagged phosphine **28**. Solvent tuning principles [33, 34] were applied to optimize the FLLE for complete removal of fluorous by-products with minimal product loss. The best solvent system reported used HFE-7100/FC-72 (2:1 v/v) as the fluorous solvent and 10% H_2O in DMF as the organic solvent. Excellent product purities were obtained using FLLE with these reagents.

In addition to phenols and carboxylic acids, other nucleophiles such as sulfonamides and phthalimides have also been demonstrated to be suitable substrates in fluorous Mitsunobu reactions [28].

2.4 Reductions with Fluorous Reagents

Most common reducing reagents such as metal hydrides are not generally problematic in terms of separation from the desired products after reaction. These reagents also do not lend themselves to being readily rendered fluorous by the addition of fluorous domains. There are, however, examples of reagents used in reductions where fluorous versions have been made. This has been done either to facilitate separation or recovery and reuse in those instances where the reagent is used in catalytic amounts. Fluorous compounds which are used as ligands in transition metal catalyzed reductions, for example fluorous BINOLs, will not be discussed in this chapter.

2.4.1 Fluorous Tin Hydride

Tributyltin hydride is a mild reducing agent for the reduction of halides, xanthates, and other functionalities [35, 36]. It has been widely used in synthetic chemistry despite the known problems in separating the tin residues from the desired product. Various workup conditions and alternate forms of tributyltin hydride have been developed to overcome these purification issues. In 1996 Curran reported the synthesis and reactivity of fluorous tin hydride **29a** as a fluorous alternative using FLLE as the fluorous separation method [37]. This was the first example of a fluorous stoichiometric reagent and clearly demonstrated the value of fluorous reagents in solving difficult separation issues. Tin hydride **29a** has all the hallmarks of later fluorous reagents and scavengers including solution phase chemistry, similar reactivity to the original nonfluorous reagent, and ease of separation through fluorous means.

Fluorous tin hydrides **29a** and **29b** (Fig. 14) have been demonstrated in many of the same reactions as tributyltin hydride including reduction of halides,

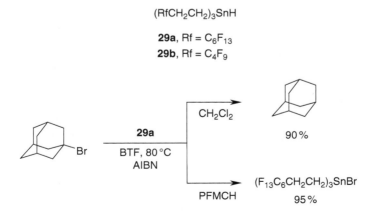

Fig. 14 Fluorous tin hydride reduction of adamantyl bromide

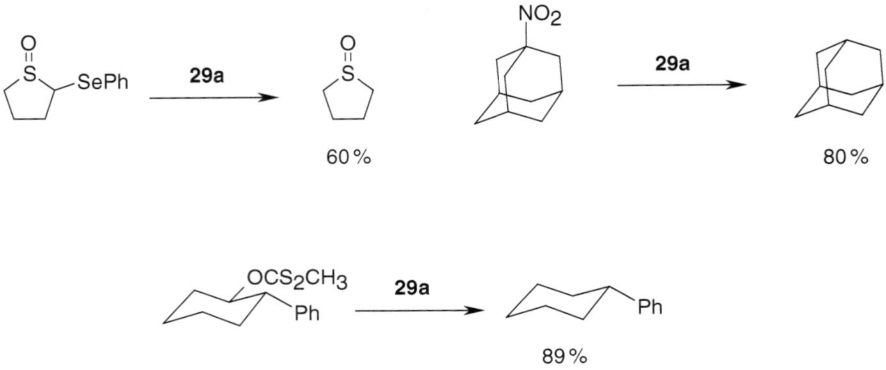

Fig. 15 Other reductions conducted using stoichiometric fluorous tin hydride

phenylselenides, tertiary nitros, and xanthates [38, 39]. In addition it has been used in hydrostannylations and reductive radical additions and cyclizations. Catalytic tin hydride conditions can also be used where NaCNBH$_3$ is the stoichiometric reductant. In all of these instances the tin residues can be removed using FLLE or FSPE. In halide reductions the resultant fluorous tin halide has been recovered, reduced back to tin hydride **29** and reused.

2.4.2 Fluorous Oxazaborolidine

The Corey–Bakshi–Shibata (CBS) reduction has, since its introduction in 1987, been a very powerful method by which to reduce ketones to chiral alcohols enantioselectively [40, 41]. The catalyst is formed by reaction of prolinol derived precatalyst **30** with either borane or a borate ester to form the oxazaborolidine catalyst **31**. Reduction is generally conducted using 10 mol% of **30** with borane-THF of borane-DMS as the stoichiometric reductant. The relatively high catalyst loadings has led to a variety of polymer-bound or other phase tag CBS catalysts being reported to facilitate recovery [42–46].

In 2005 Soós and coworkers introduced a fluorous CBS precatalyst **32** (Fig. 16) which was found to give comparable results to **30** [47]. The researchers developed a standard protocol using borane-THF with FSPE as the fluorous separation method. They reported near quantitative recovery of **32** which was recycled three times. Overall reaction yields and ees were similar to that observed using nonfluorous CBS catalysis. Chu et al. used **32** in a fluorous biphasic system with the catalyst immobilized in a hydrofluoroether [48]. A protocol which extracted the alcohol product from the fluorous solvent was developed, after which more substrate and borane could be introduced. The fluorous phase containing the fluorous immobilized **32** was used up to eight times.

A different fluorous prolinol, **33**, was prepared by Funabiki and used as a thermomorphic CBS precatalyst. Reaction yields and ees were lower than that

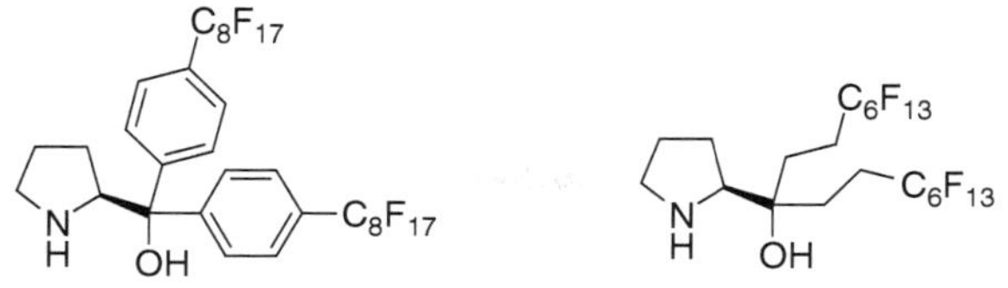

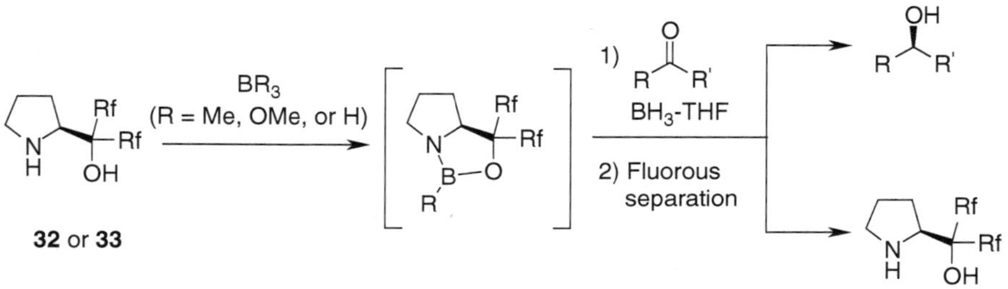

Fig. 16 Fluorous oxazaborolidine catalysts

found using **32**, but **33** (Fig. 17) was recovered by simple precipitation at low temperature and filtration. No fluorous solid or liquid phase was required [49].

2.5 Oxidations with Fluorous Reagents

Nontransition metal based oxidizing agents are frequently used in synthetic chemistry due to the mild, selective nature. The reduced forms of the oxidizing agent can sometimes create separation issues. Fluorous versions of these oxidizing agents can therefore lead to reagents that obviate the need for time-consuming purifications. Examples include hypervalent iodonium salts, amine oxides, and sulfur and selenium reagents.

2.5.1 Fluorous Hypervalent Iodine Compounds

Hypervalent iodonium species have found wide applicability in organic chemistry as mild, nonmetallic based reagents, most often used in oxidations. Not surprisingly various phase tagged versions of hypervalent iodines have been made including fluorous versions in order to facilitate purification after the reaction. In 2003 Gladysz reported the preparation of a series of fluorous modified (diacetoxy) iodobenzenes (F-DAIB) and their application to the oxidation of hydroquinones to quinones [50]. These were predominantly double or triple fluorous tagged

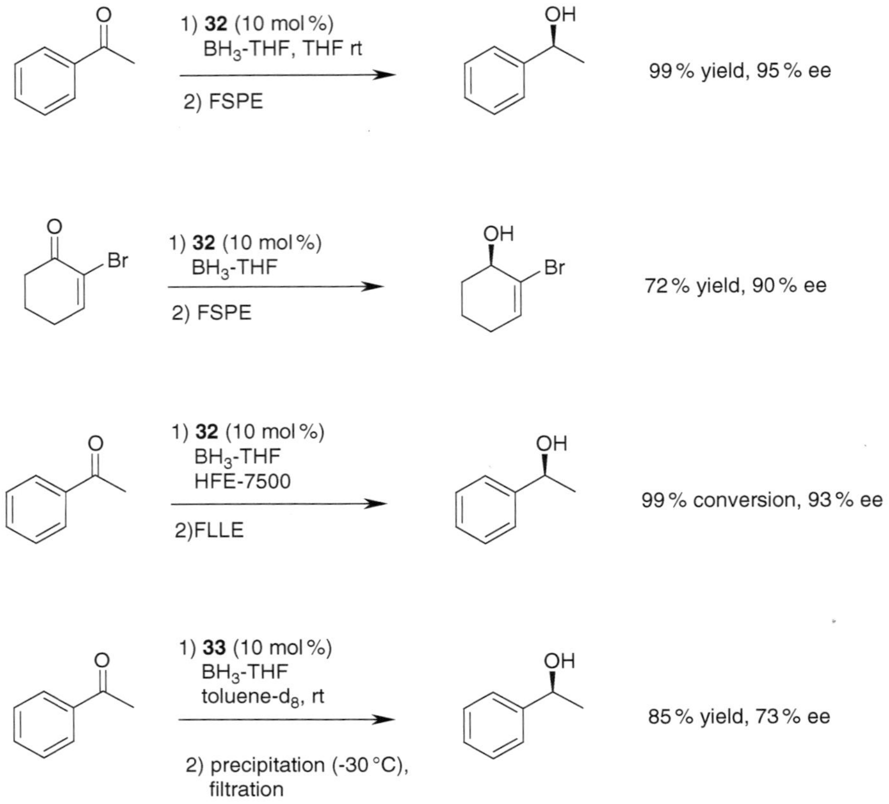

Fig 17 Example fluorous CBS reductions

molecules, exemplified by compounds **34–37** (Fig. 18) so that the resultant aryl iodine by-products could then be recovered by FLLE.

Oxidations were reported to occur with high yield and the resultant aryl iodide by-products were recovered in near quantitative yield. These could then be reoxidized to F-DIAB.

A "light" version of F-DAIB, **38**, (Fig. 19) containing a single fluorous chain was reported by Lindsley and Zhao to effect oxidative dimerization of 4-alkoxy substituted phenols [51]. The fluorous iodobenzene was separated from the product by FSPE.

Gladysz and Tesevic later showed that bis(trifluoroacetoxy) perfluoroalkyliodines, **39**, are effective at oxidizing hydroquinones to quinones and secondary alcohols to ketones [52, 53]. Once again the authors chose to use fluorous biphasic conditions for separation and recycling, although the Rf_{10} version could be readily precipitated and recovered by filtration. Recently, Miura et al. also described the oxidation of secondary alcohols to ketones and benzylic alcohols to benzaldehydes

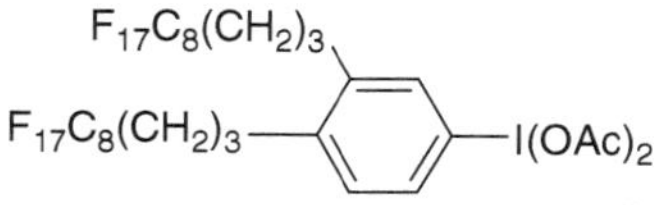

Fig. 18 F-DAIB compounds prepared by Gladysz

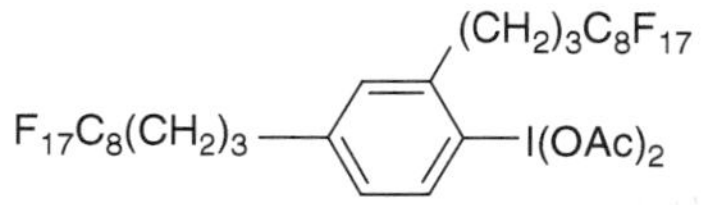

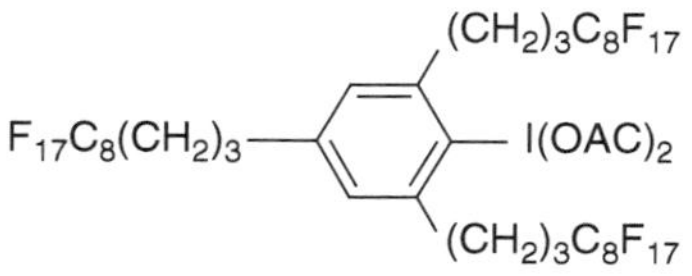

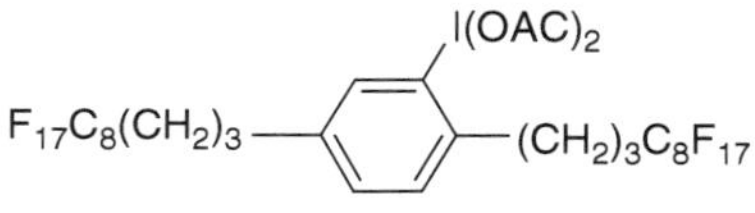

using a fluorous modified hypervalent iodine compound **40**, a fluorous version of *o*-iodoxybenzoic acid [54]. In this instance **40** was used in catalytic quantities with Oxone as the stoichiometric oxidant.

2.5.2 Fluorous Sulfur and Selenium Compounds

Sulfur and selenium based reagents and catalysts can affect a number of oxidations including alcohol oxidations, olefin epoxidation, and Baeyer–Villager oxidations among others. The toxicity of selenium and the malodorous nature of sulfur is a barrier to more widespread use. Fluorous variants of these useful oxidizing reagents have been developed and were, along with tin reagents, some of the first fluorous tagged reagents.

In 1999, Knochel described the synthesis and use of phenyl selenide **41** in the epoxidation of olefins [55]. Under the reaction conditions described, H_2O_2 was the stoichiometric oxidant with 5 mol% of **41** (Fig. 20) employed relative to the substrate. It was presumed that the fluorous selenide was first oxidized to the selenic acid then to the selenic peracid which catalyzes the reaction. The fluorous species were then recovered by FLLE and recycled up to ten times with little loss of activity. Sheldon and coworkers described a similar fluorous aryl selenide **42** which they also used with H_2O_2 in Baeyer–Villager oxidations and oxidations of benzaldehydes to benzoic acids [56]. Crich and Zou later used fluorous selenic acid **43** as a catalyst for allylic oxidation using iodoxybenzene as the stoichiometric oxidant [57]. The authors used continuous fluorous liquid extraction for removal of the fluorous residues post-reaction. It is interesting to note the differing products, i.e., epoxidation vs allylic oxidation, which are accessed depending on the exact fluorous selenium species and oxidant used. The selenium by-products were isolated as the

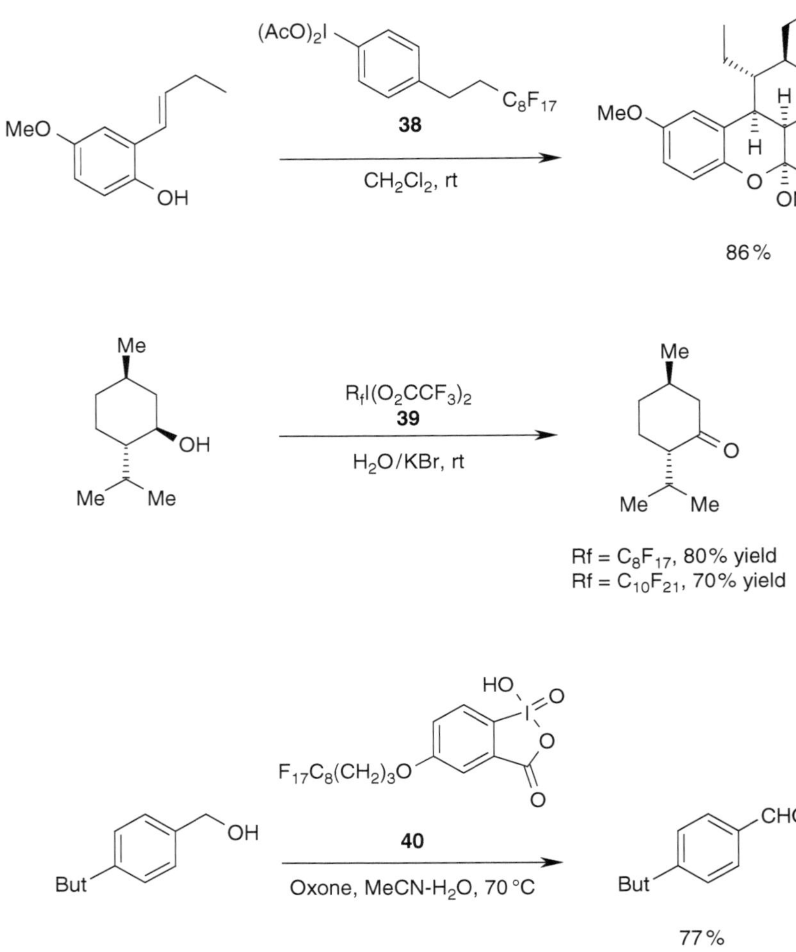

Fig. 19 Oxidation reactions with fluorous hypervalent iodonium reagents

diselenide **44** which could also be used as the precatalyst. The authors later extended the reaction to the oxidation of benzylic carbons and α-keto carbons [58].

Crich also prepared fluorous sulfoxides **45** as a DMSO equivalent for use in Swern oxidations [59]. The Swern oxidation is one of the milder and more selective oxidation methods available, but suffers from the generation of DMS which is malodorous. Sulfoxide **45** (Fig. 21), and its corresponding sulfide, is a white odorless solid and was demonstrated to oxidize primary and secondary alcohols to aldehydes and ketones under standard conditions. The fluorous DMS by-product was then recovered by continuous FLLE.

Chemical Applications of Fluorous Reagents and Scavengers

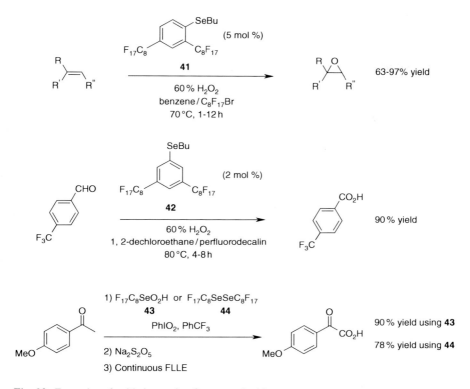

Fig. 20 Examples of oxidations using fluorous selenides

2.5.3 Fluorous TEMPO

Nitroxyl radicals such as 2,2,6,6-tetramethylpiperidine-1-oxyl (TEMPO) have found utility in catalytic oxidation of alcohols. Various stoichiometric oxidants, including peroxy acids, bleach, oxygen, and hypervalent iodoniums, have been used in these reactions. The removal and recycling of the catalytic TEMPO has been of interest and the usual phase tagging strategies have been employed including fluorous tags.

The first reported fluorous TEMPO analog was from Pozzi et al. who used a triazine linker to attach fluorous chains to a TEMPO core for use in fluorous biphasic catalysis [60, 61]. The researchers used either bleach or DAIB as the stoichiometric oxidant and found that a number of primary and secondary alcohols were oxidized to their corresponding carbonyl compounds in high yield and conversion in the presence of **46** (Fig. 22). The catalyst could be recovered by FLLE and reused up to five times. The same authors later reported other fluorous TEMPO derivatives which could be used in similar fashion and recovered using either FLLE

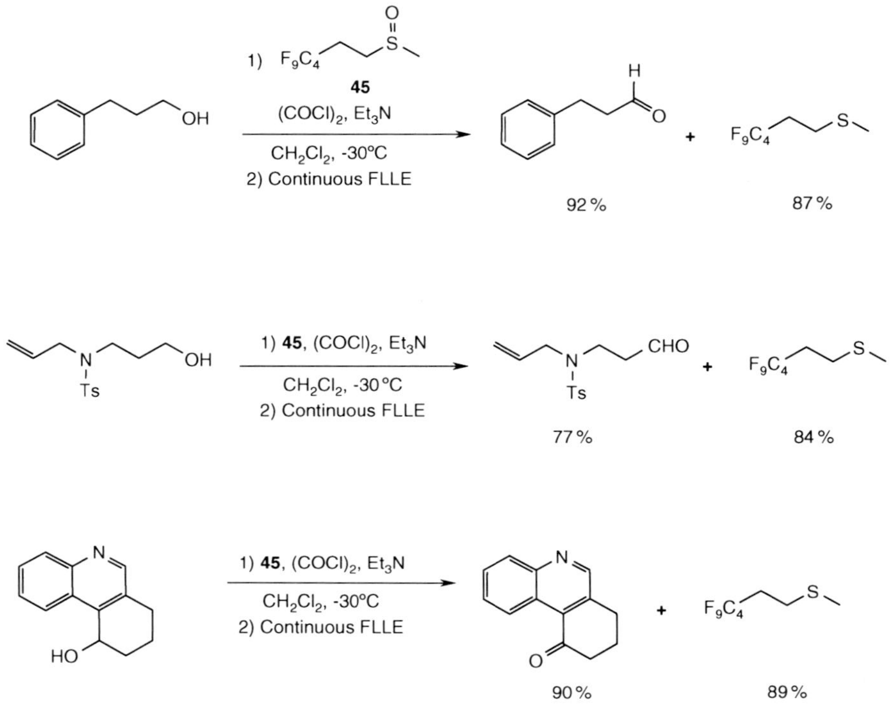

Fig. 21 Fluorous Swern reaction

or FSPE, although the recovery and recycling with compound **46** using FLLE was superior to FSPE [62].

Dobbs, meanwhile, has reported several different fluorous tagged TEMPO derivatives which were separated using FSPE. These compounds, as exemplified by **47a** and **47b**, were used to oxidize primary and secondary alcohols to aldehydes and ketones using either bleach or Oxone as the stoichiometric oxidant [63, 64].

Other fluorous tagged TEMPOs introduced by Reiser were prepared through click chemistry to provide fluorous reagents using triazoles as a linker [65, 66]. Light fluorous TEMPO **48** was first reported and was recovered using standard silica gel SPE. The heavy fluorous TEMPO **49**, also formed using click chemistry, was found to be insoluble in most solvent systems yet retained good catalytic activity similar to that of TEMPO itself and four to five times higher than that reported for polymer-supported TEMPO. Catalyst **49** was used in the oxidation of alcohols to carbonyls using either bleach or oxygen as the terminal oxidant and was easily recovered by filtration.

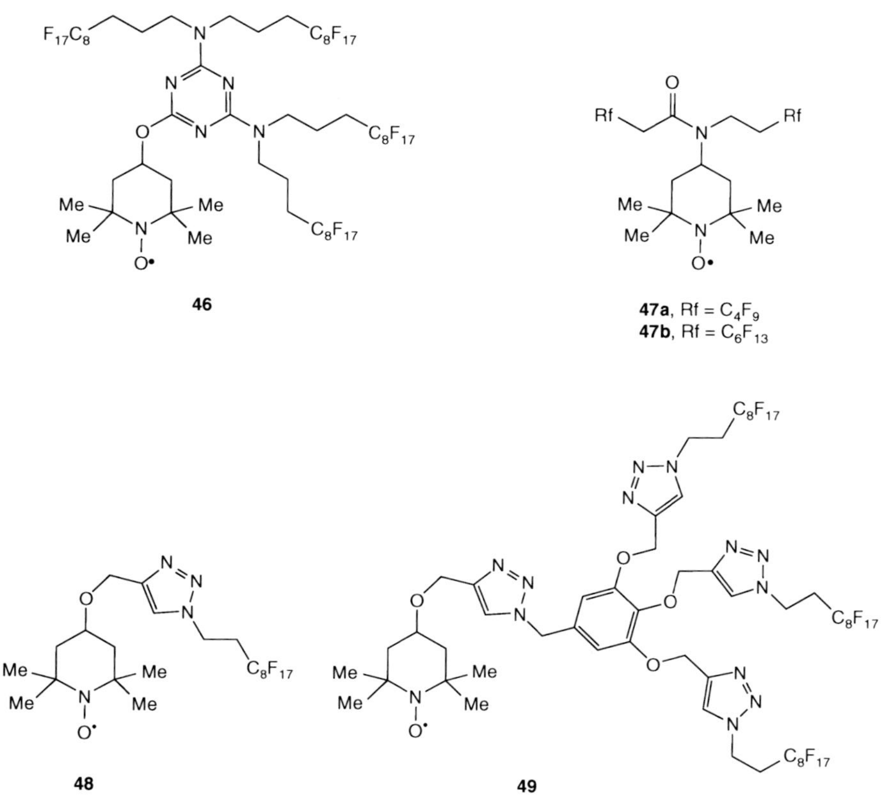

Fig. 22 Fluorous TEMPO reagents

2.6 Miscellaneous Transformations

2.6.1 Carbon–Carbon Bond Formation Using Fluorous Tin Reagents

Various other fluorous trialkyltin reagents have been reported. The common thread in each of these is that the separation of the tin residues after reaction is facilitated by fluorous separation methods. For example, fluorous allyl tin **50** (Fig. 23) has been prepared and used in radical allylations, thermal allylations, and radical carbonylative allylations [67–69]. Fluorous tin hydride **29a** has been used in radical carbonylations and hydroxymethylations [70, 71]. Other fluorous tin reagents that have been prepared and used in reactions typical of trialkyltins include fluorous tin azide [72] and fluorous aryl tins [73–75], each of which has been used in reactions typical of the corresponding trialkyltin.

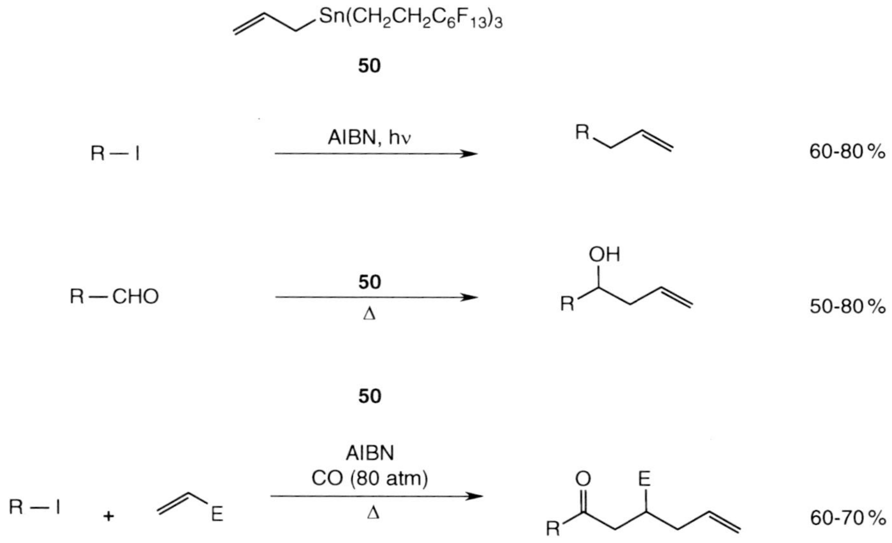

Fig. 23 Fluorous allyltin reactions

2.6.2 Thionation with Fluorous Lawesson's Reagent

The conversion of a carbonyl to a thiocarbonyl is a common procedure for preparation of sulfur containing compounds. Lawesson's reagent, **51**, (Fig. 24) has emerged as the reagent of choice for this transformation due to its generality and efficiency. The downside of **51** is that separation of the reagent by-products can be quite difficult requiring laborious effort.

In response Soós developed a fluorous Lawesson's reagent, **52**, which is of similar reactivity to **51**, but whose residues can be removed by FSPE [76]. Subsequent to that initial report, Soós and Dembinski reported a different fluorous analog of Lawesson's reagent, **53**, and found that it also had similar reactivity to **51** [77]. In either case, the fluorous by-products of the reaction were readily removed by FSPE to yield pure product. A variety of different thiocarbonyls and heterocycles were formed using these reagents.

2.6.3 Deoxy-Fluorination with F-DFMBA

Deoxy-fluorination is a selective method for the introduction of fluorines and is commonly conducted using reagents such as diethylaminosulfur trifluoride (DAST). The introduction of individual fluorines to molecules is particularly important in medicinal chemistry where fluorine is used to modify the activity and physical properties of drug candidates. Hara et al. described *N,N*-diethyl-α, α-difluoro-3-methylbenzylamine (DFMBA) as a versatile deoxy-fluorinating agent

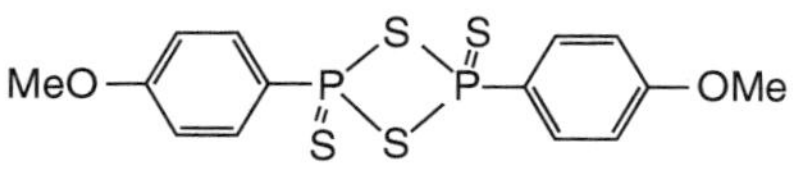

Chemical Applications of Fluorous Reagents and Scavengers

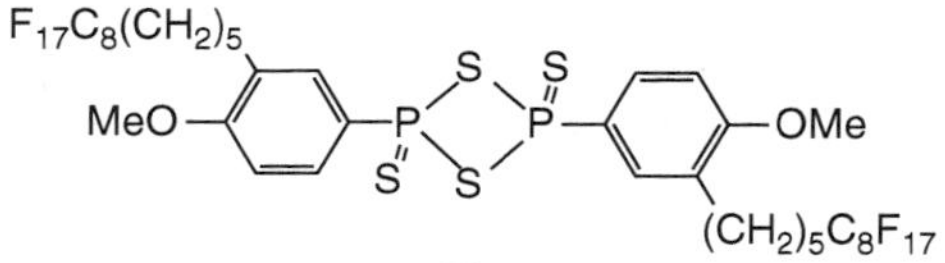

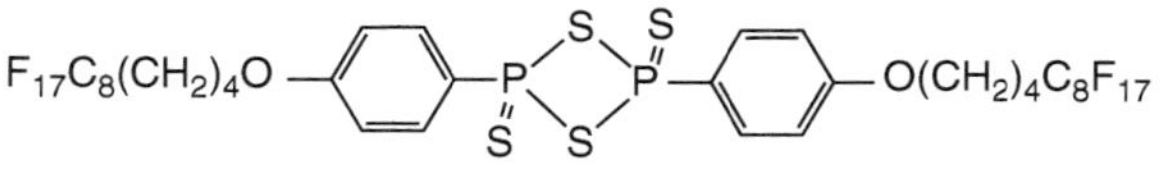

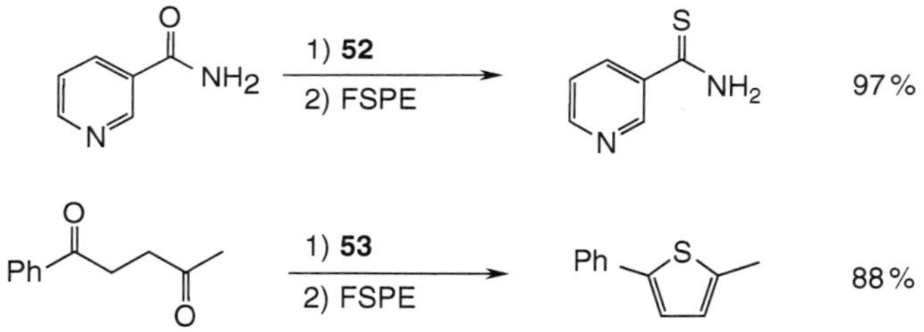

Fig. 24 Fluorous Lawesson's reagent and example reactions

[78, 79]. They went on to prepare F-DFMBA, **54** (Fig. 25), as a fluorous version of DFMBA. The authors demonstrated the use of **54** on primary alcohols and 1,2-diols. The amide by-product **55** could then be recovered and recycled to **55** using Et$_3$N-3HF. The FLLE separation did result in some loss of product to the fluorous phase, but the product itself was isolated in good yield and excellent purity in the organic phase.

3 Fluorous Scavengers

The use of phase tagged scavengers is a common method for the removal of unwanted by-products or excess reagents. Solid-phase scavenging techniques have been widely used as have fluorous techniques. Fluorous scavengers confer specific advantages over solid-phase scavengers including solution phase kinetics, control of stoichiometry, ease of scalability, and larger range of compatible reaction solvents. This section will not go into the details of these advantages which can be found elsewhere [51, 80], but will rather be a survey of the different types of fluorous scavengers that have been reported.

Fluorous scavengers can be generally grouped into nucleophilic scavengers or electrophilic scavengers, although other types of scavengers have been described.

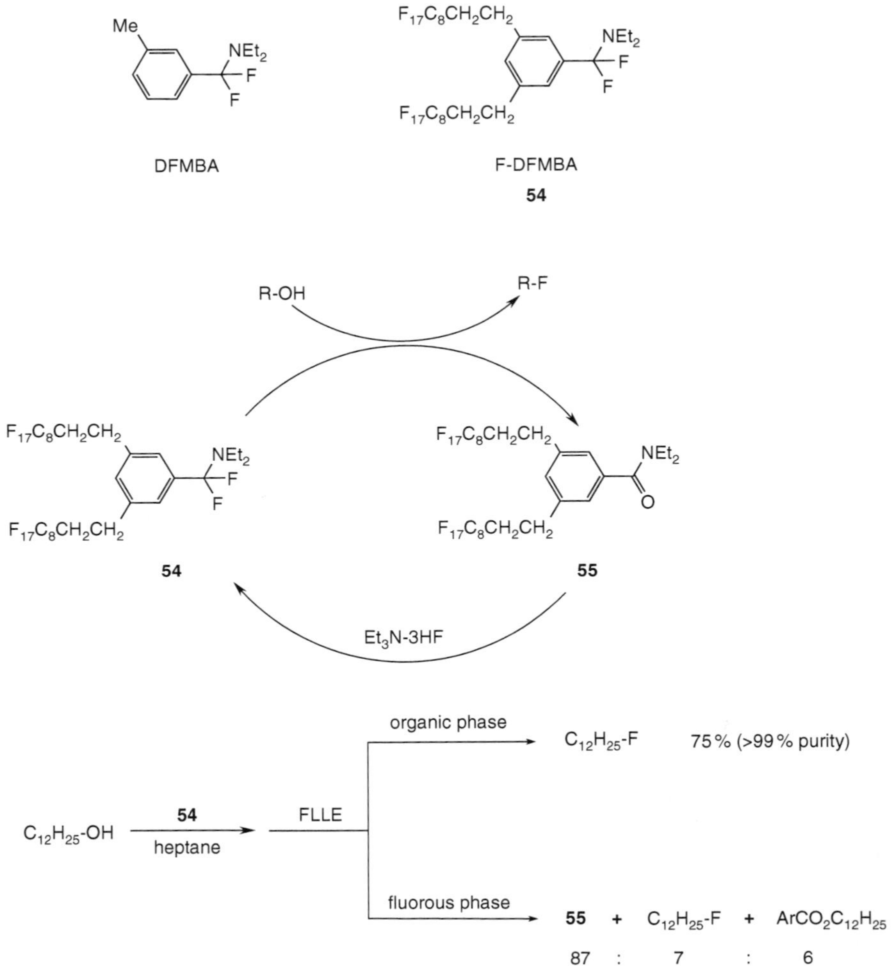

Fig. 25 F-DFMBA reaction scheme and example

As with other fluorous strategies, the separation methods used include FLLE, FSPE, and thermomorphic methods for separating the desired products from the fluorous tagged by-products.

3.1 Nucleophilic Fluorous Scavengers

Nucleophilic scavengers react with remaining amounts of excess electrophile after reaction. A number of nucleophilic fluorous scavengers have been described, both in "heavy" and "light" fluorous versions, which can then be removed using either FLLE or FSPE as appropriate.

3.1.1 Fluorous Amines

The scavenging of isocyanates with fluorous amines was one of the first examples of fluorous scavenging that was reported by Curran in 1999 [81]. In this instance the amine was "heavy" fluorous amine **56** (Fig. 26) containing six fluorous chains which, upon reaction with isocyanates, formed fluorous ureas that were separated from the desired reaction products by FLLE. Three small urea libraries were produced using the optimized reaction, scavenging, and FLLE protocols.

Lindsley et al. later reported using "light" fluorous tethered amine **57**, which was used to scavenge a number of different electrophiles such as acid chlorides, isocyanates, and sulfonyl chlorides [82, 83]. The researchers found that effective separation of the scavenged species could be conducted by FSPE. Commercially available **58** has also been used in a similar fashion.

3.1.2 Fluorous Thiols

Thiols react readily with a number of electrophiles including activated halides, epoxides, α-bromocarbonyls, and Michael acceptors. Fluorous tagged thiols have

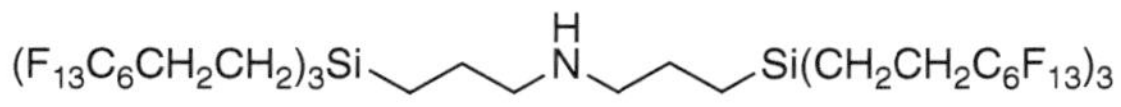

56

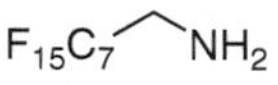

57

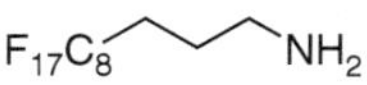

58

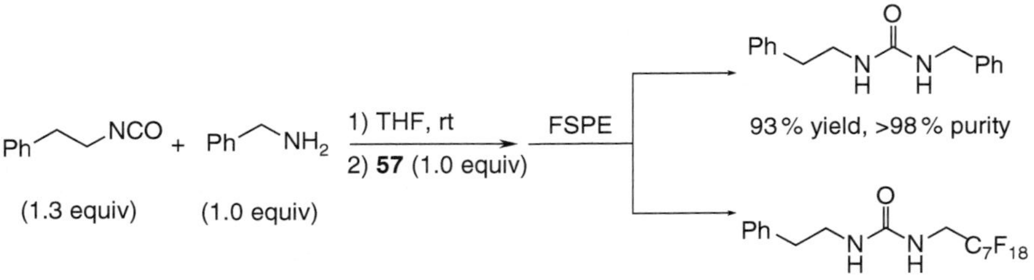

Fig. 26 Fluorous amines as nucleophilic scavengers

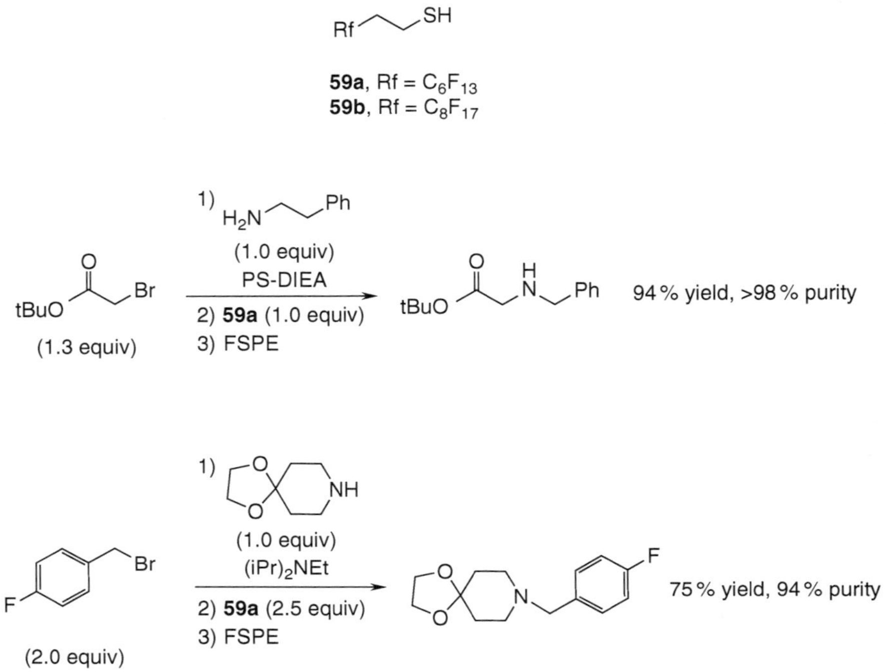

Fig. 27 Fluorous thiols as nucleophilic scavengers

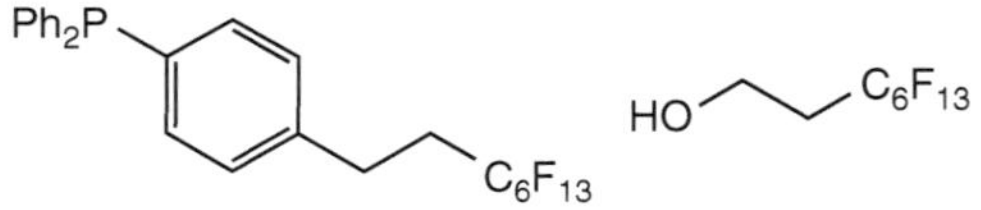

Fig. 28 Other fluorous scavengers

been used as scavengers for each of these classes of compounds [82–84]. The fluorous thiol that has been used can either be a C_6F_{13} or C_8F_{17} supported thiol (Fig. 27) **59a** and **59b**. The product of the fluorous thiol and electrophile can then be separated from the desired product by FSPE.

3.1.3 Other Nucleophilic Fluorous Scavengers

Unpublished results from Merck Research Laboratories used other fluorous nucleophiles (Fig. 28) as scavengers, specifically fluorous phosphines for the scavenging of activated halides and fluorous alcohols for the scavenging of silylating agents such as trialkylsilyl chlorides and triflates [51]. Experimental details are not available, but it is known that FSPE was used to effect fluorous separation.

3.2 Electrophilic Fluorous Scavengers

Analogous to nucleophilic scavengers, electrophilic fluorous reagents have also been used for the scavenging of nucleophilic reagents used in small molecule synthesis. In contrast to nucleophilic scavengers, however, fluorous electrophilic scavengers have for the most part been "light" fluorous compounds utilizing FSPE as the fluorous separation mode.

3.2.1 Fluorous Isocyanate and Isatoic Anhydride

The use of excess amine for amide forming reactions and aminations is a common technique in small molecule synthesis. After the reaction, remaining amine can be scavenged using fluorous isocyanate **60** (Fig. 29) or fluorous isatoic anhydride **61** [85]. Either **60** or **61** may be used to scavenge excess amine used for reaction with isocyanates or thioisocyanates, while **61** was demonstrated to be a suitable scavenger for use after epoxide opening by amines. In each of these, FSPE was used to separate the product from the scavenger and scavenger product.

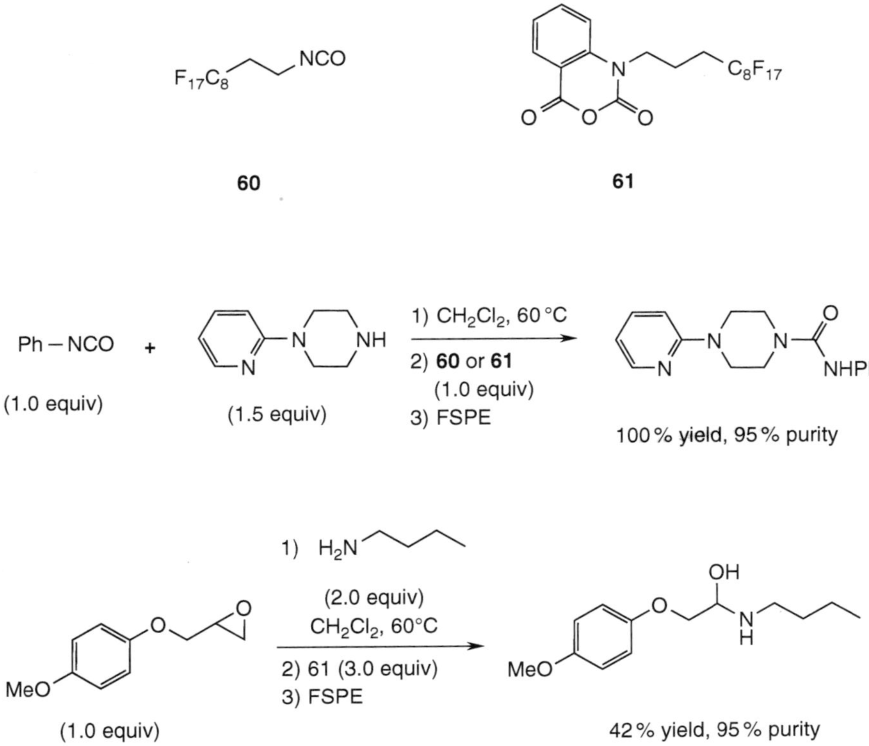

Fig. 29 Fluorous isocyanate and isatoic anhydride in amine scavenging

A particularly useful application of amine scavenging using **60** (Fig. 30) was demonstrated in the preparation of ^{35}S labeled sulfonamides [86]. Due to the high value of the isotope labeled sulfonyl chloride, the amine is often used in large excess. The resultant ^{35}S labeled sulfonamide often requires multiple HPLC purifications due to the large amount of amine present. Fluorous scavenging followed by FSPE provided a quick and efficient method for the removal of excess reagents.

3.2.2 Fluorous 2,4-Dichloro-1,3,5-Triazines

Lu and Zhang have described the preparation and use of two fluorous 2,4-dichloro-1,3,5-triazines (F-DCTs) **62** and **63** (Fig. 31) as electrophilic reagents for the

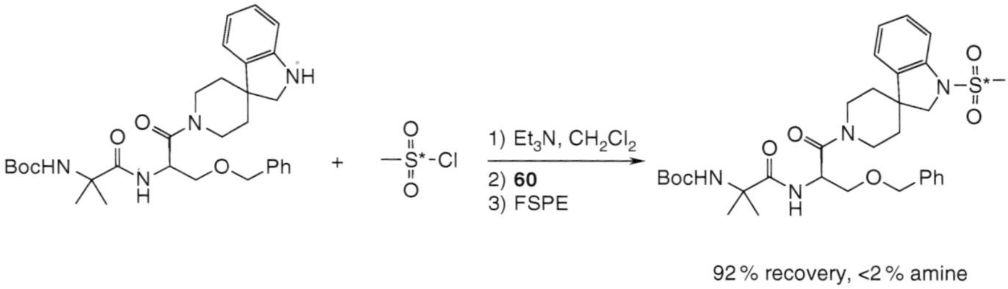

Fig. 30 Isotope labeled sulfonamide preparation with fluorous scavenging

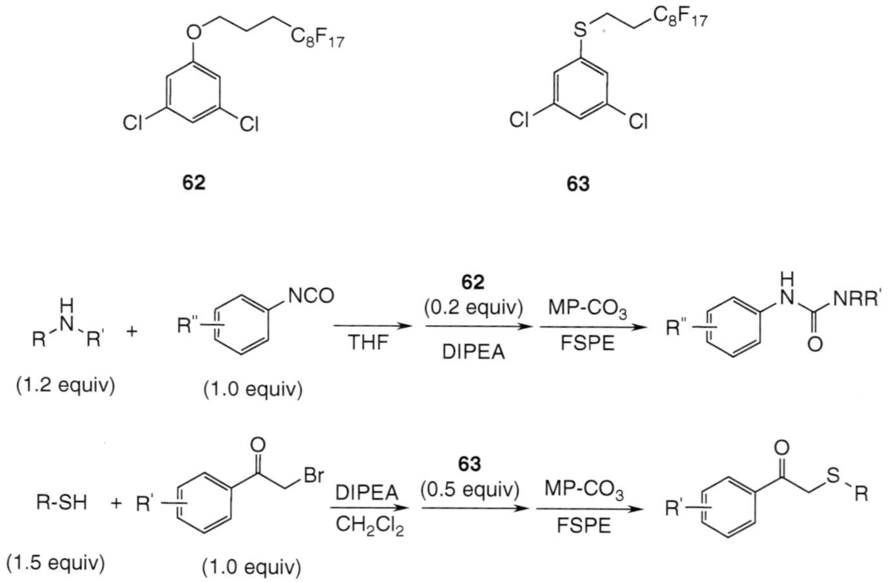

Fig. 31 F-DCT compounds and example reactions

scavenging of amines and thiols [87]. The authors found that F-DCT was an effective scavenger which could be used in lower stoichiometry due to the two reactive sites on the triazine. Two different 18-member libraries were produced using these scavengers. In addition, it was demonstrated that FSPE using scavenger **62** could be automated using a RapidTrace SPE station equipped with FSPE cartridges.

3.2.3 Other Fluorous Electrophilic Scavengers

In addition to the electrophilic scavengers noted above, there have been other fluorous compounds which have been used as electrophilic scavengers. For the scavenging of amines, fluorous acid chlorides, sulfonyl chlorides, sulfonic acids, and primary iodides have all been used in conjunction with FSPE [82, 83, 88]. Fluorous sulfonyl chlorides have also been demonstrated to scavenge alkoxides effectively, while fluorous epoxides have been used for the scavenging of amines, thiols, thiophenols, and organometallic reagents [82, 83].

3.3 Miscellaneous Fluorous Scavengers

Beyond the two major classes of fluorous scavengers described, other fluorous scavengers for specific types of reactions have also been reported. These scavengers demonstrate that, due to the solution-phase solubility of fluorous compounds and the high degree of chemical compatibility, many reactions that may not be suitable for solid-phase scavenging can incorporate fluorous scavenging to remove excess or unreacted starting materials.

3.3.1 Fluorous Tin Scavengers

In addition to the use of fluorous tin hydride **29a** (Fig. 32) in radical reduction reactions as described in Sect. 2.4.1, **29a** has also been utilized as a scavenger of excess olefins and alkynes through hydrostannylation [39]. Curran et al. demonstrated the scavenging of these compounds following Giese radical reactions, Diels–Alder reactions, and nitrile oxide cycloadditions. In each case, the excess **29a** and the fluorous hydrostannylated reagents were separated from the desired product by FLLE.

3.3.2 Fluorous Dienophile Scavenger

Werner and Curran have developed a series of fluorous dienophiles **64–66** (Fig. 33) for use as scavengers following Diels–Alder reaction. For each Diels–Alder

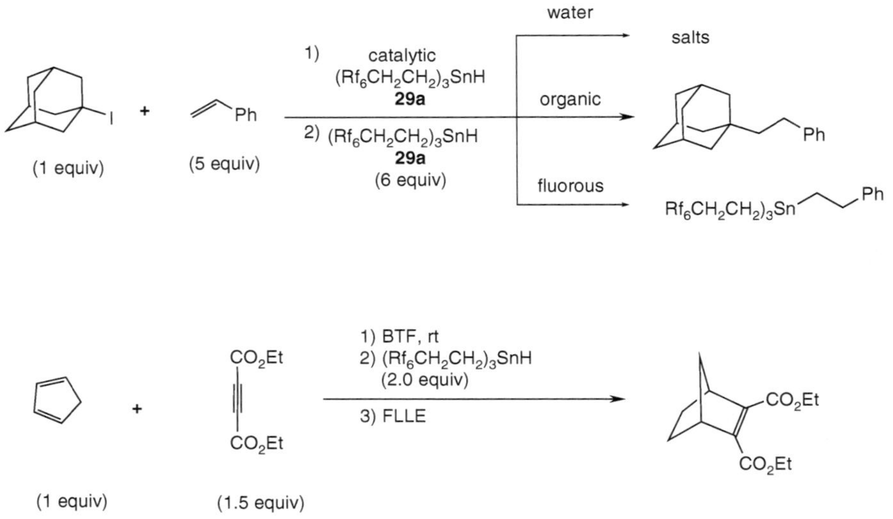

Fig. 32 Scavenging of alkenes and alkynes with fluorous tin hydride

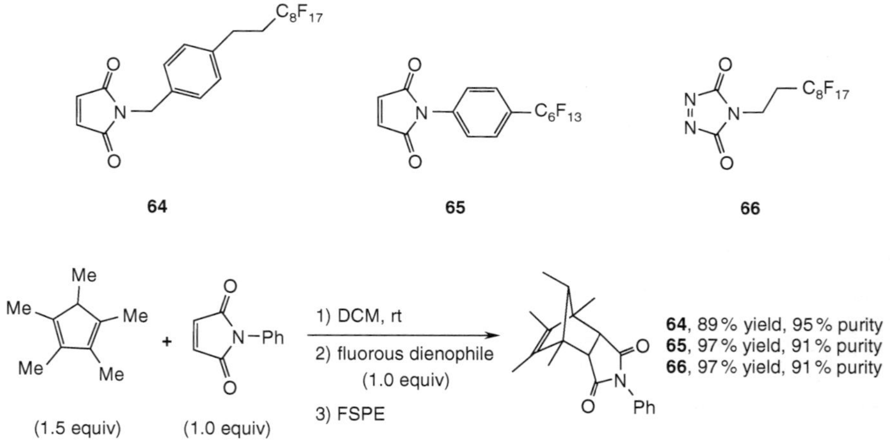

Fig. 33 Scavenging with fluorous dienophiles

reaction, the excess dienophile was scavenged with one of the fluorous dienophiles. The desired nonfluorous Diels–Alder adduct was isolated in good to excellent purity using FSPE. The fluorous Diels–Alder adducts and excess scavenger were retained on the fluorous silica gel [89].

3.3.3 Fluorous Ligand Capture

Fluorous scavengers have also been used for metal ligands which are used in excess in the formation of metal complexes. Valliant has reported the synthesis of fluorous

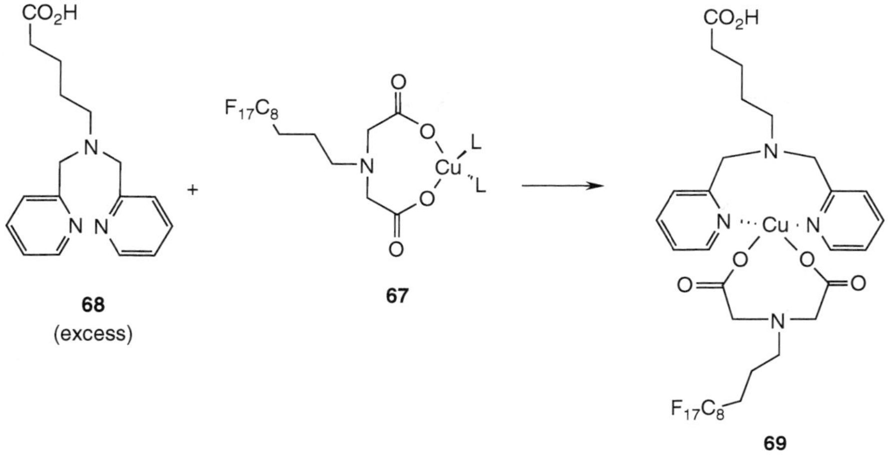

Fig. 34 Fluorous ligand scavenger for bispyridyl valeric acid.

copper complex **67** as just such a scavenger [90]. Complex **67** (Fig. 34) was then shown to incorporate effectively bispyridyl valeric acid ligand **68** to form new complex **69** which was retained on fluorous silica gel. The authors then applied **67** to the synthesis of a gamma-emitting ^{99m}Tc complex in a scavenging application which was termed fluorous ligand capture (FLC).

^{99m}Tc is the most widely used radionuclide in diagnostic medicine used in SPECT imaging. Most commonly a ^{99m}Tc radioimaging agent is formed by complexation with a vector ligand. The vector ligand is generally used in large excess in order to utilize fully the ^{99m}Tc. This, however, presents an issue with uncomplexed vector ligand competing with the ^{99m}Tc ligand complex for binding with the target. Removal of excess vector ligand therefore leads to higher quality radioimaging agent.

Valliant complexed bispyridyl modified peptide vector **70** (Fig. 35) with [^{99m}Tc(CO)$_3$(OH$_2$)$_3$]$^+$ to form a mixture of putative imaging agent **71** and **70**. Fluorous complex **67** was then added to scavenge excess **70**, forming new complex **72**. After FSPE, HPLC analysis showed that 95% of **70** had been removed from **71** which was recovered in the fluorophobic wash of the FSPE.

4 Conclusions

Fluorous methods have significant advantages over other methods for the phase tagging of reagents and scavengers and their use in synthesis. These advantages include solution phase kinetics, ease of separation, and complete stoichiometric control. One of the most advantageous aspects, however, is the large range of chemistries that can be conducted and that are compatible with fluorous domains. As seen in this survey, the range of reactions includes oxidation, reduction, acidic,

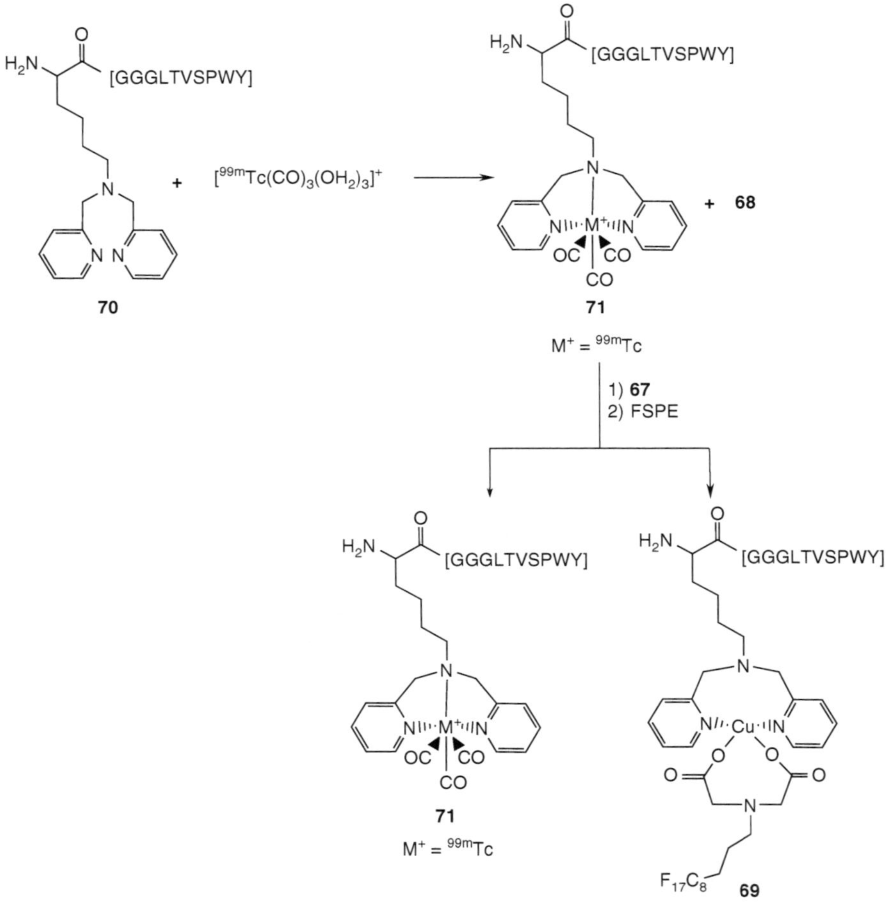

Fig. 35 Fluorous ligand capture (FLC) for the preparation of 71

basic, radical, and Lewis acidic conditions, while a host of transformations using those various conditions have been demonstrated. This provides researchers with considerable freedom to design new fluorous reagents and scavengers to meet their specific requirements so as to exploit the selectivity and utility of fluorous separations.

References

1. Curran DP (2000) Fluorous techniques for the synthesis of organic molecules: a unified strategy for reaction and separation. In: Stoddard F, Reinhoudt D, Shibasaki M (eds) Stimulating concepts in chemistry. Wiley-VCH, New York
2. Curran DP (2001) Fluorous reverse phase silica gel. A new tool for preparative separations in synthetic organic and organofluorine chemistry. Synlett:1488–1496

3. Curran DP (2006) Synthetic applications of fluorous solid-phase extraction (F-SPE). Tetrahedron 62:11837–11865
4. Palomo C, Aizpurua JM, Loinaz I, Fernandez-Berridi MJ, Irusta L (2001) Scavenging of fluorinated N,N'-dialkylureas by hydrogen binding: a novel separation method for fluorous synthesis. Org Lett 3:2361–2364
5. Del Pozo C, Keller AI, Nagashima T, Curran DP (2007) Amide bond formation with a new fluorous carbodiimide: separation by reverse fluorous solid-phase extraction. Org Lett 9:4167
6. Markowicz MW, Dembinski R (2004) Fluorous coupling reagents: application of 2-chloro-4,6-bis[(heptadecafluorononyl)oxy]-1,3,5-triazine in peptide synthesis. Synthesis:80–86
7. Zhang W, Lu Y, Nagashima T (2005) Plate-to-plate fluorous solid-phase extraction for solution-phase parallel synthesis. J Combi Chem 7:893–897
8. Matsugi M, Hasegawa M, Sadachika D, Okamoto S, Tomioka M, Ikeya Y, Masuyama A, Mori Y (2007) Preparation and condensation reactions of a new light-fluorous Mukaiyama reagent: reliable purification with fluorous solid phase extraction for esters and amides. Tetrahedron Lett 48:4147–4150
9. Matsugi M, Suganuma M, Yoshida S, Hasebe S, Kunda Y, Hagihara K, Oka S (2008) An alternative and facile purification procedure of amidation and esterification reactions using a medium fluorous Mukaiyama reagent. Tetrahedron Lett 49:6573–6574
10. Matsugi M, Nakamura S, Kunda Y, Sugiyama Y, Shioiri T (2009) Pronounced rate enhancements in condensation reactions attributed to the fluorous tag in modified Mukaiyama reagents. Tetrahedron Lett 51:133–135
11. Otera J (2004) Toward ideal (trans)esterification by use of fluorous distannoxane catalysts. Acc Chem Res 37:288–296
12. Xiang J, Toyoshima S, Orita A, Otera J (2001) A practical and green chemical process: fluoroalkyldistannoxane-catalyzed biphasic transesterification. Angew Chem Int Ed 40:3670–3672
13. Orita J, Man-e S, Otera J (2006) Fluorophilicity switch by solvation. J Am Chem Soc 128:4182–4183
14. Bucher B, Curran DP (2000) Selective sulfonylation of 1,2-diols and derivatives catalyzed by a recoverable fluorous tin oxide. Tetrahedron Lett:9617–9621
15. Beeler AB, Acquilano DE, Su Q, Yan F, Roth BL, Panek JS, Porco JA (2005) Synthesis of a library of complex macrodiolides employing cyclodimerization of hydroxy esters. J Comb Chem 7:673–681
16. Kesavan S, Su Q, Shao J, Porco JA, Panek JS (2005) Enantioselective synthesis of linear polypropionate arrays using anthracene-tagged organosilanes. Org Lett 7:4435–4438
17. Mikami K, Mikami Y, Matsuzawa H, Matsumoto Y, Nishikido J, Yamamoto F, Nakajima H (2002) Lanthanide catalysts with tris(perfluorooctanesulfonyl)methide and bis(perfluorooctanesulfonyl)amide ponytails: recyclable Lewis acid catalysts in fluorous phases or as solids. Tetrahedron 58:4015–4021
18. Hao X, Yoshida A, Nishikido J (2004) Recyclable and selective Lewis acid catalysts for transesterification and direct esterification in a fluorous biphase system: tin(IV) and hafnium (IV) bis(perfluorooctanesulfonyl)amide complexes. Tetrahedron Lett 45:781–785
19. Yoshida A, Hao X, Nishikido J (2003) Development of the continuous-flow reaction system based on the Lewis acid-catalysed reactions in a fluorous biphasic system. Green Chem 5:554–557
20. For a review see Cai C, Yi W-B, Zhang W, Shen M-G, Hong M, Zeng L-Y (2009) Fluorous Lewis acids and phase transfer catalysts. Mol Divers 14:209–239
21. Hartwig JF (1998) Transition metal catalyzed synthesis of arylamines and aryl ethers from aryl halides and triflates: scope and mechanism. Angew Chem Int Ed 37:2046–2067
22. Hartwig JF (2000) Palladium-catalyzed amination of aryl halides and sulfonates. In: Ricci A (ed) Modern amination methods. Wiley-VCH, Weinheim, pp 195–262
23. Cioffi CL, Berlin ML, Herr RJ (2004) Convenient palladium-catalyzed preparation of primary anilines using a fluorous benzophenone imine reagent. Synlett:841–845

24. Trabanco AA, Vega JA, Fernandez MA (2007) Fluorous-tagged carbamates for the pd-catalyzed amination of aryl halides. J Org Chem 72:8146–8148
25. Dandapani S, Curran DP (2004) Separation-friendly mitsunobu reactions: a microcosm of recent developments in separation strategies. Chem Eur J 10:3130–3138
26. Dembinski R (2004) Recent advances in the Mitsunobu reaction: modified reagents and the quest for chromatography-free separation. Eur J Org Chem:2763–2772
27. Dandapani S, Curran DP (2002) Fluorous Mitsunobu reagents and reactions. Tetrahedron 58:3855–3864
28. Dandapani S, Curran DP (2004) Second generation fluorous dead reagents have expanded scope in the mitsunobu reaction and retain convenient separation features. J Org Chem 69:8751–8757
29. Jogireddy R, Dakas P-Y, Valot G, Barluenga S, Winssinger N (2009) Synthesis of a resorcylic acid lactone (RAL) library using fluorous-mixture synthesis and profile of its selectivity against a panel of kinases. Chem-Eur J 15:11498–11506
30. Chu Q, Henry C, Curran DP (2008) Second-generation tags for fluorous chemistry exemplified with a new fluorous Mitsunobu reagent. Org Lett 10:2453–2456
31. Kiss LE, Kovesdi I, Rábai J (2001) An improved design of fluorophilic molecules: prediction of the ln P fluorous partition coefficient, fluorophilicity, using 3D QSAR descriptors and neural networks. J Fluorine Chem 108:95–109
32. Curran DP, Bajpai R, Sanger E (2006) Purification of fluorous Mitsunobu reactions by liquid-liquid extraction. Adv Synth Catal 348:1621–1624
33. Yu MS, Curran DP, Nagashima T (2005) Increasing fluorous partition coefficients by solvent tuning. Org Lett 7:3677–3680
34. Chu Q, Yu MS, Curran DP (2007) New fluorous/organic biphasic systems achieved by solvent tuning. Tetrahedron 63:9890–9895
35. Kuivila HG (1968) Organotin hydrides and organic free radicals. Acc Chem Res 1:299
36. Neumann WP (1987) Tri-n-butyltin hydride as reagent in organic synthesis. Synthesis:665
37. Curran DP, Hadida S (1996) Tris(2-(perfluorohexyl)ethyl)tin hydride: a new fluorous reagent for use in traditional organic synthesis and liquid phase combinatorial synthesis. J Am Chem Soc 118:2531–2532
38. Hadida S, Super MS, Beckman EJ, Curran DP (1997) Radical reactions with alkyl and fluoroalkyl (fluorous) tin hydride reagents in supercritical CO_2. J Amer Chem Soc 119:7406–7407
39. Curran DP, Hadida S, Kim SY, Luo ZY (1999) Fluorous tin hydrides: a new family of reagents for use and reuse in radical reactions. J Am Chem Soc 121:6607–6615
40. Corey EJ, Bakshi RK, Shibata S (1987) Highly enantioselective borane reduction of ketones catalyzed by chiral oxazaborolidines. Mechanism and synthetic implications. J Am Chem Soc 109:5551–5553
41. Corey EJ, Bakshi RK, Shibata S, Chen CP, Singh VK (1987) A stable and easily prepared catalyst for the enantioselective reduction of ketones. Applications to multistep syntheses. J Am Chem Soc 109:7925–7926
42. Franot C, Stone GB, Engeli P, Spöndlin C, Waldvogel E (1995) A polymer-bound oxazaborolidine catalyst: enantioselective borane reductions of ketones. Tetrahedron-Asymmetry 6:2755–2766
43. Price MD, Sui JK, Kurth MJ, Schore NE (2002) Oxazaborolidines as functional monomers: ketone reduction using polymer-supported Corey, Bakshi, and Shibata catalysts. J Org Chem 67:8086–8089
44. Kell RJ, Hodge P, Snedden P, Watson D (2003) Towards more chemically robust polymer-supported chiral catalysts: α, α-diphenyl-L-prolinol based catalysts for the reduction of prochiral ketones with borane. Org Biomol Chem 1:3238–3243
45. Degni S, Wilén C-E, Rosling A (2004) Highly catalytic enantioselective reduction of aromatic ketones using chiral polymer-supported Corey, Bakshi, and Shibata catalysts. Tetrahedron-Asymmetry 15:1495–1499

46. Schunicht C, Biffis A, Wullf G (2000) Microgel-supported oxazaborolidines: novel catalysts for enantioselective reductions. Tetrahedron 56:1693–1699
47. Dalicsek Z, Pollreisz F, Gómóry A, Soós T (2005) Recoverable fluorous CBS methodology for asymmetric reduction of ketones. Org Lett 7:3243–3246
48. Chu Q, Yu MS, Curran DP (2008) CBS reductions with a fluorous prolinol immobilized in a hydrofluoroether solvent. Org Lett 10:749–752
49. Goushi S, Funabiki K, Ohta M, Hatano K, Matsui M (2007) Novel fluorous prolinol as a precatalyst for catalytic asymmetric borane reduction of various ketones. Tetrahedron 63:4061–4066
50. Rocaboy C, Gladysz JA (2003) Convenient syntheses of fluorous aryl iodides and hypervalent iodine compounds: ArI(L)n reagents that are recoverable by simple liquid/liquid biphase workups, and applications in oxidations of hydroquinones. Chem Eur J 9:88–95
51. Lindsley CW, Zhao Z (2004) Fluorous scavengers. In: Gladys JA, Curran DP, Horváth IT (eds) Handbook of fluorous chemistry. Wiley-VCH, New York, pp 236–246
52. Tesevic V, Gladysz JA (2005) An easily accessed class of recyclable hypervalent iodide reagents for functional group oxidations: bis(trifluoroacetate) adducts of fluorous alkyl iodides, $CF_3(CF_2)_{n-1}I(OCOCF_3)_2$. Green Chem 7:833–836
53. Tesevic V, Gladysz JA (2006) Oxidations of secondary alcohols to ketones using easily recyclable bis(trifluoroacetate) adducts of fluorous alkyl iodides, $CF_3(CF_2)_{n-1}I(OCOCF_3)_2$. J Org Chem 71:7433–7440
54. Miura T, Nakashima K, Tada N, Itoh A (2010) An effective and catalytic oxidation using recyclable fluorous IBX. Chem Commun 47:1875–1877
55. Betzemeier B, Lhermitte F, Knochel P, (1999) A selenium catalyzed epoxidation in perfluorinated solvents with hydrogen peroxide. Synlett 489–491
56. ten Brink G-J, Vis JM, Arends IWCE, Sheldon RA (2002) Selenium catalysed oxidations with aqueous hydrogen peroxide. Part 3: oxidation of carbonyl compounds under mono/bi/triphasic conditions. Tetrahedron 58:3977–3983
57. Crich D, Zou Y (2004) Catalytic allylic oxidation with a recyclable, fluorous seleninic acid. Org Lett 6:775–777
58. Crich D, Zou Y (2005) Catalytic oxidation adjacent to carbonyl groups and at benzylic positions with a fluorous seleninic acid in the presence of iodoxybenzene. J Org Chem 70:3309–3311
59. Crich D, Neelamkavil S (2001) Fluorous Swern reaction. J Am Chem Soc 123:7449–7450
60. Pozzi G, Cavazzini M, Hozczknecht O, Quici S, Shepperson I (2005) Synthesis and catalytic activity of a fluorous-tagged TEMPO radical. Tetrahedron Lett 45:4249–4251
61. Holczknecht O, Pozzi G, Quici S (2006) Fluorous TEMPO: an efficient mediator for the aerobic oxidation of alcohols to carbonyl compounds. QSAR Comb Sci 25:736–741
62. Holczknecht O, Cavazzini M, Quici S, Shepperson I, Pozzi G (2005) Selective oxidation of alcohols to carbonyl compounds mediated by fluorous-tagged TEMPO radicals. Adv Synth Catal 347:677–688
63. Dobbs AP, Penny MJ, Jones P (2008) Novel light-fluorous TEMPO reagents and their application in oxidation reactions. Tetrahedron Lett 49:6955–6958
64. Dobbs AP, Jones P, Penny MJ, Rigby SE (2009) Light-fluorous TEMPO: reagent, spin trap and stable free radical. Tetrahedron 65:5271–5277
65. Gheorghe A, Cuevas-Yanez E, Horn J, Bannwarth W, Narsaiah B, Reiser O (2006) A facile strategy to a new fluorous-tagged, immobilized TEMPO catalyst using a click reaction, and its catalytic activity. Synlett 17:2767–2770
66. Gheorghe A, Chinnusamy T, Cuesvas-Yañez E, Hilgers P, Reiser O (2008) Combination of perfluoroalkyl and triazole moieties: a new recovery strategy for TEMPO. Org Lett 10:4171–4174
67. Curran DP, Lou Z, Degenkolb P (1998) "Propylene spaced" allyl tin reagents: a new class of fluorous tin reagents for allylations under radical and metal-catalyzed conditions. Bioorg Med Chem Lett 8:2403–2408

68. Curran DP, Hadida S, He M (1997) Thermal allylations of aldehydes with a fluorous allylstannane. Separation of organic and fluorous products by solid phase extraction with fluorous reverse phase silica gel. J Org Chem 62:6714–6715
69. Ryu I, Nigumo T, Minakata S, Komatsu M (1999) Radical carbonylations with fluorous allyltin reagents. Tetrahedron Lett 40:2367–2370
70. Ryu I (2004) Radical carbonylations using fluorous tin reagents: convenient workup and facile recycle of the reagents. In: Gladys JA, Curran DP, Horváth IT (eds) Handbook of fluorous chemistry. Wiley-VCH, New York, pp 182–190
71. Ryu I, Niguma T, Minakata S, Komatsu M, Hadida S, Curran DP (1997) Hydroxymethylation of organic halides. Evaluation of a catalytic system involving a fluorous tin hydride reagent for radical carbonylation. Tetrahedron Lett 38:7883–7886
72. Curran DP, Hadida S, Kim S-Y (1999) Tris(2-perfluorohexylethyl) tin azide: a new reagent for preparation of 5-substituted tetrazoles from nitriles with purification by fluorous organic liquid-liquid extraction. Tetrahedron 55:8997–9006
73. Curran DP, Hoshino M (1996) Stille couplings with fluorous tin reactants: attractive features for preparative organic synthesis and liquid-phase combinatorial synthesis. J Org Chem 61:6480–6481
74. Hoshino M, Degenkolb P, Curran DP (1997) Palladium-catalyzed Stille couplings with fluorous tin reactants. J Org Chem 62:8341–8349
75. Osswald T, Schneider S, Wang S, Bannwarth W (2001) Stille couplings in supercritical CO2 catalyzed with perfluoro-tagged and un-tagged Pd complexes. Tetrahedron Lett 42:2965–2967
76. Kaleta Z, Tárkányi G, Gömöry A, Kálmán F, Nagy T, Soós T (2006) Synthesis and application of a fluorous Lawesson's reagent: convenient chromatography-free product purification. Org Lett 8:1093–1095
77. Kaleta Z, Makowski BT, Soós T, Dembinski R (2006) Thionation using fluorous Lawesson's reagent. Org Lett 8:1625–1628
78. Kobayashi S, Yoneda A, Fukuhara T, Hara S (2004) Selective synthesis of fluorinated carbohydrates using N, N-diethyl-α, α-difluoro-(m-methylbenzyl)amine. Tetrahedron Lett 45:1287–1289
79. Kobayashi S, Yoneda A, Fukuhara T, Hara S (2004) Tetrahedron 60:6932–6930
80. Chen CH-T, Zhang W (2005) Fluorous reagents and scavengers versus solid-supported reagents and scavengers, a reaction rate and kinetic comparison. Mol Divers 9:353–359
81. Linclau B, Sing AK, Curran DP (1999) Organic-fluorous phase switches: a fluorous amine scavenger for purification in solution phase parallel synthesis. J Org Chem 64:2835–2842
82. Lindsley CW, Zhao Z, Leister WH (2002) Fluorous-tethered quenching reagents for solution phase parallel synthesis. Tetrahedron Lett 43:4225–4228
83. Lindsley CW, Zhao Z, Leister WH, Strauss KA (2002) Fluorous-tethered amine bases for organic and parallel synthesis: scope and limitations. Tetrahedron Lett 43:6319–6323
84. Zhang W, Curran DP, Chen CH-T (2002) Use of fluorous silica gel to separate fluorous thiol quenching derivatives in solution-phase parallel synthesis. Tetrahedron 58:3871–3875
85. Zhang W, Chen CH-T, Nagashima T (2003) Fluorous electrophilic scavengers for solution-phase parallel synthesis. Tetrahedron Lett 44:2065–2068
86. Zhang AS, Elmore CS, Egan MA, Mellilo DG, Dean DC (2005) Use of fluorous and solid-phase electrophiles as scavengers for excess amine in the preparation of sulfur-35 labelled radioligands. J Label Compd Radiopharm 48:203–208
87. Lu Y, Zhang W (2006) Fluorous 2,4-dichloro-1,3,5-triazines (F-DCTs) as nucleophile scavengers. QSAR Comb Sci 8:728–731
88. Baslé E, Jean M, Gouault N, Renault J, Uriac P (2007) Fluorous scavenger for parallel preparation of tertiary sulfonamides leading to secondary amines. Tetrahedron 48:8138–8140
89. Werner S, Curran DP (2003) Fluorous dienophiles are powerful diene scavengers in Diels-Alder reactions. Org Lett 5:3293–3296
90. Hicks JW, Harrington LE, Valliant JF (2011) Fluorous ligand capture (FLC): a chemoselective solution-phase strategy for isolating ^{99m}Tc-labelled compounds in high effective specific activity. Chem Commun. doi:10.1039/c1cc11079a

Fluorous Methods for the Synthesis of Peptides and Oligonucleotides

Bruhaspathy Miriyala

Abstract The non-covalent affinity of a perfluoro chain towards similar has been exploited by many to separate fluorous tagged compounds from non-fluorous compounds by F-SPE or F-LLE. This purification strategy found its application across diverse fields including peptide and oligonucleotide synthesis where even slight inefficient couplings result in deletion sequences that are often difficult to remove from the target sequence. Two commonly employed strategies to address this problem involve end-tagging the target sequence or capping the deletion sequences with fluorous tags. Solution phase syntheses using soluble fluorous supports are easier and quicker. These approaches are reviewed here in detail.

Keywords Capping · F-SPE · Fluorous tag · Fluorous peptide synthesis · Oligonucleotide synthesis

Contents

1 Introduction 106
2 Peptide Synthesis 108
  2.1 Solid Phase Synthesis 110
  2.2 Solution Phase Synthesis 117
3 Oligonucleotides 123
  3.1 Solid Phase Synthesis 125
  3.2 Solution Phase Synthesis 130
4 Conclusions 131
References 132

B. Miriyala (✉)
Department of Chemistry, Chevron Science Center, University of Pittsburgh, 219 Parkman Avenue, Pittsburgh, PA 15260, USA
e-mail: yehaibru@gmail.com; bru@pitt.edu

1 Introduction

The term "Fluorous" represents a perfluorocarbon chain (C_nF_{2n+1}) which is much like a hydrocarbon except that all the hydrogens are replaced with fluorines. As a result, a new class of organic compounds emerged that have very different physical and chemical properties. Notable amongst those is the ability of these compounds to veer away from both water and organic solvents but have a strong noncovalent interaction with fluorous compounds or solvents. This unique property has been first exploited by Horváth and Rábai [1] and led the way for further demonstrations by others [2] to phase separate fluorous tagged compounds from nonfluorous compounds either by extracting with a fluorous solvent (fluorous liquid–liquid extraction, F-LLE) or by using a fluorous stationary phase (fluorous solid phase extraction, F-SPE or F-HPLC) [3].

A typical fluorous tagged compound has three components (Fig. 1) attached covalently in a linear fashion. (1) A perfluorocarbon group (R_f) one end of which is linked to (2) a hydrocarbon spacer (2–3 carbon length) to shield high electronegativity of the R_f group. This spacer arm in turn is attached to (3) the organic compound through a reactive functional group. Depending on the type of application, the fluorous tag along with the spacer is attached to the organic portion either temporarily (as a protecting group that can be removed after the main purpose of the fluorous tag is served) or permanently. The length of the perfluorocarbon tag and the number of tags is dependent on the type of application for which it is being used.

In F-LLE the test mixture containing fluorous and nonfluorous compounds is taken in a fluorous or fluorophilic solvent (e.g., FC-72, PFMC, HFE-7100, HFE-7500) and the nonfluorous materials are washed into either an organic solvent or water (Fig. 2a). The fluorous compounds, by virtue of having a higher partition coefficient in fluorous solvents, remain in the fluorous medium. In the case of F-SPE (Fig. 2b), typically the hydroxyl groups present at the surface of silica are derivatized with fluorous tags and packed into a small cartridge (F-SPE). The sorbent is also packed as an HPLC column (F-HPLC) (Fig. 2c). So, when a mixture of fluorous and nonfluorous materials is passed through this stationary phase, the fluorous compounds adsorb to the sorbent and get retained while the nonfluorous compounds elute out (Fig. 3). After a brief washing step with aqueous and/or organic solvents (fluorophobic) to exclude all nonfluorous stuff, the fluorous compounds are eluted with a fluorophilic solvent. Typically heavy fluorous tags, where net fluorine content exceeds 40% of the total molecular

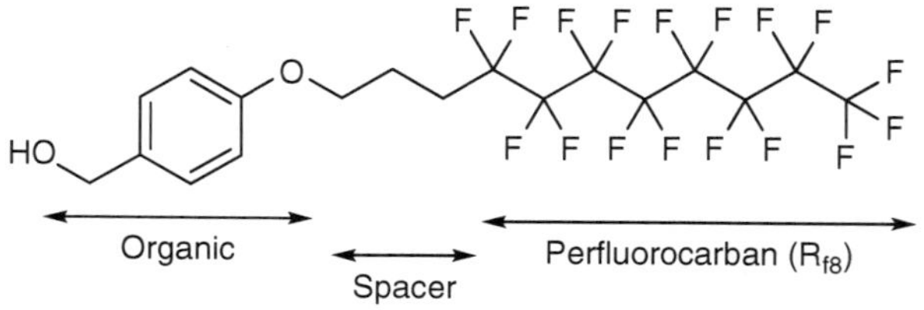

Fig. 1 General structure of a fluorous tagged compound. A C_8F_{17} (R_{f8}) perfluorocarbon group is shown as an example

Fluorous Methods for the Synthesis of Peptides and Oligonucleotides

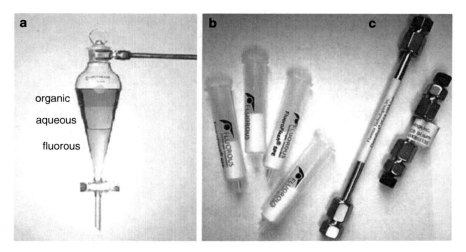

Fig. 2 Fluorous purification using (**a**) fluorous solvent – LLE or (**b**) fluorous sorbent packed as F-SPE cartridge or (**c**) F-HPLC column

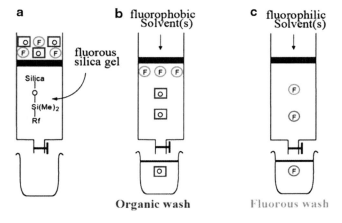

Fig. 3 Schematic diagram of a typical F-SPE purification. (**a**) Mixture loaded on top of fluorous silica cartridge is first washed with (**b**) aqueous/organic solvent mix to remove nonfluorous material and (**c**) fluorous compound eluted with fluorophilic solvents

weight, are purified by LLE and the light fluorous tags (net fluorine content <40%) by F-SPE.

Since the introduction of this concept, numerous studies have been undertaken by chemists that resulted in the development of a huge collection of fluorous reagents, catalysts, protecting groups, scavengers, etc. [4]. Most of these involved the utility of fluorous reagents to small molecule organic synthesis. The development of fluorous protecting groups led biochemists to use these in biomolecule synthesis and their purification. The biomolecules, especially peptides, oligonucleotides, and oligosaccharides, are oligomers/biopolymers that are repeating units of monomers in a defined sequence. So, their synthesis involves the repeated coupling of their respective monomer units to the growing oligomer. The synthesis of these biopolymers could be carried out either in solution much like any organic reaction or on a solid support. Because of the complexity and the number of reactions involved in their synthesis, solid phase synthesis (SPS) has become the preferred choice for many. In the context of this review, fluorous methods used in the synthesis and purification of peptides and oligonucleotides are discussed.

2 Peptide Synthesis

In solid phase peptide synthesis (SPPS) [5], the peptide is synthesized from the carboxyl (C) terminus to the amino (N) terminus (Fig. 4a) following the same direction seen in nature (protein synthesis). Thus, the entire peptide chemistry is based on protecting the amino group and side chains of the incoming amino acid and the peptide chain on the solid support leaving the carboxyl group free to form a peptide bond with the amino group of the previous amino acid (Fig. 4b). Based on the type of amino protecting group used, "Boc" or "Fmoc" chemistry could be used for SPPS.

The amino acid at the C-terminus is first immobilized covalently through a linker to a solid support that is typically made of a polystyrene resin (loading) (Fig. 5). The amino protecting group is then deprotected to expose the amino group (deprotection). The carboxyl group of the next amino acid is activated and coupled to the previously deprotected free amino group (coupling). Any unreacted amino groups

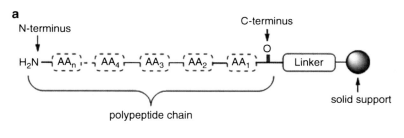

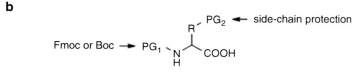

Fig. 4 Components of (**a**) solid phase peptide synthesis and (**b**) amino acid building block

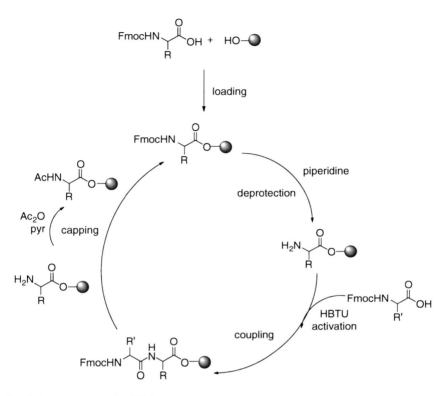

Fig. 5 General scheme for SPPS

are then capped using acetic anhydride (capping). This iteration involving deprotection, activation, coupling, and capping are repeated until the target sequence of peptide is reached. Finally, the peptide is cleaved from the resin under acidic conditions and purified by reverse-phase HPLC (RP-HPLC).

The recipe for a successful synthesis of peptide lies in its sequence. This in turn translates to the extent of inefficient couplings or deprotection of amino acids. As the formation of each peptide bond is pivotal, the efficiency of each coupling determines the yield and purity of the final peptide. There are many factors that influence the coupling step. Even a small reduction in the coupling or deprotection efficiencies would result in a mixture of failed or deletion sequences along with the final peptide (Fig. 6). This is by far the major limitation of SPPS. This makes the subsequent purification very tricky and time consuming. To this end, various groups have looked at fluorous tagging as an alternate approach to facilitate purification.

Fluorous applications to peptide synthesis and purification can be broadly categorized into solid phase or solution phase synthesis.

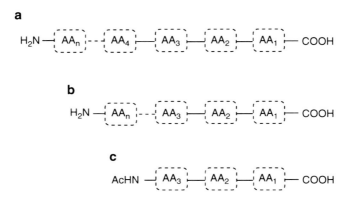

Fig. 6 Deletion sequences during SPPS. (**a**) Target sequence. (**b**) Deletion sequence without acetate capping. (**c**) Deletion sequence with acetate capping

2.1 Solid Phase Synthesis

Since its introduction in 1963 by Merrifield, solid phase peptide synthesis has undergone numerous changes with respect to the type of linkers, solid supports, amino acid protecting groups, solvents, and reagents used. There still remains, however, the issue of deletion sequences that accumulate during the course of synthesis and interfere with purification of the final product (Fig. 6). Until there is a clear understanding of what determines the inefficient coupling of a given residue in a given sequence, one should continue to find ways to minimize the complexity of purification.

The introduction of fluorous chemistry to peptide synthesis addresses this limitation. The basic concept of fluorous chemistry is to tag a given compound with a perfluorocarbon group and in turn use its high partition in a fluorous solvent or fluorous stationery phase to achieve separation from other nonfluorous compounds. Therefore, the first step would be to fluorous tag the target peptide, and only the target peptide, or tag all the deletion sequences while not tagging the target peptide. These are achieved through tagging the amino terminus of a full sequence (fluorous end-tagging) or capping the unreacted amino ends of would-be deletion sequence peptides after coupling, but prior to deprotection (fluorous capping).

2.1.1 Fluorous End Tagging

In the case of fluorous end tagging (Fig. 7), each coupling step of the SPPS is followed by standard acetate capping of the unreacted amines to truncate any deletion sequences. The final target peptide is then fluorous end-tagged at the N-terminus residue. The fluorous end-tagged peptide can be separated from the nonfluorous deletion sequences using a fluorous separation. The fluorous tag can

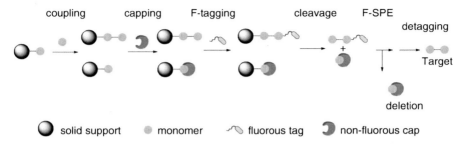

Fig. 7 Schematic representation of fluorous tagging the amino terminus of the full length peptide followed by F-SPE isolation

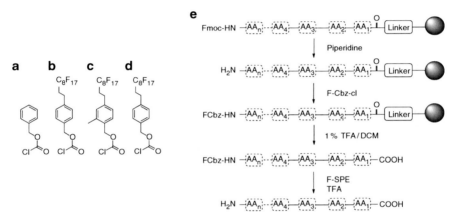

Fig. 8 Structure of (**a**) Z-Cl, (**b**) F-ZCl, (**c**) FMZ-Cl, (**d**) FEZ-Cl, and (**e**) scheme for tagging [7]

be introduced either after the final coupling and deprotection or during the final coupling by using a pretagged final amino acid [6].

A number of different fluorous tags have been used for fluorous end tagging.

F-Cbz

Filippov *et al.* [7] were first to demonstrate the utility of fluorous tagging reagents in solid phase peptide synthesis. They described the synthesis of three fluorous tagged benzyloxycarbonyl (F-Cbz) groups that vary in their lability to acid catalyzed hydrolysis (Fig. 8). In order to prevent F-Cbz hydrolysis during cleavage from the resin, a 4-hydroxymethyl-3-methoxyphenoxybutyric acid linker (HMPB), which can be selectively hydrolyzed while leaving the F-Cbz tag intact, was used to synthesize a series of peptides of varying length (7-, 19-, and 22-mer). The peptides were synthesized on 4-methylbenzhydrylamine (MBHA) resin derivatized with HMPB linker following the standard Fmoc chemistry involving acetate capping. At the end, Fmoc was deprotected and F-Cbz-chloroformate was used to protect the amino terminus. The peptide was cleaved from the resin using 1% TFA in DCM and the resulting F-Cbz tagged partially protected peptide was separated by F-HPLC from the nonfluorous capped sequences. The fluorous tagged Cbz enabled the peptides to retain much longer compared to nonfluorous impurities. The F-Cbz was then removed by TFA hydrolysis to provide the final desired peptide.

F-Msc

The acid lability of the F-Cbz tag required judicious choice of linker and side chain protecting groups which placed some limitations on its use. This prompted the same group to propose an alternate base labile fluorous tagged methylsulfonylmethoxycarbonyl (F-Msc) group for protecting the amino terminus (Fig. 9) [8]. Using standard Wang and Rink amide resins, they repeated similar SPPS of another set of peptides that varied from 7 to 35 residues in length. The 35-mer was earlier reported to be riddled with multiple incomplete couplings. Fmoc was replaced by F-Msc similar to that in Fig. 8e and the peptides cleaved from the resin. The fluorous tagged peptides were subjected to F-SPE and F-HPLC purification and detagged. Both methods were shown to be successful (Table 1) with the longer 35-mer requiring F-HPLC. The lower fluorine content of the 35-mer resulted in insufficient retention when using F-SPE.

F-Fmoc

Instead of replacing the N-terminal Fmoc with other fluorous tagged groups, it is easy for one to envisage the use of fluorous-Fmoc (F-Fmoc, Fig. 10). The reporting of a synthetic strategy [9] to fluorous tag Fmoc has enabled the group (Miriyala B, Montanari V, Yu MS, unpublished work) at Fluorous Technologies Inc. to conduct

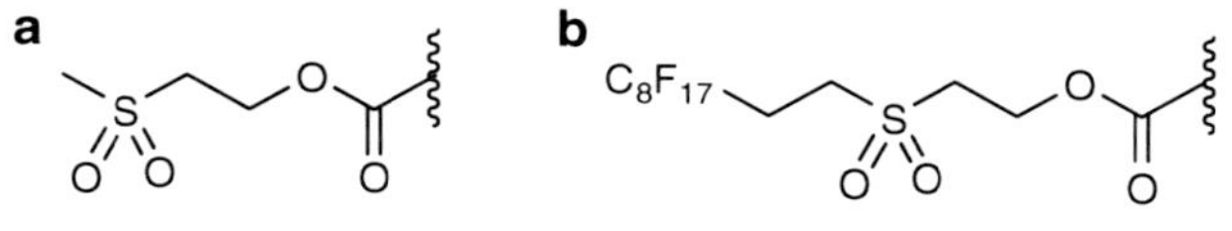

Fig. 9 Structure of (**a**) Msc and (**b**) its fluorous version – F-Msc [8]

Table 1 Details of peptide sequence and their purification yields by F-HPLC and F-SPE [8]

Peptide sequence	F-HPLC		F-SPE	
	Yield (%)	Purity (%)	Yield (%)	Purity (%)
GCCSLPPCALNNPDYC	37	98	59	91
RQIKIWFQNRRMKWKK	10	94	7	72
LSELDDRADALQAGFSQFESSAAKLKRKYWWKNLK	21	99	–	–

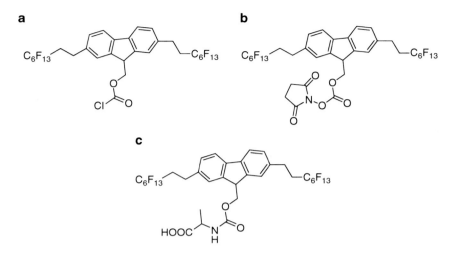

Fig. 10 Structure of (**a**) Fmoc-Cl, (**b**) F-Fmoc-OSu, and (**c**) F-Fmoc-Ala-OH [6]

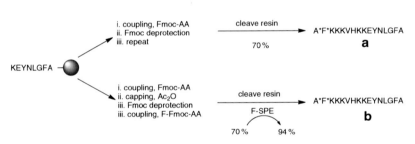

Fig. 11 Synthetic strategy for the synthesis of (**a**) nonfluorous and (**b**) F-Fmoc-tagged peptides. *denotes the point of deletion

similar SPPS studies using F-Fmoc protected alanine (Fig. 10c). Commercially available F-Fmoc-chloroformate (Fig. 10a) or F-Fmoc-*O*-succinimide ester (Fig. 10b) [6] were used to tag amino acids.

As shown in Fig. 11, a lysine rich 17-mer was used as the target sequence. Standard Fmoc chemistry with acetate capping was used throughout and the N-terminal alanine was coupled as its F-Fmoc derivative. For better visualization of the effect of deletion sequences to the final purification, artificial insertion of deletions was done at two places wherein 10% of the resin was separated from the main batch prior to coupling and added back in time for it to be acetate capped. After cleaving from the resin, the crude mixture consisting of fluorous tagged target peptide (RT 26 min, Fig. 12a) and the nonfluorous deletion sequences (RT 10–15 min) were easily separated by F-SPE with a simple washing step (Fig. 12b). This resulted in the increase of purity of the target peptide from 70% to 94%. It is

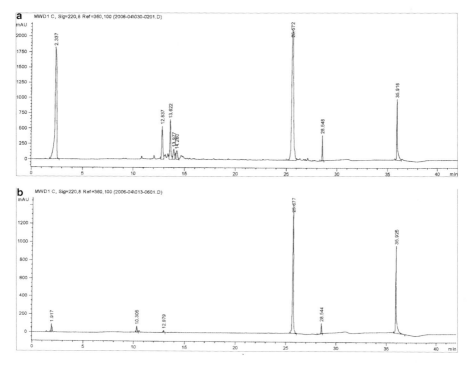

Fig. 12 Reverse phase HPLC of F-Fmoc-tagged peptide (**a**) before and (**b**) after F-SPE

also evident from the chromatographs (RP-HPLC) that the addition of a fluorous tag enables it to be retained longer on a reverse phase HPLC column, making it easier to separate from deletion sequences. F-Fmoc was subsequently removed in solution and F-SPE was again used to purify the now nonfluorous target peptide (data not shown). Alternatively, during the course of first F-SPE after the washing step to remove deletion sequences, F-Fmoc was deprotected on the cartridge and the deprotected peptide was eluted (data not shown). This strategy was employed to synthesize and purify longer peptides of up to 50 amino acids.

2.1.2 Fluorous Capping

A reverse strategy is used to separate the deletion sequences from the target sequence. The deletion sequences are capped with a fluorous capping reagent while the target peptide being nonfluorous is eluted from the cartridge during the washing step (Fig. 13).

The same study described above for F-Fmoc end tagging also attempted to cap the deletion sequences with fluorous tags (Miriyala B, Montanari V, Yu MS, unpublished work) (Fig. 14). Instead of acetate capping, a fluorous N-hydroxy

Fluorous Methods for the Synthesis of Peptides and Oligonucleotides

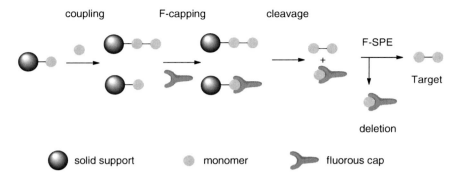

Fig. 13 Schematic representation of fluorous end capping the amino terminus of the deletion sequence followed by F-SPE isolation

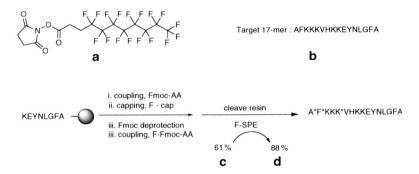

Fig. 14 (a) Fluorous capping reagent R_{f8}-OSu, (b) target sequence, and (c) synthetic strategy for the synthesis of non-fluorous target peptide (b, d) with fluorous capping (a). *denotes the point of fluorous capping

succinimide ester (Fig. 14a) was used to end-cap ~10% of the resin at three different sites in the sequence. The nonfluorous target peptide (Fig. 14b) was then purified from the fluorous tagged deletion peptides by a simple washing step during F-SPE. By virtue of having the hydrophobic fluorous tags the deletion sequences are retained longer (RT 18–22 min) than the nonfluorous target peptide (RT 13 min) in a reverse phase stationery phase (Fig. 15a). By F-SPE, the target peptide was easily enriched from 61% to 88% purity (Fig. 15b).

Montanari and Kumar [10, 11] were the first to demonstrate the concept of fluorous capping to SPPS using both Boc and Fmoc chemistries. A novel fluorous (R_{f7}) tagged trivalent iodonium reagent (Fig. 16a) was used to cap the failed sequences (Fig. 16b). The researchers produced mixtures artificially high in deletion sequences to demonstrate the technique. The shorter target sequences of 10-mer and 14-mer (Fig. 16d, f) were isolated by a simple water wash while the longer 21-mer (Fig. 15e) was purified by F-SPE.

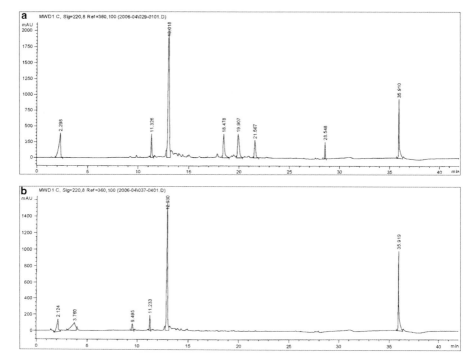

Fig. 15 Reverse phase HPLC (**a**) before and (**b**) after F-SPE. The peak with retention time of 13.0 min represents the target peptide. The peaks whose RT is 18–22 min represent the fluorous end capped deletion sequences

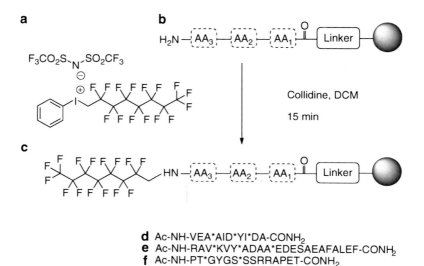

d Ac-NH-VEA*AID*YI*DA-CONH$_2$
e Ac-NH-RAV*KVY*ADAA*EDESAEAFALEF-CONH$_2$
f Ac-NH-PT*GYGS*SSRRAPET-CONH$_2$

Fig. 16 Fluorous capping reagent (**a**) used to cap the deletion sequence (**b**). The resulting deletion sequences for the synthesis of peptides (**d**) and (**f**) were removed by water precipitation and peptide (**e**) was isolated by F-SPE [10, 11]. *denotes the site of fluorous-capping

2.2 Solution Phase Synthesis

Even though the SPPS is popular in terms of speed and efficiency, it has its limitations. Solution phase syntheses offer better reaction kinetics, the ability to analyze intermediates, and the option to purify intermediates as needed. Solution phase synthesis is also more desirable for large scale synthesis of shorter sequences. At the same time, solution phase synthesis has its share of limitations. The isolation of the product at each step is labor intensive. Fluorous chemistry overcomes these limitations. The fluorous tag replaces the solid phase resin. The amino acids can be conveniently fluorous tagged as amine protecting groups (F-Cbz/Z, F-Boc, Froc) or carboxyl protecting groups (F-TMSE). During the reaction, the fluorous-support with the growing peptide is in solution. As in solid phase synthesis, excess nonfluorous reactants (amino acids) can be used to drive the reaction to completion. After the reaction is over, F-SPE (for light fluorous tags), F-LLE (for heavy fluorous tags), or a solvent switch from fluorophilic to flurophobic enable the easy isolation of fluorous materials. Even though fluorous supports have been used for this purpose, it appears that it may be better suited for synthesis of short peptides or derivatives.

2.2.1 Light Fluorous Tags

F-Cbz

Based on earlier findings [7] about the utility of fluorous tagging the peptides in solid phase synthesis with F-Cbz group, Curran's group [12] demonstrated its use in solution. By following a similar synthetic strategy, F-Cbz-O-succinimide ester with one R_{f6} (Fig. 17a) or R_{f8} (Fig. 17b) were made and used to tag naturally occurring L-amino acids (except Arg and His). The carboxyl group was in turn coupled to various amines used in excess (Fig. 17c). The fluorous tagged product was easily

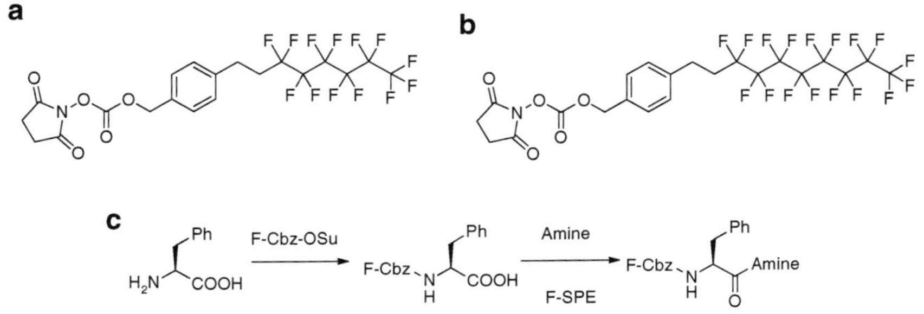

Fig. 17 Structure of F-Cbz-OSu with (**a**) C_6F_{13} or (**b**) C_8F_{17}. (**c**) L-Phenylalanine was tagged with F-Cbz, reacted with excess amine, and isolated by F-SPE [7]

isolated from the reaction mixture by F-SPE using *FluoroFlash*® cartridge. The F-Cbz group was easily removed by standard hydrogenolysis. They also validated the fluorous utility to mixture synthesis by fluorous tagging the D-amino acids. In both cases, F-Cbz behaved similar to the regular Cbz.

F-Boc

Fluorous version of *tert*-butyloxycarbonyl (Boc) was first reported by Curran [13]. Although they described the synthesis of four different versions varying in number of fluorous chains and length of spacer, the one with single R_{f8} (Fig. 18b) was used to protect a set of amino acids. However, these were not used for synthesis of peptides *per se*, but instead as a 96-member library paving the way for the utility of fluorous tagging to combinatorial and high throughput synthesis (Fig. 18c). F-SPE was used to clean up the fluorous intermediates. It appears that the removal of F-Boc group is slightly slower than the regular Boc. As a side note, for the purification of a smaller library, an F-HPLC column was also used to separate fluorous tagged from nonfluorous compounds. The fluorous HPLC column is much like a regular reverse phase column except that the stationary phase consists of fluorous derived matrix. By F-HPLC, it was observed that the retention time for fluorous compounds was marginally longer than that of the corresponding nonfluorous version.

Froc

Fluorous based trichloroethoxycarbonyl (Froc, Fig. 19b) was developed as an alternative support to solid-phase and used to protect the amino groups in peptide and carbohydrate synthesis [14]. Like its nonfluorous counterpart (Troc, Fig. 19a), Froc is stable to mild acids and bases but is selectively removed under reducing conditions (Zn/AcOH/Et$_3$N). Froc-chloride was used to protect commercially available amino esters. After hydrolyzing the ester under basic conditions it was coupled

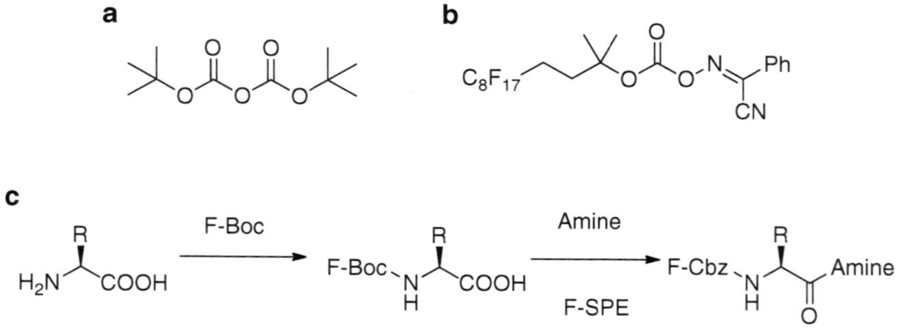

Fig. 18 Structures of (**a**) Boc anhydride and (**b**) F-Boc reagent. (**c**) Reaction scheme for F-Boc application [13]

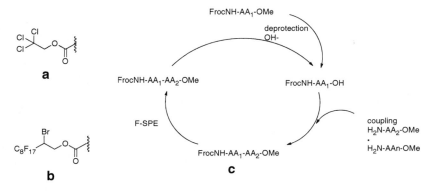

Fig. 19 Structure of (**a**) Troc and (**b**) Froc. (**c**) Synthetic scheme using Froc [14]

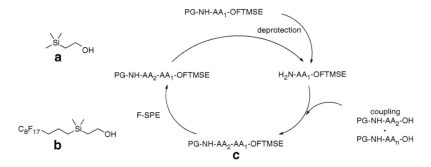

Fig. 20 Structures of (**a**) TMSE and (**b**) F-TMSE. (**c**) Synthetic scheme using F-TMSE. PG represents either Fmoc or Boc [15]

to another amino ester. Gly–Gly dipeptide and a bioactive Arg–Gly–Asp tripeptide were both prepared in high yield (Fig. 19c). At each step excess reactants were used to push the reaction to completion and the fluorous peptide was purified by F-SPE. Unlike SPPS, the peptides were synthesized from the N-terminus to the C-terminus.

F-TMSE

Fustero et al. [15] devised a new fluorous tag (F-TMSE, Fig. 20b) that is analogous to 2-(trimethylsilyl) ethanol (Fig. 20a) for the protection of carboxylic acids. This tag consisting of one R_{f8} was used to synthesize tripeptides (C- to N-terminus) consisting of α- and β-amino acids. At the end of synthesis F-TMSE was selectively removed by *trans*-esterification with TBAF that prevented racemization. All the fluorous intermediates were purified by F-SPE to remove excess nonfluorous impurities. The same strategy was further used to synthesize modified fluorinated retropeptides that are easily epimerized by exposure to acids and bases and therefore difficult to prepare by SPPS (Fig. 20c). The diastereoselectivity of aza-Michael reaction was maintained as in nonfluorous reactions and was superior to solid phase synthesis.

2.2.2 Heavy Fluorous Tags

F-Teoc

For scale-up purposes, F-SPE may be less desirable than F-LLE, requiring greater fluorine content. Nakamura *et al.* [16] had replaced all the three methyls attached to silicon in 2-(trimethylsilyl) ethoxycarbonyl with perfluorodecyl groups (Fig. 21a). The presence of three R_{f8} groups (heavy fluorous tag) rendered the purification of fluorous intermediates from organic impurities by a simple liquid–liquid extraction with fluorous solvent (FC-72). The resulting F-Teoc group was used as an amine protecting group for the solution phase synthesis of thiazole (Fig. 21b) and methyloxazole (Fig. 21c) based amino acids as well as a naturally occurring macrolactam – Bistratamide H with little racemization. F-Teoc was selectively removed by TBAF after the linear peptide was made and cyclized (Fig. 21e). The fluorous fragment was also shown to be recycled.

F-Cbz

Another version of a perfluoro-tagged benzyloxycarbonyl protecting group was reported by Schwinn *et al.* [17] This is similar to that described above (Fig. 17) except that the single R_{f8} is replaced by three R_{f6} in the form of tris[2-(perfluorohexyl)ethyl]silyl group (Fig. 22) that makes it amenable to liquid–liquid

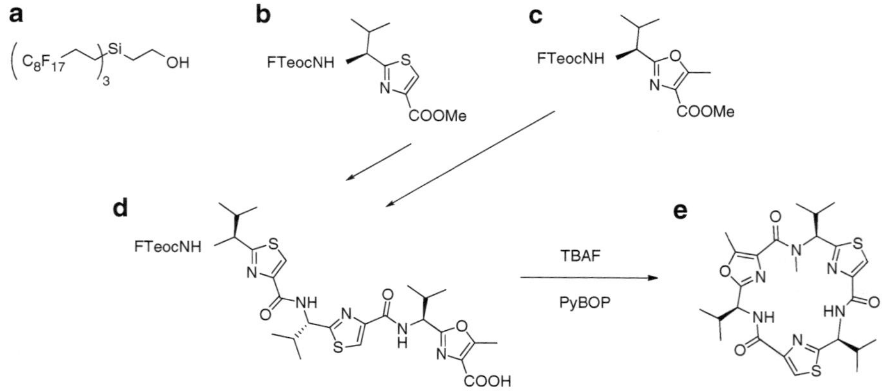

Fig. 21 Structures of F-Teoc (**a**) and F-Teoc protected heterocyclic amino acids (**b, c**) that were used to synthesize the linear peptide (**d**). Bistratamide H (**e**) was obtained upon cyclization of linear peptide (**d**) [16]

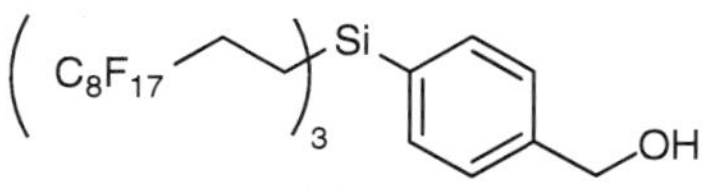

Fig. 22 Structure of F-Cbz [17]

extraction with a perfluorinated solvent. By protecting the amino group of an amino acid, this fluorous group facilitated synthesis of a small library of quinazoline-2,4-diones. The last step involved a base mediated simultaneous detagging and cyclization. The final product, being nonfluorous, was extracted using an organic solvent. During all other steps fluorinated solvent was used to separate the fluorous from organic reactants.

Hfa and Hfb

In SPPS the solid support serves three functions. It acts as carboxyl protecting group for the C-terminus amino acid, helps modify the C-terminus (amide or acid), and acts as an insoluble matrix during filtrations. Studies by Mizuno *et al.* [18–20] have shown that a heavy fluorous tag could perform similar functions when used as an anchor for the growing peptide chain. Fluorous solution phase syntheses offer improved reaction kinetics and stoichiometries compared to SPPS. Based on the functional group present, two types of fluorous support each having six R_{f8} groups have been reported – Hfa (amine, Fig. 23a) and Hfb (carboxylic acid, Fig. 23b).

Similar to the presence of linkers between solid support and peptide chain that determines the nature of C-terminus of the peptide, four linkers have been reported for the fluorous supports. A 4-hydroxylmethylphenoxyacetyl (HMPA, Fig. 23d), 4-(4-hydroxylmethyl-3methoxyphenoxy)-butyl (HMPB, Fig. 22f), or *tert*-butyl type linkers (Fig. 23e) were attached to Hfa or Hfb to synthesize peptides with C-terminal acid. Rink type linker was used to yield carboxamides at the C-end of the peptides (Fig. 23c). HMPB linker being very acid labile results in protected peptide acid.

Standard Fmoc chemistry was used to synthesize biologically active peptides – thyrotropin-releasing hormone (TRH), a tripeptide, and leu-enkephalin, a pentapeptide. In the case of TRH synthesis where Hfb was used with Rink amide linker, the Fmoc deprotection was done using 5% piperidine in FC-72/DMF (1:1) while 40% diethylamine in acetonitrile/FC72 (1:1) was used when HMPA-Hfa was involved. Like any solution phase synthesis, the reactions were monitored chromatographically or spectroscopically and separation of the fluorous supported peptide from organic impurities was conducted by F-LLE. Seven steps were involved in the synthesis of TRH (64% yield) with only one chromatographic purification step at the end of synthesis. Nine steps with single chromatographic purification were involved for leu-enkephalin (70% yield). However, there appear some limitations associated with these fluorous supports. There were some solubility issues when the net fluorine content fell to <40% of the whole molecular weight which makes these supports unattractive for synthesizing longer peptides (>6-mer).

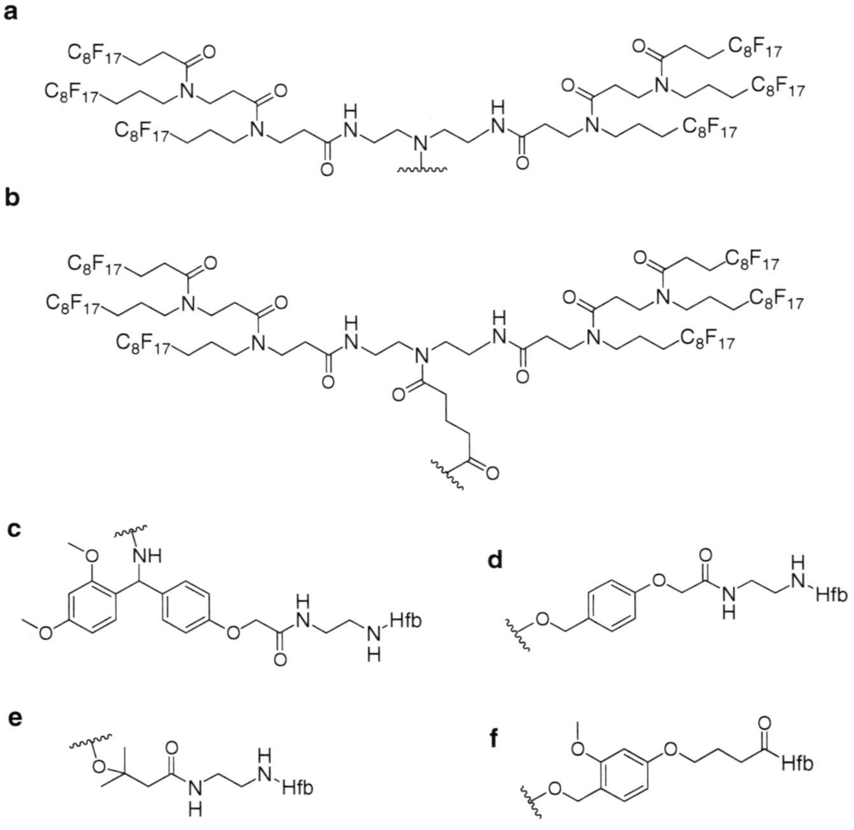

Fig. 23 Structures of (**a**) Hfa, (**b**) Hfb, (**c**) Hfb-Rink amide, (**d**) Hfb-HMPA, (**e**) Hfb-*tert*-butyl, and (**f**) Hfa-HMPB [18–20]

Also, it appears that Fmoc deprotection conditions need to be standardized for each type of linker.

F-Fmoc

Matsugi *et al.* [21] came up with a synthetic strategy to fluorous tag Fmoc. Fluorene (Fig. 24a) was derivatized to include two R_{f4} or R_{f6} tags (Fig. 24b, c). The use of two tags would enable the fluorine content to be high enough for longer peptide syntheses. Also, the difference between the number of fluorines enables its application to fluorously mixture syntheses. Simple dipeptides were made in solution and purified by F-SPE in good yields.

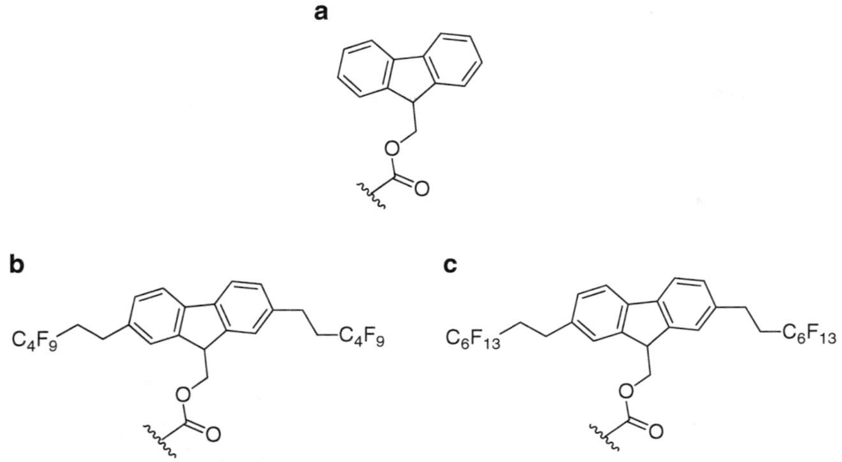

Fig. 24 Structures of (**a**) Fmoc, (**b**) F-Fmoc with two R_{f4} groups, or (**c**) two R_{f6} groups [9]

3 Oligonucleotides

The advent of the use of phosphoramidite monomers in conjunction with tetrazole chemistry in 1983 [21] revolutionized the solid phase synthesis of oligonucleotides. The speed and efficiency of coupling were greatly improved, thereby enabling the rapid synthesis of even long oligonucleotides (>100). A phosphoramidite monomer consists of a nucleotide whose heterocyclic amine, 5'-OH and phosphite are protected by benzoyl, dimethoxytrityl (DMT) and cyanoethyl and diisopropylamine respectively.

The solid phase synthesis of oligonucleotides is loosely based on that of peptides (Fig. 25). The direction of peptide synthesis is from carboxyl end (C) to amino end (N) while the oligonucleotides are synthesized from the 3'-hydroxyl end to the 5'-hydroxyl end. Similar to the immobilization of the C-terminal amino acid on a solid support in peptide synthesis, the nucleotide at the 3'-end is covalently attached to the resin and the subsequent nucleotides are attached in the 5' direction. The "Boc" or "Fmoc" group is represented by a hyper-acid labile dimethoxytrityl (DMT) group that protects the 5'-hydroxyl. Each cycle consists of deprotection, coupling, oxidation, and capping. The DMT group is removed to expose the 5'-hydroxyl end. The incoming monomer is first activated by tetrazole that replaces diisopropylamine. It is then coupled to the free 5'-end through a weak phosphite linkage which is subsequently oxidized (iodine/water/pyridine) to a more stable phosphate linkage. Acetate capping is then used to block any unreacted 5'-ends. This cycle of steps is repeated until the required numbers of nucleotides are coupled. The oligonucleotide is then cleaved from the solid support with concomitant deprotection by ammonolysis.

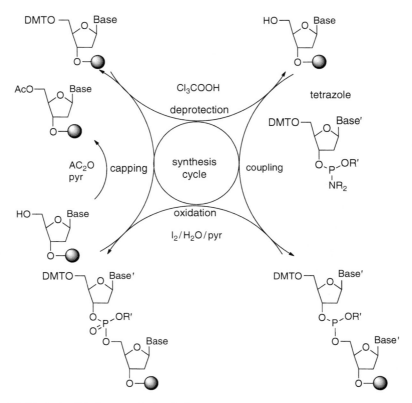

Fig. 25 Typical solid phase synthesis of oligonucleotides. Controlled pore glass (cpg) is shown here as the solid support

As in peptide synthesis, the coupling efficiency varies by the type of monomer, its position, is and the sequence of the oligonucleotide. Attaining coupling efficiency of >99% per coupling is much more common in oligonucleotide synthesis than in peptides. However, since synthetic oligonucleotides (25–28-mer) are on average longer than peptides, significant amounts of deletion sequences can accumulate. For example, for a 25-mer, a drop in coupling efficiency from 99.5 to 99% results in an increase in impurities from 10 to 21% [22, 23]. Moreover, synthesis of long oligonucleotides (60–100-mer) is quite common, resulting in dramatic increases of deletion sequences requiring tedious purification steps. At present, most custom oligonucleotide synthesis facilities do not offer purity guarantees on oligonucleotides longer than 60-mers. Applications like site-directed mutagenesis, cloning, and protein binding assays require the oligonucleotides to be of the highest purity. Owing to the presence of charges, the oligonucleotides are much more hydrophilic in nature. This greatly complicates their purification from the deletion sequences by HPLC. Polyacrylamide gel electrophoresis (PAGE) is another tool commonly used to separate the deletions from the target sequence. Low recoveries from the gel are reported to be a factor. Delaying the removal of the hydrophobic DMT referred to as "DMT-on" or "Trityl-on" from 5'-end until after purification using a reverse phase cartridge has been employed by some [24] to separate the hydrophilic acetate capped deletions from DMT-protected target sequence for shorter oligonucleotides (30–40-mer). With further increases in the overall chain length of the oligonucleotide, however, the influence of DMT on lipophilicity is diminished, resulting in very low yields. In addition, the presence of ammonia in the crude oligonucleotide mixture interferes with its adsorption to the stationery phase requiring multiple loading and decreased recoveries.

The emergence of fluorous tags in the early 2000s resulted in various groups focusing in this direction to overcome the above-mentioned shortcomings associated with oligonucleotide synthesis and purification. Similar to peptide synthesis, the fluorous applications to oligonucleotide synthesis could also be categorized into solid phase or solution phase. Solid phase approaches focused mainly on fluorous tagging at the 5'-end towards the end of the synthesis, although fluorous capping has also been reported. For solution phase synthesis, heavy fluorous based PMBs have been used as the support on which oligonucleotide is built. Some of these studies are discussed below.

3.1 Solid Phase Synthesis

As in the case of peptide synthesis, solid phase synthesis of oligonucleotides continues to be the most preferred because of the simplicity, efficiency, and automation. Controlled pore glass (CPG) or cross-linked polystyrene based resin are commonly used as the solid support.

3.1.1 5'-Tagging

In 2005, two independent groups [25, 26] demonstrated the utility of fluorous tagging at the 5'-OH end. The synthesis of oligonucleotides was conducted on a solid support up to the $n-1$ nucleotide as in a regular phosphoramidite based synthesis. The DMT protecting group for the last nucleotide was chemically modified to incorporate either a single or double fluorous tag (Fig. 26).

In order to decrease the sensitivity to acid, the dimethoxytrityl group of nucleotide phosphoramidite was replaced with either a fluorous tagged monomethoxytrityl (F-MMT, Fig. 26b) or fluorous tagged trityl (F-Trt, Fig. 26a) group. In each case, two phenyls of the trityl were derivatized with a perfluoro chain of eight carbons (C_8F_{17} or R_{f8}) with an ethylene spacer. Following the standard activation, the fluorous tagged nucleotide was then coupled to the 5'-end of a [deoxythymidine]$_{20}$ or oligonucleotides ($n-1$) of 30 residues (Table 2d). An artificial mixture made up of fluorous tagged target oligonucleotide thus obtained along with some of the acetate capped deletion sequences varying in length from 9–30 (Table 2a–c) was

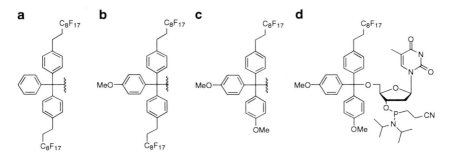

Fig. 26 Structures of (a) F-Trt, (b) F-MMT, (c) F-DMT, and (d) F-DMT protected thymidine [25, 26]

Table 2 Details of oligomer deletion sequences (a, b, c), F-MMT tagged target sequence (d), and deprotected target sequence (e) [25]

	Oligonucleotide sequence
a	5′-d(ACA GGA CCT)-3′
b	5′-d(AAC ACA GGA CCT)-3′
c	5′-d(TCG AAC ACA GGA CCT)-3′
d	5′-F-MMT-d(TCG AAC ACA GGA CCT TCG AAC ACA GGA CCT)-3′
e	5′-d(TCG AAC ACA GGA CCT TCG AAC ACA GGA CCT)-3′

then subjected to F-SPE purification using fluorous tagged silica gel as sorbent (Fig. 27a). The fluorous tagged oligonucleotide was easily retained on the fluorous silica gel (Fig. 27c) while the nonfluorous impurities were washed through the cartridge (Fig. 27b). Brief acidic exposure of the sorbent resulted in the release and recovery of the desired oligonucleotide from the cartridge while the F-MMT continued to be retained (Table 2e, Fig. 27c).

Pearson et al. [26] preferred to retain the dimethoxytrityl group but derivatized the unsubstituted phenyl with a single chain of C_8F_{17} (F-DMT, Fig. 26c). This modified F-DMT was used to protect 5′-OH of the 5′-end nucleotide and coupled using standard conditions to oligonucleotides of chain length 30, 50, 75, or 100 nucleotides (Table 3). At the end of each synthesis, the oligonucleotides were deprotected and cleaved off the solid support and loaded directly over fluorous sorbent that was immobilized on a base resistant polymer (F-SPE, *Fluoro-Pak*[TM], Fig. 28a). In each case, the F-DMT capped target oligonucleotide was well retained on the fluorous sorbent and was easily separated from their respective deletion sequences with simple washing (Fig. 28b–d). The desired full length oligomer was then eluted from the cartridge after on-column F-DMT deprotection (Fig. 28e). Unprecedented recoveries (70–100%) as well as near quantitative yields for such long oligonucleotides are also reported. It is interesting to note that these highly specific fluorous–fluorous interactions do not lose their identity even when they are dwarfed by the nonfluorous content (1:75 in the case of 100-mer). Without

Fluorous Methods for the Synthesis of Peptides and Oligonucleotides

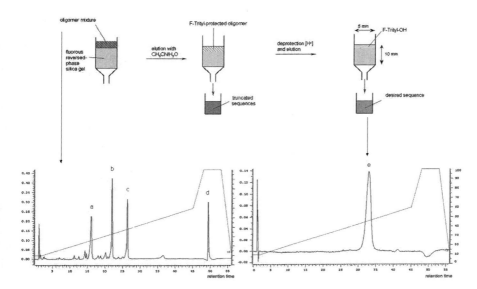

Fig. 27 Schematic representation of F-SPE (**a**) and anion-exchange HPLC of mixture before (**b**) and after F-SPE purification (**c**). Pre-mixture consists of non-fluorous deletion oligomers (**a–c**, sequence in Table 2) and FMMT protected target oligomer (**d**). After on-cartridge F-MMT deprotection and elution target oligonucleotide (**e**) is obtained. Reproduced with permission from John Wiley & Sons [25]

Table 3 Details of the oligonucleotide sequences and their purification yield after F-SPE [26]

	Sequence	Recovery (%)
30-mer	TTTCTCTAGACAATTGTGCAATACGTCTTT	75–88
50-mer	TTTCTGTTGACAATTTATCATCGGTCGTATAATGTGGAATTGGTCTTT	60–100
75-mer	TTTCTGGTTAAGGTGTGTATATGCTCGGCTACTAATTAGTGAGTATTATTCTCGCTACTATTAACAGTTGTCTTT	100
100-mer	TTTCTGGTTAAGGTGTGTGTAATATGCTCAGCTACTAATTAACAGTTGTCTAAGCTGGTTAACGTGAGTAATATGATCAGCTACTATTTAACAGTTGTCTTT	76–100

adapting to any newer techniques, the authors have demonstrated a single-pass loading method to separate selectively the deletion sequences from the desired oligonucleotide with near quantitative recoveries which are not possible with traditional DMT-on reverse-phase purification. At present these F-MMT and F-DMT based phosphoramidites are available commercially [27].

This successful demonstration of the fluorous affinity purification method using a single fluorous chain to the purification of very long oligonucleotide is of great relevance.

For the preparation of DNA based multiporphyrin arrays (Fig. 29), F-DMT protected porphyrin-phosphoramidite nucleotide was used to end-tag the

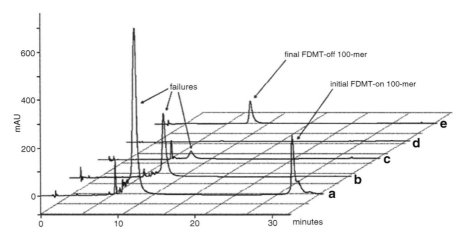

Fig. 28 Reverse phase HPLC analysis of (**a**) crude mixture, (**b**) eluate from first loading, (**c**) failure wash #1, (**d**) failure wash #2, and (**e**) target 100-mer after on-cartridge detritylation. Reproduced with permission from American Chemical Society [26]

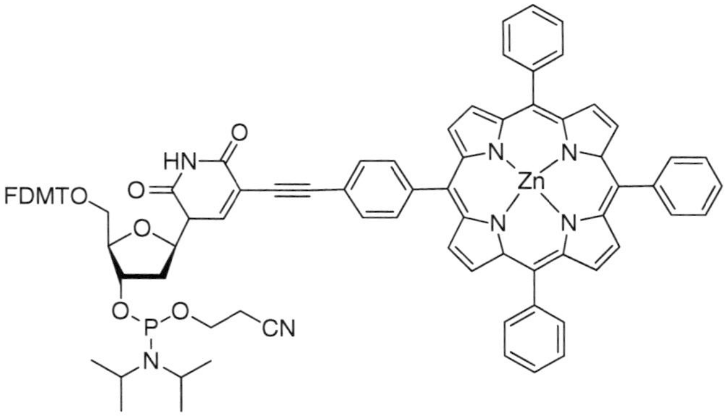

Fig. 29 Structure of porphyrin-phosphoramidite nucleotide [28]

oligonucleotide during the last synthesis cycle [28]. These complex oligonucleotides were finally purified, exploiting the fluorous tag using F-SPE. According to the authors, the F-SPE method of purification was simpler, cleaner, and quicker compared to the traditionally used PAGE and RP-HPLC.

In yet another instance [29] where the acid-sensitivity of the DMT group was a concern, it was reported that a fluorous silyl protected phosphoramidite (Fig. 30a) could also be used during the final synthesis cycle to end-tag the oligonucleotide. By fluorous tagging relatively short oligomers of 5–19 (Fig. 30b), this group showed that it is possible to separate the nonfluorous deletion sequences from the fluorous-tagged target sequence using just water. The high hydrophobicity of the fluorous protecting group would have played a significant role in overshadowing the

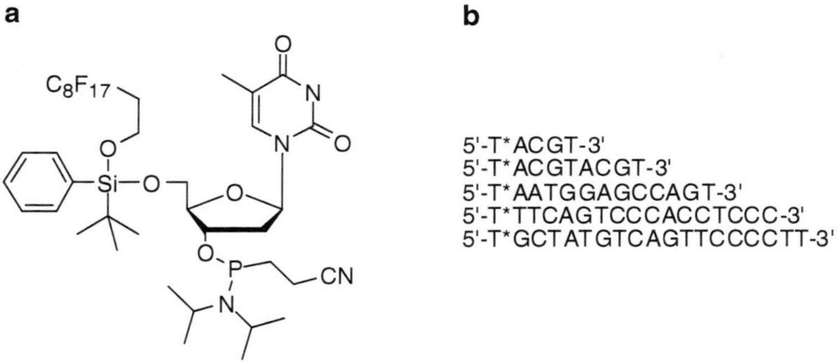

Fig. 30 (a) Structure of fluorous silyl protected phosphoramidite thymidine and (b) the sequences of oligomers. T* represents a [29]

hydrophilicity of the oligo, thereby precipitating out of water. Once isolated from impurities, the fluorous tag was removed (TBAF/THF) and the mixture was subjected to fluorous liquid–liquid extraction (F-LLE) using FC-72 and water. The target oligonucleotide which is now nonfluorous was isolated from the water layer.

Besides the advantages of fluorous tags to purification, recent studies indicate that fluorous tagging has also found other useful applications in molecular biology like in microarraying and nanoparticle based DNA detection systems [30]. The fluorous tagged oligonucleotides bring a new dimension to the field of microarrays wherein the immobilization does not involve any traditional covalent or charge based immobilization. The fluorous tagged oligonucleotides or other biomolecules can be immobilized on the glass slide by simple adsorption, but in a display-specific orientation. For most of the *in-vitro* applications, a variety of commercially available reagents are now available for permanent fluorous tagging of the oligonucleotides and peptides [6, 27].

3.1.2 Capping

In addition to fluorous tagging the 5′-end of the desired sequence of oligonucleotide and separating that from the nonfluorous impurities by a simple F-SPE, the exact opposite could also be done [31]. In this patent application, the inventors developed a fluorous phosphoramidite capping reagent (Fig. 31). Instead of using acetic anhydride for capping, this reagent is activated similarly as a DMT-nucleotide and is added to phosphitylate the unreacted 5′-hydroxyl group of the growing oligo-T (15-mer) chain. At the end of the synthesis, the mixture consisting of nonfluorous DMT-capped desired oligonucleotide and fluorous deletion sequences (Fig. 32a) is subjected to F-SPE using Pearson's conditions [26]. Only the nonfluorous desired oligonucleotide elutes out (Fig. 32b) which is then deprotected and cleaned-up by HPLC. The retained fluorous impurities could be eluted later if needed (Fig. 32c).

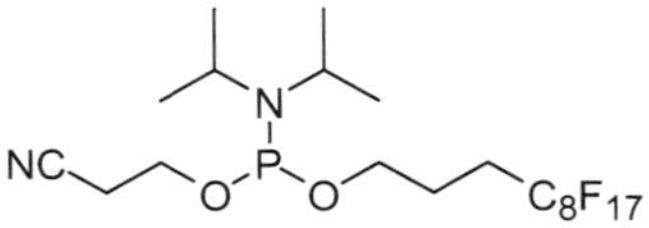

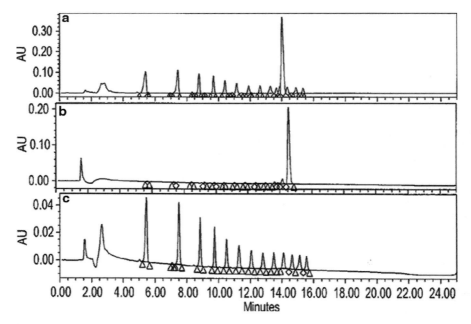

Fig. 31 Structure of fluorous capping reagent [31]

Fig. 32 Reverse-phase HPLC analysis of F-SPE fractions: (**a**) mixture of fluorous deletions and nonfluorous target oligonucleotide before loading, (**b**) wash eluent that correspond to target oligomer, and (**c**) fluorous capped deletions after elution. Reproduced with permission from Roche Molecular Systems, Inc. [31]

3.2 Solution Phase Synthesis

Solution phase synthesis definitely has some advantages over solid-phase in terms of scale-up, amount of materials, favorable reaction kinetics, wide range of chemical manipulations, and easy analysis of the reaction progression at every phase, thereby greatly reducing the unwanted deletion sequences. Solution phase synthesis of oligonucleotides, however, is seldom used because of the labor intensive steps involved at every point of the synthesis. With a better understanding of the partitioning of fluorous materials between fluorophilic, organic, and aqueous phases, it is possible to replace the solid phase support with fluorous support and overcome this limitation. The fluorous support consisting of a number of fluorous chains (heavy) is suitably derivatized with a linker so that the 3′-end nucleotide could be loaded. The fluorous-support with the growing oligonucleotide is conveniently

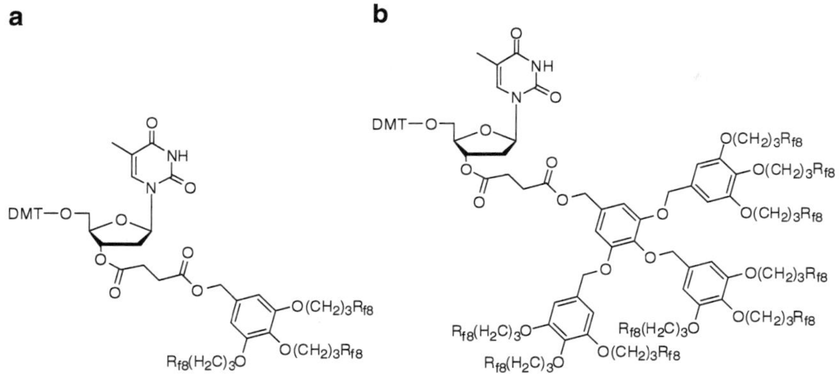

Fig. 33 Structure of fluorous supports for solution phase synthesis of oligonucleotides [32]

switched between organic and fluorous phases during reaction and purification respectively until the desired length is reached. Wada *et al.* [32] have invented 3,4,5-triperfluoroalkoxy benzyl alcohol based fluorous supports that could be used to synthesize oligonucleotides in solution. Depending on the type of application one could choose a support that has three R_{f8} chains (Fig. 33a) or nine R_{f8} chains (Fig. 33b). The 3′-OH of the nucleotide at the 3′-end of the oligonucleotide is attached to fluorous support through a linker. As a proof-of-concept, oligonucleotides of three bases long have been synthesized on these fluorous supports using standard DMT protected nucleotides. However, in the absence of further reports in this area it is difficult to assess its applicability to general oligonucleotide synthesis.

4 Conclusions

The discovery of fluorous protecting groups has indeed added another dimension to the synthesis and purification of biopolymers. Fluorous chemistry offers high compatibility to existing solid phase synthetic approaches to biopolymers, requiring few changes to the standard chemistry currently practiced. In addition, fluorous supports offer the option of using solution phase synthetic approaches. Much like in affinity chromatography, fluorous separations provide solutions to the long-existing problem of deletion sequences associated with these biomolecules' synthesis. This sample enrichment method could be used in conjunction with HPLC for studies involving very pure materials. In addition to the synthesis, the future looks promising for fluorous tags as they are now increasingly being used in other biological applications such as microarrays and proteomics [33–35].

Acknowledgments I would like to express my gratitude to Prof. Dennis Curran for all the support and guidance during the course of my journey through fluorous chemistry and preparation of this manuscript. I would also like to thank Fluorous Technologies Inc. especially Dr. Marvin S. Yu for introducing me to this fascinating field and for his valuable inputs during the preparation of this chapter.

References

1. Horváth IT, Rábai J (1994) Facile separation without water: fluorous biphase hydroformylation of olefins. Science 266:72–76
2. Gladysz JA, Curran DP, Horvath IT (2004) Handbook of fluorous chemistry. Wiley-VCH, Germany
3. Curran DP, Hadida S, He M (1997) Thermal allylations of aldehydes with a fluorous allylstannane. Separation of organic and fluorous products by solid phase extraction with fluorous reverse phase silica gel. J Org Chem 62:6714–6715
4. Wei Z (2009) Fluorous linker-facilitated chemical synthesis. Chem Rev 109:749–795
5. Merrifield RB (1963) Solid phase peptide synthesis. I. The synthesis of a tetrapeptide. J Am Chem Soc 85:2149–2154
6. Fluorous Technologies Inc. 970 William Pitt Way, Pittsburgh, PA 15238. http://www.fluorous.com
7. Filippov DV, van Zoelen DJ, Oldfield SP, van der Marel GA, Overkleeft HS, Drijfhout JW, van Boom JH (2002) Use of benzyloxycarbonyl (Z)-based fluorophilic tagging reagents in the purification of synthetic peptides. Tetrahedron Lett 43:7809–7812
8. de Visser PC, van Helden M, Filippov DV, van der Marel GA, Drijfhout JW, van Boom JH, Noort D, Overkleeft HS (2003) A novel, base-labile fluorous amine protecting group: synthesis and use as a tag in the purification of synthetic peptides. Tetrahedron Lett 44:9013–9016
9. Matsugi M, Yamanaka K, Inomata I, Takekoshi N, Hasegawa M, Curran DP (2006) Synthesis of fluorous-Fmoc reagents and purification of protected dipeptides with fluorous solid phase extraction. QSAR Comb Sci 25:713–715
10. Montanari V, Kumar K (2004) Just add water: a new fluorous capping reagent for facile purification of peptides synthesized on the solid phase. J Am Chem Soc 126:9528–9529
11. Montanari V, Kumar K (2006) A fluorous capping strategy for Fmoc-based automated and manual solid-phase peptide synthesis. Eur J Org Chem 874–877
12. Curran DP, Amatore M, Guthrie D, Campbell M, Go E, Luo Z (2003) Synthesis and reactions of fluorous carbobenzyloxy (F-Cbz) derivatives of α-amino acids. J Org Chem 68:4643–4647
13. Luo Z, Williams J, Roger W, Read RW, Dennis P, Curran DP (2001) Fluorous Boc (FBoc) carbamates: new amine protecting groups for use in fluorous synthesis. J Org Chem 66:4261–4266
14. Manzoni L, Castelli R (2006) Froc: a new fluorous protective group for peptide and oligosaccharide synthesis. Org Lett 8:955–957
15. Fustero S, Sancho AG, Chiva G, Sanz-Cervera JF, del Pozo C, Acena JL (2006) Fluorous (trimethylsilyl)ethanol: a new reagent for carboxylic acid tagging and protection in peptide synthesis. J Org Chem 71:3299–3302
16. Nakamura Y, Okumura K, Kojima M, Takeuchi S (2005) An expeditious synthesis of bistratamide H using a new fluorous protecting group. Tetrahedron Lett 47:239–243
17. Schwinn D, Bannwarth W (2002) Perfluoro-tagged benzyloxycarbonyl protecting group and its application in fluorous biphasic systems. Helv Chim Acta 85:255–264
18. Mizuno M, Goto K, Miura T, Hosaka D, Inazu T (2003) A novel peptide synthesis using fluorous chemistry. Chem Commun 972–973

19. Mizuno M, Goto K, Miura T, Matsuura T, Inazu T (2004) Peptide synthesis on fluorous support. Tetrahedron Lett 45:3425–3428
20. Mizuno M, Goto K, Miura T, Inazu T (2006) Rapid oligosaccharide and peptide synthesis on a recyclable fluorous support. QSAR Comb Sci 25:742–752
21. McBride LJ, Caruthers MH (1983) An investigation of several deoxynucleoside phosphoramidites useful for synthesizing deoxyoligonucleotides. Tetrahedron Lett 24:245–248
22. Temsamani J, Kubert M, Agrawal S (1995) Sequence identity of the n-1 product of a synthetic oligonucleotide. Nucleic Acids Res 23:1841–1844
23. Hecker KH, Rill RL (1998) Error analysis of chemically synthesized polynucleotides. Biotechniques 24:256–260
24. Seliger H, Holupirek M, Görtz H-H (1978) Solid phase oligonucleotide synthesis with affinity-chromatographic separation of the product. Tetrahedron Lett 24:2115–2118
25. Beller C, Bannwarth W (2005) Noncovalent attachment of nucleotides by fluorous-fluorous interactions: application to a simple purification principle for synthetic DNA fragments. Helv Chim Acta 88:171–179
26. Pearson WH, Berry DA, Stoy P, Jung K-Y, Sercel AD (2005) Fluorous affinity purification of oligonucleotides. J Org Chem 70:7114–7122
27. Berry & Associates Inc. 2434 Bishop Circle East, Dexter, MI 48130 USA. http://www.berryassoc.com
28. Fendt LA, Bouamaied I, Thoni S, Amiot N, Stulz E (2007) DNA as supramolecular scaffold for porphyrin arrays on the nanometer scale. J Am Chem Soc 129:15319–15329
29. Tripathi S, Misra K, Sanghvi YS (2005) Fluorous silyl protecting group for 5′-hydroxyl protection of oligonucleotides. Org Prep Procedures Int 37:257–263
30. Hong M, Zhou X, Lu Z, Zhu J (2009) Nanoparticle-based fluorous-tag-driven DNA detection. Angew Chem Int Ed 48:9503–9506
31. Gupta AP, Will SG (2008) Compounds and methods for synthesis and purification of oligonucleotides. WO2008077600A1
32. Wada T, Narita R, Kato Y, Saigo K (2005) Fluorous supports and processes for production of oligonucleotide derivatives with the same. WO2005070859A1
33. Vegas AJ, Koehler AN (2010) Detecting protein–small molecule interactions using fluorous small-molecule microarrays. Methods in Molecular Biology 669:43–55
34. Go EP, Uritboonthai W, Apon JV, Trauger SA, Nordstrom A, O'Maille G, Brittain SM, Peters EC, Siuzdak G (2007) Selective metabolite and peptide capture/mass detection using fluorous affinity tags. J. Proteome Res. 6:1492–1499
35. Kim JK, Lee JR, Kang JW, Lee SJ, Shin GC, Yeo WS, Kim KH, Park HS, Kim KP (2011). Selective enrichment and mass spectrometric identification of nitrated peptides using fluorinated carbon tags. Anal. Chem. 83:157–163

Fluorous Organic Hybrid Solvents for Non-Fluorous Organic Synthesis

Hiroshi Matsubara and Ilhyong Ryu

Abstract The rapid progress in fluorous chemistry shed the light on the use of fluorous-organic hybrid solvents for fluorous reactions; however, these hybrid solvents also have good potentials as solvents for ordinary organic synthesis. This chapter will survey the state of the art of the fluorous organic hybrid solvents as green substitutes for traditional organic solvents.

Keywords Fluorous chemistry · Green chemistry · Organic synthesis · Reaction solvent

Contents

1 Introduction 136
  1.1 Benzotrifluoride 137
  1.2 Trifluoromethylcyclohexane 143
  1.3 1,1,1,3,3-Pentafluorobutane (Solkane 365mfc) 144
  1.4 1H,1H,2H,2H-Perfluorooctyl 1,3-dimethylbutyl ether (F-626) .. 145
  1.5 Methyl Perfluorobutyl Ether (Novec 7100) 147
  1.6 Perfluorotriethylamine 147
  1.7 N-(1H,1H,2H,2H,3H,3H-Perfluorononanyl)-N-methyl formamide (F-DMF) 148
2 Conclusions 149
References 150

Abbreviations

AIBN   2,2′-Azobisisobutyronitrile
Bmim   Butylmethylimidazolium

H. Matsubara and I. Ryu (✉)
Department of Chemistry, Graduate School of Science, Osaka Prefecture University, Sakai, Osaka 599-8531, Japan
e-mail: ryu@c.s.osakafu-u.ac.jp

BTF	Benzotrifluoride (trifluoromethylbenzene α,α,α−trifluorotoluene)
DCE	1,2-Dichloroethane
E_T value	A scale for polarity of a solvent based on negatively solvatochromism of the pyridinium betaine dye
FC-72	Perfluorohexanes
F-SPE	Fluorous solid phase extraction
mpg-C_3N_4	Mesoporous carbon nitride
NHPI	N-Hydroxyphthalimide
NBS	N-Bromosuccinimide
PCC	Pyridinium chlorochromate
PDC	Pyridinium dichromate
PDE 4	Phosphodiesterase 4
PFD	Perfluorodecalin
SPC	Sodium percarbonate
S-PTPA	N-Phthaloyl-(S)-phenylalaninate
TBPA	Tris(p-bromophenyl)aminium hexachloroantimonate
TEA	Triethylamine
TFA	Trifluoroacetic acid
TfOH	Trifluoromethanesulfonic acid
TMS	Trimethylsilyl
TPP	Tetraphenylporphyrin
TTMSS	Tris(trimethylsilyl)silane

1 Introduction

The rapid evolution of fluorous tag methods in organic synthesis, to which a tremendous amount of attention of organic chemists is directed (comprehensive review of fluorous chemistry, for example [1]), has concurrently triggered expansion of the repertoire of fluorous solvents (fluorinated organic solvents), since fluorous tag reagents require fluorous solvents to dissolve themselves. Perfluorocarbons, such as perfluorohexanes (FC-72^{TM}), are immiscible with organic solvents at room temperature with the exception of some low molecular weight solvents such as pentane and ether. Thus, when perfluorocarbons are mixed with organic solvents, the lighter organic phase takes a position at the upper layer, with the denser fluorous phase at the lower layer. This organophobic nature is successfully applied to multi-layer synthetic strategy, the phase-vanishing method, which employs a fluorous phase as a liquid membrane to connect reagent layers and substrate layers [2]. On the other hand, there are many cases where such a mixed solvent system becomes homogeneous upon heating. This thermomorphic behavior in a fluorous/organic binary system was applied beautifully by Horváth and Rávai in landmark work in fluorous chemistry in 1994, which achieved facile biphasic separation of catalysts and products in Rh-catalyzed hydroformylation reaction of alkenes [3]. Nowadays, thermomorphic behaviors are among the most attractive

Table 1 Physical data of fluorous organic hybrid solvents

Solvent	Formula	BP (°C)[a]	MP (°C)	n_D	d
1: BTF	$C_6H_5CF_3$	102[b]	−29[c]	1.415[d]	1.190[e]
2: TFMC	$c\text{-}C_6H_{11}CF_3$	106[f]	−39[g]	1.374[f]	1.077[f]
3: Vertrel XF	$CF_3CHFCHFCF_2CF_3$	55[h]	−80[h]	−[i]	1.580[h]
4: Solkane 365 mfc	$CF_3CH_2CF_2CH_3$	40[j]	−36[j]	1.282[k]	1.267[k]
5: F-626	$C_4F_9(CH_2)_2OCH(CH_3)\text{-}CH_2CH(CH_3)_2$	214[l]	−110[l,m]	1.342[l]	1.354[l]
6: Novec 7100	$CH_3OC_4F_9$	61[n]	−135[n]	1.270[n]	1.520[n]
7: Novec 7200	$C_2H_5C_4F_9$	76[n]	−138[n]	1.280[n]	1.430[n]
8: PF-TEA	$N(C_2F_5)_3$	72[o]	−145[p]	1.262[q]	1.736[q]
9: F-DMF	$C_6F_{13}(CH_2)_3 N(CH_3)\text{-}CHO$	110[r] (0.75 mmHg)	−38[m,r]	1.359[r]	1.563[r]
10: FC-72	C_6F_{14}	58–60[s]	−82[t]	1.252[u]	1.670[t]

[a]At 760 mmHg, [b][41], [c][5], [d][42], [e][43], [f][44], [g][45], [h][46], [i]Not measured, [j][47], [k][48], [l][6], [m]Transition temperature to glassy state, [n]http://www.mmm.co.jp/emsd/product/pdt_-1_-1.html, [o][49], [p][50], [q][51], [r][39], [s][52], [t][53], [u][54]

aspects of fluorous chemistry [4]. On the other hand, the challenge to explore the *green* potentials of fluorous reaction media as substitutes for toxic organic solvents is now growing to constitute another important aspect of fluorous chemistry. Fluorous-organic hybrid solvents, such as benzotrifluoride (BTF) [5] and fluorous ether F-626 [6] have already found steady use in non-fluorous organic reactions. This chapter focuses on how fluorous organic hybrid solvents are useful for ordinary organic synthesis. As for BTF, since there is an excellent review [7] by the Curran group from 1999 in *Topics in Current Chemistry*, we only refer to some marked progress ever since. Some physical properties of fluorous-organic hybrid solvents, hydrofluorocarbons **1–4**, fluorous ethers **5–7**, fluorous amine **8**, and fluorous DMF **9**, as well as a typical fluorous solvent, perfluorohexanes (FC-72) **10**, are summarized in Table 1.

1.1 Benzotrifluoride

BTF **1** is a clear, low viscosity liquid with a boiling point of 102 °C, a melting point of −29 °C, and a density of 1.190 g mL^{-1}. Estimating from the absorption spectra of pyridinium *N*-phenolate betaine dye, BTF is slightly more polar than THF and slightly less polar than dichloromethane (Table 2) [8].

BTF is miscible with common organic solvents and able to dissolve many organic compounds, which encourages its consideration as a solvent for organic synthesis. In 1997, Ogawa and Curran reported that dichloromethane is replaceable by BTF in several reactions [5]. Since then, BTF has attracted the continuous attention of organic chemists as a green substitute for organic solvents. Thus far BTF has been used in a wide range of reactions including thermal reactions, radical reactions, Lewis-acid reactions, oxidations and reductions, phase-transfer reactions,

Table 2 Solvent E_T values of BTF and other organic solvents

	THF	BTF	CH_2Cl_2	DMF	CH_3CN
E_T^{30} (kcal mol^{-1})	37.4 (37.4)	39.3	41.0 (40.7)	43.3 (43.2)	45.7 (45.6)
E_T^N	0.207 (0.207)	0.265	0.318 (0.309)	0.389 (0.386)	0.463 (0.460)

Values taken from [55] are shown in parentheses

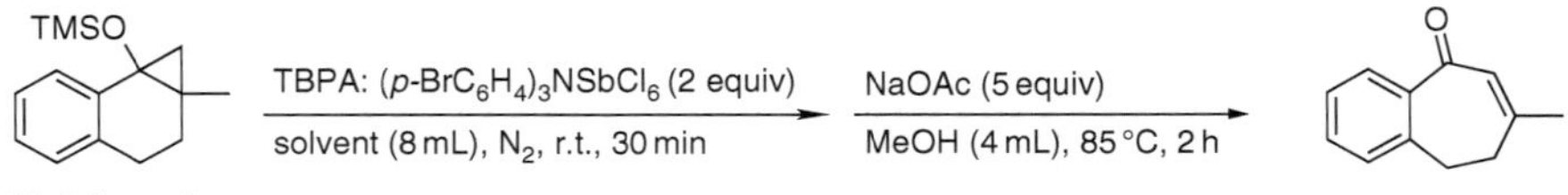

solvent	yield (%)
BTF	67 (89)[a]
CH_2Cl_2	79
CH_3CN	43

Scheme 1

and transition metal catalyzed reactions. Following up an excellent review published in 1999 [7], in this section we have focused on recent applications of BTF. Inspection of the applications of BTF to synthetic chemistry reveals that studies can be classified into two types: (1) BTF can be used as an alternative solvent to harmful organic solvents such as dichloromethane or carbon tetrachloride and (2) it affords the best results in yield or selectivity among various organic solvents examined.

Hasegawa et al. carried out TTMSS-mediated radical reaction of organic halides using BTF as a solvent [8]. BTF exhibited an efficiency similar to dichloromethane in a ring-expansion reaction of a siloxycyclopropane **11** by tris(p-bromophenyl) aminium hexachloroantimonate (TBPA) (Scheme 1) [9]. Hasegawa and his co-workers also developed a binary solvent system comprised of BTF and ionic liquids for a similar oxidative ring-opening reaction of siloxycyclopropanes by $FeCl_3$ [10]. Aza-Prins cyclization reactions in a similar biphasic system (BTF and $FeCl_3$-$BminFeCl_4$) gave piperidine derivatives in good yields (Scheme 2) [11].

Easton and his co-workers performed aromatic chlorination of ω-phenylalkylamine and ω-phenylalkylamide with chlorine gas in BTF as a solvent (Scheme 3) [12]. The rate constant of chlorination of 3-phenylpropionamide (**13**) in BTF was determined to be 8.8×10^{-3} M^{-1} s^{-1}, which is slightly larger than that in carbon tetrachloride (1.3×10^{-3} M^{-1} s^{-1}) and two orders of magnitude larger than that in acetic acid. Ortho/para selectivity for chlorination in acetic acid was higher than that in BTF and carbon tetrachloride.

Muzart et al. examined chromium(VI) catalyzed benzylic oxidation reactions with *tert*-butyl hydroperoxide or sodium percarbonate (SPC: $Na_2CO_3 \cdot 1.5H_2O_2$) as oxidant. Benzylic alcohols and active methylene compounds were oxidized in the presence of catalytic amounts of CrO_3 [13] or pyridinium dichromate (PDC: $(C_5H_5NH^+)_2 \cdot Cr_2O_7^{2-}$) [14] to give the corresponding ketones. As shown in Table 3,

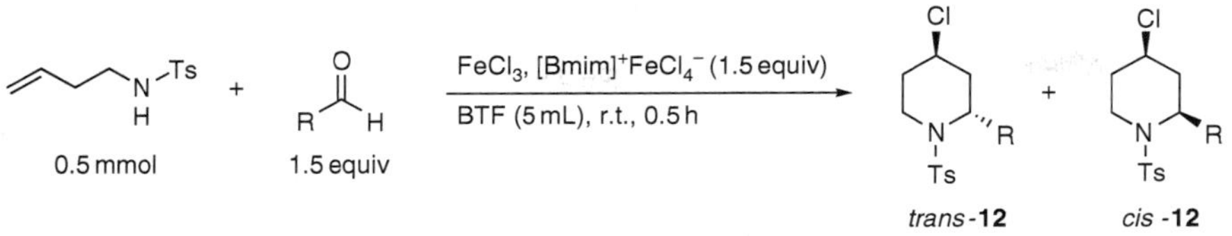

Scheme 2

R	yield (trans-12 : cis-12) (%)
Ph	93 (89 : 11)
p-MeC$_6$H$_4$	72 (89 : 11)
p-ClC$_6$H$_4$	67 (85 : 15)
p-O$_2$NC$_6$H$_4$	80 (90 : 10)

Scheme 3

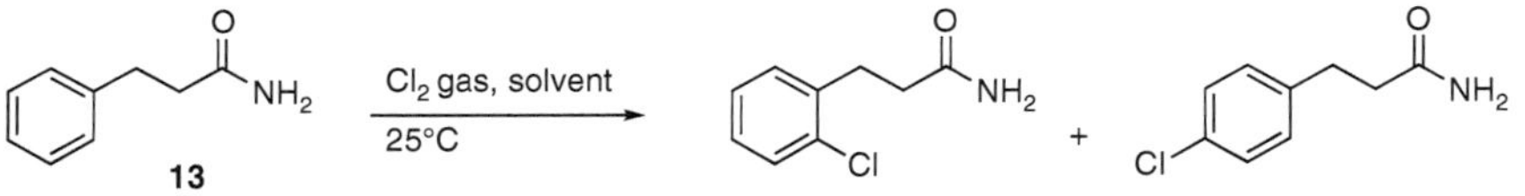

solvent	product ratio ortho/para	rate constant k/s^{-1} M^{-1}
BTF	55 : 45	8.8 × 10^{-3}
CCl$_4$	55 : 45	1.3 × 10^{-3}
AcOH	61 : 39	4.1 × 10^{-5}

Rate constants k were calculated with listed k' and concentration (M) taken from the literature.

isolated yields of the product in BTF were constantly better than those in dichloromethane and dichloroethane.

Dudley et al. developed a novel benzylation reagent, 2-benzyloxy-1-methylpyridinium triflate (**14**) [15]. Among various solvents examined for benzylation of alcohols using this reagent, BTF was found to be the choice (Scheme 4). Some studies [16–18] of total syntheses successfully employed this reagent and BTF for the benzylation steps. Benzylation also proceeded when Dudley's reagent was prepared in situ (Scheme 5) [19].

BTF provides the best outcomes in some asymmetric reactions. Hashimoto et al. carried out intramolecular 1,3-dipolar cycloaddition of α-diazo ketones in the presence of chiral dirhodium(II) carboxylate catalyst **15** and found that BTF was the best solvent for this reaction in terms of yield and enantioselectivity (Scheme 6) [20].

BTF also acts as a superior solvent in asymmetric synthesis using organocatalysts. Maruoka et al. reported an *exo*-selective asymmetric Diels–Alder reaction catalyzed by an organocatalyst of diamine salt (R)-**16** [21, 22]. As shown in Scheme 7, among seven solvents examined, BTF gave the best results in terms of reaction rate and stereoselectivity.

Table 3 Chromium-catalyzed oxidation of benzylic compounds

Entry	Substrate	Product	Reaction condition[a]	Isolated yield (%) BTF	Chlorinated solvent[b]
1	indanol	indanone	A B	81 96	77 83
2	fluorenol	fluorenone	A B	94 99	87 83
3	1-phenylpropan-1-ol	propiophenone	A B	97 63	83 60
4	indane	indanone	A B	92 38	60 30
5	tetralin	tetralone	A B	77 41	43 −[c]

[a]Condition A: substrate (1 mmol), CrO_3 (0.05 mmol, 5 mol%), solvent (6 mL). 70% t-BuOOH (0.64 mmol, 4 equiv.), rt, 24 h. Condition B: substrate (1 mmol), pyridinium dichromate (PDC, 0.1 equiv.), sodium percarbonate (SPC, 4 equiv.), solvent (6 mL). methyltrialkyl (C_8C_{10}) ammonium chloride (Adogen 464, 0.2 equiv.), solvent (19 mL), reflux, 24 h
[b]1,2-Dichlorethane (condition A), dichloromethane (condition B)
[c]Not performed

Mesoporous carbon nitride (mpg-C_3N_4) polymer was used as a metal-free photocatalyst to activate O_2 for oxidation reactions [23]. In oxidation of benzyl alcohol to benzaldehyde using this catalyst under irradiation of visible light, BTF gave a better conversion and selectivity of this reaction to toluene and acetonitrile (Scheme 8).

In ruthenium(III) tetraphenylporphyrin chloride **17** catalyzed oxidation reactions, BTF gave better results than toluene and acetonitrile due to less over-oxidation to give benzoic acid (Scheme 9) [24].

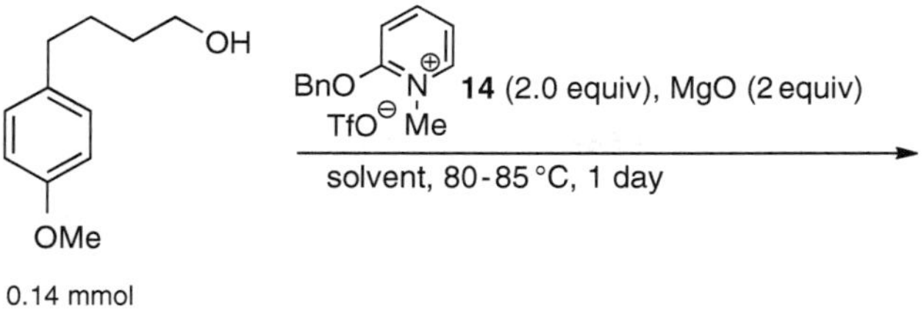

Scheme 4

solvent	yield (%)
BTF	> 95
toluene	91
DCE	67

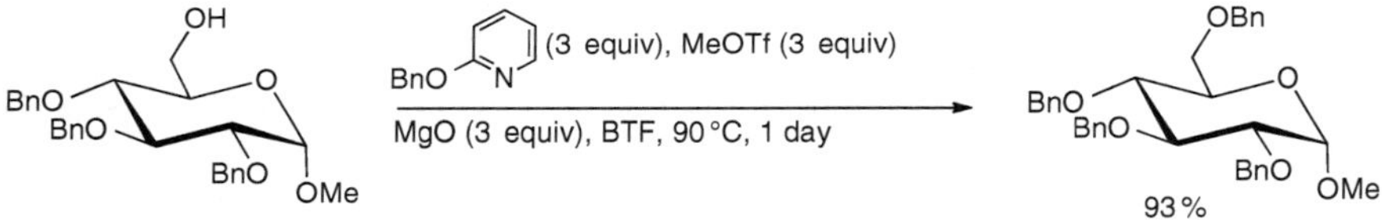

Scheme 5

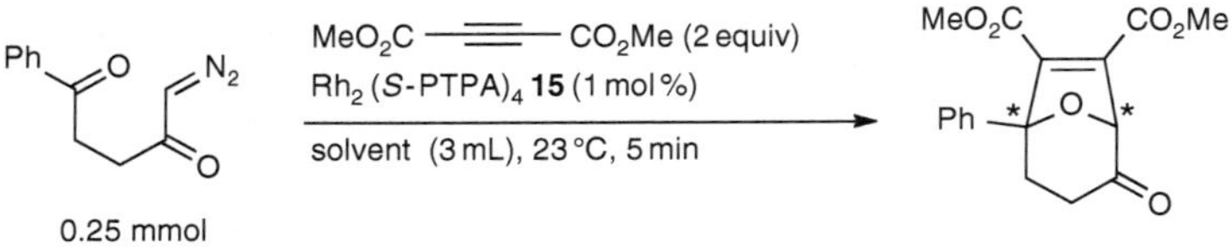

Scheme 6

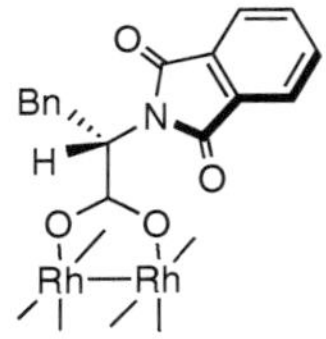

solvent	yield (%)	% ee
BTF	81	60
C$_6$H$_6$	64	59
CH$_2$Cl$_2$	79	20
Et$_2$O	63	29

Yanagisawa et al. reported that BTF and dichloromethane were good solvents for hydrocyanation of styrene-type compounds (Scheme 10) [25]. Utilizing BTF as a solvent, they achieved total synthesis of PDE 4 inhibitor.

Yorimitsu, Nakamura and their co-workers reported that ultra-rapid preparation of ^{15}O labeled deoxyglucose for positron emission tomography (PET) was achieved using a mixture of BTF/perfluorodecalin (PFD)/2-butanol as solvents, which appears to be a very good solvent system for both molecular oxygen and unprotected glucose. Starting from 6-iodo-2-deoxy-D-glucose (18), 6-[^{15}O]-2-deoxy-

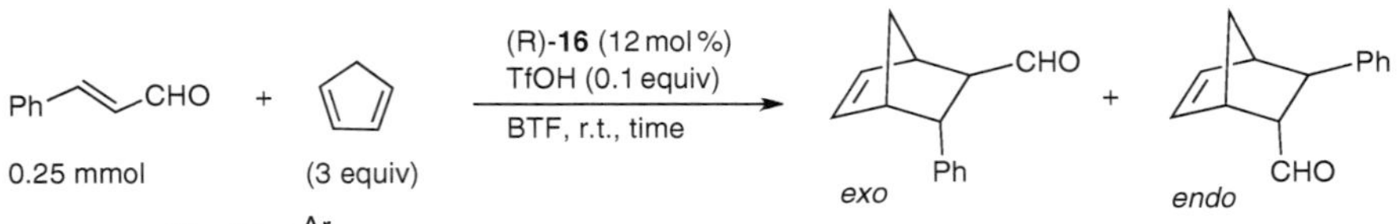

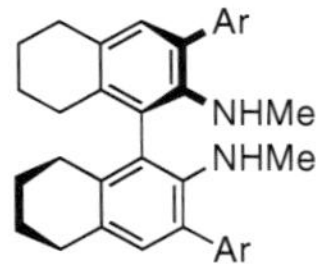

Scheme 7

solvent	time (h)	yield [%] (exo/endo)	ee (%) exo	ee (%) endo
BTF	9	89 (9.5:1)	79	71
CH$_2$Cl$_2$	20	87 (6.5:1)	72	68
MeOH	20	97 (6.5:1)	75	64
DMF	20	56 (5.1:1)	60	45
THF	20	80 (7.1:1)	72	68
hexane	20	93 (4.6:1)	74	74
toluene	14	93 (6.4:1)	78	76

(R)-**16**: Ar = 4-^{t}BuC$_6$H$_4$

Scheme 8

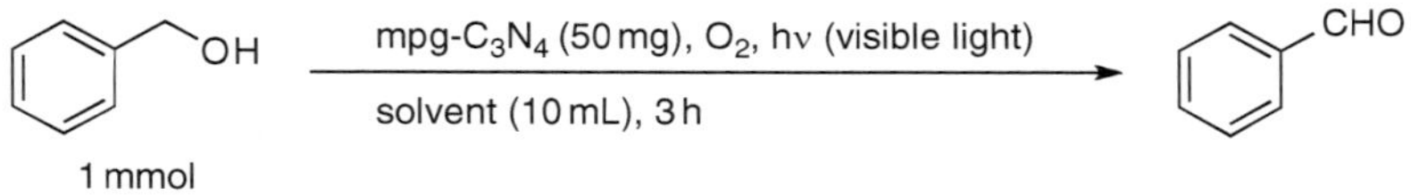

solvent	temp.(°C)	O$_2$ press.(bar)	conv.(%)	select.(%)
BTF	100	8	57	> 99
toluene	100	8	40	> 99
CH$_3$CN	100	8	70	68
FC-72	25	1	46	> 99

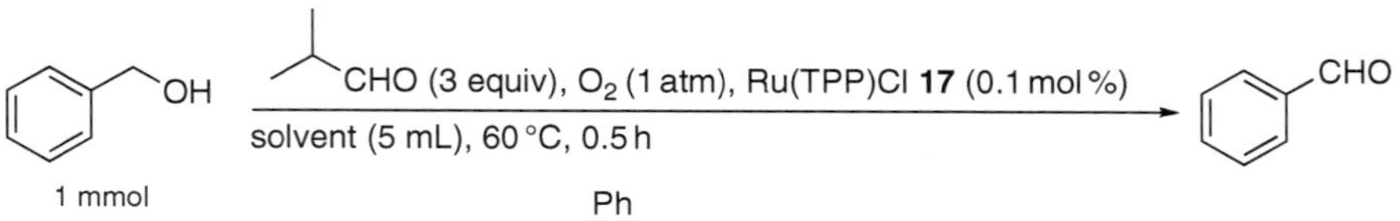

Scheme 9

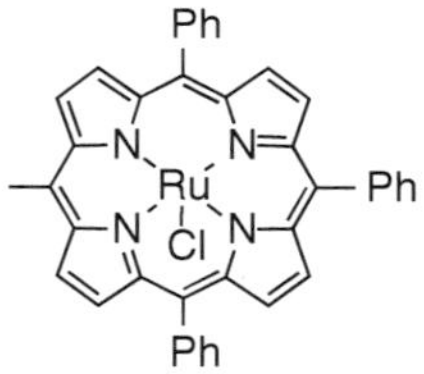

solvent	yield (%)
BTF	> 99
toluene	61(12)
CH$_3$CN	18(8)

Yield(%) of benzoic acid is shown in parentheses.

D-glucose was synthesized in 7 min in such a solvent system (Scheme 11) [26]. Aerobic oxidation of cyclohexane catalyzed by *N*-hydoxyphthalimide (NHPI) was successfully carried out using BTF as a solvent [56].

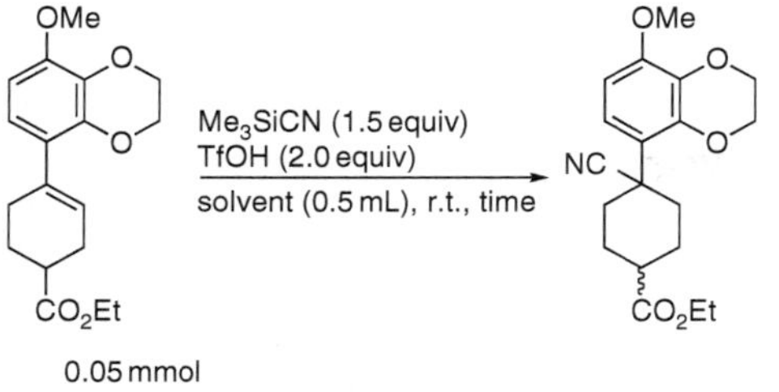
OMe
Me3SiCN (1.5 equiv)
TfOH (2.0 equiv)
solvent (0.5 mL), r.t., time
NC
CO2Et
0.05 mmol

solvent	time (h)	yield (%)	*cis/trans*
BTF	2	54	(63/37)
DMSO	72	0	
EtOH	24	0	
EtOAc	24	3	(52/48)
CH_3CN	5	14	(65/35)
TFA	9	23	(66/34)
hexane	2	29	(63/37)
toluene	2	36	(62/38)
CH_2Cl_2	2	56	(63/37)

Scheme 10

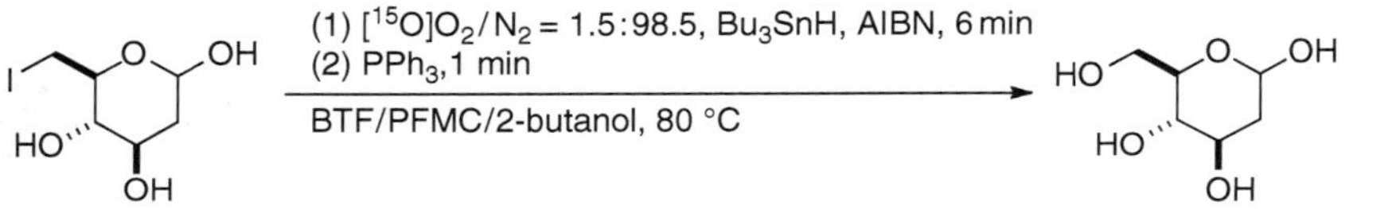
(1) [15O]O2/N2 = 1.5:98.5, Bu3SnH, AIBN, 6 min
(2) PPh3,1 min
BTF/PFMC/2-butanol, 80 °C

18

6-[^{15}O]-2-deoxy-D-glucose

Scheme 11

1.2 Trifluoromethylcyclohexane

Trifluoromethylcyclohexane (TFMC, **2**) was tested as a substitute for dichloromethane [27] in PCC (pyridinium chlorochromate: $(C_5H_5NH)^+ClCrO_3^-$) oxidation and carbon tetrachloride for allylic bromination with NBS (Scheme 12). In PCC oxidation with TFMC, however, the reaction was sluggish, requiring heating at 70 °C for 15 h, whereas concise filtrative work-up was possible since TFMC did not dissolve chromium residues. On the other hand, TFMC worked equally well with CCl_4 in allylic bromination of cyclohexene.

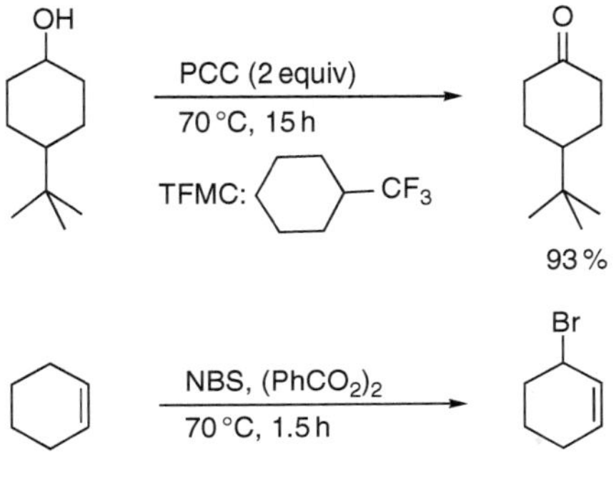
PCC (2 equiv)
70 °C, 15 h
TFMC:
CF3
93 %
NBS, (PhCO2)2
70 °C, 1.5 h
Br

solvent	yield
TFMC	46 %
CCl_4	45 %

Scheme 12

1.3 1,1,1,3,3-Pentafluorobutane (Solkane 365mfc)

1,1,1,3,3-Pentafluorobutane (**4**), the boiling point of which is 40 °C, is supplied from Solvay Fluorides as the trade name of Solkane 365mfc. Solkane 365mfc was examined for a solvent of trifluoromethylation using TMSCF$_3$ by Shibata and coworkers (Scheme 13) [28]. Bases, such as LiOAc and P(*t*-Bu)$_3$, which are frequently employed for reactions in DMF and THF, were not effective in Solkane 365mfc. However, the use of a catalytic amount of KOH or CsF in Solkane 365mfc gave the desired products in good yields. The same group also found that TBAF worked well in Solkane 365mfc for fluoride anion induced nucleophilic ring-opening reaction of an azirizine ring. Solkane 365mfc acts as a good solvent for catalyst-free Friedel–Crafts alkylation of indoles with ethyl trifluoropyruvate and glyoxylate [29]. While catalyst-free Friedel–Crafts alkylation of indole with ethyl trifluoropyruvate proceeds sluggishly in ether, the use of Solakane 365mfc led the

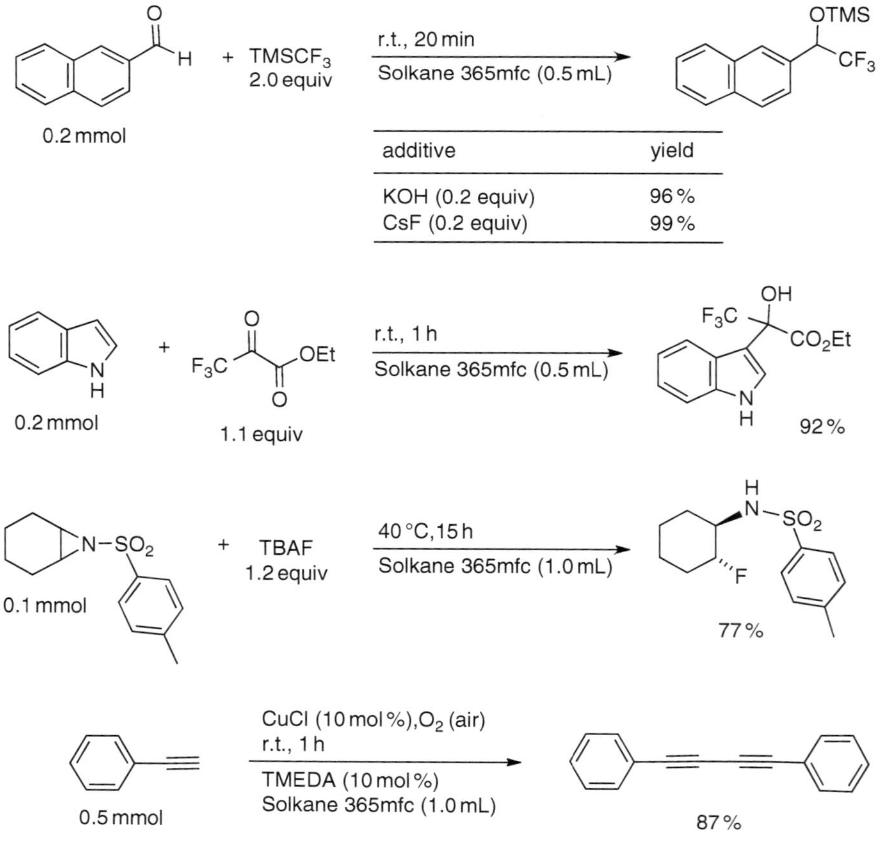

additive	yield
KOH (0.2 equiv)	96 %
CsF (0.2 equiv)	99 %

Scheme 13

reaction to complete in 1 h [30]. The Glaser type homo-coupling of terminal alkynes was carried out using Solkane 365 mfc as solvent [31]. Blended Solkane (93/7 mixture of Solkane 365mfc and 227 (1,1,1,2,3,3,3-heptafluoropropane), which has no flash point, worked equally well as Solkane 365 mfc.

1.4 1H,1H,2H,2H-Perfluorooctyl 1,3-dimethylbutyl ether (F-626)

F-626 **5** (1H,1H,2H,2H-perfluorooctyl-1,3-dimethylbutyl ether) is a fluorous/organic hybrid ether solvent (bp 214 °C, glass transition −110 °C), developed by the Kao Corporation [32]. F-626 is a colorless, clear, slightly viscous liquid, and is miscible with common organic solvents but hardly soluble in water. The approximate partition coefficients of F-626 associated with biphasic treatment are shown in Table 4 together with the data of F-DMF **9** and BTF **1** for comparison. It shows that the majority of F-626 is distributed in the FC-72 phase except for the cases of chloroform and ethyl acetate. Since BTF is preferentially distributed in the organic solvents, F-626 is more fluorous than BTF. The highly fluorous nature of F-626 allows its easy recovery from the reaction mixture via a fluorous/organic biphasic treatment. F-626 can be used as a solvent for both fluorous and non-fluorous reactions, such as fluorous tin hydride based radical reaction [6], Mizoroki–Heck reaction using fluorous Pd catalyst, LAH reduction, and Pd/C catalyzed hydrogenation (Scheme 14) [6, 33].

Standard high temperature reactions, for example, the Vilsmeier formylation, the Wolff–Kishner reduction, the Diels–Alder reaction, and retro-aldol reaction, are carried out using high boiling organic solvents such as diethylene glycol, DMF, o-dicholorobenzene, and diphenyl ether. In these reactions, generally separation of high boiling organic solvents is done by aqueous treatment and/or distillation. In contrast, F-626 is easily removable from the reaction mixture by fluorous/organic biphasic workup [34]. The following examples demonstrate that F-626 can be used

Table 4 Approximate partition coefficient of BTF, F-626, and F-DMF

	Organic solvent/FC-72		
	BFT[a]	F-626[a]	F-DMF[b]
CH$_3$CN	1/0.13	1/7.3	1/0.078
MeOH	1/0.21	1/3.8	1/0.05
C$_6$H$_6$	1/0.18	1/1.6	1/1.13
Cyclohexane	–[c]	–[c]	1/8.30
Acetone	1/0.08	1/1.1	1/0.10
AcOEt	1/0.13	1/0.85	1/0.20
CHCl$_3$	1/0.16	1/0.85	1/0.13

[a] Taken from [6]
[b] Taken from [39]
[c] Not determined

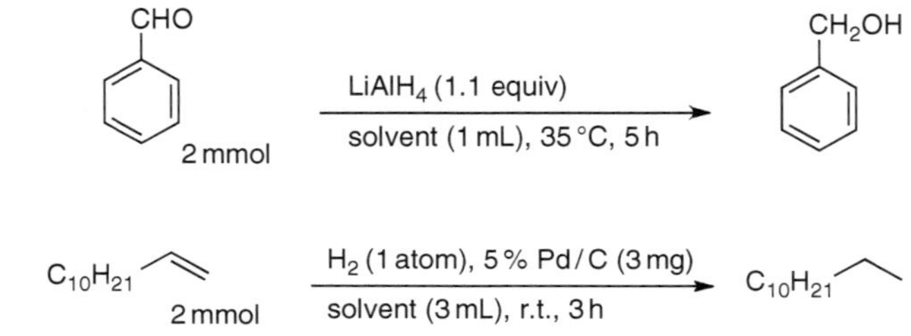

Scheme 14

Reaction 1: PhCHO (2 mmol) + LiAlH₄ (1.1 equiv), solvent (1 mL), 35 °C, 5 h → PhCH₂OH

solvent	yield
F-626	93%
ether	96%

Reaction 2: $C_{10}H_{21}CH=CH_2$ (2 mmol) + H₂ (1 atom), 5% Pd/C (3 mg), solvent (3 mL), r.t., 3 h → $C_{10}H_{21}CH_2CH_3$

solvent	yield
F-626	98%
ether	95%

Scheme 15

Reaction 1: 3,4-dichlorobenzaldehyde (2.5 mmol) + N₂H₄·H₂O (1 mL) + KOH (2.0 equiv), 110 °C, 2 h, then 200 °C, 6 h, solvent (2 mL) → 3,4-dichlorotoluene

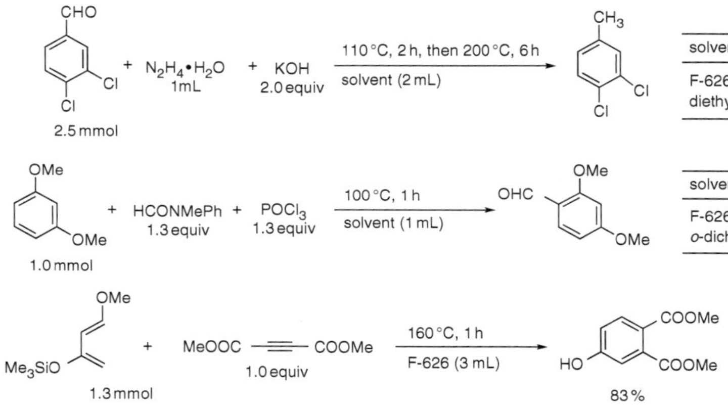

solvent	yield
F-626	89%
diethylene glycol	89%

Reaction 2: 1,3-dimethoxybenzene (1.0 mmol) + HCONMePh (1.3 equiv) + POCl₃ (1.3 equiv), 100 °C, 1 h, solvent (1 mL) → 2,4-dimethoxybenzaldehyde

solvent	yield
F-626	83%
o-dichlorobenzene	60%

Reaction 3: Me₃SiO-CH=CH-CH=CH-OMe (1.3 mmol) + MeOOC-C≡C-COOMe (1.0 equiv), 160 °C, 1 h, F-626 (3 mL) → 4-hydroxyphthalate dimethyl ester derivative, 83%

as an easily removed and reusable solvent for the Vilsmeier formylation, the Wolff–Kishner reduction, and the Diels–Alder reaction (Scheme 15) [6].

The thermal retro-aldol reaction is a well known reaction and can be frequently used for the preparation of some natural product analogs and for structure confirmation. F-626 was demonstrated to be an excellent reaction medium for the thermal retro-aldol reaction, which is easily separable from the products by fluorous/organic biphasic treatment. In the following example (Scheme 16) [35], the result without solvent gave a 3/1 mixture of the desired retro-aldol and undesired dehydration products while the use of ionic liquid [Bmim]NTf₂ as a solvent gave an even worse result for retro-aldol reaction, indeed a 50/50 mixture of desired and undesired products. F-626 worked quite well for the selective thermal retro-aldol reaction. The procedure including workup is detailed in Scheme 17. While fluorous ether solvent F-626 does not dissolve the starting aldol at room temperature, upon heating one layer results. After cooling, biphase workup using acetonitrile and FC-72, followed by purification using silica gel chromatography, gave the desired ketone in 94% yield with a ratio of 95/5 while F-626 was recovered in 98%.

solvent	yield (**20a**/**20b** ratio)
neat	47 % (75 : 25)
[Bmim]NTf_2 (0.5 mL)	65 % (50 : 50)
F-626 (4 mL)	98 % (95 : 5)

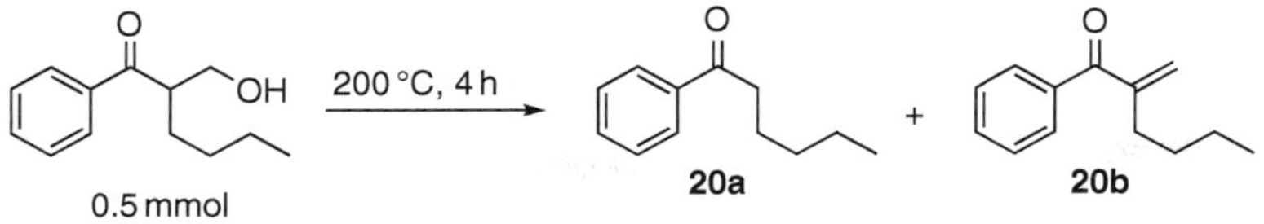

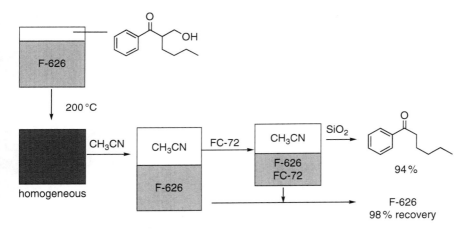

Scheme 16

Scheme 17

1.5 Methyl Perfluorobutyl Ether (Novec 7100)

The fluorous organic hybrid ether solvents Novec 7100 **6** (isomeric mixtures of methyl perfluorobutyl ether, $C_4F_9OCH_3$) and Novec 7200 **7** (isomeric mixtures of ethyl perfluorobutyl ether, $C_4F_9OC_2H_5$) are commercially available from 3M Ltd. Novec-7100 was successfully used as a co-solvent for electrophilic fluorination of aryl Grignard reagents with *N*-fluoro-2,4,6-trimethylpyridinium tetrafluoroborate (**19**) (Scheme 18) [36].

1.6 Perfluorotriethylamine

Perfluorotriethylamine (PF-TEA, **8**) was used as the solvent for some Lewis acid induced reactions such as the Hosomi–Sakurai and the Friedel–Crafts reactions

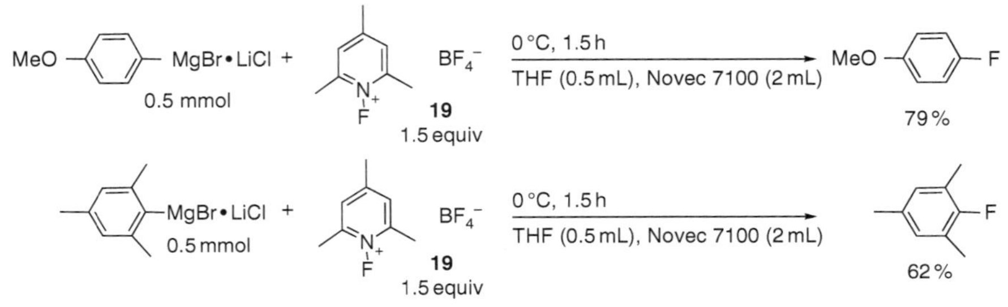

Scheme 18

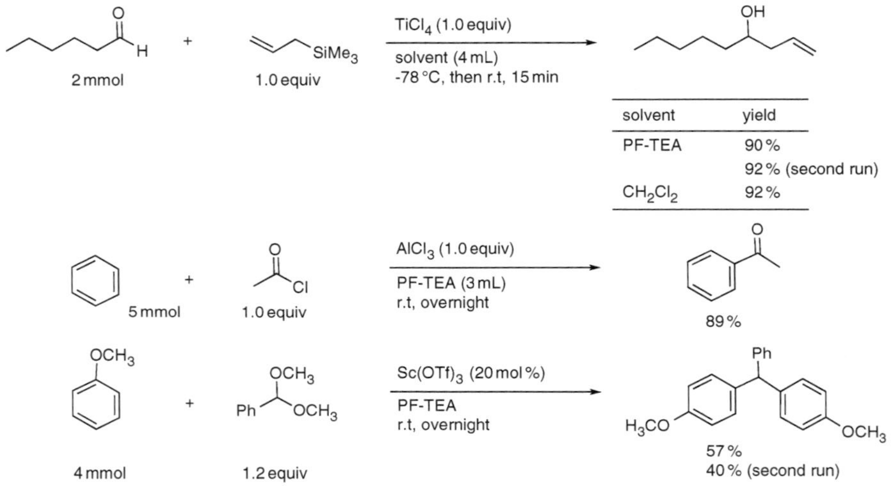

Scheme 19

(Scheme 19) [37, 38]. Interestingly, despite the existence of an amine structure, PF-TEA does not react with Lewis acids. After the reaction, PF-TEA can be recovered by triphasic (organic/aqueous/fluorous) workup and reused.

1.7 N-(*1*H,*1*H,*2*H,*2*H,*3*H,*3*H-*Perfluorononanyl*)-N-*methyl formamide (F-DMF)*

Fluorous DMF, (*N*-(1*H*,1*H*,2*H*,2*H*,3*H*,3*H*-perfluorononanyl)-*N*-methyl formamide) (**9**), is a colorless, slightly viscous liquid with a boiling point of 110 °C at 0.75 Torr, and a density of 1.544 g/cm [3] (25 °C). F-DMF does not freeze at −35 °C but transforms to a glassy state at −40 °C. F-DMF is miscible with a wide range of organic solvents, such as hexane, benzene, chloroform, ether, acetone, ethyl acetate, and ethanol; however, it is hardly soluble in cyclohexane and water.

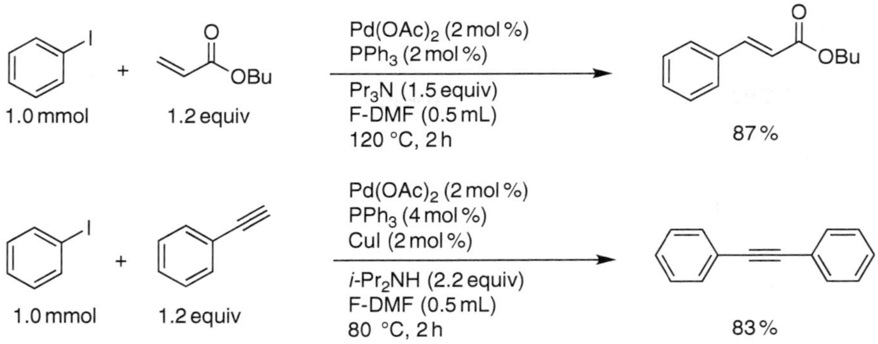

Scheme 20

The approximate partition coefficients of F-DMF are listed in Table 4. For biphasic separation a combination of cyclohexane and FC-72 is recommended.

F-DMF was tested for the Mizoroki–Heck and the Sonogashira reactions (Scheme 20) [39]. These cross-coupling reactions proceeded smoothly to give the products in good yields. Recycling and reusing of catalyst was achieved by organic/fluorous biphasic workup in these reactions.

2 Conclusions

BTF has been used in synthetic chemistry not only as an alternative of harmful chlorine-containing solvents but also as an option on optimizing reaction conditions, often providing superior results in yields or selectivity to classical organic solvents. Apart from BTF, the use of other newcomer fluorous hybrid solvents has just started. However hydrofluorocarbons such as Solkane and Vertrel appear promising as a green substitute for toxic chlorohydrocarbons. Fluorous ether solvents, such as F-626 and Novec, exhibit more fluorous character than BTF. Nevertheless, they can be employed for a variety of organic syntheses thus far carried out with organic solvents. Advantage of hybrid solvents having strong fluorous character is its ease of recovery by fluorous/organic biphasic workup or fluorous solid phase extraction (F-SPE). F-DMF is unique since, unlike DMF, it is not miscible with water, while the high polarity is retained. Needless to say, finding the most suitable solvent to give highest conversion and selectivity is essential and in this regard we still need a repertoire in designed solvents. For this purpose, it should be noted that Chu, Yu, and Curran extensively investigated solvent polarity and fluorophilicity/phobicity [40], which is quite useful for the design of fluorous/organic biphasic systems with mixed hydrofluoroethers (HFEs, RfOR). This information is also useful for the design of blend solvents for synthetic chemistry. Having diversity in solvents, organic synthesis would be more extensive, and the role of fluorous hybrid solvents should be quite large.

References

1. Horváth I, Gladysz J, Curran DP (2004) Eds. Handbook of fluorous chemistry, Wiley-VCH, Weinheim
2. Ryu I, Matsubara H, Nakamura H, Curran DP (2008) Phase-vanishing methods based on fluorous phase screen. A simple way for efficient execution of organic synthesis. Chem Rec 8:351
3. Horváth I, Rábai J (1994) Facile catalyst separation without water: fluorous biphase hydroformylation of olefins. Science 266:72
4. Curran DP (1998) Strategy-level separations in organic synthesis: from planning to practice. Angew Chem Int Ed Engl 37:1174
5. Ogawa A, Curran DP (1997) Benzotrifluoride: a useful alternative solvent for organic reactions currently conducted in dichloromethane and related solvents. J Org Chem 62:450
6. Matsubara H, Yasuda S, Sugiyama H, Ryu I, Fujii Y, Kita K (2002) A new fluorous/organic amphiphilic ether solvent, F-626: execution of fluorous and high temperature classical reactions with convenient biphase workup to separate product from high boiling solvent. Tetrahedron 58:4071
7. Maul JJ, Ostrowski PJ, Ublacker GA, Linclau B, Curran DP (1999) Benzotrifluoride and derivatives: useful solvents for organic synthesis and fluorous synthesis. Top Curr Chem 206:79
8. Hasegawa E, Ogawa Y, Kakinuma K, Tsuchida H, Tosaka E, Takizawa S, Muraoka H, Saikawa T (2008) Tris(trimethylsilyl)silane promoted radical reaction and electron-transfer reaction in benztrifluoride. Tetrahedron 64:7724
9. Hasegawa E, Kakinuma K, Yanaki T, Komata S (2009) In situ generated tris(p-bromophenyl) amine radical cation promoted electron transfer reaction of cyclopropyl silyl ethers. Tetrahedron 65:10876
10. Tsuchida H, Hasegawa E (2010) Novel biphasic reaction system of ferric chloride dissolved imidazolium hexafluorophosphate and benzotrifluoride: application to electron transfer reaction of cyclopropyl silyl ethers. Tetrahedron 66:3447
11. Hasegawa E, Hiroi N, Osawa C, Tayama E, Iwamoto H (2010) Application of biphasic reaction procedure using ferric chloride dissolved in an imidazolium salt and benzotrifluoride (Felm-BTF procedure) to aza-Prins cyclization reaction. Tetrahedron Lett 51:6535
12. O'Connel JL, Simpson JS, Dumanski PG, Simpson GW, Easton CJ (2006) Aromatic chlorination of ω-phenylalkylamines and ω-phenylalkylamides in carbon tetrachloride and α, α, α-trifluorotoluene. Org Biomol Chem 4:2716
13. Boitsov S, Riahi A, Muzart J (2000) Chromium(IV) oxide-catalysed oxidations by *tert*-butyl hydroperoxide using benzotrifluoride as solvent. C R Acad Sci Paris Sér IIc Chimie 3:747
14. Delaval N, Bouquillon S, Hénin F, Muzart J (1999) Use of benzotrifluoride as solvent for chromium-catalysed oxidations with sodium percarbonate. J Chem Res (S):286
15. Poon KWC, Dudley GB (2006) Mix-and-heat benzylation of alcohols using a bench-stable pyridinium salt. J Org Chem 71:3923
16. Caubert V, Massé J, Retailleau P, Langlois N (2007) Stereoselective formal synthesis of the potent proteasome inhibitor: salinosporamide A. Tetrahedron Lett 48:381
17. Legeay J-C, Langlois N (2007) Nitron [2+3]-cycloaddition in stereocontrolled synthesis of a potent proteasome inhibitor: (−)-omuralide. J Org Chem 72:10108
18. Schmidt JP, Beltrán-Rodil S, Cox RJ, McAllister GD, Reid M, Taylor RJK (2007) The first synthesis of the ABC-ring system of 'Upenamide'. Org Lett 9:4041
19. Lopez SS, Dudley GB (2008) Convenient method for preparing benzyl ethers and esters using 2-benzyloxypyridine. Beilstein J Org Chem 4:44
20. Kitagaki S, Anada M, Kataoka O, Matsuno K, Umeda C, Watanabe N, Hashimoto S (1999) Enantiocontrol in tandem carbonyl ylide formation and intermolecular 1,3-dipolar cycloaddition of α−diazo ketones mediated by chiral dirhodium(II) carboxylate catalyst. J Am Chem Soc 121:1417

21. Kano T, Tanaka Y, Maruoka K (2006) *exo*-Selective asymmetric Diels–Alder reaction catalyzed by diamine salts as organocatalyst. Org Lett 8:2687
22. Kano T, Tanaka Y, Maruoka K (2007) exo-Selective asymmetric Diels–Alder reaction catalyzed by diamine salts as organocatalysts. Chem Asian J 2:1161
23. Su F, Mathew SC, Lipner G, Fu X, Antonietti M, Blechert S, Wang X (2010) mpg-C$_3$N$_4$-catalyzed selective oxidation of alcohols using O$_2$ and visible light. J Am Chem Soc 132:16299
24. Ji H-B, Yuan Q-L, Zhou X-T, Pei L-X, Wang L-F (2007) Highly efficient selective oxidation of alcohols to carbonyl compounds catalyzed by ruthenium(III) *meso*-tetraphenylphorphyrin chloride in the presence of molecular oxygen. Bioorg Med Chem Lett 17:6364
25. Yanagisawa A, Nezu T, Mohri S (2009) Brϕnsted acid-promoted hydrocyanation of arylalkenes. Org Lett 11:5286
26. Yorimitsu H, Murakami Y, Takamatsu H, Nishimura S, Nakamura E (2005) Ultra-rapid synthesis of ^{15}O-labeled 2-deoxy-D-glucose for positron emission tomography (PET). Angew Chem Int Ed Engl 44:2708
27. Legros J, Crousse B, Bonnet-Delpon D, Bégué J–P, Maruta M (2002) Trifluoromethycyclohexane as a new solvent? Limits of use. Tetrahedron 58:4067
28. Kusuda A, Kawai H, Nakamura S, Shibata N (2009) Solkane® 356mfc is an environmentally benign alternative solvent for trifluoromethylation reactions. Green Chem 11:1733
29. Török B, Abid M, London G, Esquibel J, Török M, Mhadgt SC, Yan P, Prakash GKS (2005) Highly enantioselective organicatalytic hydroxyalkylation of indoles with ethyl trifluoropyruvate. Angew Chem Int Ed Engl 44:3086
30. Xu X–H, Kusuda A, Tokunaga E, Shibata N (2011) Catalyst-free and catalytic Friedel–Crafts alkylations of indoles in Solkane® 356mfc, an alternative solvent. Green Chem 13:46
31. Kusuda A, Xu X–H, Wang X, Tokunaga E, Shibata N (2011) Organic reaction in Solkane® 356mfc: homocoupling reaction of terminal alkynes. Green Chem 13:843
32. Fujii Y, Furugaki H, Kajihara Y, Kita K, Morimoto H, Uno M (2000) U. S. Patent 6011071
33. Fukuyama T, Arai M, Matsubara H, Ryu I (2004) Mizoroki–Heck arylation of α,β-unsaturated acids with a hybrid fluorous ether, F-626: facile filtrative separation of products and efficient recycling of a reaction medium containing a catalyst. J Org Chem 69:8105
34. Curran DP (2000) In: Stoddart F, Reinhoudt D, Shibasaki M (eds) Stimulating concept in chemistry. Wiely-VCH, New York, p 25
35. Fukuyama T, Kawamoto T, Okamura T, Denichoux A, Ryu I (2010) Thermal retro-aldol reaction using fluorous ether F-626 as a reaction medium. Synlett:2193
36. Anbarasan P, Neumann H, Beller M (2010) Efficient synthesis of aryl fluorides. Angew Chem Int Ed Engl 49:2219
37. Nakano H, Kitazume T (1999) Organic reactions without an organic medium. Green Chem 1:21
38. Nakano H, Kitazume T (1999) Friedel-Crafts reaction in fluorous fluids. Green Chem 1:179
39. Matsubara H, Maeda L, Ryu I (2005) Preparation of fluorous DMF solvents and their use for some Pd-catalyzed cross-coupling reactions. Chem Lett 34:1548
40. Chu Q, Yu MS, Curran DP (2007) New fluorous/organic biphasic systems achieved by solvent tuning. Tetrahedron 63:9890
41. Baldwin JE, Smith RA (1967) Cycloadditions. XII. The relative reactivity of carbethoxycarbene and carbethoxynitrene in cycloadditions with aromatics. J Am Chem Soc 89:1886
42. Fialkov YA, Moklyachuk LI, Kremlev MM, Yagupol'skii LM (1980) Fluorination of aromatic carboxylic acids by sulfur tetrafluoride. XIII. Fluorinat benzoic acid and its derivatives in hydrogen fluoride. J Org Chem USSR (Engl Transl) 16:1269
43. Rudenko AP, Sperkach VS, Timoshenko AN, Yagupol'skii LM (1981) The elastic properties of trifluoromethylbenzene along equlibrium curve. Russ J Phys Chem (Engl Transl) 55:591
44. Roberts JD, Webb RL, McElhill EA (1950) The electrical effect of the trifluoromethyl group. J Am Chem Soc 72:408
45. Ban B, Chachaty C, Renaud M, Fourme R (1971) Molecular motion and phase transition in solid perfluoromethylcyclohexane. Can J Chem 49:2953

46. Tochigi K, Kikuchi C, Kurihara K, Ochi K, Mizukado J, Otake K (2005) vapor−liquid equilibrium data for the binary decafluoropentane (HFC-43-10meec) + heptane, decafluoropentane + butyl ethyl ether, octafluorobutane (HFC-338pcc) + butyl ethyl ether, and heptafluoro propyl methyl ether (HFE-347mcc) + HFC-338pcc systems at 101.3 kPa. J Chem Eng Data 50:784
47. Frohn H-J, Giesen M, Welting D, Bardin VV (2010) Bis(perfluoroorganyl)bromonium salts [(RF)$_2$Br]Y (RF = aryl, alkenyl, and alkynyl). J Fluorine Chem 131:922
48. Henne AL, Hinkamp JB (1945) The synthesis and directed chlorination of 2,2-difluorobutane. J Am Chem Soc 67:1194
49. Felling KW, Lagow RJ (2003) An efficient high-yield synthesis for perfluorinated tertiary alkyl amines. J Fluorine Chem 123:233
50. Bispen TA, Mikhailova TV, Moldavskii DD, Furin GG, Shkul'tetskaya LV (1996) Purification of perfluorinated organic compounds. Russ J Appl Chem 69:96
51. Haszeldine RN (1951) Perfluoro-*tert*-amines. J Chem Soc 102
52. Banks RE, Burgess JE, Haszeldine RN (1965) Heterocyclic polyfluoro-compounds. Part VIII. Perfluoro-(2-, 3-, and 4-methylpyridine) and tetrafluoroisonicotinic acid. J Chem Soc: 2720
53. Dunlap RD, Murphy CJ Jr, Bedford RG (1958) Some physical properties of perfluoro-n-hexane. J Am Chem Soc 80:83
54. Brice TJ, Coon RI (1953) The effects of structure on the viscosities of perfluoroalkyl ethers and amines. J Am Chem Soc 75:2921
55. Reichardt C (1994) Solvatochromic dyes as solvent polarity indicators. Chem Rev 94:2319
56. Guha SK, Obora Y, Ishihara D, Matsubara H, Ryu I, Ishii Y (2008) Aerobic oxidation of cyclohexane using *N*-hydroxyphthlimide bearing fluoroalkyl chain. Adv Synth Catal 350:1323

Fluorous Catalysis: From the Origin to Recent Advances

Jean-Marc Vincent

Abstract Among the various strategies developed in the last two decades to recycle catalysts, fluorous catalysis has emerged as one of the most powerful approaches as it combines the advantages of homogeneous catalysis for reactivity (molecular catalysts, most often reactions conducted in one-phase homogeneous conditions) and heterogeneous catalysis for catalyst recovery (liquid/liquid- or solid/liquid-phase separation protocols). Of particular interest is the general character of this approach and the variety and efficiency of separation protocols available to recover catalysts.

Keywords Fluorous catalysis · Fluorous chemistry · Green chemistry · Perfluorocarbons · Purification procedures · Separation techniques

Contents

1 Introduction 154
2 Liquid/Liquid Phase Separation: Fluorous Biphasic Catalysis 156
   2.1 Metal-Based Catalysis 156
   2.2 Organocatalysis and Biocatalysis 162
3 Solid/Liquid Phase Separation 164
   3.1 Fluorous Solid-Phase Extraction 164
   3.2 Supported Fluorous Phase Catalysis 165
   3.3 Precipitation 167
4 Conclusion 169
References 170

J.-M. Vincent (✉)
Institute of Molecular Sciences, University of Bordeaux, UMR CNRS 5255, 351, Cours de la Libération, 33405 Talence Cedex, France
e-mail: jm.vincent@ism.u-bordeaux1.fr

1 Introduction

Most chemical reactions, whether they are conducted in industry on large scales or in smaller scales in research and development laboratories, require a catalyst to proceed efficiently. Triggered by both environmental and economic concerns, as catalysis very often deals with potentially toxic and/or costly metal complexes, the search for recoverable catalysts has become a major concern of modern chemistry [1]. As outlined by Gladysz [2], a recoverable catalyst would, ideally, satisfy an exacting set of criteria related to its preparation (low cost, easy synthesis and handling, being non-toxic and hazard-free), its reactivity (no activation, giving 100% product yield, fast kinetics at ambient temperature, low loading), and its recovery (quantitative by decantation or filtration). Beside the environmental issue, routinely used automated synthesis applied to drug discovery has pushed toward the development of adapted synthetic methodologies for the rapid and efficient purification of the product while being able to recycle the catalyst if needed. Fluorous catalysis emerged in 1994 [3] with the publication of the Fluorous Biphasic Catalysis (FBC) concept first applied to the rhodium(I) catalyzed hydroformylation of alkenes by Horváth and Rábai in *Science* magazine. Since this seminal work, there has been an enormous interest in developing catalyst separation/recovery procedures exploiting the unique physico-chemical properties of the perfluoroalkyl tags in association with perfluorocarbons (PFC) or fluorous supports, i.e., fluorous silica and perfluoropolymers [4–7]. A selection of fluorous ligands, organocatalysts/phase transfer catalysts developed within the past 17 years, is shown in Schemes 1 and 2.

The variety of fluorous ligands now available illustrates the fact that essentially all important catalytic processes exist in a fluorous version. Among the many strategies developed in recent years for the recovery of catalysts [1], fluorous catalysis is undoubtedly one of the most powerful and general approaches for the following reasons: (1) fluorous catalysis deals with molecular catalysts displaying structures and reactivity similar to their non-fluorous analogs; (2) in most cases reactions are conducted in optimal homogeneous conditions for reactivity while a range of separation strategies based on liquid/liquid or solid/liquid separation procedures are available; (3) the modification of ligands/catalysts by the fluorous tags, can be achieved by straightforward reactions from commercially available fluorous reactants.

The objective of this chapter is not to provide an exhaustive list of fluorous catalysts and related reactions, but rather to highlight the variety of conceptually innovative catalyst recovery procedures developed so far which exploit the specific properties of the perfluoroalkyl chains. Each type of separation/recovery procedure will be illustrated by one or several examples of representative catalytic processes.

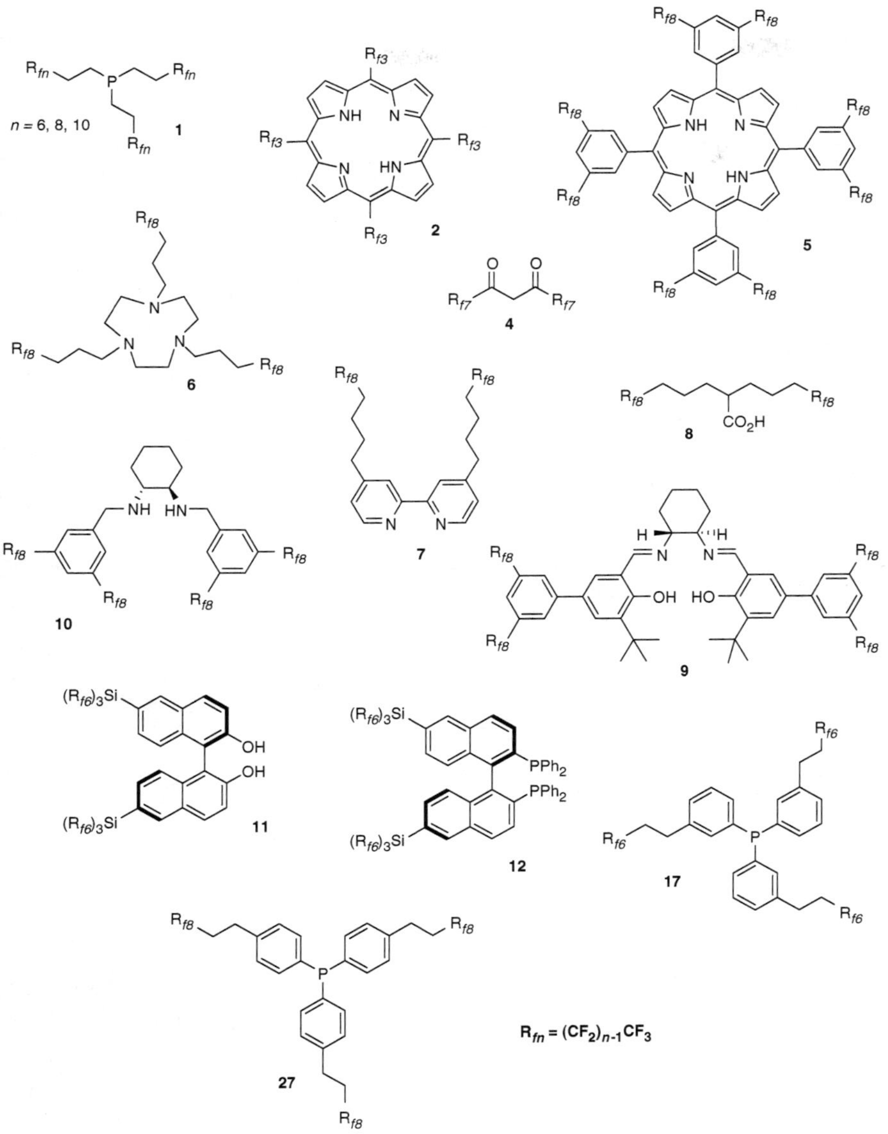

Scheme 1 Structures of some fluorous ligands synthesized since 1994

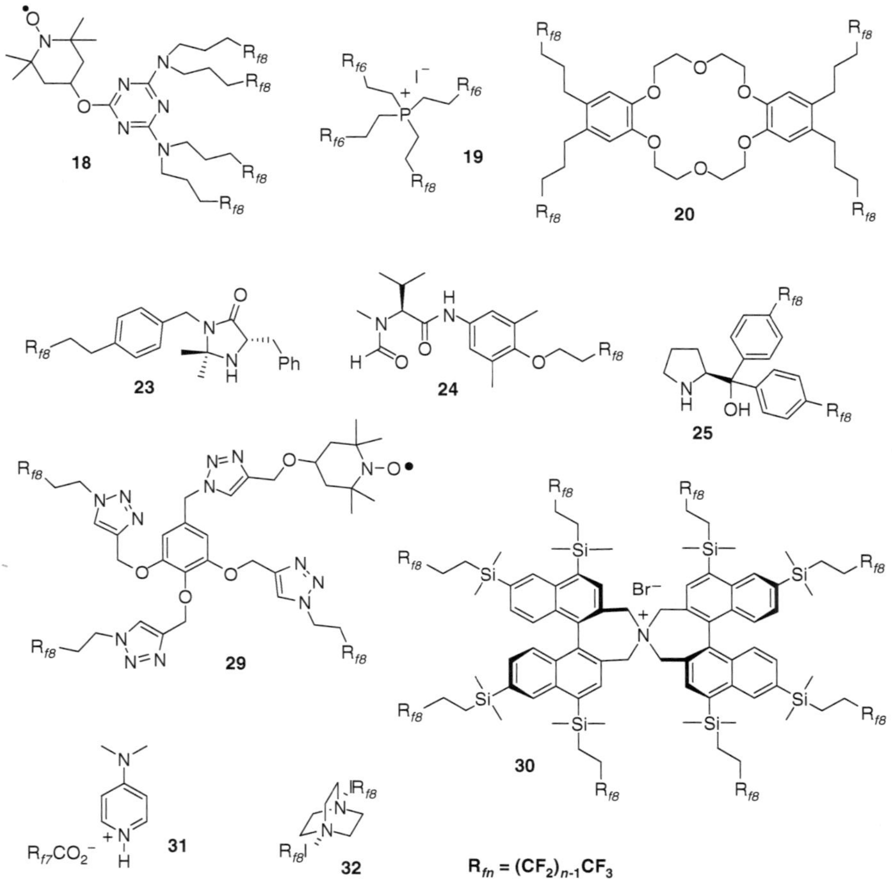

Scheme 2 Structures of some fluorous organocatalysts synthesized in recent years

2 Liquid/Liquid Phase Separation: Fluorous Biphasic Catalysis

2.1 Metal-Based Catalysis

The FBC concept, reported by Horváth and Rábai in 1994 [3][1], was first applied to the rhodium(I) catalyzed hydroformylation of alkenes. The thermomorphic properties of PFC and hydrocarbons (HC) were exploited to run the

---

[1] In 1991 Vogt M defended a PhD thesis entitled "The application of perfluorinated polyethers for the immobilization of homogeneous catalysts," Rheinisch-Wesfälischen Technischen Hochschule, Aachen, Germany. Results from this work were published in 1999. See [8].

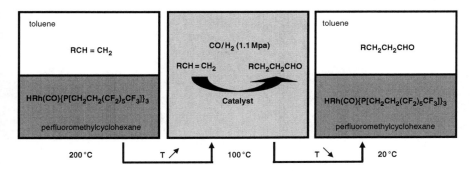

Scheme 3 Rhodium-catalyzed hydroformylation of alkenes under fluorous biphasic conditions (adapted from [3])

hydroformylation reactions under optimal homogeneous conditions at elevated temperatures, while the fluorous-phase soluble catalyst was efficiently recovered at the end of the reaction by simple liquid/liquid phase separation which occurred at ambient temperature (Scheme 3). Their work clearly demonstrated that minimal leaching of the fluorous catalyst into the organic phase was possible when the highly fluorinated phosphines (1) (Scheme 1) are used to ensure a high partition coefficient of the catalyst for the fluorous phase [3, 9]. An Rh loss as low as 4.2% was reported after nine consecutive runs and a *turnover number* of 35,000 (TON: mole of aldehyde/mole of Rh). It was also shown that the introduction of two or three methylene groups before the fluorous chain was necessary to isolate the coordinating atom from the electron-withdrawing effect of the perfluoroalkyl chain to ensure a reactivity similar to the unmodified catalyst [9, 10]. For further details and developments on fluorous hydroformylations, see the chapter by Horváth and Zhao in this issue.

Since this pioneering work, the FBC protocol has been applied with success to important rhodium-catalyzed reactions such as hydroboration [11], hydrogenation [12], or hydrosilylation [13] of alkenes. Two chapters in this issue are devoted to fluorous hydrogenations and hydrosilylations. Interestingly, a variety of perfluoroalkylated phosphines are now commercially available whilst all kind of fluorous monodentate, bidentate, or chiral phosphines can be prepared via reported procedures [14, 15].

Because of the specific properties of PFCs, such as the chemical inertness and the capacity to solubilize gases such as O_2 in higher concentrations compared to organic solvents and water, the FBC concept rapidly emerged as a highly appealing protocol for catalyzed oxidation reactions [16–18]. A very interesting early example of FBC for oxidations was reported by DiMagno, Dussault et al. They showed that the fluorous porphyrin (2) combined with a PFC was a very efficient photosensitizer for singlet oxygenation of alkenes [19]. The described FBC, applied to the hydroperoxidation of **3** (Scheme 4), is particularly appealing for the following reasons: (1) due to both the electron-withdrawing effect of the ponytails and the

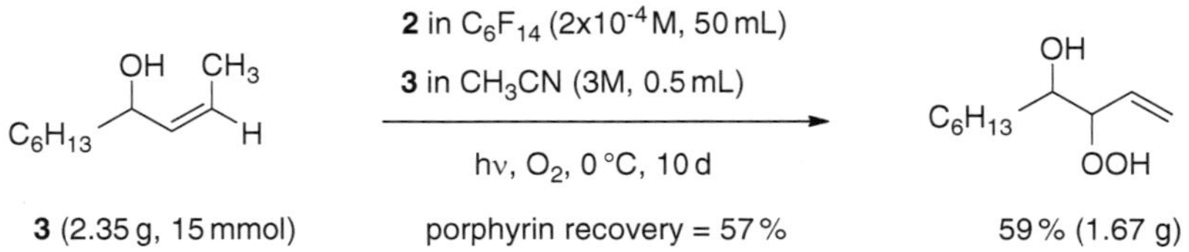

Scheme 4

Reactant: C₆H₁₃-CH(OH)-CH=CH-CH₃ (with OH and CH₃ labels), labeled **3** (2.35 g, 15 mmol)

Conditions: **2** in C₆F₁₄ (2×10⁻⁴ M, 50 mL); **3** in CH₃CN (3M, 0.5 mL); hv, O₂, 0 °C, 10 d; porphyrin recovery = 57%

Product: C₆H₁₃-CH(OH)-CH(OOH)-CH=CH₂, 59% (1.67 g)

Scheme 4 Practical photooxygenation conducted under FBC (adapted from [19])

physical segregation of the catalyst from the hydroperoxide that forms, the catalyst was found to be much more stable compared to a non-fluorinated porphyrin employed in one-phase organic solvent; (2) recovery and recycling of **2** was possible by straightforward decantation; (3) the lifetime of $^1O_2$ in PFCs (~100 ms) is exceptionally high compared to organic solvents (for instance 54.4 μs in CH_3CN).

Following this work, a range of efficient fluorous biphasic protocols for the oxidation of alkenes [20–24], alcohols [25, 26], or sulfides [20, 27, 28] has been reported through the use of a variety of fluorous ligands such as the diketonate (**4**) [20, 23], the porphyrin (**5**) [21, 27], other nitrogenated macrocycles such as the triazacyclononane (**6**) [22, 24], the bipyridine (**7**) [25], or carboxylates such as (**8**) [26]. It is important to bear in mind that the key parameter to achieve efficient recycling is the partition coefficient of the catalyst between the PFC and the HC phase. In the case of highly polar cationic transition metal complexes, the use of highly fluorophilic counter-anions such as fluorous tetraphenylborates [29], or coordinating fluorous carboxylates [22, 26, 30], proved to be essential to ensure a high solubility and partitioning into the PFCs.

Particular attention has been paid to the recovery of costly chiral catalysts [7, 31]. Among major achievements in FBC, one can cite the development by Pozzi et al. of highly effective and recyclable chiral(salen)manganese epoxidation catalysts based on highly fluorophilic ligands such as (**9**) (Scheme 1) [30]. Using these crowded second-generation (salen)manganese complexes, highly effective and enantioselective epoxidations for a range of alkenes were achieved, with the best results obtained with a biphasic *n*-perfluorooctane/CH₃CN solvent system at 100 °C (Scheme 5) in the presence of PhIO/pyridine *N*-oxide (PNO). Under such conditions, good to excellent epoxide yields (68–98%,) and *ee*s (50–92%) were obtained with values being close to those reported by Janda and Reger using gel-type resins. Moreover, the catalysts were recycled twice without significant drop in enantioselectivities and epoxide yields. Only after the fourth run was a decrease observed.

Other enantioselective reactions carried out in fluorous media include the reduction of ketones using Ir catalysts bearing chiral fluorous diamines/diimines such as **10** (Scheme 1) [32], and the asymmetric formation of carbon–carbon bond catalyzed by various chiral fluorous Ti-BINOL (**11**) [33], Pd-BINAP (**12**) [34], or Rh-prolinate [35] complexes. In most cases yields and enantioselectivities

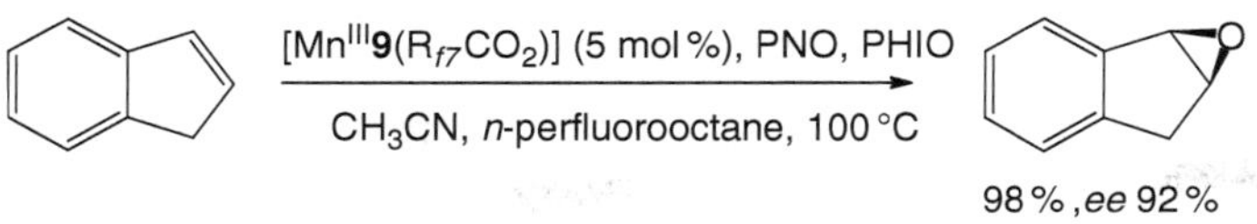

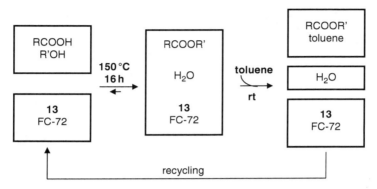

Scheme 5 Asymmetric epoxidation of indene under fluorous biphasic conditions (adapted from [30])

Scheme 6 Esterification under FBC conditions catalyzed by **13** (adapted from [37])

matching those observed for the non-fluorous analogs were obtained while liquid/liquid phase-separation led, in general, to efficient catalyst or ligand recovery.

Metal-catalyzed carbon–carbon bond forming reactions [36] and esterification [37]/transesterification [38] reactions have also been transposed into the fluorous world. Otera and coworkers have developed a particularly efficient fluorous biphasic esterification methodology catalyzed by the highly fluorophilic fluoroalkyldistannoxane [{Cl(C$_6$F$_{13}$C$_2$H$_4$)$_2$SnOSn(C$_2$H$_4$C$_6$F$_{13}$)$_2$Cl}] (**13**) [37]. Interestingly, the extremely low solubility of water in PFCs liberated during the esterification was exploited to shift the reaction equilibrium towards the ester formation, thus requiring no extra dehydration protocol (Scheme 6). Accordingly, virtually 100% yields of esters were obtained starting from equimolar amounts of acids and alcohols in the presence of 3–5 mol% catalyst. Moreover, quantitative yields were attained in ten runs with only 5% catalyst loss after the tenth run. Overall, almost complete atom efficiency was realized, making this methodology a green esterification process.

Nishikido and coworkers reported on a practical fluorous biphasic process for small and bench-scale esterifications catalyzed by lanthanide(III) bis(perfluorooctansulfonyl)amide [39]. They developed a continuous-flow apparatus consisting of a reactor with a mechanical stirrer and a separate decanter where phase-separation occurs (Scheme 7). Using a semi-industrial set up (reactor of 500 mL) the acetylation of cyclohexanol by acetic anhydride proceeded successfully for more than 500 h with high TON (~10,000) attained for the catalyst whilst very low leaching of Yb was detected [39]. This catalytic system also proved to be highly efficient for the Baeyer–Villiger oxidations and Friedel–Craft acylations [40]. Overall, this work

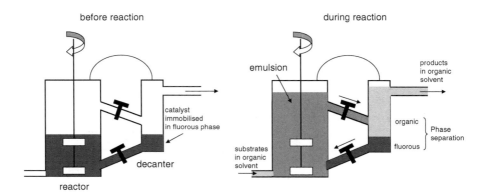

Scheme 7 Continuous-flow reaction model based on FBC (adapted from [39])

Scheme 8 Structure of a second generation Grubbs catalyst supported on fluorous polyacrylate (adapted from [41])

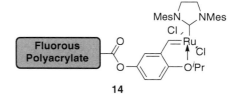

clearly demonstrated the potential of such continuous-flow/fluorous biphase systems for industrial applications.

An interesting contribution was made by Yao and Zhang who immobilized the second generation Grubbs catalyst on a fluorous polyacrylate polymer bearing a bidentate isopropoxystyrene ligand (Scheme 8) [41]. The supported catalyst (**14**) was typically employed in homogeneous PhCF$_3$/CH$_2$Cl$_2$ solvent system at 50 °C, and was shown to exhibit an excellent activity in ring-closing olefin metathesis for a range of substrates. Efficient recovery of **14** under an activated form was achieved by extraction with FC-72.

Conceptually interesting results were reported by Gladysz and Corrêa da Costa in 2006 showing that fluorous solvents can be used in a different manner, i.e., not for catalyst/product separation purposes, but to activate the catalyst [42]. This new concept of fluorous phase-transfer activation of catalysts was applied to alkene metathesis. The principle of catalyst activation is depicted in Scheme 9. Fluorous second generation Grubbs catalysts, such as the ruthenium complex (**15**), have been prepared. The fluorous phosphines possessing a high thermodynamic affinity for PFCs, it was hypothesized that they could be effectively scavenged into the fluorous phase, thus providing an activated non-fluorophilic catalyst partitioning mostly in the organic phase containing the reactant. They showed that the reaction rates were indeed much faster when the reactions were conducted in the presence of a fluorous solvent. For instance, the cyclopentene (Scheme 9) was formed in 74% yield after 2 h in the presence of perfluoro(2-butyltetrahydrofuran), whilst only 6% was obtained using DCM only.

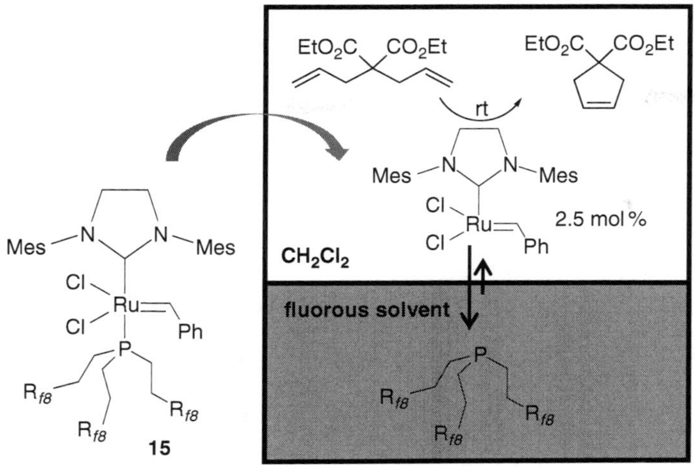

Scheme 9 Principle of fluorous phase-transfer catalyst activation applied to alkene metathesis (adapted from [42])

Scheme 10 Structure of the fluorous ionic liquid (**16**)

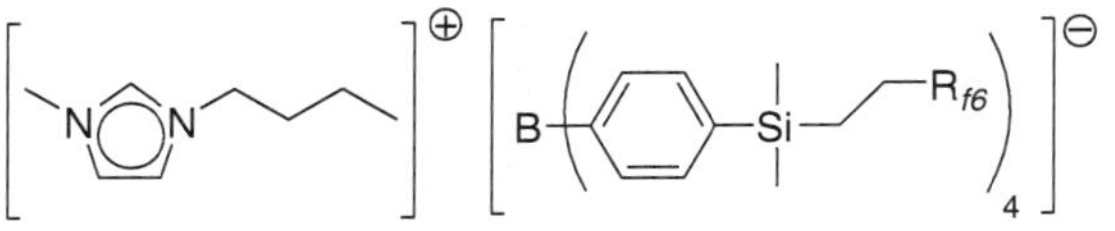

The fluorous ionic liquid (**16**) (Scheme 10) developed by Deelman and co-workers exhibits characteristics of both fluorous solvents and ionic liquids [43]. Of particular interest is the increased solubility of apolar substrates in **16** compared to classical ionic liquids, and also that it forms a monophasic system with organic solvent at high temperature but a biphasic one at room temperature. This fluorous ionic liquid was successfully applied to the hydrosilylation of 1-octene using a fluorous Wilkinson's catalyst. Excellent catalytic activity was observed over 15 reaction cycles, with 92% retention of catalytic activity per cycle (Scheme 10).

It is important to emphasize that fluorous ligands have been widely employed to carry out catalytic processes in supercritical fluids, in particular in supercritical carbon dioxide (scCO$_2$). Indeed, fluorous compounds in general exhibit excellent solubilities in such apolar media. For instance, using the fluorous phosphines (**17**) (Scheme 1), Leitner and coworkers conducted very efficient rhodium-catalyzed hydroformylations in scCO$_2$ [44], while the bipyridine (**7**) has been employed by Matyjazewsky and co-workers to perform copper-catalyzed Atom Transfer Radical Polymerization (ATRP) reactions in scCO$_2$ [45]. Biocatalysis can be conducted in scCO$_2$ as well, employing fluorous enzymes solubilized thanks to the hydrophobic ion pairing (HIP) strategy (see Sect. 2.2) developed by Thomas and co-workers [46].

Finally, fluorous catalysis can be coupled to microwave heating, thus speeding-up the reaction kinetics while facilitating the elimination/recovery of the catalyst [47].

2.2 Organocatalysis and Biocatalysis

In the last 10 years, organocatalysis has experienced a tremendous development [48]. It is thus not surprising that a variety of fluorous catalysts have been designed in order to facilitate their purification/recycling, particularly when considering sophisticated chiral organocatalysts. Three chapters in this book will describe in detail the achievements in the area of fluorous organocatalysis. The fluorous TEMPO (**18**) (Scheme 2) developed by Pozzi and coworkers proved to be an excellent catalyst for oxidation of alcohols to carbonyl compounds in combination with various stoichiometric oxidants such as NaOCl, trichloroisocyanuric acid (TCCA), and [bis(acetoxy)iodo]benzene (BAIB) [49]. Reactions were typically conducted in DCM while direct extraction by perfluoro-(1,3-dimethyl)cyclohexane at the end of the reaction led to excellent catalyst recovery (95–100%). Phase-transfer catalysts such as the phosphonium salt (**19**) developed by Gladysz and co-workers [50], and the crown ether (**20**) developed by Fish, Pozzi and co-workers [51], have been employed in multiphasic conditions in conjunction with PFCs. The phosphonium salt (**19**) (10 mol%) efficiently catalyzes the halide substitution of fluorous alkyl halides ($R_{f8}(CH_2)_nX$; $n = 2, 3$; X=Cl, Br, I) under biphasic PFC/water conditions with the aqueous phase containing a sodium or potassium salt. These results demonstrated, for the first time, that ionic displacement can occur in the least polar existing solvents. The conversion of 1-bromooctane into its iodo derivative was efficiently catalyzed by **20** (2 mol%) in a multiphasic catalytic system in which the bromooctane is the organic phase, while the catalyst is solubilized in a PFC in the presence of solid KI. The separation of the product from the catalyst is easily carried out through a liquid/liquid decantation to isolate the product, followed by a filtration to eliminate the inorganic salts and recover **20** in a pure form.

PFCs have also been employed as solvents for enzymatic reactions with the objective of developing straightforward purification procedures for biocatalytic processes [52]. The liquid/liquid system PFC/hexane was considered as a particularly attractive solvent combination for two main reasons: (1) enzymes are known to retain their activity in hexane at rather high temperature (40–60 °C); (2) such solvent combination is biphasic below 20–25 °C while becoming monophasic at temperatures slightly above 20–25 °C, thus providing smooth conditions for enzymatic reactions. In 2002, O'Hagan and co-workers reported for the first time that biocatalysis could be conducted under FBC protocol [53]. Lipase-mediated transesterifications between esters and long chain polyfluorinated alcohols (Scheme 11) were performed at 40 °C in a one phase hexane/perfluorohexane solvent system while the enzyme was used as heterogeneous catalyst. After ~50% conversion the reactions were stopped, the enzyme removed by filtration, and the

Fluorous Catalysis: From the Origin to Recent Advances 163

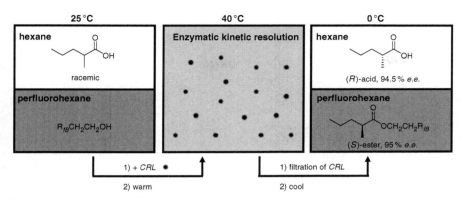

Scheme 11 Kinetic resolution of a racemic acid catalyzed by *Candida Rugosa Lipase* (CRL) using a fluorous biphasic system (adapted from [53])

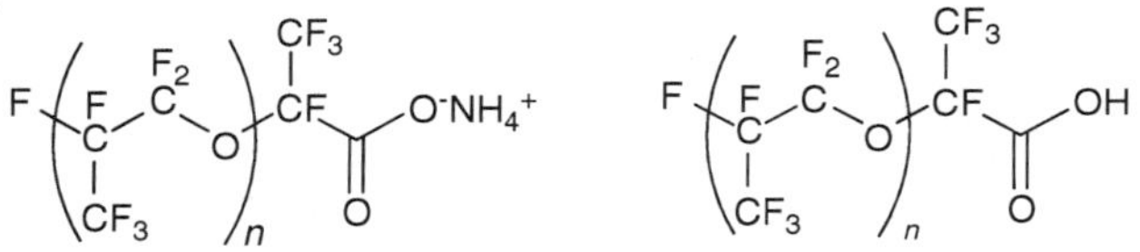

KDP 4606
(mw ~ 1400; n ~ 9)

Krytox 157 FSL
(mw ~ 2500; n ~ 17)

Scheme 12 Structures of the fluorous carboxylate KDP 4606 and carboxylic acid Krytox 157 FSL

reaction mixture cooled to 0 °C leading to efficient partitioning of the enantioenriched unreacted (R)-acid in the organic phase, while the fluorous (S)-ester product was mostly recovered in the fluorous phase. Interestingly, an improved stereoselectivity has been noticed compared to previous studies conducted in hexane only. Moreover, it has been shown that this fluorous biphasic process can be applied on a preparative scale. Starting from 6 g of the racemic acid, 1.97 g (66%) of the (S)-acid (96% e.e.) was isolated after hydrolysis of the corresponding ester, whilst 1.81 g (57%) of unreacted (R)-acid (79% e.e.) was obtained.

In 2007, Thomas and co-workers demonstrated that enzymatic reactions could proceed efficiently in fully homogeneous fluorous conditions i.e., with a soluble enzyme [54]. Proteins such as cytochrome c or α-chymotrypsin (CMT) were very effectively transferred from an aqueous phase into perfluoromethylcyclohexane through HIP with the highly fluorophilic perfluoropolyether carboxylate KDP and carboxylic acids Krytox (Scheme 12), affording concentrated limpid solutions of the proteins in the fluorous solvent (up to 20 mg of cytochrome c per mL). Transesterifications were then conducted at 40 °C in a fully homogeneous hexane/PFMC solution. At the end of the reaction, cooling the reaction mixture led to phase separation, allowing straightforward recycling of the enzyme, which was shown to retain its activity during four reaction cycles in fluorous biphasic conditions.

3 Solid/Liquid Phase Separation

3.1 Fluorous Solid-Phase Extraction

Because PFCs are rather expensive solvents and potentially environmentally persistent chemicals, much effort has been devoted to the development of recovery procedures avoiding the use of PFCs. Fluorous molecules can be separated from non-fluorous compounds by Fluorous Solid-Phase Extraction (FSPE) using fluorous silica-gel as the stationary phase which is typically silica grafted with perfluorooctylethyl chains [55, 56]. Interestingly, molecules bearing only one perfluoroalkyl chain, called light fluorous compounds, are very efficiently separated from the non-fluorous molecules on the fluorous support. The non-fluorous compounds are typically eluted first using an MeOH/H$_2$O (10/1) mixture as eluent, while the more retained fluorous compounds are eluted using more fluorophilic solvents such as pure MeOH. This approach has the advantage of employing catalysts modified by fewer fluorous tags, thus exhibiting enhanced solubility in conventional organic solvent and reactivity similar to the native catalyst. Moreover, as stated above, no PFC is required for the separation/recovery step. Using this approach Stuart and coworkers succeeded in recycling three times a fluorous tagged complex, [Ni{F$_{13}$C$_6$C(O)CHC(O)C$_6$F$_{13}$}$_2$], employed as Lewis acid catalyst in the reaction between β-diketones and ethyl cyanoformate [57]. Curran and coworkers prepared light fluorous analogs of the first- and second-generation Grubbs–Hoveyda metathesis catalysts (**21**) and (**22**) (Scheme 13) [58, 59]. These catalysts react similarly to the original one with reactions typically run in CH$_2$Cl$_2$. More importantly, these catalysts have been recovered in high yield and purity after straightforward elution of the reaction mixture on fluorous silica and reused up to five times with no decrease in activity.

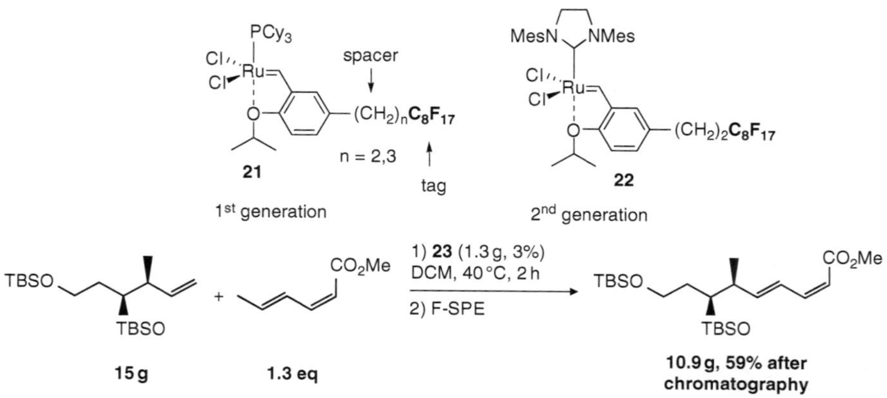

Scheme 13 Application of "light fluorous" Grubbs-Hoveyda metathesis catalysts to the synthesis of an intermediate of the dictyostatin anticancer agent (adapted from [59])

Other chiral catalysts/ligands have been recovered/recycled through FSPE protocols [60]. Among them a number of light fluorous chiral organocatalysts such as the imidazolidine (**23**) [61], the diamide (**24**) [62], or the prolinol (**25**) [63] (Scheme 2) have recently been reported. For instance, the fluorous prolinol (**25**) catalyzes the reduction of ketones in high yields (72–99%) and enantioselectivities (71–95%) through the intermediate of a fluorous oxazaborolidine which is generated in situ [63].

3.2 Supported Fluorous Phase Catalysis

The adsorption of fluorous catalysts onto fluorous solid supports such as fluorous silica or Teflon represents another very attractive strategy for the efficient and straightforward recycling of catalysts under PFC-free conditions. Bannwarth et al. [64, 65] and Biffis et al. [66] first independently reported that fluorous catalysts which are readily adsorbed from organic solutions onto the surface of fluorous silica can then be used as conventional supported catalyst with the advantage of easy separation from the reaction product by filtration or decantation. This methodology was successfully applied to Suzuki–Miyaura and Sonogashira cross-coupling reactions (Scheme 14) catalyzed by Pd complexes such as **26** bearing the fluorous phosphine (**27**) (Scheme 1) [64, 65], and silylation of alcohols catalyzed by fluorous Rh-carboxylate complexes [65]. The supported catalysts were shown to retain a high reactivity while being efficiently recycled with low metal leaching (~2%). Nishikido et al. showed that fluorous Lewis acids such as $Sn[N(SO_2C_8F_{17})_2]$ and Sc $[C(SO_2C_4F_9)_3]_3$ immobilized on fluorous silica were effective catalysts for Baeyer–Villiger and Diels–Alder reactions carried out in water [67]. The catalysts (5 mol% loading), recovered by simple filtration, were shown to retain their activity

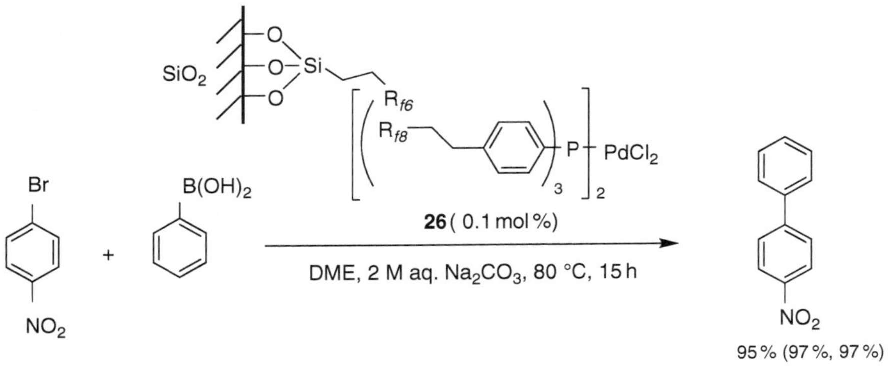

Scheme 14 Suzuki reaction catalyzed by a fluorous catalyst adsorbed on fluorous silica. The numbers in parenthesis are the yields of the second and third runs with recycled catalyst (adapted from [64])

over four reaction cycles. Curran and coworkers showed that the fluorous silica supported catalysts can be employed in a slightly different manner [58]. Indeed, by simply switching the solvent polarity the catalyst can be driven onto or extracted off the fluorous silica gel. For example, the second generation fluorous Grubbs–Hoveyda catalyst (Scheme 13) adsorbed on fluorous silica does not leach from the support when suspended in 80% MeOH/H$_2$O mixture because of its pronounced hydrophobic character. However, the complex was readily extracted from the silica when washed with a good solubilizing solvent such as dichloromethane. Metathesis reactions may then be performed in the following manner. The substrate and silica-supported catalyst (5 mol% loading) was reacted in refluxing DCM in optimal homogeneous conditions. After reaction the DCM was evaporated, MeOH/water was added and the resulting slurry filtered to separate the product from the supported catalyst. The recovered complex was reused for five consecutive runs with no significant decrease in activity.

In their quest to render the catalyst recycling process even more practical, Gladysz et al. discovered that Teflon tape can play the role of the fluorous support, providing a smart an effective means to deliver/recover the catalyst into/from solution with the additional attractive feature of greatly facilitating the handling of catalyst at low loadings [68, 69]. Thus, when a hot (55 °C) homogeneous solution of the fluorous phosphine-Rh complex (**28**) (Scheme 15) in dibutyl ether containing Teflon tape was cooled, adsorption on the tape occurred instead of precipitation. Hydrosilylation of ketones was then conducted using the reaction sequence depicted in Scheme 15. The recovered catalyst (resting state) was reused with the same protocol for three consecutive runs with no decrease in catalytic activity (>96% yields and similar reaction rates) showing that no deactivation and minor metal loss occurred under these conditions. More recently, Gladysz and co-workers have shown that Teflon fibers (commercial name "Gore-Rastex") were more efficient than Teflon tape for catalyst adsorption because of higher surface area [69].

Hope, Stuart et al. compared the catalytic activity and recycling efficiency of a fluorous Wilkinson's catalyst adsorbed on micro- and meso-porous silica or powdered Teflon [70]. Hydrogenations of styrene were run in toluene at 63.5 °C in the

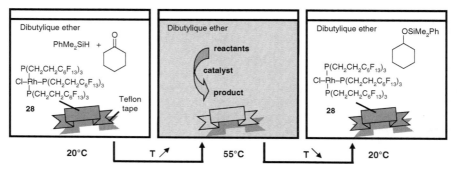

Scheme 15 Delivery/recovery protocol of a hydrosilylation Rh catalyst immobilized on Teflon tape (adapted from [68])

presence of small amounts of perfluoro-1,3-dimethylcyclohexane (~5 vol%). They showed that while Rh leaching was about one order of magnitude lower with the fluorous meso-porous silica (~0.5 ppm/cycle) compared to powdered Teflon powder (~5.95 ppm/cycle), the catalyst deactivation was more pronounced with the silica support, the reaction rates decreasing from 84.9 to 51.8 and 38.1 mmol dm^{-3} h^{-1} in three consecutive runs. The catalyst decomposition was ascribed to possible reaction of the metal with the free hydroxy groups of silica surface. With the powdered Teflon the reaction rates in three consecutive runs were found to be 71.8, 62.7, and 59.4 mmol dm^{-3} h^{-1}.

Interestingly, Stuart et al. developed perfluoroalkylated polystyrene resins for use as insoluble supports for fluorous catalysts [71]. A range of perfluoroalkylated polystyrene beads was synthesized by radical-initiated suspension polymerization of 4-(perfluoro-n-octyl)styrene, styrene and 1,4-divinylbenzene (Scheme 16). The resulting fluorinated polystyrene beads exhibit limited swelling behavior in organic solvents. The selected polymers for the preliminary catalytic studies were those prepared from DVB/R$_{f8}$-styrene/styrene with ratios of 5–10/60/35–30 as they represent the best compromise considering the catalysis efficacy, the fluorine content, and the mechanical stability. The palladium complex (**26**) (Scheme 14) and a fluorous Wilkinson's complex were adsorbed on the polymer beads and the catalytic efficiency of the resulting heterogeneous supports assessed in Suzuki–Miyaura and styrene hydrogenation reactions. For both reactions the catalysts supported on polystyrene proved to be more active and much more stable compared to the same catalysts adsorbed on fluorous silica. The results were particularly impressive for the hydrogenations for which no decrease of the reaction rates was observed in three runs while an Rh leaching as low as 0.04–0.12% was measured after each run. Overall, this makes such fluorous polystyrenes very attractive supports for catalyst recycling procedures.

3.3 Precipitation

The fluorous catalysts, because of their limited solubility in conventional organic solvents at ambient temperature or below, can eventually be separated from the reaction mixture by precipitation and thus recovered by simple filtration with no need for a fluorous solvent or support. Moreover, fluorous molecules were shown to

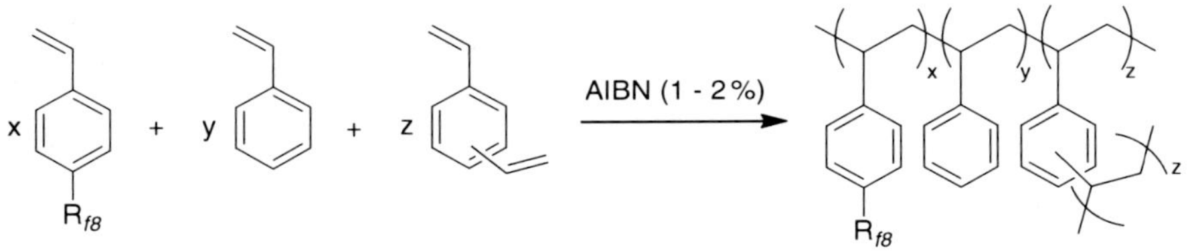

Scheme 16 Synthesis of a fluorous polystyrene (adapted from [71])

exhibit large temperature-dependent solubilities which allowed reactions to be run under homogeneous conditions at elevated temperatures. The solubility-based thermomorphic properties were effectively exploited by Sheldon et al. for the recycling of the perfluoroheptadecan-9-one ($C_8F_{17}C(O)C_8F_{17}$) which was employed as epoxidation organocatalyst in the presence of H_2O_2 [72]. Reactions were conducted in refluxing EtOAc/1,2-dichloroethane mixture, while cooling of the reaction mixture in an ice bath led to catalyst crystallization which was recovered by filtration in 92% yield. The rather general character of the solid/liquid thermomorphic properties of fluorous compounds in organic solvents, in particular their large temperature-dependant solubility, was clearly established by Gladysz et al. [73–75]. For example, the fluorous phosphine, $P[(CH_2)_2(CF_2)_7CF_3]_3$, exhibited a 600-fold solubility increase in *n*-octane between $-20\ °C$ and $80\ °C$. Such phosphines were employed as catalysts in addition reactions between alcohols and methyl propiolate under homogeneous conditions in *n*-octane at $65\ °C$ and recovered in 90–95% yields by precipitation at $-30\ °C$ (Scheme 17) [73, 75]. Yamomoto and coworkers also demonstrated the utility of this concept in direct amide condensation, acylation, and acetylation reactions, as well as the Mukaiyama aldol reaction catalyzed by fluorous acids [76, 77]. Fluorous transition metal catalysts have also been recycled through the PFC-free thermomorphic mode. For example, Contel, Vincent, Fish et al. reported on the oxidation of benzylic alcohols catalyzed by fluorous copper(I)–carboxylate complexes in the presence of TEMPO/O_2 [26].

Fluorous compounds, because of their lower solubility and amphiphilic character, can favor the formation of emulsions in HC/water mixtures. Reiser et al. have prepared the fluorous "click" TEMPO derivative (**29**) (Scheme 2) and have shown that this compound promotes the formation of an emulsion in a DCM/water mixture [78]. Under heterogeneous DCM/water conditions **29** (0.2 mol%) proved to be highly reactive and selective for the oxidation of alcohols, while its recycling was achieved by simple filtration followed by washings. No loss of activity was noticed after four runs.

The chiral fluorous PT catalyst (**30**) (Scheme 2) developed by Maruoka and co-workers was also employed as a heterogeneous catalyst (3 mol%) in a toluene/

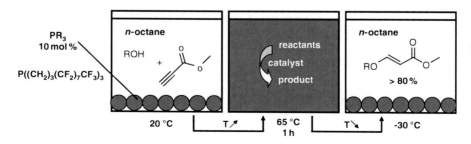

Scheme 17 Phosphine-catalyzed addition reaction using a PFC-free thermomorphic mode (adapted from [73])

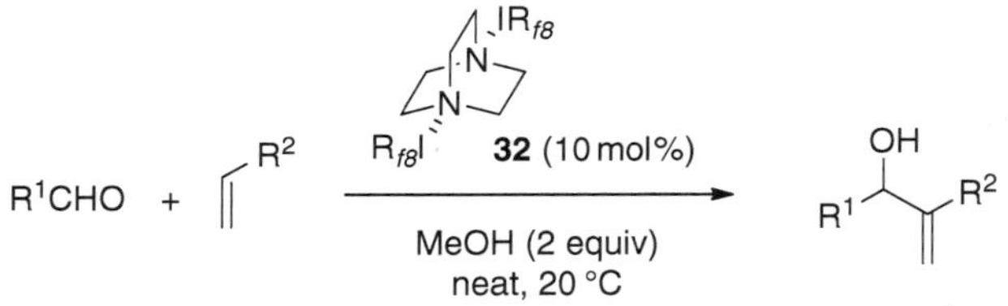

R^1CHO + ⟋⟍R^2 →[R_{f8}I—N⟨N⟩—IR_{f8} **32** (10 mol%) / MeOH (2 equiv) / neat, 20 °C] R^1—CH(OH)—C(=CH$_2$)R^2

Scheme 18 Morita-Bayis-Hillman reactions catalyzed by a supramolecular fluorous DABCO (adapted from [81])

aqueous KOH biphasic system [79]. These conditions were successfully applied to enantioselective α-alkylations of glycine derivatives. Straightforward recovery of the catalyst was achieved by extraction of the reaction mixture with FC-72. The catalyst has been reused twice with no observed loss of yields and enantioselectivities.

Recently Legros and co-workers demonstrated that non-covalent fluorous tagging could provide a simple but powerful strategy for catalyst recycling. The concept was first demonstrated for alcohol acylations catalyzed by the fluorous DMAP salt (**31**) (Scheme 2) [80]. Circumventing the synthetic and reactivity drawbacks that can be associated with covalent modification of DMAP, **31** was prepared in >98% yield by simply mixing the DMAP and the carboxylic acid followed by recrystallization. Alcohol acylation reactions were conducted in neat conditions at 25 °C in the presence of **31** (10 mol%) and acetic or isobutyric anhydrides. Using these conditions a range of esters were prepared in good to excellent yields. At the end of the reaction, after evaporation of the generated carboxylic acid, the catalyst is precipitated by addition of toluene or pentane and quantitatively recovered (>99%) by filtration. The same group also reported that halogen bonding was a valuable supramolecular interaction to tag catalysts [81]. The fluorous DABCO derivative (**32**) (Scheme 2) is prepared by reacting $C_8F_{17}I$ with DABCO in DCM. Under these conditions **32** immediately precipitates and is isolated in 94% yield by filtration. The fluorous DABCO proved to be a good catalyst in Morita–Bayis-Hillman reactions between aldehydes and Michael acceptors using the conditions shown in Scheme 18. Importantly, addition of DCM and acetonitrile at the end of the reaction leads to the precipitation of pure **32** which is separated by filtration with a typical 90% recovery.

4 Conclusion

As described above, since the discovery of the FBC concept there has been a creative and continuous effort in the quest for innovative, efficient, and practical procedures for the recovery of catalysts. Of course these methodologies complement or compete with a range of methods employing catalysts immobilized on insoluble or soluble polymers, grafted onto dendrimers, using other tagging approaches, or employing liquid phases such as water, ionic liquids, or supercritical CO_2 as the non-organic phase. It is thus very important when facing such a large array of potential methods, in order to assess whether a methodology is the most suitable for a given reaction, to compare as accurately as possible a set of parameters including: easiness of preparation and handling, cost, reactivity, and stability of the catalyst in the reaction conditions, as well as the efficiency of the recovery procedure.

Acknowledgements I wish to warmly thank all my coworkers who actively participated to the fluorous chemistry program. The CNRS, the University of Bordeaux, the Région Aquitaine and the Ministère de l'Enseignement Supérieur et de la Recherche are gratefully acknowledge for their financial support.

References

1. Gladysz JA (ed) (2002) Recoverable catalysts and reagents. Special issue of Chem Rev 102:3215–3892
2. Gladysz JA (2001) Recoverable catalysts. Ultimate goals, criteria of evaluation, and the green chemistry interface. Pure Appl Chem 73:1319–1324
3. Horváth IT, Rábai J (1994) Facile catalyst separation without water: fluorous biphase hydroformylation of olefins. Science 266:72–75
4. Gladysz JA, Curran DP, Horváth IT (eds) (2004) Handbook of fluorous chemistry. Wiley-VCH, Weinheim, Germany
5. Fish RH (1999) Fluorous biphasic catalysis: a new paradigm for the separation of homogeneous catalysts from their reaction substrates and products. Chem Eur J 5:1677–1680
6. Dobbs AP, Kimberley MR (2002) Fluorous phase chemistry: a new industrial technology. J Fluor Chem 118:3–17
7. Pozzi G, Cavazzini M, Quici S, Maillard D, Sinou D (2002) Chiral fluorous catalysts: synthesis and purposes. J Mol Catal A Chem 182–183:455–461
8. Keim W, Vogt M, Wasserscheid P, Drießen-Hölscher B (1999) Perfluorinated polyethers for the immobilisation of homogeneous nickel catalysts. J Mol Catal A Chem 139:171–175
9. Horváth IT, Kiss G, Cook RA, Bond JE, Stevens PA, Rábai J, Mozeleski E (1998) Molecular engineering in homogeneous catalysis: one-phase catalysis coupled with biphase catalyst separation. The fluorous-soluble $HRh(CO)\{P[CH_2CH_2(CF_2)_5CF_3]_3\}$ hydroformylation system. J Am Chem Soc 120:3133–3143
10. Jiao H, Le Stang S, Soós T, Meier R, Kowski K, Rademacher P, Jafarpour L, Hamard J-B, Nolan SP, Gladysz JA (2002) How to insulate a reactive site from a perfluoroalkyl group: photoelectron spectroscopy, calorimetric, and computational studies of long-range electronic effects in fluorous phosphines $P((CH_2)_m(CF_2)_7CF_3)_3$. J Am Chem Soc 124:1516–1523
11. Juliette JJJ, Rutherford D, Horváth IT, Gladysz JA (1999) Transition metal catalysis in fluorous media: practical application of a new immobilization principle to rhodium-catalyzed hydroborations of alkenes and alkynes. J Am Chem Soc 121:2696–2704
12. Richter B, Spek AL, van Koten G, Deelman B-J (2000) Fluorous versions of Wilkinson's catalyst. Hydrogenation of 1-alkenes and recycling by fluorous biphasic separation. J Am Chem Soc 122:3945–3951
13. de Wolf E, Speets EA, Deelman B-J, van Koten G (2001) Recycling of rhodium-based hydrosilylation catalysts; a fluorous approach. Organometallics 20:3686–3690

14. Hope EG, Stuart AM (2004) Synthesis of perfluoralkylated phosphines. In: Gladysz JA, Curran DP, Horváth IT (eds) Handbook of fluorous chemistry. Wiley-VCH, Weinheim, Germany, pp 247–256
15. Gudmunsen D, Hope EG, Paige DR, Stuart A (2009) Coordination chemistry of perfluoroalkylated phosphorus(III) ligands. J Fluor Chem 130:942–950
16. Crich D, Zhou Y (2004) Recyclable oxidation reagents. In: Gladysz JA, Curran DP, Horváth IT (eds) Handbook of fluorous chemistry. Wiley-VCH, Weinheim, Germany, pp 202–222
17. Pozzi G, Quici S (2004) Fluorous nitrogen ligands for oxidation reactions. In: Gladysz JA, Curran DP, Horváth IT (eds) Handbook of fluorous chemistry. Wiley-VCH, Weinheim, Germany, pp 290–298
18. Vincent J-M, Lastécouères D, Contel M, Laguna M, Fish RH (2004) Synthesis of fluorous nitrogen ligands and their metal complexes as precatalysts for applications in alkane, alkenes, and alcohol oxidations, and atom transfer radical reactions. In: Gladysz JA, Curran DP, Horváth IT (eds) Handbook of fluorous chemistry. Wiley-VCH, Weinheim, Germany, pp 298–305
19. DiMagno SG, Dussault PH, Schultz JA (1996) Fluorous biphasic singlet oxygenation with a perfluoroalkylated photosensitizer. J Am Chem Soc 118:5312–5313
20. Klement I, Lütjens H, Knochel P (1997) Transition metal catalyzed oxidations in perfluorinated solvents. Angew Chem Int Ed Engl 36:1454–1457
21. Pozzi G, Montanari F, Quici S (1997) Cobalt tetraarylporphyrin-catalysed epoxidation of alkenes by dioxygen and 2-methylpropanal under fluorous biphasic conditions. Chem Commun 69–70
22. Vincent J-M, Rabion A, Yachandra VK, Fish RH (1997) Fluorous biphasic catalysis: complexation of 1,4,7-[C$_8$F$_{17}$(CH$_2$)$_3$]$_3$-1,4,7-triazacyclononane with [M(C$_8$F$_{17}$(CH$_2$)$_2$CO$_2$)$_2$] (M=Mn, Co) to provide perfluoroheptane-soluble catalysts for alkane and alkene functionalization in the presence of t-BuOOH and O$_2$. Angew Chem Int Ed Engl 36:2346–2349
23. Betzemeier B, Lhermitte F, Knochel P (1998) Wacker oxidation of alkenes using a fluorous biphasic system. A mild preparation of polyfunctional ketones. Tetrahedron Lett 39:6667–6670
24. Vincent J-M, Rabion A, Yachandra VK, Fish RH (2001) Fluorous biphasic catalysis. 2. Synthesis of a variety of fluoroponytailed ligands and complexation with [Mn(C$_8$F$_{17}$(CH$_2$)$_2$CO$_2$)$_2$] and its Co^{2+} analog: demonstration of a perfluoroheptane soluble precatalyst for alkane and alkene functionalization in the presence of t-butyl hydroperoxide and oxygen gas. Can J Chem 79:888–895
25. Ragagnin G, Betzemeier B, Quici S, Knochel P (2002) Copper-catalysed aerobic oxidation of alcohols using fluorous biphasic catalysis. Tetrahedron 58:3985–3991
26. Contel M, Villuendas PR, Fernández-Gallardo J, Alonso PJ, Vincent JM, Fish RH (2005) Fluorocarbon soluble copper(II) carboxylate complexes with nonfluoroponytailed nitrogen ligands as precatalysts for the oxidation of alkenols and alcohols under fluorous biphasic or thermomorphic modes: structural and mechanistic aspects. Inorg Chem 44:9771–9778
27. Colonna S, Gaggero N, Montanari F, Pozzi G, Quici S (2001) Fluorous biphasic catalytic oxidation of sulfides by molecular oxygen/2,2-dimethylpropanal. Eur J Org Chem 181–186
28. Cavazzini M, Pozzi G, Quici S, Shepperson I (2003) Fluorous biphasic oxidation of sulfides catalysed by (salen)manganese(III) complexes. J Mol Catal A Chem 204–205:433–441
29. van den Broeke J, de Wolf E, Deelman B-J, van Koten G (2003) Enhanced hydrogenation activity and recycling of cationic rhodium diphosphine complexes through the use of highly fluorous and weakly-coordinating tetraphenylborate anions. Adv Synth Catal 345:625–635
30. Cavazzini M, Manfredi A, Montanari F, Quici S, Pozzi G (2001) Asymmetric epoxidation of alkenes in fluorinated media catalyzed by second-generation fluorous chiral (salen)manganese complexes. Eur J Org Chem 4639–4649
31. Sinou D (2004) Enantioselective catalysis: biphasic conditions. In: Gladysz JA, Curran DP, Horváth IT (eds) Handbook of fluorous chemistry. Wiley-VCH, Weinheim, Germany, pp 306–315

32. Maillard D, Pozzi G, Quici S, Sinou D (2002) Asymmetric hydrogen transfer reduction of ketones using chiral perfluorinated diimines and diamines. Tetrahedron 58:3971–3976
33. Nakamura Y, Takeuchi S, Okumura K, Ohgo Y, Curran DP (2002) Recyclable fluorous chiral ligands and catalysts: asymmetric addition of diethylzinc to aromatic aldehydes catalyzed by fluorous BINOL-Ti complexes. Tetrahedron 58:3963–3969
34. Nakamura Y, Takeuchi S, Zhang S, Okumura K, Ohgo Y (2002) Preparation of a fluorous chiral BINAP and application to an asymmetric Heck reaction. Tetrahedron Lett 43:3053–3056
35. Biffis A, Braga M, Cadamuro S, Tubaro C, Basato M (2005) A fluorous chiral dirhodium(II) complex as a recyclable asymmetric catalyst. Org Lett 7:1841–1844
36. Schneider S, Tzschucke CC, Bannwarth W (2004) Metal-catalyzed carbon-carbon bond forming reactions in fluorous biphasic system. In: Gladysz JA, Curran DP, Horváth IT (eds) Handbook of fluorous chemistry. Wiley-VCH, Weinheim, Germany, pp 257–271
37. Xiang J, Orita A, Otera J (2002) Fluorous biphasic esterification directed towards ultimate atom efficiency. Angew Chem Int Ed Engl 41:4117–4119
38. Xiang J, Toyoshima S, Orita A, Otera J (2001) A practical and green chemical process: fluoroalkyldistannoxane-catalyzed biphasic transesterification. Angew Chem Int Ed Engl 40:3670–3672
39. Yoshida A, Hao X, Nishikido J (2003) Development of the continuous-flow reaction system based on the Lewis acid-catalysed reactions in a fluorous biphasic system. Green Chem 5:554–557
40. Hao X, Yoshida A, Nishikido J (2006) Metal bis(perfluorooctansulfonyl)amides as highly efficient Lewis acid catalysts for fluorous biphase organic reactions. J Fluor Chem 127:193–199
41. Yao Q, Zhang Y (2004) Poly(fluoroalkyl acrylate)-bound ruthenium carbene complex: a fluorous and recyclable catalyst for ring-closing olefin metathesis. J Am Chem Soc 126:74–75
42. Corrêa da Costa R, Gladysz JA (2006) Fluorous phase-transfer activation of catalysts: application of a new rate-enhancement strategy to alkene metathesis. Chem Commun 2619–2621
43. Van den Broeke J, Winter F, Deelman B-J, van Koten G (2002) A highly fluorous room-temperature ionic liquid exhibiting fluorous biphasic behavior and its use in catalyst recycling. Org Lett 4:3851–3854
44. Koch D, Leitner W (1998) Rhodium-catalyzed hydroformylation in supercritical carbon dioxide. J Am Chem Soc 120:13398–13404
45. Xia J, Johnson T, Gaynor SG, Matyjaszewski K, DeSimone J (1999) Atom transfer radical polymerization in supercritical carbon dioxide. Macromolecules 32:4802–4805
46. Benaissi K, Poliakoff M, Thomas NR (2010) Solubilisation of α-chimotrypsin by hydrophobic ion pairing in fluorous systems and supercritical carbon dioxide and demonstration of efficient enzyme recycling. Green Chem 12:54–59
47. Olofsson K, Larhed M (2004) Microwave-assisted fluorous chemistry. In: Gladysz JA, Curran DP, Horváth IT (eds) Handbook of fluorous chemistry. Wiley-VCH, Weinheim, Germany, pp 359–365
48. List B (ed) (2007) Organocatalysis. Special issue of Chem Rev 107:5413–5883
49. Holczknecht O, Cavazzini M, Quici S, Shepperson I, Pozzi G (2005) Selective oxidation of alcohols to carbonyl compounds mediated by fluorous-tagged TEMPO radicals. Adv Synth Catal 347:677–688
50. Consorti CS, Jurisch M, Gladysz JA (2007) Ionic transformations in extremely nonpolar fluorous media: phase transfer catalysis of halide substitution reactions. Org Lett 9:2309–2312
51. Pozzi G, Quici S, Fish RH (2008) Perfluorocarbon soluble crown ethers as phase transfer catalysts. Adv Synth Catal 350:2425–2436
52. Hobbs HR, Thomas NR (2007) Biocatalysis in supercritical fluids, in fluorous solvents, and under solvent-free conditions. Chem Rev 107:2786–2820
53. Beier P, O'Hagan D (2002) Enantiomeric partitioning using fluorous biphase methodology for lipase-mediated (trans)esterifications. Chem Commun 1680–1681

54. Hobbs HR, Kirke HM, Poliakoff M, Thomas NR (2007) Homogeneous biocatalysis in both fluorous biphasic and supercritical carbon dioxide systems. Angew Chem Int Ed Engl 46:7860–7863
55. Krafft M-P, Jeannaux F, Le Blanc M, Riess JG, Berthod A (1988) Highly fluorinated stationary phases for analysis of perfluorinated solutes by reverse-phase high-performance liquid chromatography. Anal Chem 60:1969–1972
56. Curran DP (2001) Fluorous reverse phase silica gel. A new tool for preparative separations in synthetic organic and organofluorine chemistry. Synlett 1488–1496
57. Croxtall B, Hope EG, Stuart AM (2003) Separation, recovery and recycling of a fluorous-tagged nickel catalyst using fluorous solid-phase extraction. Chem Commun 2430–2431
58. Matsugi M, Curran DP (2005) Synthesis, reaction, and recycle of light fluorous Grubbs-Hoveyda catalyst for alkene metathesis. J Org Chem 70:1636–1642
59. Moura-Letts G, Curran DP (2007) Selective synthesis of (2Z,4E)-dienyl esters by ene-diene cross metathesis. Org Lett 9:5–8
60. Takeuchi S, Nakamura Y (2004) Enantioselective catalysis in non-biphasic conditions. In: Gladysz JA, Curran DP, Horváth IT (eds) Handbook of fluorous chemistry. Wiley-VCH, Weinheim, Germany, pp 316–322
61. Chu Q, Zhang W, Curran DP (2006) A recyclable fluorous organocatalyst for Diels-Alder reactions. Tetrahedron Lett 47:9287–9290
62. Malkov AV, Figlus M, Stoncius S, Kocovsky P (2007) Organocatalysis with a fluorous tag: asymmetric reduction of imines with trichlorosilane catalyzed by amino acid-derived formamides. J Org Chem 72:1315–1325
63. Dalicsek Z, Pollreisz F, Gömöry A, Soos T (2005) Recoverable fluorous CBS methodology for asymmetric reduction of ketones. Org Lett 7:3243–3246
64. Tzschucke CC, Markert C, Glatz H, Bannwarth W (2002) Fluorous biphasic catalysis without perfluorinated solvents: application to Pd-mediated Suzuki and Sonogashira couplings. Angew Chem Int Ed Engl 41:4500–4503
65. Tzschucke CC, Bannwarth W (2004) Fluorous-silica-supported perfluoro-tagged palladium complexes catalyze Suzuki couplings in water. Helv Chim Acta 87:2882–2889
66. Biffis A, Zecca M, Basato M (2003) A green protocol for the silylation of alcohols using bonded fluorous phase catalysis. Green Chem 5:170–173
67. Yamazaki O, Hao X, Yoshida A, Nishikido J (2003) Fluorous reverse-phase silica gel-supported Lewis acids as recyclable catalysts in water. Tetrahedron Lett 44:8791–8795
68. Dinh LV, Gladysz JA (2005) "Catalyst on a tape"-teflon: a new delivery and recovery method for homogeneous fluorous catalysts. Angew Chem Int Ed Engl 44:4095–4097
69. Siedel FO, Gladysz JA (2008) Catalysis of intramolecular Morita-Baylis-Hillman and Raulut-Currier reactions by fluorous phosphines; facile recovery by liquid/solid organic/fluorous biphase protocols involving precipitation, Teflon® Tape, and Gore-Rastex® fiber. Adv Synth Catal 350:2443–2449
70. Hope EG, Sherrington J, Stuart AM (2006) Supported fluorous phase catalysis on PTFE, fluoroalkylated micro- and meso-porous silica. Adv Synth Catal 348:1635–1639
71. Audic N, Dyer PW, Hope EG, Stuart AM, Suhard S (2010) Insoluble perfluoroalkylated polymers: new solid supports for supported fluorous phase catalysis. Adv Synth Catal 352:2241–2250
72. van Vliet MCA, Arends IWCE, Sheldon RA (1999) Perfluoroheptadecan-9-one: a selective and reusable catalyst for epoxidations with hydrogen peroxide. Chem Commun 263–264
73. Wende M, Meier R, Gladysz JA (2001) Fluorous catalysis without fluorous solvents: a friendlier catalyst recovery/recycling protocol based upon thermomorphic properties and liquid/solid phase separation. J Am Chem Soc 123:11490–11491
74. Rocaboy C, Gladysz JA (2002) Highly active thermomorphic fluorous palladacycle catalyst precursors for the Heck reaction; evidence for a palladium nanoparticle pathway. Org Lett 4:1993–1996

75. Wende M, Gladysz JA (2003) Fluorous catalysis under homogeneous conditions without fluorous solvents: a "greener" catalyst recycling protocol based upon temperature-dependent solubilities and liquid/solid phase separation. J Am Chem Soc 125:5861–5872
76. Ishihara K, Kondo S, Yamamoto H (2001) 3,5-Bis(perfluorodecyl)phenylboronic acid as an easily recyclable direct amide condensation catalyst. Synlett 1371–1374
77. Ishihara K, Hasegawa K, Yamamoto H (2002) A fluorous super Brønsted acid catalyst: application to fluorous catalysis without fluorous solvents. Synlett 1299–1301
78. Gheorghe A, Chinnusamy T, Cuevas-Yanez E, Hilgers P, Reiser O (2008) Combination of perfluoroalkyl and triazole moieties: a new recovery strategy for TEMPO. Org Lett 10:4171–4174
79. Shirakawa S, Tanaka Y, Maruoka K (2004) Development of a recyclable fluorous chiral phase-transfer catalyst: application to the catalytic asymmetric synthesis of α-amino acids. Org Lett 6:1429–1431
80. Vuluga D, Legros J, Crousse B, Bonnet-Delpon D (2010) Fluorous 4-N,N-dimethylaminopyridine (DMAP) salts as simple recyclable acylation catalysts. Chem Eur J 16:1776–1779
81. Dordonne S, Crousse B, Bonnet-Delpon D, Legros J (2011) Fluorous tagging of DABCO through halogen bonding: recyclable catalyst for the Morita-Baylis-Hillman reaction. Chem Commun 47:5855–5857

Fluorous Organocatalysis

Wei Zhang

Abstract Organocatalysis is a new and fast growing research area, especially for asymmetric synthesis. Compared to metal catalysis and biocatalysis, organocatalysis has a number of unique features such metal-free, mild reaction conditions, novel mode of activations, and good structural amenability. Fluorous organocatalysis provides an efficient way for catalyst recovery. In addition to phase tag separation, fluorous chain can also be used to modify the reactivity and selectivity in the design of new catalysts.

Keywords Asymmetric synthesis · Fluorous · Green chemistry · Organocatalysis

Contents

1 Introduction 176
2 Fluorous Organocatalysis 176
  2.1 Fluorous Secondary Amines 177
  2.2 Fluorous Cinchona Alkaloids 181
  2.3 Fluorous Thioureas 183
  2.4 Fluorous 1, 2-Diamine-Derivatives 183
  2.5 Other Fluorous Catalysts 185
3 Summary 188
References 189

Abbreviations

Boc   *tert*-Butyloxycarbonyl
BTF   Benzotrifluoride

W. Zhang (✉)
Department of Chemistry, University of Massachusetts Boston, 100 Morrissey Boulevard, Boston, MA 02125, USA
e-mail: wei2.zhang@umb.edu

CBS	Corey-Bakshi-Shibata
CBZ	Carboxybenzyl
DABCO	1,4-Diazabicyclo[2.2.2]octane
DBU	Diazabicyclo[5.4.0]undec-7-ene
DIBAL	Diisobutylaluminum hydride
DMAP	4-N,N-Dimethylaminopyridine
F-LLE	Fluorous liquid–liquid extraction
F-SPE	Fluorous solid-phase extraction
HFIP	Hexafluoroisopropanol
NCS	N-Chlorosuccinimide

1 Introduction

Transition metal catalysis, organocatalysis, and biocatalysis are three major catalytic methods for organic synthesis. Organocatalysis is a newcomer but has generated great momentum. It has advantages of new activation mechanisms, low cost, less sensitive to moisture and oxygen, flexible for structural modification, and good availability from nature sources [1, 2]. Metal-free catalysis is also attractive in the development of green reaction processes [3]. However, organocatalysis usually requires 10–20 mol% catalyst loading, which is significantly higher than metal catalysis. The development of more efficient organocatalysts and catalyst recovery are two major challenges in this field.

Polymers [4], ionic liquids [5], fluorous tags [6], and other supporting materials have been developed to address the separation issue in organocatalysis. Different from solid-supported organocatalysis, fluorous organocatalysis is conducted in solution. The homogeneous environment could provide more consistent and predictable results. In the development of fluorous biphasic catalysis, heavy fluorous tags with high fluorine content have been used to make fluorous ligands [7–10]. Most organocatalysts described in this chapter have light fluorous tags such as C_8F_{17}. They can be recovered by fluorous solid-phase extraction (F-SPE) [11] or by precipitation/filtration from fluorophobic solvents such as water and hexane. Light fluorous organocatalysis avoids using environmentally persistent perfluorinated solvents in both reaction and separation processes. Since the first examples of fluorous organocatalysis reported in 2001 [12–14], over a dozen new fluorous organocatalysts have been developed and applied for a wide range of organic transformations, especially for asymmetric reactions [15].

2 Fluorous Organocatalysis

Attachment of a perfluorinated chain at an appropriate position of normal organocatalysts is a general approach to make fluorous organocatalysts. The perfluorinated chain has a strong electron-withdrawing power. It could have positive or negative impact on the attached catalysts. Depending on the nature of the catalyst, the electronic effect could be shielded by introducing a $-(CH_2)_n-$ (usually $n = 2$) spacer, or could be utilized to modify the reactivity and selectivity of the catalyst without using the spacer.

A number of fluorous Lewis acids have been developed for organocatalysis. It has been reviewed by a recent paper [16] and is not be covered by this chapter.

2.1 Fluorous Secondary Amines

Natural and synthetic secondary amines have been extensively studied in the development of new organocatalysts [17]. The best known example is L-proline which has been used to promote transformations such as aldol, Mannich, cycloaddition, and amination reactions. Synthetically modified secondary amines have even broader utilities. Some representative fluorous secondary amine organocatalysts are shown in Scheme 1. The fluorous tag in catalysts 1 and 2 is attached

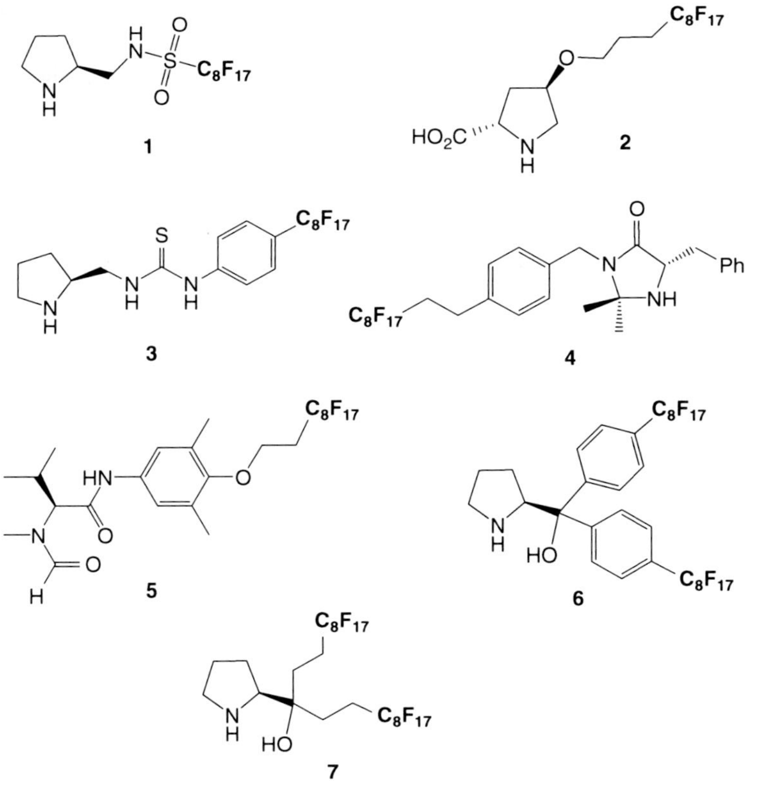

Scheme 1 Fluorous secondary amine organocatalysts

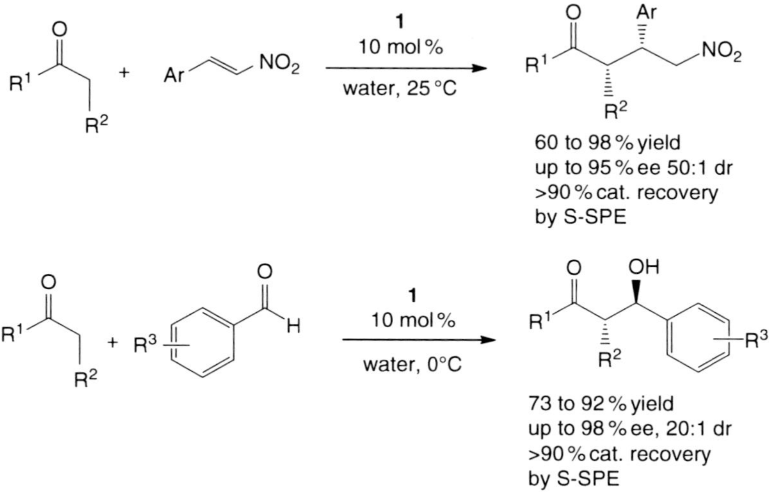

Scheme 2

to a pyrrolidine ring through a sulfonamide or ether functional group. Catalyst **3** has pyrrolidine and thiourea bifuncational groups. Its utility is discussed in Sect. 2.3. Catalyst **4** is a fluorous version of MacMillan imidazolidinone catalyst. Catalyst **5** is an acyclic amide. Pyrolinol derivatives **6** and **7**, so far, are the most studied fluorous catalysts.

Catalyst **1** was prepared by the condensation of CBZ-protected (*S*)-2-aminomethylpyrrolidine with perfluorobutanesulfonyl fluoride followed by CBZ deprotection. It has been used for Michael and aldol reactions (Scheme 2) [18, 19]. The reactions were carried out using 10 mol% of the catalyst in water to give the condensation products in excellent yield and enantioselectivity. C_4F_9 is a relatively short fluorous tag, but the attached catalyst can still be efficiently separated by F-SPE and reused six to seven times without a significant loss of catalytic reactivity.

Catalyst **2** was prepared through a multi-step synthesis using 4-hydroxy-L-proline as a starting material [20]. This catalyst has been successfully used for aldol and allylic oxidation reactions (Scheme 3). The aldol reaction was conducted in benzotrifluoride (BTF) at room temperature for 2 days. The allylic oxidation reaction was carried out in reflux hexafluoroisopropanol (HFIP) for 2 h. In both cases, the catalyst was recovered by precipitating out from petroleum ether followed by filtration.

A fluorous version of MacMillan imidazolidinone catalyst **4** was prepared by the cycloaddition of a fluorous amino amide with acetone [21]. This catalyst has been compared side-by-side with the original MacMillan catalyst for a Diels–Alder reaction (Scheme 4). The fluorous catalyst **4** has slightly better diastereo- and enantioselectivity than the non-fluorous counterpart. More importantly, it was recovered by F-SPE in 84% yield with 99% purity, whereas the non-fluorous imidazolidinone catalyst was recovered by acid–base work up in 65% yield and only 74% purity.

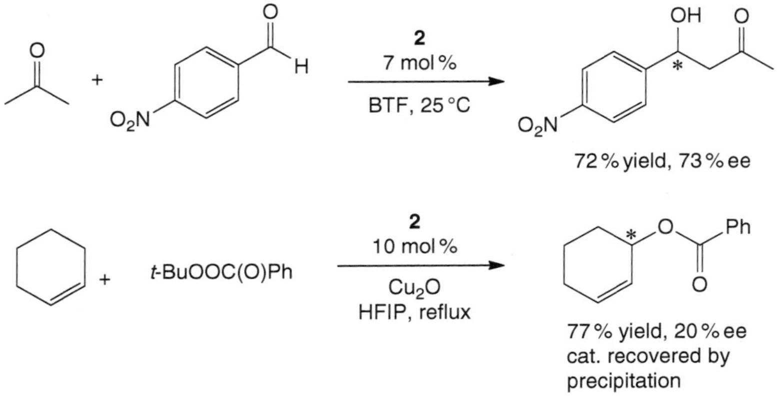

Scheme 3

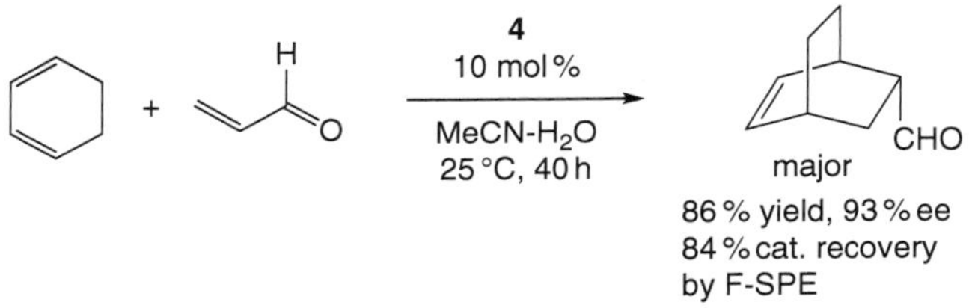

Scheme 4

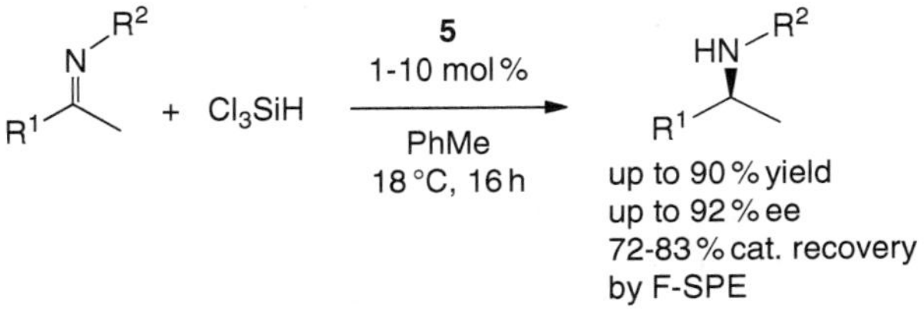

Scheme 5

Fluorous organocatalyst **5** was prepared through amide coupling reaction of a chiral amino acid with a fluorous aniline [22]. It has been tested for catalytic asymmetric reduction of imines with trichlorosilane (Scheme 5). The catalyst was separated by F-SPE and tested for recycling four times.

There are several reports on the development of diphenylpyrolinol **6** for diethylzinc addition and Corey-Bakshi-Shibata (CBS) reduction of benzaldehydes and acetophenones. This catalyst could be prepared by the Grignard reaction of N-protected proline methyl ester with 4-bromophenylmagnesium bromide followed by the formation of cyclic carbamate, Ullmann-type reaction to introduce two C_8F_{17} tags, and the reduction of carbamate with diisobutylaluminum hydride (DIBAL) [23].

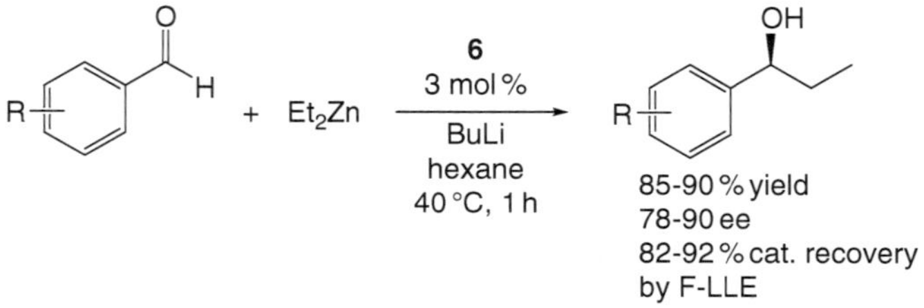

Scheme 6

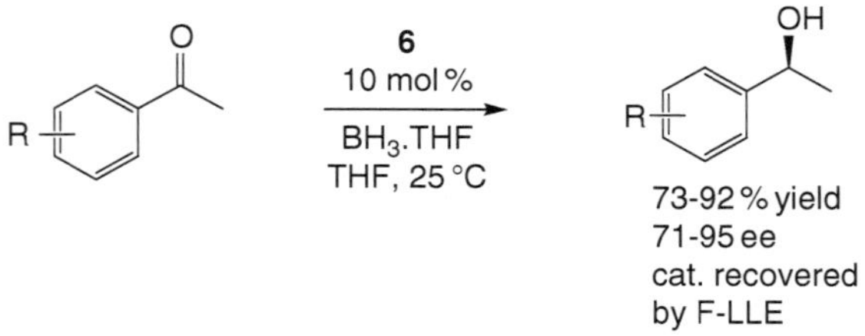

Ar-CHO + Et$_2$Zn →[6, 3 mol%; BuLi; hexane; 40 °C, 1 h]→ Ar-CH(OH)Et

85–90% yield
78–90 ee
82–92% cat. recovery by F-LLE

Scheme 7

Ar-COCH$_3$ →[6, 10 mol%; BH$_3$·THF; THF, 25 °C]→ Ar-CH(OH)CH$_3$

73–92% yield
71–95 ee
cat. recovered by F-LLE

The diethylzinc addition of benzaldehydes catalyzed by **6** gave good yields with high enantioselectivity (Scheme 6) [23]. The catalyst was recovered by fluorous liquid-liquid extraction (F-LLE) with perfluorohexane (FC-72). It could be recycled nine times with 82–92% catalyst recoveries.

Pyrolinol **6** has been used as a precatalyst to form a complex with BH$_3$·THF for CBS reduction of acetophenones (Scheme 7) [24]. Catalyst recycling could be achieved by F-SPE. Different reaction solvent and precatalyst recovery methods have been developed. A complex of catalyst **6** and BH$_3$·THF in HFE-7500 (C$_3$F$_7$CF(OC$_2$H$_5$)CF(CF$_3$)$_2$) for the reduction of acetophenone gave the product in quantitative conversion and 92% ee [25]. The product was extracted by organic solvents such as MeCN and DMSO. The catalyst retained in the HFE-7500 layer was recycled six times. The conversion was greater than 99% in each run, but the ee was gradually reduced from 92% to 45%. Another precatalyst recovery method has been introduced using 1:1 MeOH-H$_2$O for aqueous-organic extraction in a U-tube continuous extraction setup [26]. It was considered a practical and cost-saving device which is useful for other kind of fluorous reactions. In addition to diphenylpyrolinol **6**, the dialkylpyrolinol **7** has also been used as a precatalyst for CBS reductions [27]. It was recovered by cooling the reaction mixture and then filtering out the precipitate.

Diphenylpyrolinol silyl ether **6a** has been used for Michael additions (Scheme 8) [28]. The reactions were conducted in BTF at room temperature to give good product yield and high diastereo- and enantioselectivity. The catalyst in was recovered by F-SPE and recycled six times without significantly loss of activity.

Pyrolinols **6** and **7** have been used for catalytic epoxidation of α,β-unsaturated ketones (Scheme 9) [29]. The reaction catalyzed by pyrolinol **7** gave the product

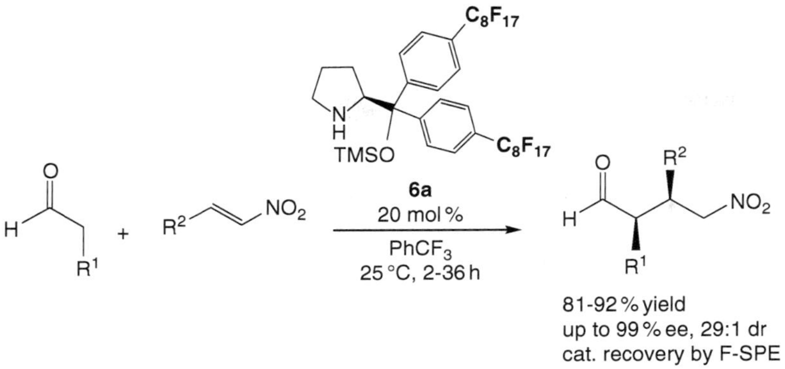

Fluorous Organocatalysis

Scheme 8

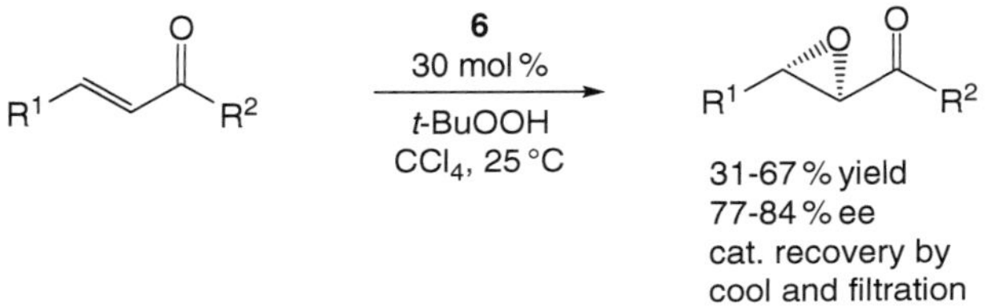

Scheme 9

81-92% yield
up to 99% ee, 29:1 dr
cat. recovery by F-SPE

31-67% yield
77-84% ee
cat. recovery by
cool and filtration

with low (10%) enantioselectivity. Much better results were obtained from the reactions using **6** as a catalyst. The catalyst recovered by cooling the reaction mixture and filtration could be recycled four times and still gave good yield and ee.

2.2 Fluorous Cinchona Alkaloids

Natural product cinchona alkaloids are an important class of organocatalysts [30]. Many modified cinchona-based catalysts have been developed to improve catalytic efficiency and enantioselectivity for reactions such as conjugated addition, amination, and cyanosilation [31]. So far only two fluorous cinchona catalysts **8** and **9** have been reported (Scheme 10). Catalyst **8** was one of the first fluorous organocatalysts reported in 2001 [13]. It was prepared by free radical addition of a fluorous alkyl chain to the vinyl group of a cinchona. This catalyst was tested for Diels–Alder reactions (Scheme 11). The reaction was conducted in fluorinated solvent BTF. Since the product was insoluble in BTF, it was easily separated from the reaction mixture by filtration. After comparison of the results obtained from the non-fluorous cinchona catalysts, it was concluded that the fluorous tag has

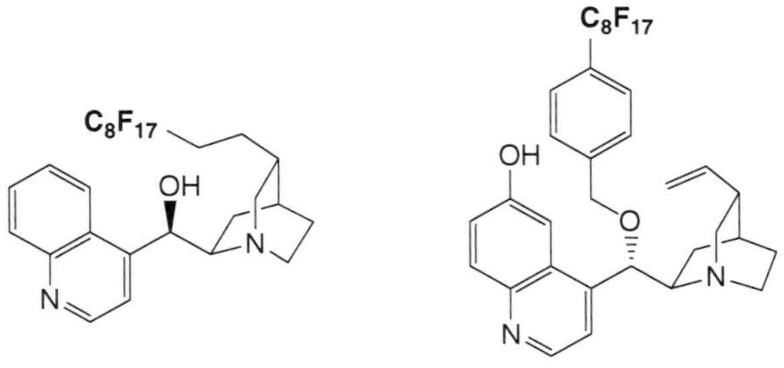

Scheme 10 Fluorous cinchona catalysts

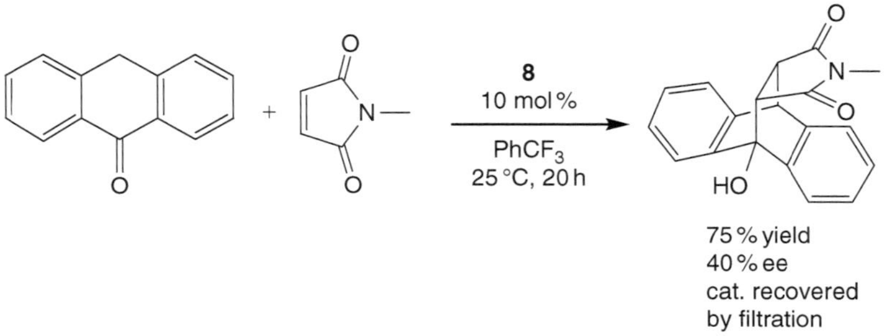

Scheme 11

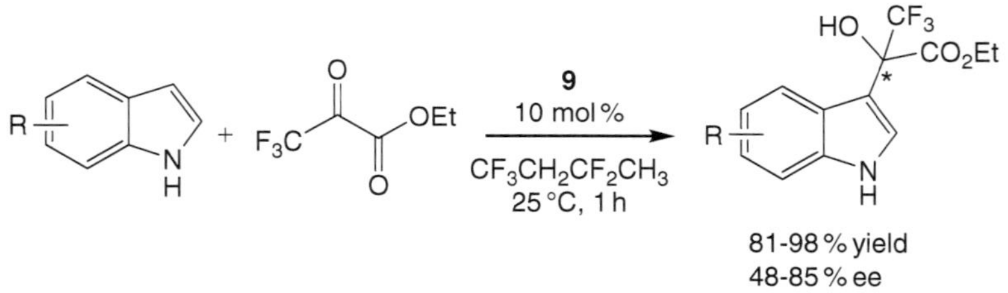

Scheme 12

a significant impact on the enantioselectivity of the catalyst, and the fluorinated solvent BTF could accelerate the reaction.

Catalyst **9** was prepared by introducing the fluorous benzyl group to the hydroxy group of the cinchona [32]. It has been tested for Friedel–Crafts reactions (Scheme 12). The reactions were conducted in $CF_3CH_2CF_2CH_3$ which is considered an environmentally benign solvent. Even though the catalyst has a fluorous tag, it was not utilized for fluorous separation. Instead, products were purified by flash chromatography and no catalyst recycling was reported in the paper.

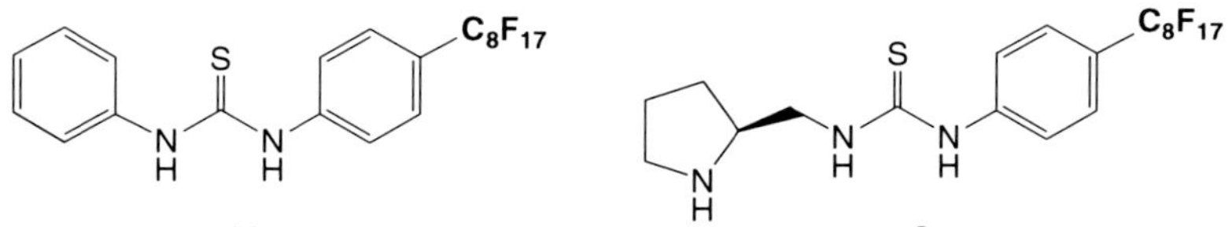

Fluorous Organocatalysis

Scheme 13 Fluorous thiourea catalysts

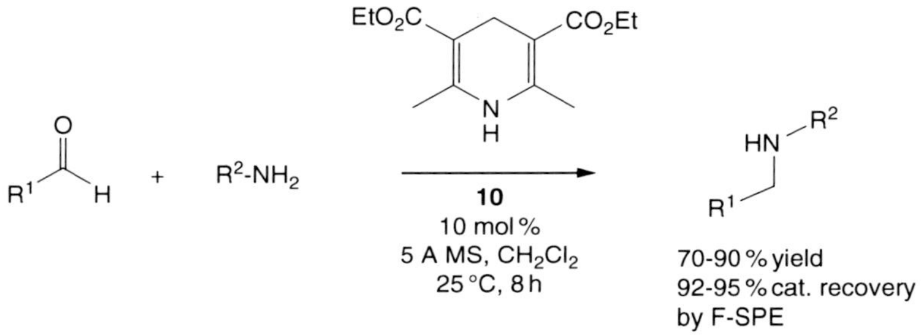

Scheme 14

2.3 Fluorous Thioureas

Thioureas are an important class of organocatalyst which partially relies on hydrogen-bonding interaction between the thiourea and the substrates for activation [33]. Two fluorous thiourea catalysts are shown in Scheme 13. Catalyst **10** is a non-chiral compound, and catalyst **3** is a chiral and bifunctional compound.

Reductive aminations can be conducted using thiourea **10** as a catalyst and Hantzsch 1,4-dihydropyridine as a reducing agent [34]. The reactions were performed in CH_2Cl_2 at room temperature to afford amine products in good to excellent yields (Scheme 14). The catalyst was isolated by F-SPE in 92–95% yield and recycled three times.

Fluorous (S)-pyrrolidine-thiourea bifunctional catalyst **3** has been developed as a new α-chlorination catalyst for aldehydes [35]. This catalyst was prepared by the condensation of 2-aminomethylpyrrolidine with phenyl chlorothioformate with a fluorous aniline. The asymmetric α-chlorinations were conducted under mild conditions to afford products in excellent yield and enantioselectivity. The catalyst was recovered by F-SPE (Scheme 15).

2.4 Fluorous 1, 2-Diamine-Derivatives

Fluorous organocatalysts **11–13** derivatized from chiral 1,2-diamines are shown in Scheme 16. They have been used for cyclopropanation, aldol, and ethylation reactions.

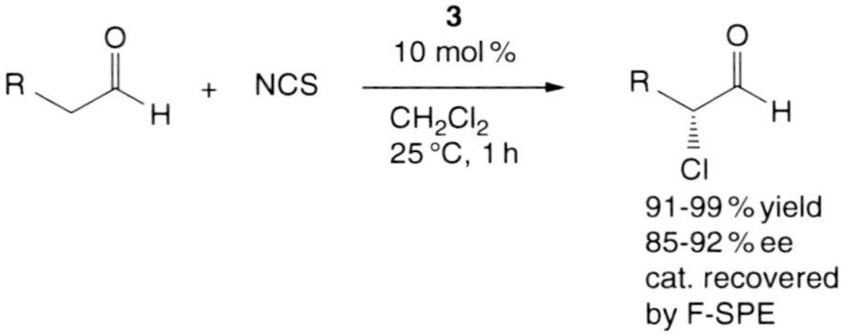

Scheme 15

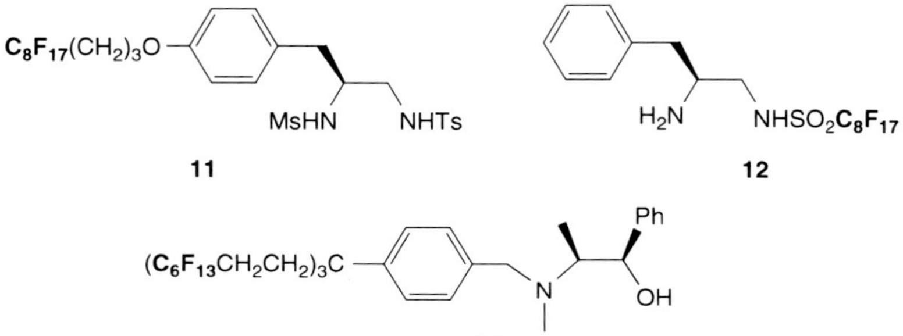

Scheme 16 Fluorous 1,2-diamine derivatized catalysts

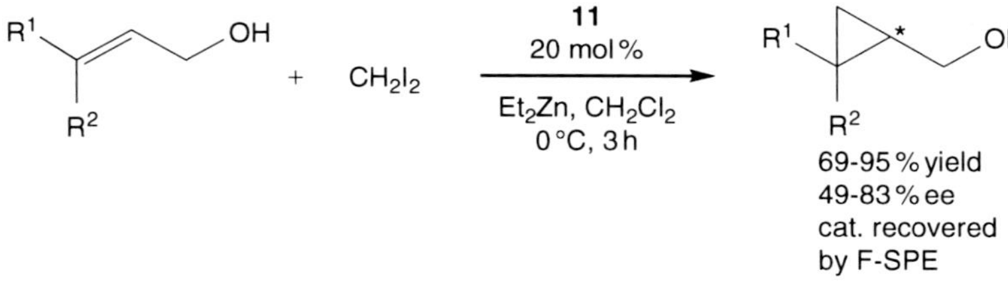

Scheme 17

Catalyst **11** was prepared by a multistep synthesis using tyrosinol as a starting material [36]. It has been used as a catalyst for cyclopropanation of allylic alcohols with CH_2I_2 and Et_2Zn (Scheme 17). The catalyst recovered by F-SPE was reused three times without significant change of reactivity and selectivity.

Fluorous sulfonamide **12** was synthesized by the reaction of a mono-Boc-protected diamine with fluorous sulfonyl fluoride followed by the Boc deprotection [37]. The aldol reaction of cyclic ketones and benzaldehydes catalyzed by **12** afforded a range of products with good diastereo- and enantioselectivity (Scheme 18). The catalyst recovered by F-SPE was reusable.

Catalyst **13** was prepared by N-alkylation of (1R,2S)-ephedrine with fluorous benzeylbromide [12]. This was one of the first fluorous organocatalysts reported in

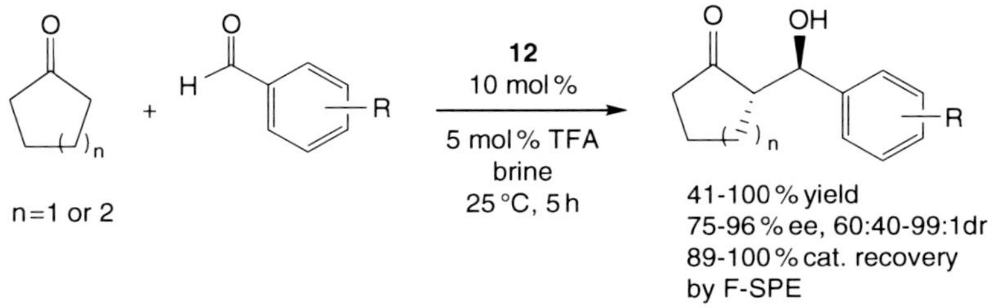

Fluorous Organocatalysis

Scheme 18

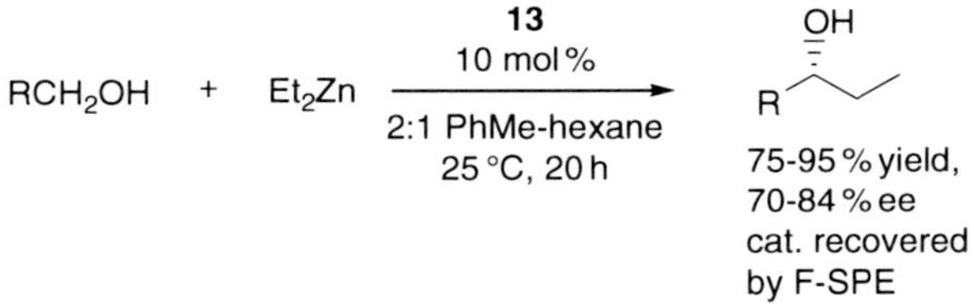

Cyclopentanone/cyclohexanone (n=1 or 2) + PhCHO (R-substituted) → β-hydroxy ketone
- **12** 10 mol%
- 5 mol% TFA
- brine
- 25 °C, 5 h
- 41–100% yield
- 75–96% ee, 60:40–99:1 dr
- 89–100% cat. recovery by F-SPE

Scheme 19

RCH$_2$OH + Et$_2$Zn → R-CH(OH)-Et
- **13** 10 mol%
- 2:1 PhMe-hexane
- 25 °C, 20 h
- 75–95% yield, 70–84% ee
- cat. recovered by F-SPE

2001 for asymmetric ethylation of benzaldehydes (Scheme 19). The catalyst was recovered by F-SPE. Since it has a heavier fluorous tag containing 39 fluorine atoms, FC-72 instead of MeOH was used as a stronger fluorophilic solvent to wash the catalyst out the cartridge. It was found that after ten cycles, catalyst **11** still performed well without significant loss of activity.

2.5 Other Fluorous Catalysts

Organocatalysts described in previous sections are mainly for asymmetric reactions. Catalysts for non-asymmetric synthesis are presented in this section.

Fluorous phenylboronic acid **14** bearing two C$_{10}$F$_{21}$ tags has been developed as a recyclable catalyst for direct amide coupling reactions (Scheme 20) [14]. The reaction of cyclohexanecarboxylic acid and benzylamine was catalyzed by 3 mol% of **14** in azeotropic reflux 1:1:1 o-xylene-toluene-perfluorodecaline to give amide product in high yield. The product in the organic layer was collected by decantation and concentration. The catalyst retained in the perfluorodecaline layer could be reused more than ten times.

Fluorous 2-iodooxybenzoic acid generated by the reaction of Oxone with **15** has been introduced as a recyclable oxidation agent to convert alcohols to ketones or aldehydes (Scheme 21) [38]. The reactions were conducted in MeNO$_2$-H$_2$O at 70°C to afford products in good yield. The catalyst separated by filtration could be reused six times. The catalyst reactivity was reduced after four runs.

Two fluorous analogs of 4-N,N-dimethylaminopyridine (DMAP) **16** and **17** have been developed as acylation catalysts (Scheme 22). Compound **16** has a fluorous tag

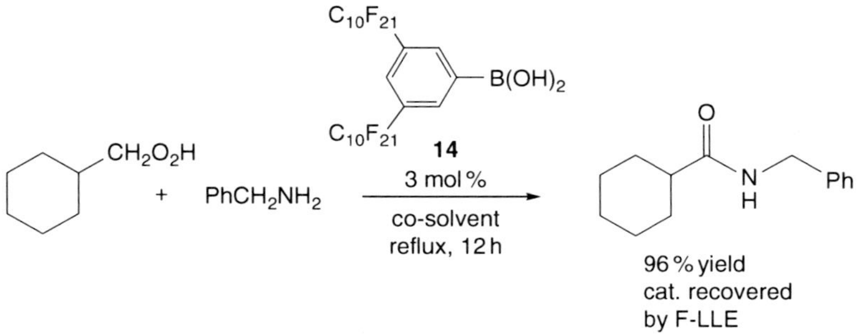

Scheme 20

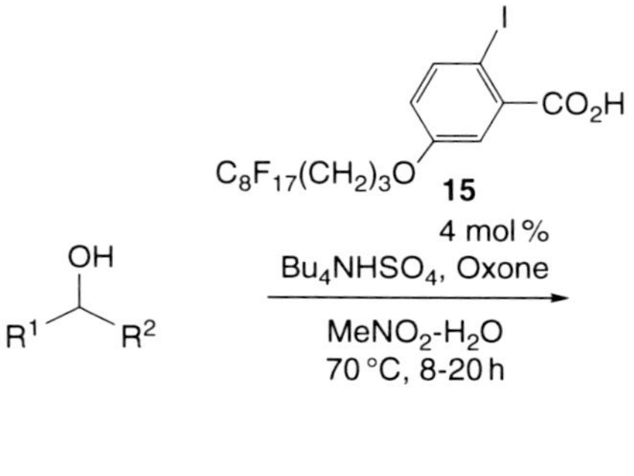

Scheme 21

on the pyrrolidine ring. It was prepared by free radical addition/cyclization of a fluorous alkyl group to an *N,N*-diallyl compound followed by 1,8-diazabicyclo[5.4.0] undec-7-ene (DBU) dehydroiodonation [39]. This catalyst efficiently promoted the acylation of secondary and tertiary alcohols with anhydrides. The catalyst was recovered by addition of hexane followed by filtration. Compound **17** is a fluorous salt which was easily prepared by mixing of a fluorous acidic with DMAP through noncovalent bonds [40]. It was used to catalyze acylation reactions under solvent- and base-free conditions. The catalyst could be fully recovered by precipitation.

Fluorous phosphine **18** has been developed for Michael addition, intramolecular Morita–Baylis–Hillman and Rauhut–Currier reactions (Scheme 23). The Michael addition was conducted in refluxing anhydrous MeCN to give 100% substrate conversion [41]. The intramolecular Morita–Baylis–Hillman reaction under similar conditions gave 85% yield after 24 h [42]. The catalyst was easily separated by cooling the reaction mixture and precipitating the catalyst onto a fluoropolymer support such as Gore–Rastex fiber or Teflon tape.

Recently, using halogen bonding as a new mechanism for organocatalysis has been reported in the literature. The first example was iodooperfluoroalkane **19**-catalyzed Hantzsch ester reduction of quinolones (Scheme 24) [43]. It was believed that the reduction reaction involves an activation of the heterocycle by halogen

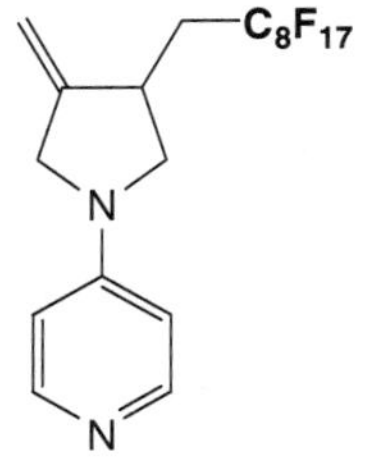

Fluorous Organocatalysis

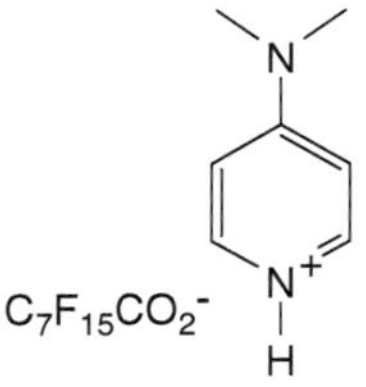

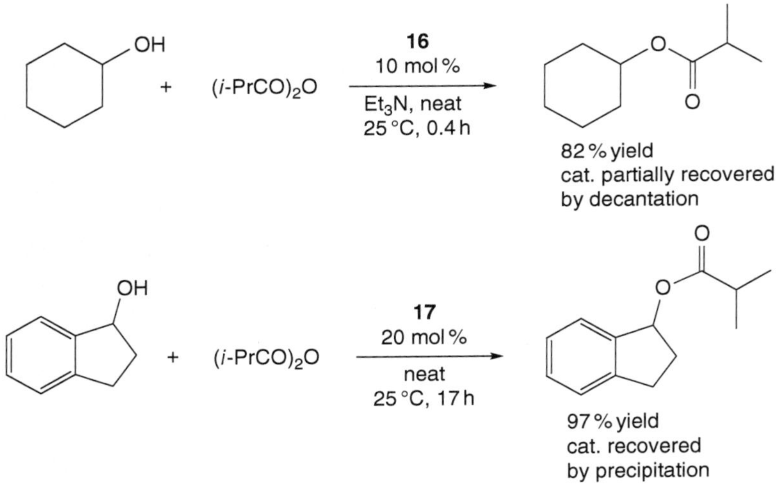

Scheme 22

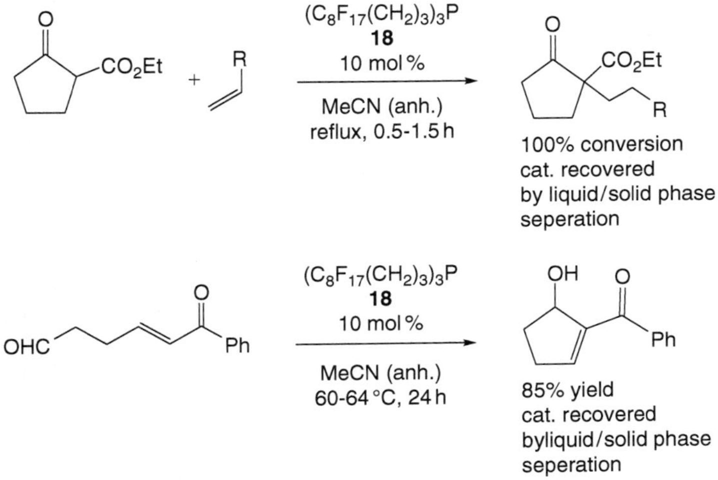

Scheme 23

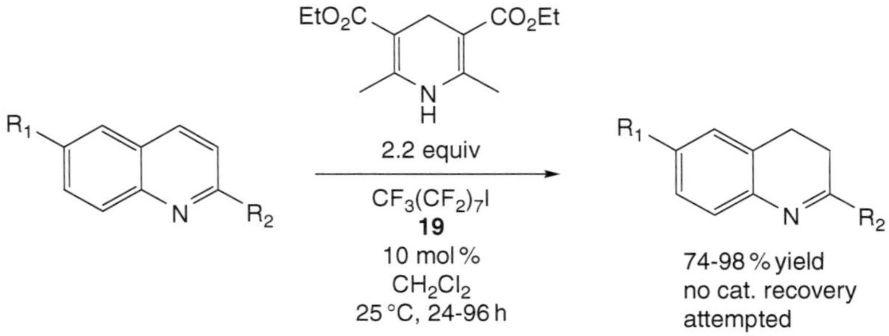

Scheme 24

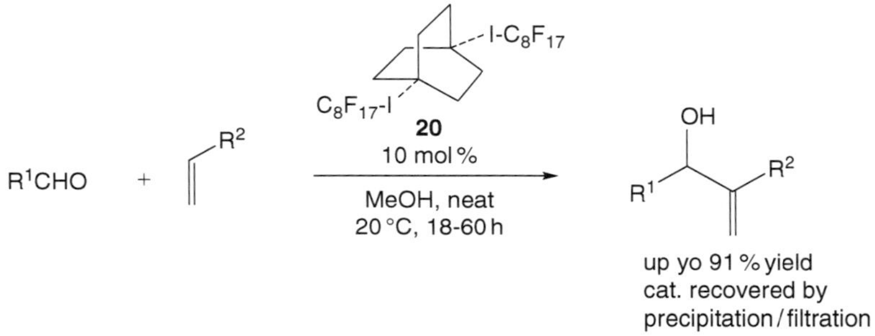

Scheme 25

bonding arising from a Hal–N interaction. In the second case, complex **20** generated from iodooperfluoroalkane and 1,4-diazabicyclo[2.2.2]octane (DABCO) through halogen bonding was used as a recyclable catalyst for the Morita–Baylis–Hillman reaction (Scheme 25) [44]. The catalyst recovered by simple precipitation and filtration could be recycled five times.

3 Summary

Attaching a fluorous tag to organocatalysts is a general strategy in the development of fluorous organocatalysts for recycling. The fluorous catalysts can be recovered by F-SPE or precipitation by cooling or addition of fluorophobic solvents such as hexane and water. The strong electron-withdrawing perfluoroalkyl group can be shielded by the –(CH$_2$)$_n$– spacer, or can be utilized for tuning the property of the catalysts. Since most fluorous catalysts presented in this chapter have relatively low fluorine content, they do not usually require fluorous solvents in the reaction and separation processes. Fluorous organocatalysis is metal-free and the catalysts are recyclable. The green chemistry and low cost advantages make it attractive in the development new catalytic reactions.

Acknowledgments University of Massachusetts Boston, Joseph P. Healey, and Proposal Development Program grants for supporting the green chemistry and fluorous organocatalysis projects are gratefully acknowledged.

References

1. Berkessel A, Groger H (2005) Asymmetric organocatalysis. Wiley-VCH, Weinheim
2. Dalko PI (2007) Enantioselective organocatalysis. Wiley-VCH, Weinheim
3. Anastas PT, Warner JC (1998) Green chemistry: theory and practice. Oxford University Press, Oxford
4. Cozzi F (2006) Immobilization of organic catalysts: when, why, and how. Adv Synth Catal 348:1367
5. de Maria PD (2008) "Nonsolvent" applications of ionic liquids in biotransformations and organocatalysis. Angew Chem Int Ed Engl 47:6960
6. Zhang W, Cai C (2008) New chemical and biological applications of fluorous technologies. Chem Commun 44:5686
7. Horvath IT (1998) Fluorous biphase chemistry. Acc Chem Res 31:641
8. de Wolf E, van Koten G, Deelman BJ (1999) Fluorous phase separation techniques in catalysis. Chem Soc Rev 28:37
9. Barthel-Rosa LP, Gladysz JA (1999) Chemistry in fluorous media: a user's guide to practical considerations in the application of fluorous catalysts and reagents. Coord Chem Rev 190–192:587
10. Pozzi G, Shepperson I (2003) Fluorous chiral ligands for novel catalytic systems. Coord Chem Rev 242:115
11. Zhang W, Curran DP (2006) Synthetic applications of fluorous solid-phase extraction (F-SPE). Tetrahedron 62:11837
12. Nakamura Y, Takeuchi S, Okumura K, Ohgo Y (2001) Enantioselective addition of diethylzinc to aldehydes catalyzed by fluorous beta-aminoalcohols. Tetrahedron 57:5565
13. Fache F, Piva O (2001) New perfluoroalkylated cinchona derivatives: synthesis and use in base-catalysed Diels-Alder reactions. Tetrahedron Lett 42:5655
14. Ishihara K, Kondo S, Yamamoto H (2001) 3,5-Bis(perfluorodecyl)phenylboronic acid as an easily recyclable direct amide condensation catalyst. Synlett 1371
15. Zhang Z, Zhang W (2011) Fluorous organocatalysis - a green approach for asymmetric synthesis. Chim Oggi 28(3):32
16. Cai C, Yi WB, Zhang W, Shen MG, Hong M, Zeng LY (2009) Fluorous Lewis acids and phase transfer catalysts. Mol Divers 13:209
17. Notz W, Tanaka F, Barbas CF (2004) Enamine-based organocatalysis with proline and diamines: the development of direct catalytic asymmetric Aldol, Mannich, Michael, and Diels-Alder reactions. Acc Chem Res 37:80
18. Zu LS, Wang J, Li H, Wang W (2006) A recyclable fluorous (S)-pyrrolidine sulfonamide promoted direct, highly enantioselective Michael addition of ketones and aldehydes to nitroolefins in water. Org Lett 8:3077
19. Zu LS, Xie HX, Li H, Wang J, Wang W (2008) Highly enantioselective aldol reactions catalyzed by a recyclable fluorous (S) pyrrolidine sulfonamide on water. Org Lett 10:1211
20. Fache F, Piva O (2003) Synthesis and applications of the first polyfluorous proline derivative. Tetrahedron Asymmetry 14:139
21. Chu QL, Zhang W, Curran DP (2006) A recyclable fluorous organocatalyst for Diels-Alder reactions. Tetrahedron Lett 47:9287

22. Malkov AV, Figlus M, Stoncius S, Kocovsky P (2007) Organocatalysis with a fluorous tag: asymmetric reduction of imines with trichlorosilane catalyzed by amino acid-derived formamides. J Org Chem 72:1315
23. Park JK, Lee HG, Bolm C, Kim BM (2005) Asymmetric diethyl- and diphenylzinc additions to aldehydes by using a fluorine-containing chiral amino alcohol: a striking temperature effect on the enantioselectivity, a minimal amino alcohol loading, and an efficient recycling of the amino alcohol. Chem Eur J 11:945
24. Dalicsek Z, Pollreisz F, Gomory A, Soos T (2005) Recoverable fluorous CBS methodology for asymmetric reduction of ketones. Org Lett 7:3243
25. Chu QL, Yu MS, Curran DP (2008) CBS reductions with a fluorous prolinol immobilized in a hydrofluoroether solvent. Org Lett 10:749
26. Dalicsek Z, Pollreisz F, Soos T (2009) Efficient separation of a trifluoromethyl substituted organocatalyst: just add water. Chem Commun 4587
27. Goushi S, Funabiki K, Ohta M, Hatano K, Matsui M (2007) Novel fluorous prolinol as a precatalyst for catalytic asymmetric borane reduction of various ketones. Tetrhedron 63:4061
28. Zu LS, Li H, Wang J, Yu XH, Wang W (2006) Highly enantioselective aldehyde-nitroolefin Michael addition reactions catalyzed by recyclable fluorous (S) diphenylpyrrolinol silyl ether. Tetrahedron Lett 47:5131
29. Cui HF, Li YW, Zheng CW, Zhao G, Zhu SS (2008) Enantioselective catalytic epoxidation of alpha, beta-enones promoted by fluorous alpha, alpha-diaryl-L-prolinols. J Fluor Chem 129:45
30. Song CE (2009) Cinchona alkaloids in synthesis and catalysis. Wiley-VCH, Weinheim
31. Tian SK, Chen YG, Hang JF, Tang L, McDaid P, Deng L (2004) Asymmetric organic catalysis with modified cinchona alkaloids. Acc Chem Res 37:621
32. Xu XH, Kusuda A, Tokunaga E, Shibata N (2011) Catalyst-free and catalytic Friedel-Crafts alkylations of indoles in Solkane (R) 365mfc, an environmentally benign alternative solvent. Green Chem 13:46
33. Zhang ZG, Schreiner PR (2009) (Thio)urea organocatalysis – what can be learnt from anion recognition? Chem Soc Rev 38:1187
34. Huang YB, Yi WB, Cai C (2010) An efficient, recoverable fluorous organocatalyst for direct reductive amination of aldehydes. J Fluor Chem 132:879
35. Wang L, Cai C, Curran DP, Zhang W (2010) Enantioselective alpha-chlorination of aldehydes with recyclable fluorous (S)-pyrrolidine-thiourea bifunctional organocatalyst. Synlett 433
36. Miura T, Itoh K, Yasaku Y, Koyata N, Murakami Y, Imai N (2008) Catalytic enantioselective cyclopropanation of allylic alcohols using recyclable fluorous disulfonamide ligand. Tetrahedron Lett 49:5813
37. Miura T, Imai K, Ina M, Tada N, Imai N, Itoh A (2010) Direct asymmetric aldol reaction with recyclable fluorous organocatalyst. Org Lett 12:1620
38. Miura T, Nakashima K, Tada N, Itoh A (2011) An effective and catalytic oxidation using recyclable fluorous IBX. Chem Commun 47:1875
39. Legros J, Crousse B, Bonnet-Delpon D (2008) Fluorous analogues of DMAP (F-DMAP): reusable organocatalysts for acylation reaction. J Fluor Chem 129:974
40. Vuluga D, Legros J, Crousse B, Bonnet-Delpon D (2010) Fluorous 4-N,N-dimethylamino-pyridine (DMAP) salts as simple recyclable acylation catalysts. Chem Eur J 16:1776
41. Gimbert C, Vallribera A, Gladysz JA, Jurisch M (2010) Thermomorphic fluorous phosphines as organocatalysts for Michael addition reactions. Tetrahedron Lett 51:4662
42. Seidel FO, Gladysz JA (2008) Catalysis of intramolecular Morita-Baylis-Hiliman and Rauhut-Currier reactions by fluorous phosphines; facile recovery by liquid/solid organic/fluorous biphase protocols involving precipitation, Teflon (R) tape, and Gore–Rastex (R) fiber. Adv Synth Catal 350:2443
43. Bruckmann A, Pena MA, Bolm C (2008) Organocatalysis through halogen-bond activation. Synlett 900
44. Dordonne S, Crousse B, Bonnet-Depon D, Legros J (2011) Fluorous tagging of DABCO through halogen bonding: recyclable catalyst for the Morita-Baylis-Hillman reaction. Chem Commun 47:5855

Thiourea Based Fluorous Organocatalyst

Yi-Bo Huang, Wen-Bin Yi, and Chun Cai

Abstract This review deals with general and significant development of the fluorous organocatalysts based on thiourea. The applications of fluorous technology are briefly discussed. The implementations of thiourea based catalysts in organic synthesis are focused on in the chapter.

Keywords Fluorous organocatalyst · Fluorous technology · Thiourea

Contents

1 Introduction 192
2 Normal Thiourea Based Organocatalysts 193
   2.1 Etter's Basic Urea 193
   2.2 Curran's Diphenyl Urea 194
   2.3 Jacobsen's Schiff Base (Thio)Urea 194
   2.4 Takemoto's Bifunctional Chiral Thiourea 195
   2.5 Tang's Chiral Bifunctional Pyrrolidine-Thiourea 196
   2.6 Wang's Bifunctional Binaphthyl-Thiourea 196
   2.7 Soós's and Connon's Bifunctional Cinchona Alkaloid-Thiourea 197
   2.8 Other Thioureas 198
3 Fluorous Thiourea Organocatalyst 198
   3.1 Schreiner's Thiourea 199
   3.2 Fluoroalkyl Tag Thiourea 199
   3.3 Thiourea Catalysis Mechanism 202
   3.4 Recycling Use of Fluorous Thiourea 202
4 Reactions Catalyzed by Fluorous Thiourea 202
   4.1 Diels–Alder Reaction 203
   4.2 Morita–Baylis–Hillman Reaction 203
   4.3 Friedel–Crafts Alkylation 204

Y.-B. Huang, W.-B. Yi, and C. Cai (✉)
Nanjing University of Science and Technology, Nanjing, P.R. China
e-mail: c.cai@mail.njust.edu.cn

4.4	Acyl-Strecker Reaction	205
4.5	Direct Reductive Amination Reaction	206
4.6	Alcoholysis of Styrene Oxides	207
4.7	α-Chlorination of Aldehydes	207
4.8	Chemoselective Oxidation Reaction	209
5	Outlook	210
References		210

1 Introduction

In the past few decades, transition metal based catalysts have attracted most of the attention in the study of catalysis, promoting many organic transformations. Meanwhile, many enzymes have also exhibited catalytic properties in biological and organic transformations. Unfortunately, not only are some transition metal catalysts toxic and expensive, but also enzyme catalyst are of narrow substrate scope, low selectivity, and low thermostability. Later, organocatalysis based on small organic molecules, where an inorganic element is not part of the active principle, has become a popular topic in chemical research [1–5]. Especially in the last few decades, organocatalysis has entered a "gold rush". Much interest has been devoted to the development of highly efficient organocatalysts for a variety of reactions.

Compared with traditional metal catalysts, organocatalysts are more eco-friendly and non-toxic, while enzyme catalysts are more stable and suitable for broad substrate application. In general, organocatalysts can be grouped mainly into three classes: (1) biomolecules, such as proline [6], cinchona alkaloids [7], certain oligopetides [8], (2) hydrogen bonding catalysts, such as bi-naphthol [9], thioureas [10], and (3) triazolium salts. Most organocatalysts display secondary amine functionality described as performing either enamine catalysis (by forming catalytic quantities of an active enamine nucleophile) or iminium catalysis (by forming catalytic quantities of activated iminium electrophiles). This catalytic mechanism is typical for covalent organocatalysis. Other organocatalysts, such as thiourea, exhibit catalytic effects based on hydrogen-bonding interaction with substrates, which could be considered as non-covalent mechanism.

Among several kinds of organocatalysts, thioureas are more diverse than other organocatalysts, and a variety of thiourea derivatives can accelerate and stereochemically alter organic transformations through double hydrogen-bonding interactions with the respective substrates.

In this review we will illustrate the development of organocatalysts based on thiourea. Then fluorous technology as a kind of powerful recyclable method is introduced in detail; this is pioneered by Curran and Zhang [11]. In addition, some applications on fluorous thiourea organocatalysts are described in detail. The introduction of the fluorous tag can improve catalytic effect as well as favor the reusability of thiourea derivatives. This strategy is well in accordance with the ideals of green chemistry.

2 Normal Thiourea Based Organocatalysts

As we all know, general mineral acid catalysts and Lewis acid catalysts have played important roles in all kinds of the organic transformations. Unfortunately, the lack of selectivity has been troublesome. Later, inspired by the natural enzymatic systems, more and more people began to explore weak acid-base/hydrogen bonding as a basis for catalyst design. Especially, some selective, robust, environmentally benign and functional-group tolerant organocatalysts have become hot topics in the research fields. Thiourea as a kind of organocatalyst has developed from the initial finding of catalytic effects to diverse structural modifications.

2.1 Etter's Basic Urea

First, Etter and coworkers [12, 13] reported all kinds of different hydrogen patterns of the organic compounds in detail, and then analyzed the host-guest relation. In their reports, general rules of hydrogen bonds included that (1) all good proton donors and acceptors are used in hydrogen bonding, (2) six-membered-ring intramolecular hydrogen bonds form in preference to intermolecular hydrogen bonds, and (3) the best proton donors and acceptors remaining after intramolecular hydrogen-bond form intermolecular hydrogen bonds.

In addition, several more rules for specific classes of functional groups are described. Furthermore, they discovered that the N–H hydrogens preferred to form three-center bonds with urea carbonyl groups C(4) [$R_2^1(6)$] (Fig. 1). Meanwhile, when there are strong *meta*-substituted electron-withdrawing substituents, like the NO_2 group on aryl rings, and the guest molecules have acceptor groups that are stronger than internally hydrogen bonded urea carbonyl oxygen, cocrystals form in $R_2^1(6)$ patterns (Fig. 2). So Etter's urea could activate the carbonyl compounds. The catalytic mechanisms of urea or thiourea are also described.

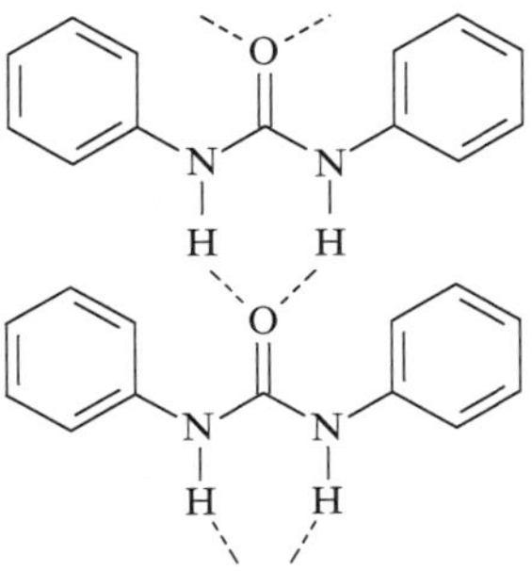

Fig. 1 The intermolecular hydrogen bond pattern of urea

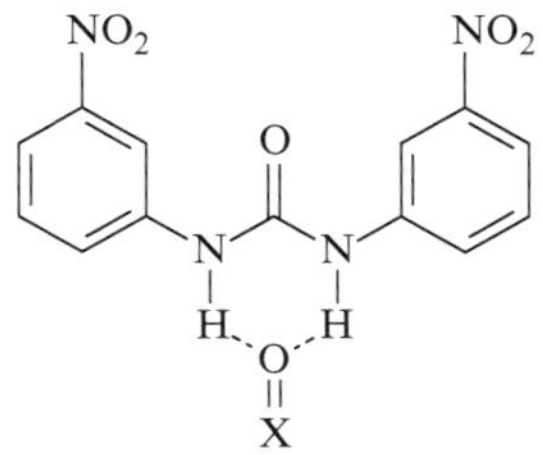

Fig. 2 The hydrogen bond interaction of X = O with urea

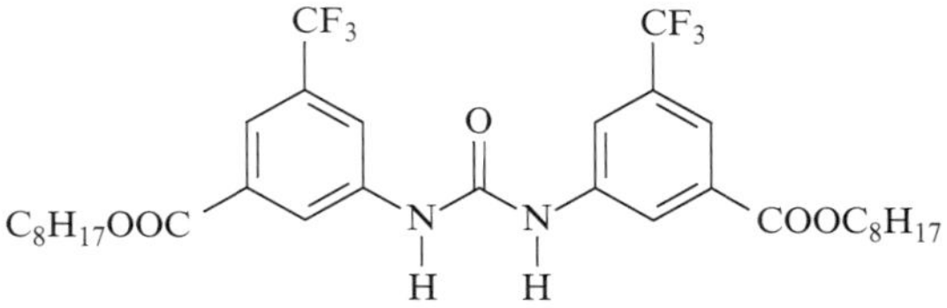

Fig. 3 The structure of Etter's urea

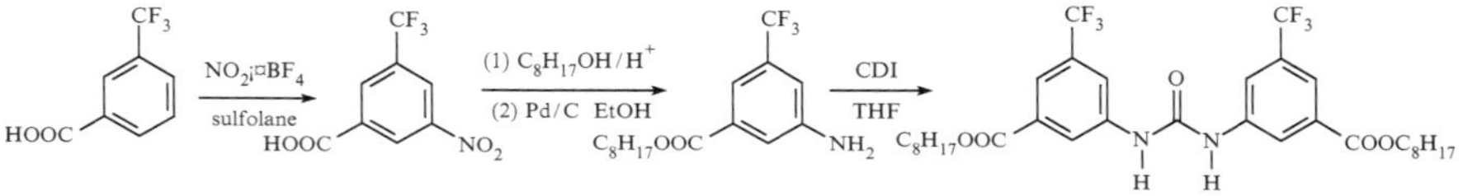

Fig. 4 The preparation of Etter's urea

2.2 Curran's Diphenyl Urea

In 1994, inspired by Etter's reports, Curran and Kuo designed a kind of novel Lewis acid urea of N,N-bis[3-(trifluoromethyl)-5-(carbooctyloxy)phenyl]urea (Fig. 3). This novel Lewis acid urea acted as an additive, which could effectively accelerate the radical allylations of 2-(phenylseleno)teterahydrothiophene oxide and 2-benzyl-5-(phenylseleno)-3-isothiazolidone 1-oxide with allyltributylatnnane [14]. In addition, the Claisen rearrangement of 6-methoxy allyl vinyl ether is catalyzed by N,N-bis[3-(trifluoromethyl)-5-(carbooctyloxy)phenyl]urea, and the corresponding evidence is presented that the urea stabilizes a dipolar transition state by hydrogen bonding [15]. Meanwhile, the detailed synthesis of Curran's urea was followed by Fig. 4. Since then, a lot of researchers began to focus on all kinds of diphenylurea catalysts. The detailed description will be introduced in the following.

2.3 Jacobsen's Schiff Base (Thio)Urea

The most remarkable work on the urea and thiourea catalysts was carried out by the Jacobsen group [16–19]. They had developed a range of chiral Schiff base ureas and thioureas, which were versatile and effective organocatalysts for asymmetric synthesis. Among these, the enantioselective thiourea catalyst (Fig. 5) could be

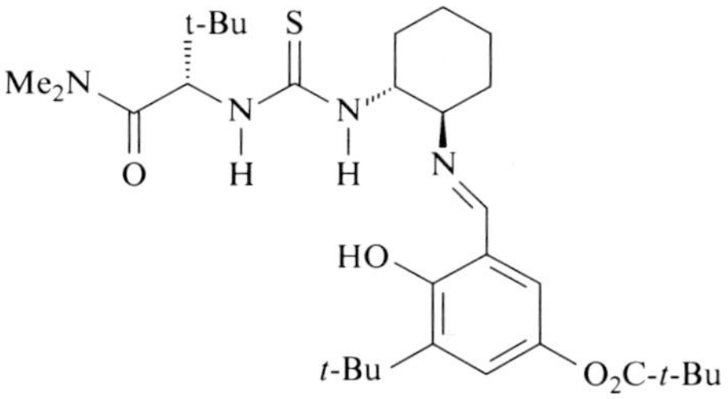

Fig. 5 The structure of Jacobsen's schiff thiourea

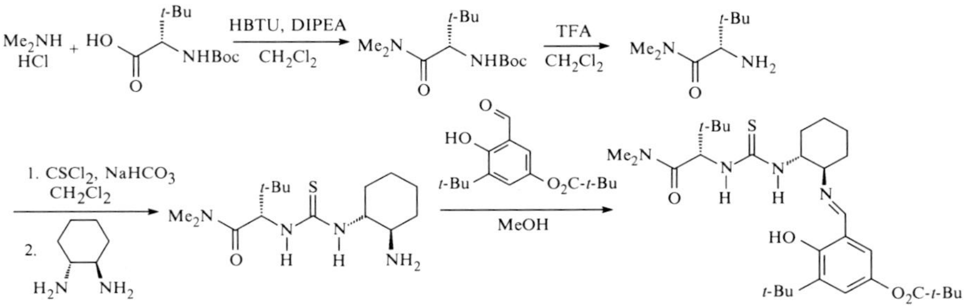

Fig. 6 The preparation of Jacobsen's schiff thiourea

successfully applied to the Stecker and the Mannich reactions. In their reports, structure-based analysis and optimization of the catalysts were thoroughly investigated. The most effective catalyst (Fig. 5) could be prepared with the following method (Fig. 6).

Later, inspired by the pioneering work on asymmetric catalysis based thiourea, more and more people were dedicating themselves to designing multifunctional chiral thioureas for enantioselective transformation. Among these thioureas, some representative ones will be elaborated in the following section.

2.4 Takemoto's Bifunctional Chiral Thiourea

In order to explore the organocatalytic scope of thiourea derivatives in depth, the Takemoto group designed a series of bifunctional thioureas with thiourea moieties and amino groups [20, 21]. Among these, a kind of thiourea catalyst (Fig. 7) was screened to be the most effective for the Michael addition reaction of malonates to nitroolefins. In particular, the chiral amino group played a key role in producing good enantioselective results. In addition, Takemoto's thiourea could be prepared with *(R,R)-N,N*-dimethyl-*trans*-diamincyclohexane and 3,5-bis(trifluoromethyl) phenyl isothiocyanate. Later, Takemoto's group reported the other novel bifunctioanl thiourea catalysts which could be obtained by modifying the amino group of Takemoto thiourea catalyst [22, 23]. The most outstanding feature of Takemoto's

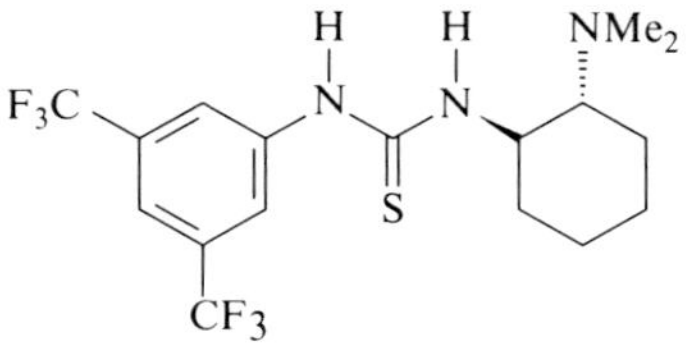

Fig. 7 The bifunctional Takemoto's catalyst

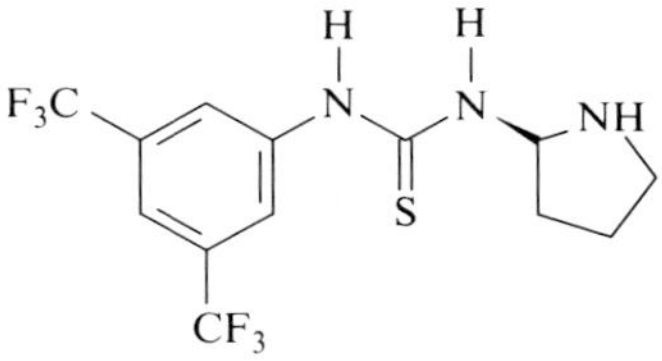

Fig. 8 The bifunctional Tang's pyrrolidine-thiourea

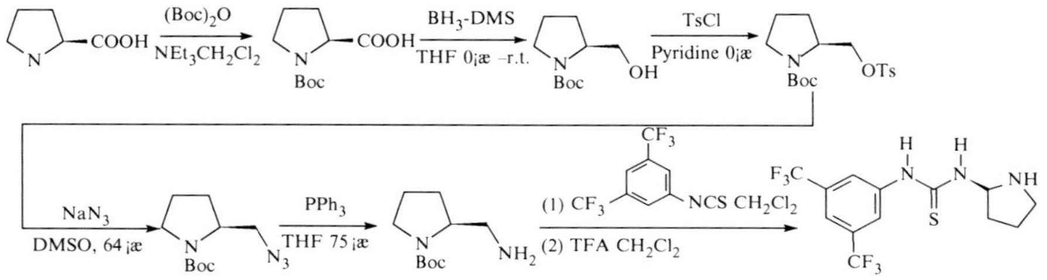

Fig. 9 The preparation of Tang's pyrrolidine-thiourea catalyst

thiourea was that the catalyst could produce dual activation of electrophile and nucleophile. Meanwhile, a lot of work on C–C bond forming reactions involving this catalyst was described by the Takemoto group [24].

2.5 Tang's Chiral Bifunctional Pyrrolidine-Thiourea

In 2006, Tang's group first reported that the combination of thiourea and pyrrolidine in a chiral scaffold could result in a potential bifunctional organocatalyst (Fig. 8), which could promote the highly enantioselective Michael addition of cyclohexanone to nitroolefins [25]. The bifunctional pyrrolidine-thiourea could be synthesized with 3,5-bis(trifluoromethyl)phenyl isothiocyanate and pyrrolidine derivative in mild conditions (Fig. 9). The pyrrolidine derivative could be prepared from L-proline according to the standard procedure [26].

2.6 Wang's Bifunctional Binaphthyl-Thiourea

With the development of thiourea based organocatalysts, many novel bifunctional thiourea catalysts were designed for a variety of organic transformations. Wang's

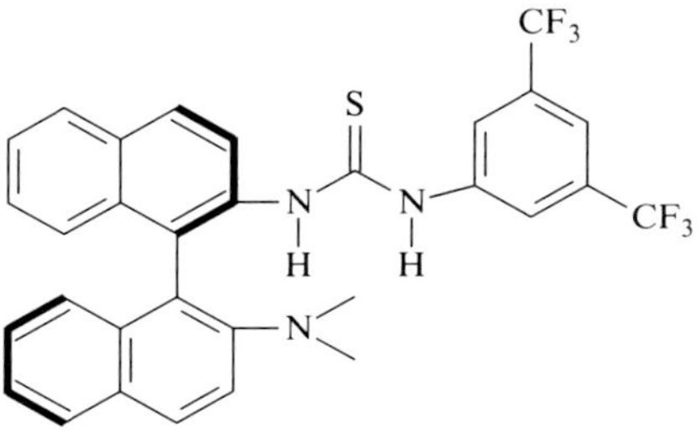

Fig. 10 The bifunctional Wang's binaphthyl-thiourea

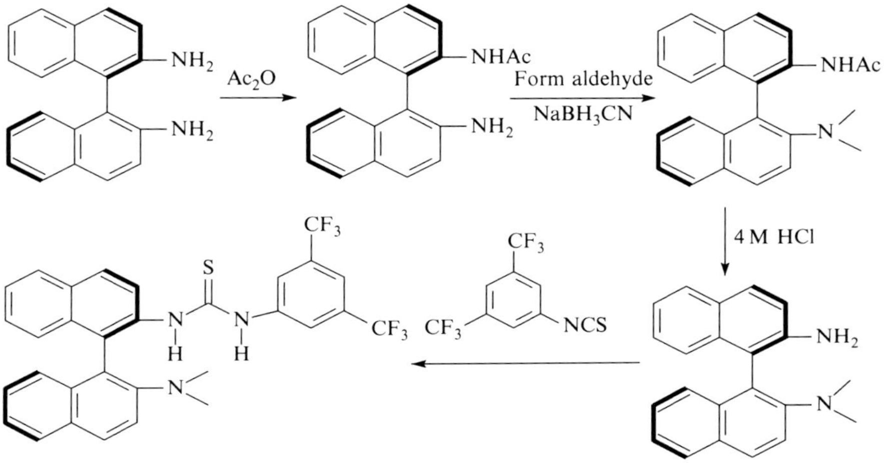

Fig. 11 The preparation of Wang's binaphthyl-thiourea

group committed themselves to develop the binaphthyl-thiourea (Fig. 10), which incorporated both a basic binaphthol moiety and a thiourea moiety. Wang's binaphthyl thiourea could be synthesized by the following methods (Fig. 11). In these reports, the bifunctional binaphthyl-thiourea could promote the asymmetric Morita–Baylis–Hillman reaction of cyclohexenone with a series of aldehydes [27]. Meanwhile, the asymmetric Michael addition of diketones to nitroalkenes could be effectively catalyzed with high yields and good enantioselectivities [28].

2.7 Soós's and Connon's Bifunctional Cinchona Alkaloid-Thiourea

Soós et al. and Connon et al. have independently investigated the use of thiourea-substituted cinchona alkaloid derivatives as bifunctional catalysts [29, 30]. The cinchona alkaloid backbone is composed of a quinuclidine moiety and a secondary chiral alcohol. After simple modification, the 9-hydroxyl could be converted into amino and other functional groups. Then the intermediate and 3,5-bis

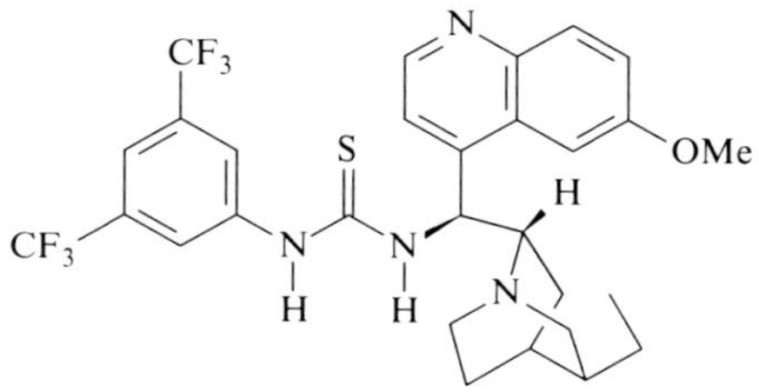

Fig. 12 The bifunctional cinchona alkaloid-thiourea

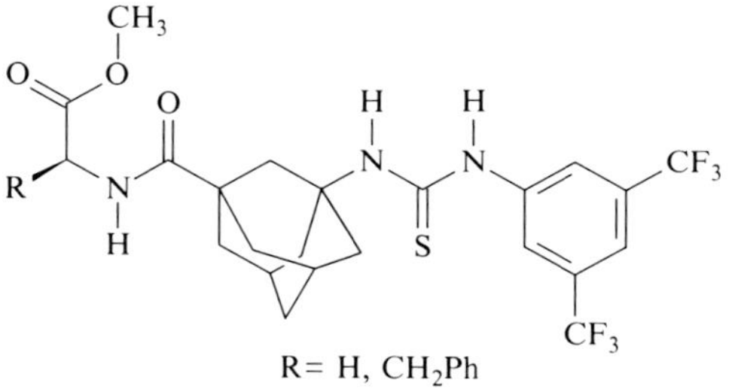

Fig. 13 Thiourea catalysts based on peptidic adamantane

R= H, CH$_2$Ph

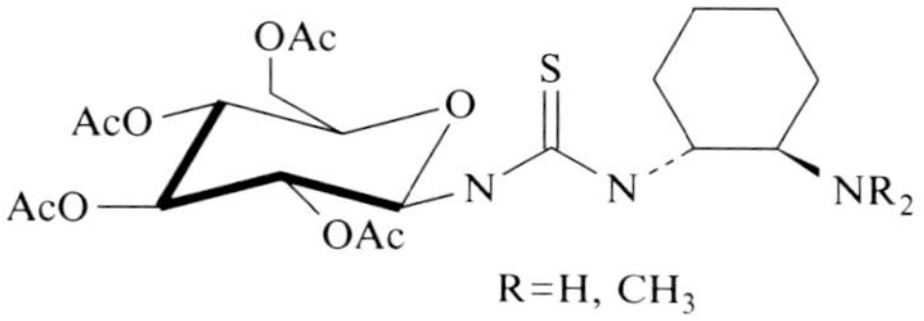

Fig. 14 The bifunctional saccharide-amine thiourea

R=H, CH$_3$

(trifluoromethyl)phenyl isothiocyanate could afford the corresponding cinchona alkaloid-thiourea (Fig. 12). The catalyst performed well during the asymmetric Michael addition reaction.

2.8 Other Thioureas

In order to recycle the thiourea catalyst, some polymer (e.g., PEG, PS) based thioureas were developed and widely used in some organic transformations [31, 32]. In addition, some novel thiourea catalysts based on peptidic adamantane (Fig. 13) [33] and saccharides (Fig. 14) [34, 35] were also designed for the application of C–C bond formation.

3 Fluorous Thiourea Organocatalyst

In the previous section we introduced several bifunctional thioureas, and their synthesis methods are also elaborated in detail. Then, with the development of organocatalysis, many researchers began to pay attention to recovery of organocatalysts.

Catalyst immobilization techniques and fluorous tag techniques emerged. Herein, we will describe some fluorous thiourea organocatalysts.

3.1 Schreiner's Thiourea

Schreiner's thiourea was a classic thiourea catalyst which was widely used in all kinds of reactions. There are four trifluoromethyl groups in the aromatic ring, which could facilitate improving catalytic activity. The experimental results showed that Schreiner's catalyst exhibits better catalytic effects than the other thioureas. The catalyst was first reported in 2002 (Fig. 15) [36], which could be prepared by a classic method (Fig. 16). Generally, a mixture of 3,5-bis(trifluoromethyl)aniline and triethylamine in THF was added into a three-necked flask. Under nitrogen atmosphere, a mixture of thiophosgene in THF was added dropwise to the stirred solution at −5 to 0 °C. After 24 h, the mixtures were separated by extraction of diethyl ether and purified by recrystallization from chloroform twice. The pure thiourea derivative catalysts could be obtained after drying in vacuo [37].

3.2 Fluoroalkyl Tag Thiourea

Inspired by Schreiner's thiourea, we decided to design some thioureas bearing fluorous tags ($C_8F_{17}^-$ group). The greatest advantage of introducing the fluoroalkyl group was that thiourea catalyst could be recovered by the fluorous solid phase extraction (FSPE) technique. Fluorous tag methodology is a component of fluorous technology, a platform for synthesis, purification, enrichment, and immobilization of molecules, which has been widely developed in the past few decades.

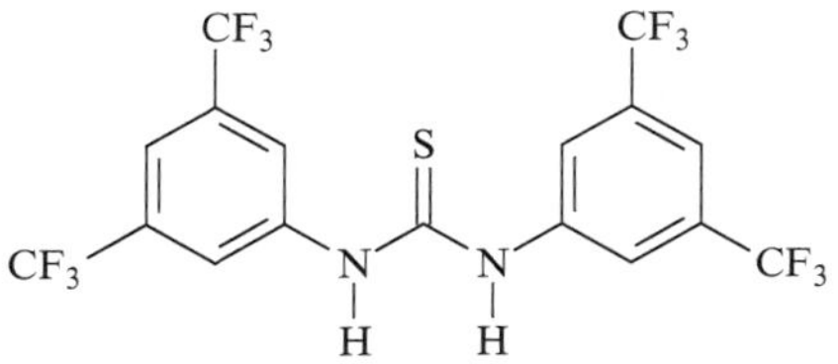

Fig. 15 The structure of Schreiner's thiourea

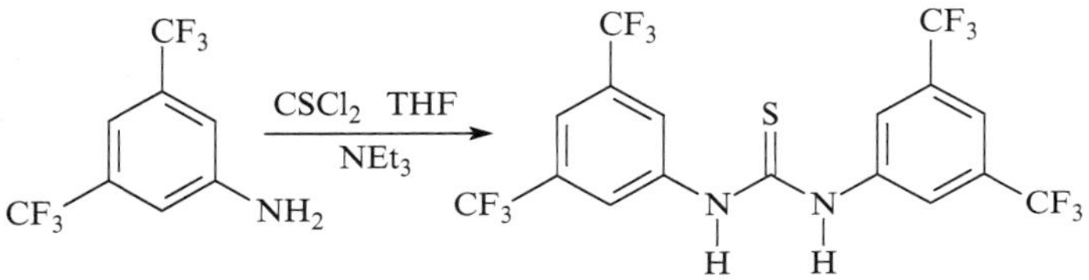

Fig. 16 The preparation of Schreiner's thiourea

3.2.1 Fluorous Tag; and Fluorous Separation

Fluorous tags and fluorous separation of molecules are both important in the application of fluorous technology[1]. Generally, fluorous separation is usually combined with fluorous tag techniques in the utilization of fluorous technology.

As far as fluorous tag techniques are concerned, some perfluorocarbon alkyl chains of varying lengths were introduced into molecules in order to promote the reaction and facilitate the separation. Fluorous tag groups such as $C_6F_{13}^-$ or $C_8F_{17}^-$ could impart special characteristics to a molecule.

Furthermore, some kinds of fluorous tags could be removed as fluorous protective groups. The fluorous Boc group is a typical example of a fluorous protecting group that is designed to be attached and removed by analogy with the standard Boc group. Such fluorous protective groups could promote rapid separation of all tagged molecules from non-tagged molecules by FSPE. Removable fluorous tags have been used extensively in small molecule parallel synthesis and in biomolecule synthesis. More recently, many researchers have focused on modifying small molecules with fluorous tag techniques for widely varied applications.

Fluorous separation as a simple protocol can be used to separate fluorous molecules and non-fluorous molecules with FSPE or fluorous liquid–liquid extraction (FLLE).

FSPE utilizes fluorous silica as the stationary phase to separate quickly all fluorous molecules from all non-fluorous molecules [11]. The fluorous solid phase is typically silica gel with a fluorocarbon bonded phase ($-SiMe_2(CH_2)_2C_8F_{17}$), and this is commercially available from Fluorous Technologies, Inc. under the trade name of Fluoro*Flash*®.

FSPE is a simple three step process. The mixture is loaded onto the silica gel using an adequate amount of organic solvent. A fluorophobic wash, e.g., 20% water in methanol, washes away all of the non-fluorous molecules while the fluorous molecules are retained on the silica gel. A fluorous wash, such as methanol or THF, is then used to wash the fluorous molecules from the stationary phase.

Compared to FSPE, FLLE was applied in earlier years. It utilizes a fluorous phase to extract fluorous molecules from an organic or aqueous solvent [38]. FLLE has been used in biphasic catalysis and small molecule synthesis. The FLLE technique is also called a "heavy" fluorous technique, since multiple fluorous chains on a molecule may be necessary to enact good separation. Liquid–liquid extractions work well when fluorous domains are relatively large. In the best cases, only a single separation is needed. With lower partition coefficients, the organic fraction is washed several times with the fluorous solvent. Since the solubilities of organic compounds in fluorous solvent are exceedingly low, the washing process can be conducted repeatedly without extractive loss of the organic product. Liquid–liquid extractive methods are typically used when the desired product is organic and

---

[1] http://www.fluorous.com

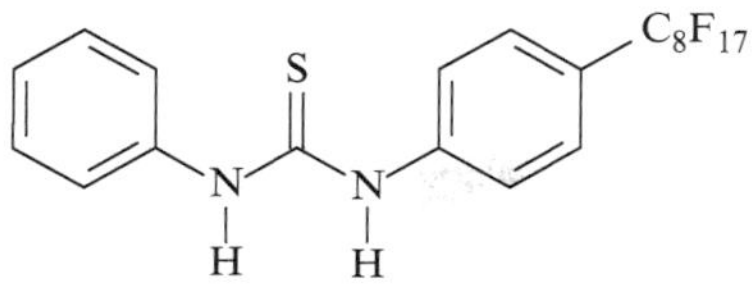

Fig. 17 The structure of 1-[4-(perfluorooctyl)phenyl]-3-phenyl thiourea

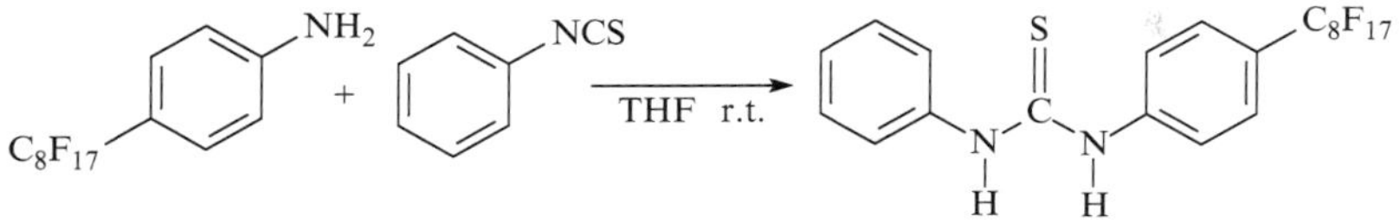

Fig. 18 The preparation of 1-[4-(perfluorooctyl)phenyl]-3-phenyl thiourea

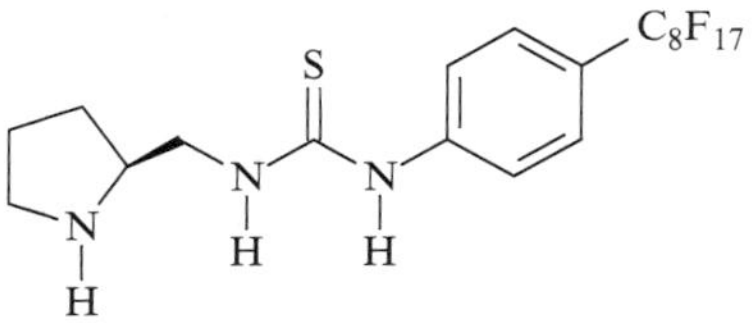

Fig. 19 The structure of (S)-pyrrolidine-thiourea

some other reaction component (reactant, reagent, catalyst, scavenged product) is fluorous.

In fact, FLLE and FSPE are two efficient separation methods which have been widely used to purify reaction mixtures.

3.2.2 Fluorous Tag Thiourea

Initially, we would have liked to synthesize the symmetric double 4-perfluorooctyl phenyl thiourea but we failed due to the troublesome post-treatment of long fluoroalkyl chains. So we have to decide to develop single 4-perfluorooctyl phenyl thiourea such as 1-[4-(perfluorooctyl)phenyl]-3-phenyl thiourea (Fig. 17). The corresponding synthetic method is as follows (Fig. 18). Moreover, the thiourea could also be applied in the direct reductive amination of aryl aldehydes and ketones [39].

Meanwhile, we have designed (S)-pyrrolidine-thiourea bifunctional organocatalyst, which contains a thiourea moiety and a pyrrolidine structure (Fig. 19). The bifunctional thiourea could be prepared by the procedure shown in Fig. 20 [40]. The N-Boc fluorous (S)-pyrrolidine-thiourea was prepared via the condensation of 4-(perfluorooctyl)aniline with phenyl chlorothioformate followed by the substitution of phenol by (S)-*tert*-butyl 2-(aminomethyl) pyrrolidine-1-carboxylate following Berkessel's method. Subsequent de-Boc in TFA/CH$_2$Cl$_2$ at 25 °C afforded the target catalyst as a yellow solid. Furthermore, the (S)-pyrrolidine-thiourea bifunctional organocatalyst was researched in the application of enantioselective α-chlorination of aldehydes [41].

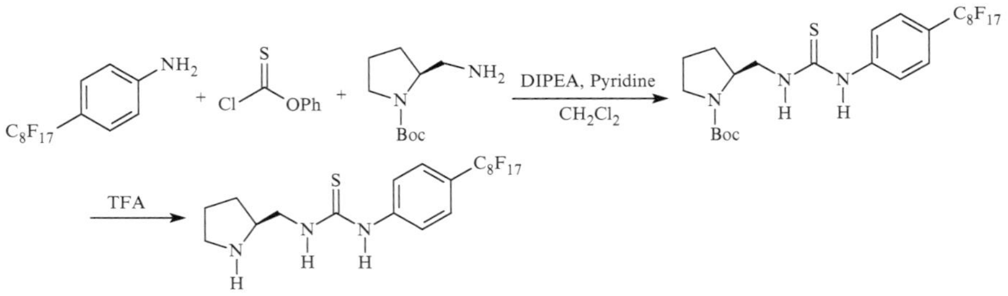

Fig. 20 The preparation of (S)-pyrrolidine-thiourea

3.3 Thiourea Catalysis Mechanism

Since 2001, several research groups have realized the potential of thiourea derivatives and developed various achiral/chiral mono- and bifunctional derivatives. During the catalytic process, thioureas and thiourea derivatives could undergo double hydrogen bond interaction with substrates such as carbonyl, nitrile, and imine, which act like poor Lewis acids [36, 42].

When the thiourea moiety contains strong electron-withdrawing groups such as 3,5-bis(trifluoromethyl) or perfluorooctyl, the catalytic activities could increase distinctly. If the thiourea catalyst incorporated other functional groups, the thiourea structure could activate hydrogen acceptor groups; meanwhile the functional groups promote some enantioselective transformations.

3.4 Recycling Use of Fluorous Thiourea

The recycling of fluorous thiourea catalysts could be achieved by FSPE methodology. Many mono- and bifunctional fluorous thioureas have been reported to be recovered over several runs for reuse. Generally, the fluorous thiourea and some non-fluorous components were dissolved in a minimum of solvent such as THF or CH_3OH, and then were loaded onto a Fluoro*Flash*® cartridge. The mixtures loading quantity depended on different quantities of Fluoro*Flash*® silica gel in the cartridge. First, 20–30% water in methanol or THF could wash away all the non-fluorous component, while the fluorous component was retained in the fluorous silica gel. Second, pure methanol or THF was used to wash the fluorous molecules from the stationary phase. After evaporating the solvent, the corresponding thiourea could be recycled for next reuse.

4 Reactions Catalyzed by Fluorous Thiourea

We have discussed the classification of various thioureas and some classic preparation methods in the previous section. Meanwhile, fluorous technology has been outlined in brief. We will now describe some important reactions catalyzed by fluorous thiourea reported by Schreiner's group and our group.

4.1 Diels–Alder Reaction

In 2003, Schreiner et al. examined the catalytic activity of substituted thioureas in a series of Diels–Alder reactions and 1,3-dipolar cycloadditions [43]. In order to screen catalyst efficiencies, Diels–Alder reaction of methyl vinyl ketone and cyclopentadiene catalyzed by thiourea derivatives (1 mol%) was examined in detail (Fig. 21). While a tenfold excess of cyclopentadiene was added, the reactions were strictly pseudo-first-order and the relative rate constants k_{rel} were determined by least-error square fits of kinetic data. The experimental results showed that the relative rate constants k_{rel} depended on the choice of catalyst. Among these thioureas, N,N'-bis(trifluoromethyl) phenyl thiourea performed the best catalytic activities. Even 1 mol% loading could increase the reaction rate by a factor of 6.0.

Meanwhile, an interesting conclusion was proposed that thiourea derivatives with rigid electron-withdrawing aromatic substituents were the most effective hydrogen-bonding catalysts for Diels–Alder reactions in this work.

4.2 Morita–Baylis–Hillman Reaction

The Morita–Baylis–Hillman reaction is a versatile carbon–carbon bond-forming reaction for the synthesis of densely functionalized compounds from aldehydes and electron-deficient alkenes in the presence of Lewis bases. The corresponding adducts had been extensively used as intermediates in organic synthesis for a variety of applications.

In 2008, Nagasawa et al. described a novel N,N'-bis(trifluoromethyl) phenyl thiourea (Fig. 22) that was more effective for the reaction of aryl aldehyde and cyclohexen-1-one than the other thioureas [44]. Meanwhile, they had developed a kind of novel chiral thiourea catalyst having two thiourea functionalities.

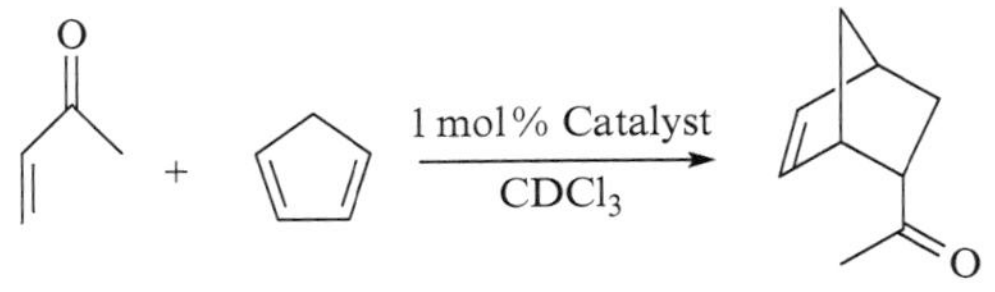

Fig. 21 Diels–Alder reaction of methyl vinyl ketone and cyclopentadiene

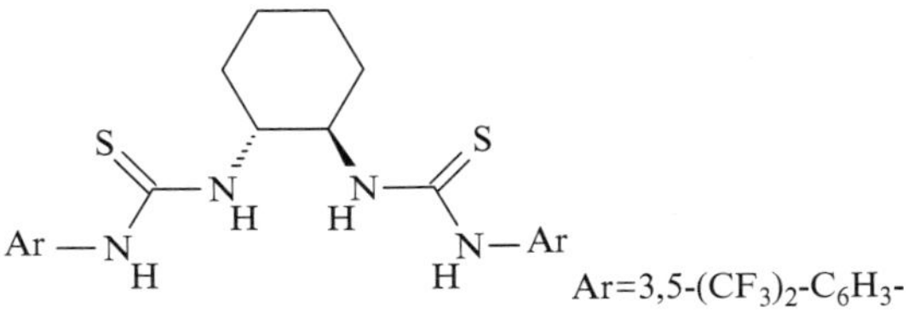

Fig. 22 The structure of N,N-bis(trifluoromethyl) phenyl thiourea

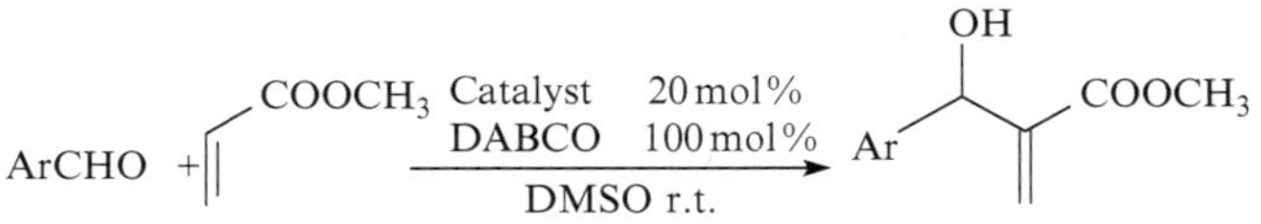

ArCHO + [CH2=CHCOOCH3] —Catalyst 20 mol%, DABCO 100 mol%, DMSO r.t.→ Ar-CH(OH)-C(=CH2)-COOCH3

Fig. 23 The MBH reaction of arylaldehydes with methyl acrylate

In our work, 1-[4-(perfluorooctyl)phenyl]-3-phenyl thiourea was also used to catalyze the MBH reaction of arylaldehydes with methyl acrylate under DABCO at room temperature [45]. In addition, 1-[4-(perfluorooctyl)phenyl]-3-phenyl thiourea could perform as well as Schreiner thiourea in the reaction (Fig. 23).

Moreover, we found that the solvent had a pronounced effect on the yield. Excellent yields were obtained when employing CH_3CN, DMSO, or DMF as reaction solvent, while for the screening of different tertiary amine bases, DABCO and DMAP exhibited better results than other bases. When catalyst loading was varied from 5 mol% to 20 mol%, the yield of MBH adduct increased gradually. But the yield was not increased when further amounts of catalyst were employed.

Under the optimized reaction conditions of 20 mol% catalyst, 5 mL DMSO, and 100 mol% DABCO, we examined the recovery of the catalyst. In each cycle, the catalyst could be recovered with high recovery of >93%. Meanwhile, a high yield of 87–91% of the model reaction could be obtained, mediated by recovered catalyst within three cycles. It showed that our fluorous thiourea catalyst could be well recycled by FSPE without any remarkable loss of reaction activity.

4.3 Friedel–Crafts Alkylation

Ricci et al. demonstrated that catalytic amounts (10 mol%) of bis-aryl(thio)ureas promoted the Friedel–Crafts alkylation with nitroolefins of aromatic and heteroaromatic N-containing derivatives [46]. In the screening of catalysts, the catalytic activities of N,N'-bis(3,5-bistrifluoromethyl) phenyl thiourea was better than N,N'-bis(3,5-bistrifluoromethyl) phenyl urea in the same loading. In addition, the reaction failed to happen in the absence of catalyst under the standard conditions.

First, in order to screen the catalyst efficiency under various conditions, they carried out the reactions in toluene as well as in the absence of solvent with catalyst loading of 10 mol% (Fig. 24). Among these, *N*-methylpyrrole reacted smoothly in toluene to give the 2-substituted product, but the reaction was disappointing in the case of the *N*-aryl derivative. In addition, aliphatic nitroolefin indicated greater reactivity than β-nitrostyrene during the Friedel–Crafts alkylation reaction.

Meanwhile, the substrate scope was extended into the indole series (Fig. 25). They reported that the reactivity series towards Michael acceptors for the indoles is 2-methylindole > indole > 1-methylindole. In the indole series, double hydrogen

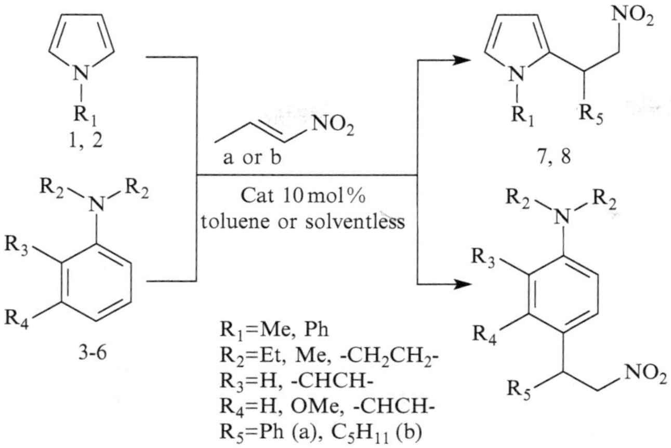

Fig. 24 The Friedel–Crafts alkylation with nitroolefins of aromatic and hetero-aromatic N-containing derivatives

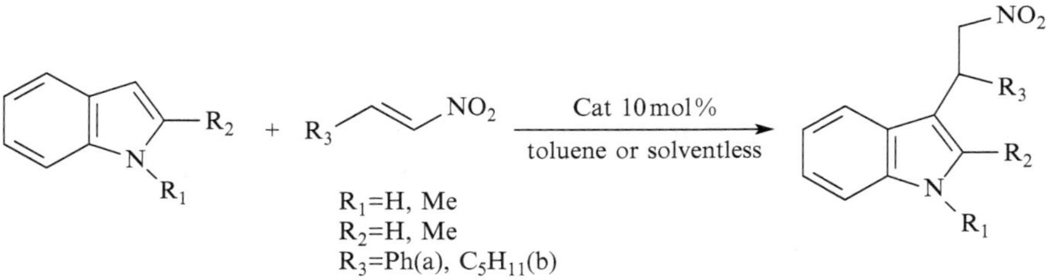

Fig. 25 The Friedel–Crafts alkylation with nitroolefins of indole derivatives

bonding catalysis offered several conspicuous advantages in terms of much milder reaction conditions and higher yields with respect to the traditional methods.

4.4 Acyl-Strecker Reaction

The Strecker reaction, as a kind of multicomponent reaction [47, 48], was often useful due to their high atom economy, selectivity, environmental friendliness, and formation of low levels of by-products. However, the Strecker reaction had some drawbacks, in particular due to the volatile and highly toxic nature of HCN. Later, trimethylsilyl cyanide (TMSCN) offered certain advantages but, due to its toxicity and high price, an alternative cyanation reagent was desirable. Acyl cyanides were not just less toxic but also had been used in acylcyanation of carbonyl compounds. In 2007, List and Chandra described an organocatalytic one-pot, three-component acyl-Strecker reaction using acetyl cyanide as a new cyanide source (Fig. 26) [49].

During the screening process of the catalyst, they found that the Schreiner thiourea catalyst turned out to be a highly efficient catalyst for this rarely used

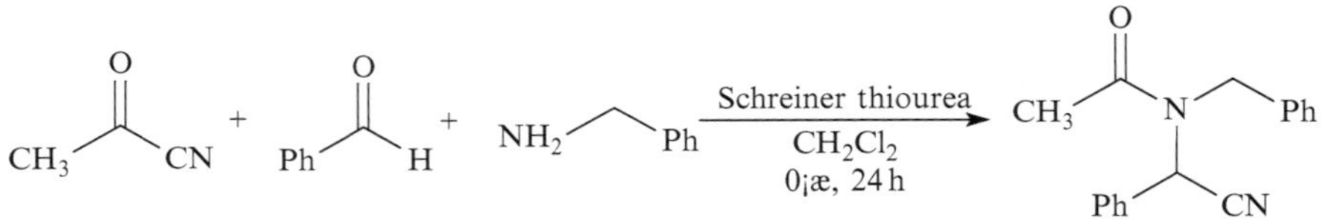

Fig. 26 The three-component acyl-Strecker reaction catalyzed by Schreiner thiourea

and yet highly atom economic reaction. In the absence of catalyst, the conversion was about 42%. When the catalyst loading varied from 1 mol% to 5 mol%, the conversion was 70–99%.

In addition, additive such as MgSO$_4$ and 5-Å molecular sieve (MS) could distinctly improve the conversion. Furthermore, a variety of amines and aldehydes was explored in order to extend the scope of reaction.

Above all, it was an attractive approach for the generation of diverse assortments of α-amido nitriles. Besides, it could be important in the application of medicinal chemistry due to its potential diversity.

4.5 Direct Reductive Amination Reaction

Reductive amination presented one of the effective methods for the synthesis of various kinds of amine [50], in which the carbonyl component was treated with amine and reductant in "one-pot" fashion. Thus, many methods had been reported to accomplish this direct process [51–53] but these methods based on Brønsted acid and Lewis acid catalysis were not suitable for sensitive, acid-labile, or polyfunctional substrates. However, people found that organocatalyst could overcome these drawbacks.

Our group carried out some work on reductive amination with fluorous thiourea. In our experiments we studied the direct reductive amination of aldehydes mediated by 1-[4-(perfluorooctyl)phenyl]-3-phenylthiourea using Hantzsch 1,4-dihydropyridine as reducing agent as shown in Fig. 27.

In the catalyst screening, fluorous thiourea of 1-[4-(perfluorooctyl)phenyl]-3-phenylthiourea performed better than other thioureas in the catalytic process. As for the model reaction of direct reductive amination of benzaldehyde and *p*-anisidine using Hantzsch 1,4-dihydropydine with the fluorous catalyst loading of 10 mol%, the yield (94%) was higher than that of the Schreiner thiourea (82%) and *N*,*N*′-diphenyl thiourea (65%).

Furthermore, we established the catalyst recycling methods using FSPE. In each run, the recovered catalyst retained its high activity with good recovery within three cycles. Furthermore, the substrate scope of different amines and aldehydes were explored in detail.

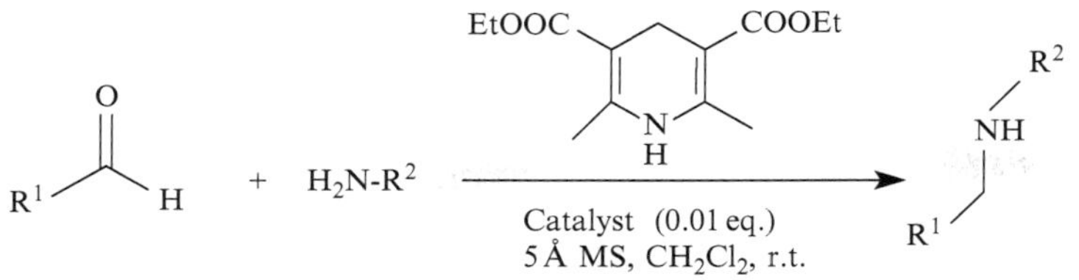

Fig. 27 The direct reductive amination of aldehydes mediated by 1-[4-(perfluorooctyl)phenyl]-3-phenylthiourea

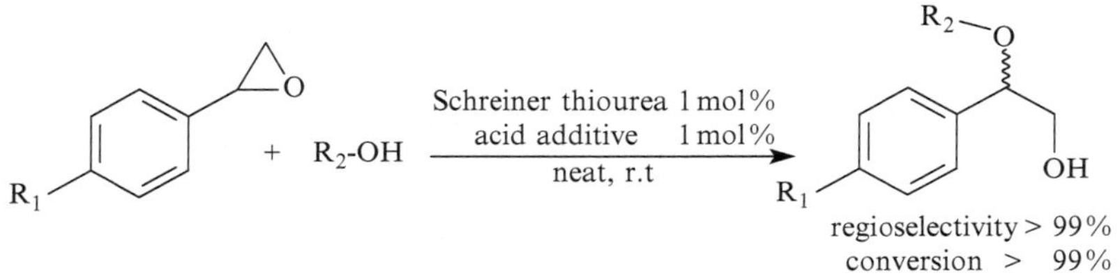

Fig. 28 The regioselective alcoholysis of styrene oxides catalyzed by Schreiner thiourea

4.6 Alcoholysis of Styrene Oxides

In 2008, Schreiner et al. presented a mild and efficient method for the completely regioselective alcoholysis of styrene oxides utilizing a cooperative Brønsted acid type organocatalytic system comprised of mandelic acid (1 mol%) and N,N'-bis-[3,5-bis-(trifluoromethyl) phenyl] thiourea (1 mol%) (Fig. 28) [54].

During the acid additive screening, aromatic acids bearing a second coordination center in the R-position (hydroxy or carbonyl) led to appreciable conversions. The blocking of the α-coordination center or removal of the aromatic system dramatically reduced the conversion rates.

In addition, various styrene oxides were readily transformed into their corresponding β-alkoxy alcohols in good to excellent yields at full conversion with mandelic acid (1 mol%) and the catalyst (1 mol%). Simple aliphatic and sterically demanding as well as unsaturated and acid-sensitive alcohols could be employed. The most important factor was that they had suggested an H-bonding-mediated cooperative Brønsted-acid mechanism, which had been proven by theoretical calculation.

4.7 α-Chlorination of Aldehydes

In 2004, the enantioselective direct α-chlorination of unbranched aldehydes was independently reported by the Jørgensen [55] and MacMillan [56] groups utilizing chiral secondary amine catalysts. Inspired by previous reports, we designed a kind

Table 1 Organocatalysts promoted α-chlorination of hydrocinnamaldehyde with NCS[a]

Entry	Catalyst (mol%)	Solvent	Time (h)	Yield (%)[b]	ee (%)[c]
1	1 (10)	CH_2Cl_2	3	99	85
2	1 (10)	$CHCl_3$	3	99	78
3	1 (10)	DCE	3	99	82
4	1 (10)	MeOH	<1 min	99[d]	0
5	1 (10)	THF	3	47	29
6	1 (5)	CH_2Cl_2	3	71	85
7	–	CH_2Cl_2	24	<5	0
8	L-Proline (10)	CH_2Cl_2	3	99	18
9	L-Prolinamide (10)	CH_2Cl_2	3	99	76
10	Thiourea (10)	CH_2Cl_2	12	52	0

[a]Reaction conditions: hydrocinnamaldehyde (0.5 mmol), NCS (0.65 mmol), solvent (1 mL), 25°C
[b]Measured by GC using benzyl methyl ether as internal standard
[c]ee determined by chiral HPLC after reduction to the corresponding alcohol
[d]Acetalization of aldehyde

of fluorous (S)-pyrrolidine-thiourea bifunctional organocatalyst **I**, which could be applied to α-chlorination of aldehydes.

In our work, a model reaction involving hydrocinnamaldehyde, NCS, and a fluorous bifunctional catalyst to optimize the reaction condition [41] was used. Excellent yields were obtained when employing CH_2Cl_2, $CHCl_3$, or DCE as a solvent, while the best ee (85%) was found in CH_2Cl_2. THF was less efficient, affording the product in 29% ee.

To our surprise, acetalization of aldehyde was found to be very quick (<1 min) when the reaction was conducted in methanol. This may be due to the in situ formation of HCl from NCS, which was accelerated by the thiourea part of the catalyst [57] (Table 1).

The effect of catalyst loading was also evaluated. We found that 10 mol% of catalyst was found to be most efficient. Other secondary amine catalysts such as L-proline and L-prolinamide were also utilized for the same reaction. Both catalysts promoted the reaction effectively, but lower ee values, 18% and 76%, were obtained, respectively. In order to prove whether the thiourea part of the catalyst can facilitate the reaction, the reaction was conducted using thiourea as catalyst. The result indicated that thiourea could promote the reaction to provide the product in 52% yield after 12 h.

In order to demonstrate the scope of the reaction, a series of aldehydes was used for chlorination reactions. As revealed in Table 2, all the reactions proceeded efficiently to furnish the products in excellent yields (91–99%) and good enantioselectivities (85–95% ee).

Table 2 Enantioselective α-chlorination: substrate scope[a]

R-CH2-CHO + NCS → (Catalyst 10 mol%, CH2Cl2 25 ¡æ) → R-CHCl-CHO

Entry	R	Time (h)	Yield (%)[b]	ee (%)[c]
1	Et	2	99	86
2	i-Pr	2	97	90
3	n-Pr	2	95	87
4	t-Bu	2	92	95
5	n-Hexyl	3	96	92
6	n-nonyl	3	95	91
7	Allyl	3	91	89
8	PhCH2	3	99	85[d]

[a]Reaction conditions: aldehyde (0.5 mmol), NCS (0.65 mmol), CH2Cl2 (1 mL)
[b]Measured by GC using benzyl methyl ether as internal standard
[c]ee determined by chiral GC
[d]ee determined by chiral HPLC after reduction to the corresponding alcohol

Moreover, we performed the catalyst recovery experiment using 0.5 mmol of hydrocinnamaldehyde and 33 mg fluorous thiourea catalyst. It was found that the catalyst could be recovered in high yield (87%) by FSPE. The second round reaction using the recovered catalyst gave a similar product yield and ee.

4.8 Chemoselective Oxidation Reaction

Sulfoxides were valuable synthetic intermediates for the production of therapeutic agents [58]. The oxidation of sulfides was undoubtedly the easiest and most direct approach for the preparation of sulfoxides. Several transition metal based methodologies have been reported to date employing different sources of oxygen donors [59]. Later, with a view to the favorable environmental impact, some green oxidative systems were developed such as hydrogen peroxide, and some organocatalysis methodologies were reported.

Among them, Lattanzi et al. [60] reported the first example of sulfoxidation with tert-butyl hydrogen peroxide (TBHP) mediated by Schreiner thiourea. Subsequently, we also carried out similar work based on hydrogen peroxide catalyzed by fluorous thiourea [61]. In our work, CH_2Cl_2 was selected as solvent for use in the reaction and the loading of fluorous thiourea catalyst was about 10 mol%. The fluorous thiourea could be recovered more than five times and be reused directly for the next run.

5 Outlook

In summary, thiourea-based organocatalysts have been developed rapidly, and have been widely used in a variety of organic transformations. Some outstanding work has been reported in recent decades. However, fluorous thiourea-based organocatalysts are in their infancy and it is very worthwhile for researchers to carry out some more work in this field.

References

1. Houk KN, List B (2004) Asymmetric organocatalysis. Acc Chem Res 37:487
2. List B (2007) Introduction: organocatalysis. Chem Rev 107:5413
3. Seayad J, List B (2005) Asymmetric organocatalysis. Org Biomol Chem 3:719
4. Dalko PI, Moisan L (2001) Enantioselective organocatalysis. Angew Chem Int Ed Engl 40:3726
5. Dalko PI, Moisan L (2004) In the golden age of organocatalysis. Angew Chem Int Ed Engl 43:5138
6. Notz W, Tanaka F, Barbas CF (2004) Enamine-based organocatalysis with proline and diamines: the development of direct catalytic asymmetric Aldol, Mannich, Michael, and Diels-Alder reactions. Acc Chem Res 37:580
7. Chen YG, Tian SK, Deng L (2000) A highly enantioselective catalytic desymmetrization of cyclic anhydrides with modified cinchona alkaloids. J Am Chem Soc 122:9542
8. Jarvo ER, Miller SJ (2002) Amino acids and peptides as asymmetric organocatalysts. Tetrahedron 58:2481
9. Yamada YMA, Ikegami S (2000) Efficient Baylis–Hillman reactions promoted by mild cooperative catalysts and their application to catalytic asymmetric synthesis. Tetrahedron Lett 41:2165
10. Schreiner PR (2003) Metal-free organocatalysis through explicit hydrogen bonding interactions. Chem Soc Rev 32:289
11. Zhang W, Curran DP (2006) Synthetic applications of fluorous solid-phase extraction (F-SPE). Tetrahedron 62:11837
12. Etter MC (1990) Encoding and decoding hydrogen-bond patterns of organic compounds. Acc Chem Res 23:120
13. Etter MC, Panunto TW (1988) 1,3-Bis(m-nitrophenyl)urea: an exceptionally good complexing agent for proton acceptors. J Am Chem Soc 110:5896
14. Curran DP, Kuo LH (1994) Structure-based analysis and optimization of a highly enantioselective catalyst for the Strecker reaction. J Org Chem 59:3259
15. Curran DP, Kuo LH (1995) Acceleration of a dipolar Claisen rearrangement by hydrogen bonding to a soluble diaryl-urea. Tetrahedron Lett 36:6647
16. Sigman MS, Jacobsen EN (1998) Schiff base catalysts for the asymmetric Strecker reaction identified and optimized from parallel synthetic libraries. J Am Chem Soc 120:4901
17. Sigman MS, Vachal P, Jacobsen EN (2000) Angew Chem Int Ed Engl 112:1336
18. Vachal P, Jacobsen EN (2002) Structure-based analysis and optimization of a highly enantioselective catalyst for the Strecker reaction. J Am Chem Soc 124:10012
19. Wenzel AG, Jacobsen EN (2002) Asymmetric catalytic Mannich reactions catalyzed by urea derivatives: enantioselective synthesis of β-aryl-β-amino acids. J Am Chem Soc 124:12964
20. Okino T, Hoashi Y, Takemoto Y (2003) Enantioselective Michael reaction of malonates to nitroolefins catalyzed by bifunctional organocatalysts. J Am Chem Soc 125:12672

21. Okino T, Hoashi Y, Furukawa T, Xu X, Takemoto Y (2005) Enantio- and diastereoselective Michael reaction of 1,3-dicarbonyl compounds to nitroolefins catalyzed by a bifunctional thiourea. J Am Chem Soc 127:119
22. Yamaoka T, Miyabe H, Takemoto Y (2007) Catalytic enantioselective Petasis-type reaction of quinolines catalyzed by a newly designed thiourea catalyst. J Am Chem Soc 129:6686
23. Inokuma T, Hoashi Y, Takemoto Y (2006) Thiourea-catalyzed asymmetric Michael addition of activated methylene compounds to α, β-unsaturated imides: dual activation of imide by intra- and intermolecular hydrogen bonding. J Am Chem Soc 128:9413
24. Miyabe H, Takemoto Y (2008) Discovery and application of asymmetric reaction by multi-functional thioureas. Bull Chem Soc Jpn 81:785
25. Cao C, Ye M, Sun X, Tang Y (2006) Pyrrolidine-thiourea as a bifunctional organocatalyst: highly enantioselective Michael addition of cyclohexanone to nitroolefins. Org Lett 8:2901
26. Dahlin N, Bϕgevig A, Adolfsson H (2004) Adv Synth Catal 346:1101
27. Wang J, Li H, Yu X, Zu L, Wang W (2005) Chiral binaphthyl-derived amine-thiourea organocatalyst-promoted asymmetric Morita-Baylis-Hillman reaction. Org Lett 7:4293
28. Wang J, Li H, Duan W, Zu L, Wang W (2005) Organocatalytic asymmetric Michael addition of 2,4-pentandione to nitroolefins. Org Lett 7:4713
29. McCooey SH, Connon SJ (2005) Urea- and thiourea-substituted cinchona alkaloid derivatives as highly efficient bifunctional organocatalysts for the asymmetric addition of malonate to nitroalkene: inversion of configuration at C9 dramatically improves catalyst performance. Angew Chem Int Ed Engl 44:6367
30. Vakulya B, Varga S, Csampai A, Soós T (2005) Highly enantioselective conjugate addition of nitromethane to chalcones using bifunctional cinchona organocatalysts. Org Lett 7:1967
31. Hideto M, Tuchida S, Yamauchi M, Takemoto Y (2006) Reaction of nitroorganic compounds using thiourea catalysts anchored to polymer support. Synthesis 19:3295
32. Mike K, Schreiner P (2007) Generally applicable organocatalytic tetrahydropyranylation of hydroxy functionalities with very low catalyst loading. Synthesis 5:779
33. Lukas W, Cabrele C, Vanejews M, Schreiner PR (2007) γ-Aminoadamantanecarboxylic acids through direct C-H bond amidations. European J Org Chem 9:1474
34. Liu K, Cui H, Nie J, Dong K et al (2007) Highly enantioselective Michael addition of aromatic ketones to nitroolefins promoted by chiral bifunctional primary amine-thiourea catalysts based on saccharides. Org Lett 9:923
35. Li X, Liu K, Ma H, Nie J et al (2008) Highly enantioselective michael addition of malonates to nitroolefins catalyzed by chiral bifunctional tertiary amine-thioureas based on saccharides. Synlett 20:3242
36. Schreiner PR, Wittkopp A (2002) H-Bonding additives act like Lewis acid catalysts. Org Lett 4:217
37. Kote M, Schreiner PR (2006) Acid-free, organocatalytic acetalization. Tetrahedron 62:434
38. Yu MS, Curran DP, Nagashima T (2005) Increasing fluorous partition coefficients by solvent tuning. Org Lett 7:3677
39. Huang YB, Yi WB, Cai C (2010) An efficient, recoverable fluorous organocatalyst for direct reductive amination of aldehydes. J Fluor Chem 131:879
40. Berkessel A, Seelig B (2009) A simplified synthesis of Takemoto's catalyst. Synlett 2113
41. Wang L, Cai C, Curran DP, Zhang W (2010) Enantioselective α-chlorination of aldehydes with recyclable fluorous (S)-pyrrolidine-thiourea bifunctional organocatalyst. Synlett 433
42. Pipko PM (2004) Activation of carbonyls by double hydrogen bonding: an emerging tool in asymmetric catalysis. Angew Chem Int Ed Engl 43:2062
43. Wittkopp A, Schreiner PR (2003) Metal-free, noncovalent catalysis of Diels–Alder reactions by neutral hydrogen bond donors in organic solvents and in water. Chemistry 9:407
44. Sohtome Y, Takemura N, Takagi R et al (2008) Thiourea-catalyzed Morita–Baylis–Hillman reaction. Tetrahedron 64:9423

45. Huang YB, Yi WB, Cai C. An efficient, recoverable fluorous organocatalyst for accelerating the DABCO-promoted Morita-Baylis-Hillman reaction. Chin Chem Lett. doi:10.1016/j.cclet.2011.03.010
46. Dessole G, Herrera RP, Ricci A (2004) H-Bonding organocatalysed Friedel-Crafts alkylation of aromatic and heteroaromatic systems with nitroolefins. Synlett 2374
47. Dömling A, Ugi I (2000) Multicomponent reactions with isocyanides. Angew Chem Int Ed Engl 39:3168
48. Ramón DJ, Yus M (2005) Asymmetric multicomponent reactions: the new frontier. Angew Chem Int Ed Engl 44:1602
49. Pan SC, List B (2007) Catalytic one-pot, three-component acyl-Strecker reaction. Synlett 318
50. Tararov VI, Kadyrov R, Riermeier TH et al (2004) Direct reductive amination versus hydrogenation of intermediates - a comparison. Adv Synth Catal 346:561
51. Gross T, Seayad AM, Ahmad M, Beller M (2002) Synthesis of primary amines: first homogeneously catalyzed reductive amination with ammonia. Org Lett 4:2055
52. Miriyala B, Bhattacharyya S, Williamson JS (2004) Tetrahedron 60:1463
53. Itoh T, Nagata K, Ishikawa H et al (2004) A selective reductive amination of aldehydes by the use of Hantzsch dihydropyridines as reductant. Tetrahedron 60:6649
54. Weil T, Kotke M, Kleiner CM, Schreiner PR (2008) Cooperative brønsted acid-type organocatalysis: alcoholysis of styrene oxides. Org Lett 10:1513
55. Halland N, Braunton A, Bachmann S et al (2004) Direct organocatalytic asymmetric α-chlorination of aldehydes. J Am Chem Soc 126:4790
56. Brochu MP, Brown SP, MacMillan DWC (2004) Direct and enantioselective organocatalytic α-chlorination of aldehydes. J Am Chem Soc 126:4108
57. Mei Y, Bentley PA, Du J (2009) NCS with thiourea as highly efficient catalysts for acetalization of aldehydes. Tetrahedron Lett 50:4199
58. Spencer CM, Faulds D (2000) Esomeprazole. Drugs 60:321
59. Russo A, Lattanzi A (2009) Hydrogen-bonding catalysis: mild and highly chemoselective oxidation of sulfides. Adv Synth Catal 351:521
60. Mba M, Prins LJ, Licini G (2007) C_3-Symmetric Ti(IV) triphenolate amino complexes as sulfoxidation catalysts with aqueous hydrogen peroxide. Org Lett 9:21
61. Huang YB, Yi WB, Cai C (2011) A recyclable fluorous thiourea organocatalyst for the chemoselective oxidation of sulfides. J Fluorine Chem. doi:10.1016/j.jfluchem.2011.05.026

Fluoroponytailed Crown Ethers and Quaternary Ammonium Salts as Solid–Liquid Phase Transfer Catalysts in Organic Synthesis

Gianluca Pozzi and Richard H. Fish

Abstract Fluorous derivatives of dibenzo-18-crown-6 ether were prepared, and then successfully applied in representative solid–liquid phase transfer catalysis reactions, which were performed in standard organic solvents, such as chlorobenzene and toluene, as well as in fluorous solvents, such as perfluoro-1,3-dimethylcyclohexane. It was clearly shown that properly designed fluoroponytailed crown ethers could promote the disintegration of the crystal lattice of alkali salts, and transfer anions from the solid surface into an apolar, non-coordinating perfluorocarbon phase, for phase transfer catalysis reactions in organic synthesis. Furthermore, 3,5-bis(perfluorooctyl)benzyl bromide and triethylamine were reacted under mild conditions to provide an analogue of the versatile phase transfer catalyst, benzyltriethylammonium chloride, containing two fluoroponytails. This fluoroponytailed quaternary ammonium salt was also successfully employed as a catalyst in a variety of organic reactions conducted under solid–liquid phase transfer catalysis conditions, without a perfluorocarbon phase. Thus, being both hydrophobic and lipophobic, fluorous crown ethers and ammonium salts, could be rapidly recovered in quantitative yields, and reused without loss of activity, over several reaction cycles.

Keywords Ammonium salts · Crown ethers · Fluorous catalysis · N-Alkylation · Nucleophilic substitution

---

G. Pozzi
CNR-Istituto di Scienze e Tecnologie Molecolari, via Golgi 19, 20133 Milano, Italy
e-mail: gianluca.pozzi@istm.cnr.it

R.H. Fish (✉)
Lawrence Berkeley National Laboratory, University of California, Berkeley, CA 94720, USA
e-mail: rhfish@lbl.gov

Contents

1 Phase Transfer Catalysis Concepts: Fluorous Crown Ethers and Quaternary
  Onium Salts 214
2 Synthesis and Solid–Liquid Phase Transfer Catalysis Reactions of Perfluorocarbon
  Soluble Crown Ether Derivatives 217
3 Fluorous Quaternary Ammonium Salts as Solid–Liquid Phase Transfer Catalysts 226
4 Conclusions 230
References 230

1 Phase Transfer Catalysis Concepts: Fluorous Crown Ethers and Quaternary Onium Salts

Crown ethers (CE) have played a significant role in many aspects of synthetic organic chemistry, and especially in phase transfer catalysis (PTC), where their ability to form specific complexes with metal cations, M^+, led to the formation of organophilic ion-pairs, $[CE-M]^+X^-$, and to the activation and transfer of anions X^- from a solid, or less frequently aqueous phase, to a liquid organic phase [1, 2]. The term PTC encompasses several different techniques for accelerating reactions between two or more reactants present in two or more phases, all characterized by operational simplicity, mild conditions, high reaction rates, high selectivity, and the utilization of inexpensive reagents [3, 4]. These techniques, that offer significant advantages over conventional procedures, have been widely applied in industry for the synthesis of pharmaceuticals, perfumes, flavorants, dyes, agricultural chemicals, monomers, polymers, and for many other applications [5, 6]. PTC processes still have tremendous potential for waste reduction and catalyst reuse. Indeed, removal of traditional phase transfer (PT) catalysts from the reaction mixture can be achieved by solvent extraction, distillation, adsorption, or simply by washing the organic phase with copious amounts of water. In most cases, such PT catalysts have not been recovered from the effluents, or, once recovered, they have not been pure enough, and have been disposed of as waste, thus increasing the process costs, and reducing the otherwise remarkable environmental benefits of the PTC approach. CE have shown higher chemical stability in comparison to other popular PT catalysts, such as ammonium and phosphonium (onium) salts, but were also found to be more expensive, and may present potential health risks. Therefore, their efficient separation and recovery from reaction mixtures, and, if possible, their recycle would be greatly desirable.

Heterogeneous PT catalysts, bound to either an insoluble polymer or an insoluble inorganic support, have been developed in order to overcome these problems [7]. Therefore, they can be easily separated from reaction products by simple filtration and then reused; but, unfortunately, their broad application has been limited by the fact that most PTC reactions are much slower with insoluble catalysts owing to mass transfer limitations. More importantly, most solid-bound catalysts have not been found to be mechanically robust enough to survive repeated reaction/separation cycles. As an alternative, immobilization of PT catalysts on soluble polymers, such as poly(ethylene glycol)s (PEGs), has been proposed [8]. Reactions were performed under standard PTC conditions, after which, selective precipitation of the supported catalyst was induced by thorough dilution of the organic phase with an additional solvent, which showed little affinity for the polymer matrix (e.g., Et$_2$O in the case of PEGs). This method showed some limitations as well; in particular, large amounts of extra solvent were required in the precipitation step, and also for the efficient washing of the crude precipitate. In addition, its applicability to catalysts other that simple ammonium salts has been not demonstrated as yet.

Fluorous biphasic catalysis (FBC), with its many variants [9, 10], has been shown to be important for the separation of catalysts and products, and it could have a beneficial impact on PTC techniques, as demonstrated by several studies that appeared in 2004–2007 describing the use of fluorous PT in typical PTC reactions (for an overview of fluorous PTC see [11]). In the first example, an enantiopure C_2-symmetric fluorous ammonium bromide developed by Maruoka and co-workers was applied to the asymmetric synthesis of both natural and unnatural α-amino acids through enantioselective liquid–liquid PTC alkylation of a protected glycine derivative [12]. Reactions were conducted at 0 °C in a 50% aqueous KOH/toluene biphasic system, where the fluorous PT catalysts formed a third solid phase. At the end of the reaction, the fluorous salt was separated from the organic products by extraction of the reaction mixture with perfluorohexane, and it was reused at least twice more, without any loss of activity and selectivity. However, the efficiency and enantioselectivity of this catalyst were clearly inferior to those exhibited by similar non-fluorous onium salts under analogous conditions. Stuart, Gladysz, and co-workers synthesized a series of symmetrically and unsymmetrically substituted phosphonium salts by quaternarization of tertiary fluorous phosphines with primary fluorous and non-fluorous alkylating agents [13, 14]. Their viability as catalysts in liquid–liquid PTC halide exchange (Finkelstein) reactions between fluorous substrates $C_8F_{17}(CH_2)_nX$ ($n = 2$ or 3, $X = I$, Br, or Cl) dissolved in either octafluorotoluene or perfluoromethyldecalin with aqueous MY (KI, NaCl, or NaBr) was demonstrated. Recovery of such fluorous onium salts by their selective, although incomplete, precipitation from the fluorous phase upon addition of hexane was also feasible, with regards to the superior solubility of the $C_8F_{17}(CH_2)_nY$ products, in the solvent mixture of hexane/perfluorocarbon [14].

The only pre-2008 literature report dealing with fluorous PT catalysts, other than onium salts, was due to Stuart and Vidal, and it was focussed on perfluoroalkylated 4,13-diaza-18-crown-6 ethers [15]. Common organic solvents, such as halocarbons, toluene, Et$_2$O, and EtOAc, were good solvents for these compounds, and their partition coefficients between perfluoro-1,3-dimethylcyclohexane (PFDMC) and organic solvents were somewhat biased towards the organic phase. It was interesting to note that only in the case of extremely polar, fluorophobic organic solvents, such as CH$_3$CN, was there a bias towards the fluorous phase. Due to this behavior, typical of light fluorous compounds, perfluoroalkylated 4,13-diaza-18-crown-6 ethers could only be tested as catalysts in reactions performed in standard organic solvents. Two substitution reactions were studied; namely, the halide exchange

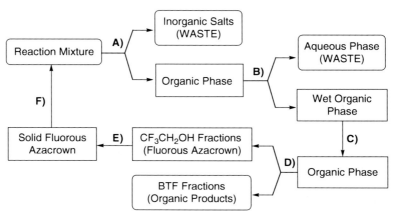

A) Filtration
B) Aqueous washing
C) Drying
D) Fluorous Solid-Phase Extraction (BTF washing followed by CF$_3$CH$_2$OH washing)
E) Evaporation of CF$_3$CH$_2$OH
F) Addition of the catalyst and solid inorganic reagent to a fresh BTF solution of organic reagent and new reaction cycle

Scheme 1 Separation and reuse of light fluorous azacrowns

reaction between 1-bromooctane and solid KI (organic solvent = (trifluoromethyl) benzene, BTF), and the aromatic nucleophilic substitution of an activated chlorobenzene derivative with solid KF (organic solvent = CH$_3$CN). The fluorinated macrocycles could be recovered from the organic phase using a multistep procedure based on fluorous solid-phase extraction (Scheme 1), and one of them could be reused six times in the Finkelstein reaction. At the end of the recycling experiments, 70% of the original catalyst quantity was recovered.

What was extremely surprising, at that time, was that despite the potential advantages associated with the use of an additional fluorous liquid phase in the reaction step [9], PTC involving organic substrates and fluorous PT catalysts dissolved in perfluorocarbons had been largely ignored, since no examples had been previously published. Therefore, we subsequently reported on the synthesis of the first perfluorocarbon soluble derivatives of dibenzo-18-crown-6 ether, **F-CE**, their solubilization of potassium anion salts, such as KI, KCN, KOMe, and KOH in the fluorous phase, and their subsequent reactivity in solid–liquid phase transfer catalysis (SL-PTC) nucleophilic substitution and oxidation reactions with aliphatic and aromatic halogen substrates, and an aromatic substituted hydrocarbon, respectively [16].

More importantly, while fluorous phosphonium salts have been readily used as PTC reagents, the corresponding fluorous quaternary ammonium salts have not been studied extensively [11]. Therefore, we recently developed a fluorous analogue of the versatile phase transfer catalyst, benzyltriethylammonium chloride

(**TEBA**), containing two fluoroponytails (**F-TEBA**), which was also successfully employed as a catalyst in a variety of organic reactions conducted under SL-PTC conditions, without a perfluorocarbon phase [17].

2 Synthesis and Solid–Liquid Phase Transfer Catalysis Reactions of Perfluorocarbon Soluble Crown Ether Derivatives

The perfluorocarbon soluble **CE** derivatives **F-CE1-3** (Scheme 2) were designed on the basis of the experience gained from our previous investigations on fluorous macrocyclic ligands [18–20]. Thus, it was realized that, besides a relatively high fluorine loading, the presence of more than two fluoroponytails was crucial for ensuring a suitable degree of fluorophilicity to the fluoroponytailed **CE** derivatives. Therefore, dibenzo-18-crown-6 ether (**DB18-C-6**) was an ideal scaffold, due to the ease of controlled functionalization of its aromatic moieties, and to their possible

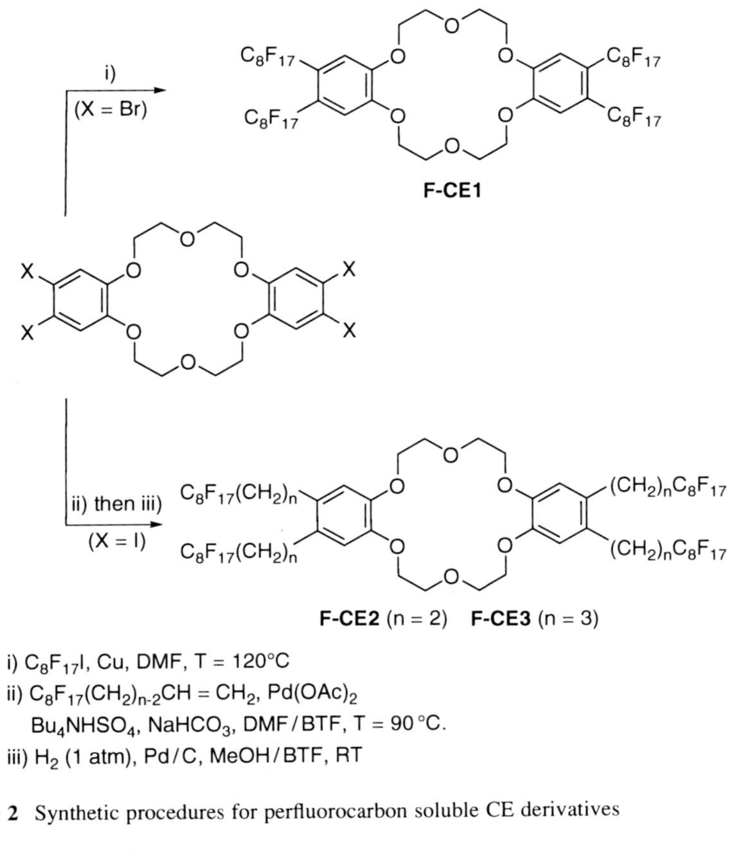
i)
(X = Br)
C_8F_{17}
F-CE1
X
ii) then iii)
(X = I)
$C_8F_{17}(CH_2)_n$
$(CH_2)_nC_8F_{17}$
F-CE2 (n = 2) F-CE3 (n = 3)
i) $C_8F_{17}I$, Cu, DMF, T = 120°C
ii) $C_8F_{17}(CH_2)_{n-2}CH=CH_2$, $Pd(OAc)_2$
Bu_4NHSO_4, $NaHCO_3$, DMF/BTF, T = 90 °C.
iii) H_2 (1 atm), Pd/C, MeOH/BTF, RT

Scheme 2 Synthetic procedures for perfluorocarbon soluble CE derivatives

shielding effect towards the electron-withdrawing effect of perfluoroalkyl substituents, which could significantly reduce the Lewis basicity of the oxygen binding sites, and thus, the utility of the resulting fluoroponytailed **CE** as a PT catalyst. We found that up to four fluoroponytails, $C_8F_{17}-$, could be directly introduced into the aromatic subunits of **DB18-C-6** by means of metal-catalyzed cross-coupling reactions, as exemplified by the synthetic pathways shown in Scheme 2.

F-CE1 (%F = 63.6%), a derivative of **DB18-C-6** bearing fluoroponytailed C_8F_{17}-substituents, directly attached to the aromatic subunits, was readily prepared in 64% yield by the copper-catalyzed perfluoroalkylation of 4,4',5,5'-tetrabromodibenzo-18-crown-6 [21] with $C_8F_{17}I$. Both **F-CE2** (%F = 60.2) and **F₁-CE3** (%F = 58.7) were prepared by Heck vinylation of 4,4',5,5'-tetraiododibenzo-18-crown-6 [22] with an appropriate perfluoroalkene (for the synthesis of 1 *H*,1 *H*,2 *H*,3 *H*,3 *H*-heptadecafluoro-1-undecene see [23]. 1 *H*,1 *H*,2 *H*-Heptadecafluoro-1-decene is a commercially available compound) under phosphine-free conditions [24], and this was followed by hydrogenation of the perfluoroalkenyl-substituted intermediates. Overall yields were 13% and 38% for **F-CE2** and **F-CE3**, respectively.

In contrast to previously reported perfluoroalkylated 4,13-diaza-18-crown-6 ethers [15], the three fluoroponytailed **DB18-C-6** derivatives were found to be almost completely insoluble in common organic solvents at room temperature, with the exception of CH_2Cl_2 and $CHCl_3$, but totally soluble in BTF, and also in perfluorocarbons at temperatures slightly > 40 °C, with partial precipitation of **F-CE2** and **F-CE3** from their 10^{-2} M solutions, after standing at room temperature for a few hours. The significant fluorophilicity exhibited by the three **F-CE** was confirmed by partition coefficient measurements between PFDMC and organic solvents, including low polar ones (Table 1). Values ranged from >98/2 to 70/30 in the case of **F-CE1** and **F-CE3** (organic solvent = CH_2Cl_2), respectively. This behavior is noteworthy also in the light of the considerable affinity for organic solvents exhibited by related fluoroponytailed **DB18-C-6** derivatives bearing only two C_nF_{2n+1}- substituents later disclosed in the literature [25].

Table 1 Partition coefficients *P* of fluorous **CE** between PFDMC and organic solvents[a]

Fluorous **CE**	Organic solvent	Partition coefficient *P*
F-CE1	Toluene	>98/2
F-CE2		96.9/3.1
F-CE3		86.5/13.5
F-CE1	CH_2Cl_2	>98/2
F-CE2		89.4/10.6
F-CE3		69.7/30.3
F-CE1	CH_3CN	>98/2
F-CE2		97.0/3.0
F-CE3		96.4/3.6

[a] $P =$ [**F-CE**]$_{PFDMC}$/[**F-CE**] $_{Organic\ solvent}$. Determined gravimetrically at $T = 20$ °C (for experimental details see [16])

The complexation ability of the new macrocyclic ligands towards alkali metal cations was evaluated by studying the extraction of the corresponding picrate salts from aqueous solutions into selected solvents, and the results are summarized in Table 2. **DB18-C-6** has been known to bind K$^+$ ions preferentially over other alkali metal cations, on the basis of the match between the cavity and the cation size. This property was maintained by its fluoroponytailed derivatives, **F-CE1-3**, as shown by the comparison of potassium and sodium picrate extraction performed in CH$_2$Cl$_2$ (Table 2, entries 1–4). Furthermore, it was also demonstrated that the two, and the better, three methylene spacer with the perfluoroponytails, was essential for efficient K$^+$ binding, since the powerful electron-withdrawing effect of a C$_8$F$_{17}$– group directly bonded to the aromatic ring inhibited cation binding to the **F-CE** oxygen array. This behavior was similar to the well-known inverse relationship between the cumulative electron-withdrawing power of the substituents, and the complexation ability of **DB18-C-6** derivatives towards potassium and sodium salts [26]. The presence of three methylene spacer units in **F-CE3** ensured a picrate extraction ability very close to that of the classical PT catalyst, **DB18-C-6**, both in a typical organic solvent (Table 2, CH$_2$Cl$_2$, entry 1 vs 4), and in the partially fluorinated solvent, BTF (Table 2, entry 5 vs 8). As previously observed in related experiments with fluorous phosphonium salts [13] or perfluoroalkylated 4,13-diaza-18-crown-6 ethers [15], the nature of the non-aqueous solvent had a critical influence on the level of potassium picrate extraction by **F-CE** (Table 2, entries 2 vs 6 and 9, 3 vs 7 and 10, 4 vs 8 and 11), and extractions in BTF proved to be far less efficient than those performed in CH$_2$Cl$_2$. Quite interestingly, the extraction ability of **DB18-C-6** was affected likewise by this change of solvent (Table 2, entry 1 vs 5), thus casting some doubts on the potential of BTF as a solvent for PTC reactions. However, the most outstanding feature of these experiments was the clear demonstration that **F-CE3**, and **F-CE2**, to a much lesser extent, retained their potassium ion complexing abilities even in an apolar, non-coordinating perfluorocarbon such as

Table 2 Picrate extraction experiments[a]

Entry	Fluorous CE	Solvent	Na$^+$	K$^+$
1	**DB18-C-6**	CH$_2$Cl$_2$	2.5	35.2
2	**F-CE1**		<1	2.5
3	**F-CE2**		1.2	20.9
4	**F-CE3**		2.3	37.3
5	**DB18-C-6**	BTF	–	12.3
6	**F-CE1**		–	n.d.
7	**F-CE2**		–	3.2
8	**F-CE3**		–	10.8
9	**F-CE1**	PFDMC	–	n.d.
10	**F-CE2**		–	1.1
11	**F-CE3**		–	30.7

n.d. Not detected
[a]Equal volumes of a 10^{-4} M aqueous solution of alkali metal picrate and a 10^{-4} M organic or fluorous solution of **CE** at $T = 25\ °C$ (for experimental details see [16])

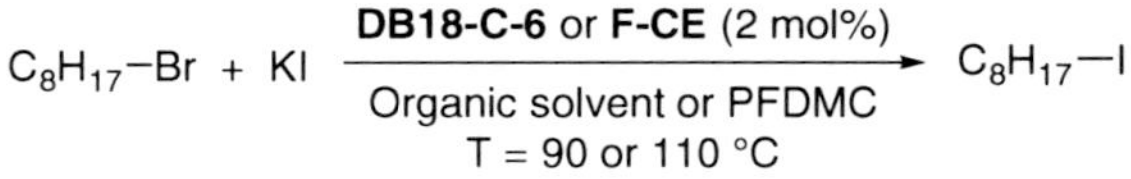

Scheme 3 Halide substitution reaction (Finkelstein reaction) under SL-PTC conditions

$$C_8H_{17}-Br + KI \xrightarrow[\text{T = 90 or 110 °C}]{\text{DB18-C-6 or F-CE (2 mol\%)}}_{\text{Organic solvent or PFDMC}} C_8H_{17}-I$$

PFDMC (Table 2, entries 10 and 11). This unprecedented behavior clearly demonstrated the ability of **F-CE** to act as significant PT catalysts in perfluorocarbons [16].

Lipophilic **CE** that were able to form complexes with alkali metal cations have been shown to be the catalysts of choice in many SL-PTC reactions [2]; namely, in reactions where a solid ionic reagent was suspended in an anhydrous organic solution containing the substrate and the PT catalyst. The latter interacted with the surface of the solid salt, thus promoting the collapse of the crystal lattice and the subsequent transfer of the anion, as a reactive ion-pair, from the surface of the solid into the liquid organic phase. This option offers specific advantages over liquid–liquid PTC; in particular, it avoids the presence of water, which may cause the reduction of the rate of the main process, lead to undesired side reactions, and complicate the isolation of the products [3]. To obtain an insight into the impact of the residual electron-withdrawing effect of the fluoroponytails and the nature of the solvent system on the efficiency of the solid–liquid PTC process, the catalytic activities of **F-CE1-3** were first investigated in the classical Finkelstein reaction between 1-bromooctane and KI (Scheme 3), and then compared to that of **DB18-C-6**.

BTF has been shown to be a credible solvent both for fluorous and purely organic molecules, while it was completely miscible with most organic solvents. Such amphiphilic behavior was clearly incompatible with the requirements of the fluorous biphasic reaction systems. However, BTF has been widely used in reactions, where fluorous reagents and/or catalysts showing limited solubility in perfluorocarbons were involved, including the halide exchange between 1-bromooctane and KI under SL-PTC conditions [15]. This model reaction was thus initially performed in the presence of 2 mol% of **F-CE1 – 3** or **DB18-C-6** in BTF at 90 and 110 °C, and the results are shown in Table 3 and Fig. 1. Independent of the temperature, the halide exchange occurred to a very limited extent in the absence of a PT catalyst (Table 3, entries 1 and 7), while the addition of **F-CE2** (Table 3, entries 5, 10), **F-CE3** (Table 3, entries 6, 11), or **DB18-C-6** (Table 3, entries 2, 10) catalyzed the formation of 1-iodooctane in high yields, with reaction rates following the order: **F-CE3** > **DB18-C-6** > **F-CE2**, both at 90 and 110 °C, although, as shown in Fig. 1, the catalytic activities of **F-CE3** and **DB18-C-6** became closer as the reaction temperature was increased. Alternatively, **F-CE1** proved to be a poor PT catalyst, both under solid–liquid (Table 3, entries 3 and 9) and liquid–liquid (Table 3, entry 4) PTC conditions, in agreement with the results obtained in the picrate extraction tests.

Further experiments performed in chlorobenzene (Table 4 and Fig. 2) demonstrated the central role of the solvent in determining the SL-PTC activity of fluoroponytailed **CE**. Indeed, chlorobenzene was known to be an excellent

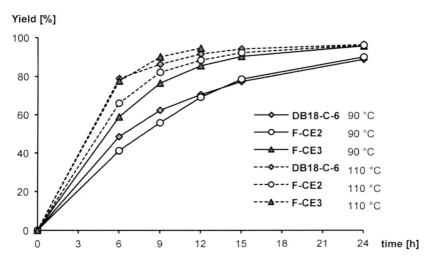

Fig. 1 Rates of formation of 1-iodooctane in SL-PTC reactions run in BTF

Table 3 Finkelstein reaction in BTF under SL-PTC conditions[a]

Entry	CE	T (°C)	t (h)	Yield (%)	TOF[b]
1	–	90	24	3	–
2	DB18-C-6		24	89	1.88
3	F-CE1		24	20	0.41
4[c]	F-CE1		24	17	0.36
5	F-CE2		24	90	1.88
6	F-CE3		15	91	3.03
7	–	110	24	3	–
8	DB18-C-6		15	94	3.14
9	F-CE1		24	21	0.45
10	F-CE2		15	92	3.08
11	F-CE3		12	95	3.94

[a]Reaction conditions: substrate = 1 mmol, KI = 5 mmol, CE = 2 mol%, solvent = 4 mL. Selectivity for $C_8H_{17}I$ > 98%
[b]TOF = mmol converted substrate/(mmol catalyst × hour)
[c]Liquid/liquid PTC. Reaction conditions: substrate = 1 mmol, KI = 5 mmol, CE = 2 mol%, solvent = 2 mL, H_2O = 0.6 mL

medium for anion promoted reactions catalyzed by standard PT catalysts [27], and in this solvent the best results were obtained with **DB18-C-6** (Table 4, entries 2 and 7). Both reaction yields and rates (Fig. 2) were superior to those observed in BTF with the same catalysts, and also with **F-CE3**. The behavior of the latter (Table 4, entries 5 and 10) was not improved by the change of solvent, while the catalytic activities of **F-CE1** (Table 4, entries 3 and 8) and **F-CE2** (Table 4, entries 4 and 9) were drastically reduced. This effect was more pronounced for reactions run at 90 °C, where, in the case of **F-CE2**, the low solubility of the fluoroponytailed CE in

Table 4 Finkelstein reaction in chlorobenzene under SL-PTC conditions[a]

Entry	CE	T (°C)	t (h)	Yield (%)	TOF[b]
1	–	90	24	1	–
2	DB18-C-6		15	97	3.24
3	F-CE1		24	7	0.15
4	F-CE2		24	32	0.68
5	F-CE3		15	92	3.05
6	–	110	24	3	–
7	DB18-C-6		9	96	5.32
8	F-CE1		24	21	0.44
9	F-CE2		24	94	1.96
10	F-CE3		12	95	3.95

[a]Reaction conditions: substrate = 1 mmol, KI = 5 mmol, CE = 2 mol%, solvent = 4 mL. Selectivity for $C_8H_{17}I$ > 98%
[b]TOF = mmol converted substrate/(mmol catalyst × hour)

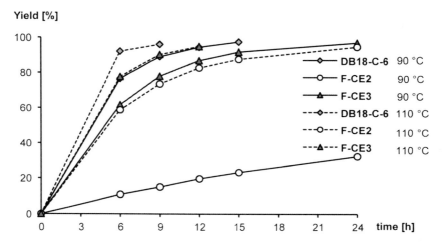

Fig. 2 Rates of formation of 1-iodooctane in SL-PTC reactions run in chlorobenzene

chlorobenzene also provided a negative effect. Nevertheless, **F-CE2** gave comparatively poorer results in chlorobenzene than in BTF, even at 110 °C. At that temperature, the fluorous catalyst was completely soluble in the reaction environment, thus highlighting the inherent influence exerted by the nature of the solvent on the outcome of the PTC process.

Based on these results, we were intrigued, at the time, by the possibility of using both a perfluorocarbon and the more promising fluorous **CE** compounds, **F-CE2** and **F-CE3**, in PTC nucleophilic substitution reactions, with KI, KCN, and KOMe, along with aromatic and aliphatic halogen substrates. In fact, the achievement of reactivity and selectivity comparable or even superior to those observed with classical PT agents, without compromising the ease of recovery and the recycling efficiency, has been a primary focus of our research on fluorous PT catalysts. These

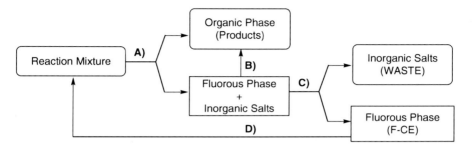

A) Phase separation
B) Organic washing
C) Filtration
D) Addition of fresh reagents to the fluorous layer and new reaction cycle

Scheme 4 Separation and reuse of perfluorocarbon soluble F-CE

goals could not have been attained without the systematic search for optimal reaction conditions and PTC techniques in association with these particular fluoroponytailed **CE** compounds. Thus far, simple liquid–liquid and solid–liquid PTC systems have been explored, but fluorous PT catalysts would also offer the opportunity to consider alternative approaches; for example, the use of a triphasic system consisting of a fluorous, organic, and inorganic phase. We expected that, in analogy to other multiphase PTC systems [28], at the end of the reaction, the immiscible fluorous phase containing **F-CE2** and **F-CE3** could be easily recovered and reused, without resorting to any specific treatment, such as fluorous solvent or solid-phase extraction techniques (Scheme 4).

The benchmark PTC experiment with PFDMC was obviously the same Finkelstein reaction previously performed under standard SL-PTC conditions. This time, the KI was found in the bottom solid phase, the **F-CE2** and **F-CE3** compounds in the PFDMC layer, while 1-bromooctane was found in the top layer. These results are summarized in Table 5.

Clearly, **F-CE2** and **F-CE3** were necessary for the SL-PTC reaction to occur, while entries 4–6 define the recycle of the **CE**, which was accomplished by simple phase separation as shown in Scheme 4, as a facile process that circumvents the necessity of any cumbersome, waste-producing, post-reaction manipulations, including the use of fluorous solid-phase materials (see Scheme 1). Interestingly, an ^{1}H NMR spectrum of the recovered solid during evaporation of the PFDMC layer, after four recycles, showed downfield shifts of the –OCH$_2$CH$_2$O- resonances from 4.0 and 4.15 ppm to 4.25 and 4.17 ppm, while the aromatic resonance shifted upfield from 6.65 to 6.1 ppm. This was similar to the literature values for the inclusion complexes of several K$^+$X$^-$ in **DB18-C-6** [29], and clearly defines the formation of [(**F-CE3**)K]$^+$ I$^-$ in a fluorous solvent. Moreover, we independently synthesized the [(**F-CE3**)K]$^+$ I$^-$ complex, which corroborated the above-mentioned NMR results. Furthermore, entries 9–12 in Table 5 allowed a pertinent example of the thermomorphic property of **F-CE3** [30], with this catalyst being

Table 5 Finkelstein reaction in PFDMC under SL-PTC conditions[a]

Entry	CE	T (°C)	t (h)	Yield (%)	TOF[b]
1	–	90	24	9	–
2	**F-CE2**		24	64	1.33
3	**F-CE3**		8 (15)	69 (95)	4.31 (3.16)
4[c]			8 (15)	68 (93)	4.31 (3.10)
5[c]			8 (15)	66 (93)	4.25 (3.10)
6[c]			8 (15)	68 (90)	4.31 (3.01)
7	–	110	24	17	–
8	**F-CE2**		24	87	1.81
9	**F-CE3**		12	96	4.01
10[d]	**F-CE3**	90	24	78	1.63
11[d, e]			24	73	1.52
12[d, e]			24	69	1.45

[a]Reaction conditions: substrate = 1 mmol, KI = 5 mmol, CE = 2 mol%, solvent = 4 mL. Selectivity for $C_8H_{17}I$ > 98%
[b]TOF = mmol converted substrate/(mmol catalyst × hour)
[c]Recycling experiment: the fluorous layer was recovered at RT from the preceding reaction, and reused in subsequent reaction cycles
[d]Solvent = toluene
[e]Recycling experiment: the solid catalyst was recovered at 0 °C from the preceding reaction, and reused; the inorganic salts were first centrifuged from the hot mixture

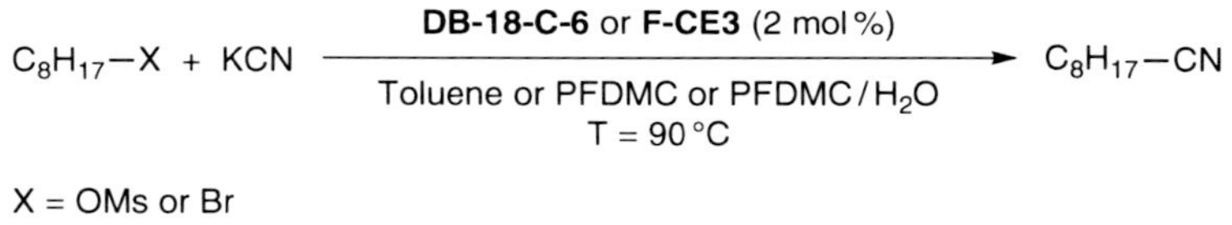

$$C_8H_{17}-X + KCN \xrightarrow[\substack{\text{Toluene or PFDMC or PFDMC/H}_2\text{O} \\ T = 90\,°C}]{\textbf{DB-18-C-6 or F-CE3 (2 mol\%)}} C_8H_{17}-CN$$

X = OMs or Br

Scheme 5 Nucleophilic substitution reaction with KCN under SL-PTC conditions

insoluble in toluene at 0 °C, while at 90 °C it was fully soluble; recycle was conducted via filtration of the ice cold suspension, after removal of the inorganic products from the hot reaction mixture. This latter thermomorphic property of **F-CE3** added a new dimension to the PTC process, and, furthermore, was accomplished without a perfluorocarbon solvent.

The nucleophilic displacement reaction of *n*-octyl methanesulfonate with solid KCN in PFDMC, to provide *n*-octyl cyanide in > 98% selectivity, was then studied (Scheme 5, X = OMs). The control experiment (Table 6, entry 1) clearly afforded less of the product in comparison to the addition of **F-CE3** (Table 6, entry 2). The less reactive 1-bromooctane (Scheme 5, X = Br) was also used as a substrate in order to demonstrate the catalytic effect of **F-CE3**. Thus, traces of *n*-octyl cyanide were formed in the absence of a catalyst (Table 6, entry 3), but also SL-PTC reactions run in the presence of either 2 mol% of **F-CE3** (Table 6, entry 4) or **DB18-C-6** (Table 6, entry 5) gave product yields of ~10%. Moreover, the rate of the SL-PTC cyanide displacement reaction of 1-bromooctane in organic solvents was

Table 6 Nucleophilic substitution reactions with KCN under SL-PTC conditions[a]

Entry	CE	X	Solvent	Yield (%)
1	–	OMs	PFDMC	35
2	F-CE3	OMs	PFDMC	71
3	–	Br	PFDMC	<1
4	F-CE3	Br	PFDMC	7
5	DB18-C-6	Br	Toluene	9
6[b]	–	Br	PFDMC/H$_2$O	<1
7[b]	F-CE3	Br	PFDMC/H$_2$O	93

[a]Reaction conditions: substrate = 0.5 mmol, KCN = 2.5 mmol, CE = 2 mol%, solvent = 2 mL, Time = 24 h. Selectivity for C$_8$H$_{17}$CN > 98%
[b]Thin-layer PTC. Reaction conditions: substrate = 0.5 mmol, KCN = 2.5 mmol, CE = 2 mol%, PFDMC = 2 mL, H$_2$O = 0.16 mL. Time = 24 h. Selectivity for C$_8$H$_{17}$CN > 98%

Scheme 6 Nucleophilic aromatic substitution reaction under SL-PTC conditions

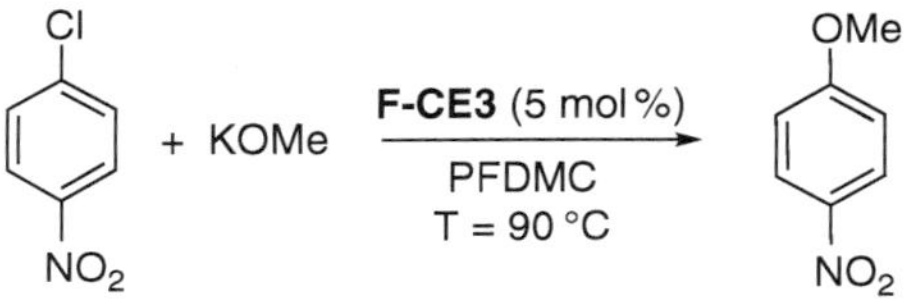

previously found to be a function of the amount of added water [31]. We had observed similar behavior for reactions in PFDMC, while the catalytic effect of **F-CE3** was clearly evident using thin-layer PTC conditions [32] (Table 6, entries 6 and 7), where the addition of a small amount of water generates a third liquid phase coating on the surface of the solid [33].

The SL-PTC of anion solubility in PFDMC was further extended to KOMe, and its nucleophilic aromatic substitution reaction (S$_N$Ar) with 4-nitrochlorobenzene (Scheme 6). Pertinently, these S$_N$Ar reactions were known to be challenging under standard conditions [34], while other competitive reactions were also known to interfere in the absence of a **CE** [35]. What was so important, as epitomized in the model reaction, was the threefold increase in yield of product, from 24%, in the absence of a **CE**, to 70%, afforded by **F-CE3** (5 mol%). As observed in common organic solvents, the addition of the fluorous **CE** selectively increased the rate of the S$_N$Ar process, and therefore it was highly beneficial with respect to the product selectivity.

It has been previously shown that carbanions of organic compounds with acidic C–H bonds can be generated in the presence of inorganic bases under liquid–liquid or SL-PTC conditions, and then oxidized with molecular oxygen [36]. The substrate most studied for the above-mentioned reaction was fluorene, which undergoes a PTC aerobic oxidation to yield fluorenone. Therefore, analogously to organic **CE** and cryptates, **F-CE3** was able to catalyze that model reaction when solid KOH was used as the base (Scheme 7). In a blank reaction performed at 60 °C in PFDMC, conversion of the substrate was as low as 30%, and fluorenone was obtained in 26% yield, while a facile conversion reached 99%, with a yield in fluorenone of 93% when **F–CE3** (5 mol%) was added to the reaction mixture.

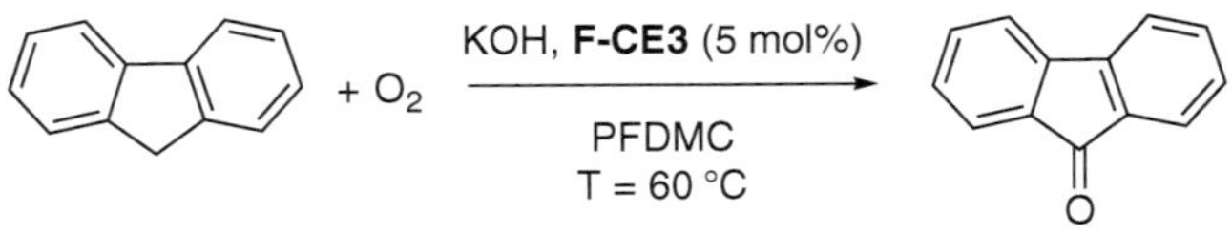

Scheme 7 Aerobic oxidation of fluorine under SL-PTC conditions

After finally establishing the feasibility and convenience of SL-PTC involving organic substrates and fluorous PT catalysts, dissolved in perfluorocarbons, specific studies addressing the manifold of mechanistic issues of this process, e.g., the way anions could be transported from the solid to the liquid phase, the nature of the reactive ion pairs formed by the fluorous PT catalysts, and the solvent phase, where the final interaction with the organic substrates takes place, can now be pursued. This can be expected to be a long-term project, as in the case of conventional PTC [5, 11, 33].

3 Fluorous Quaternary Ammonium Salts as Solid–Liquid Phase Transfer Catalysts

Fluoroponytailed PT catalysts, such as quaternary phosphonium salts $(R_f)_3R^1P^+Y^-$ ($R_f = C_nF_{2n+1}(CH_2)_n$-, $R^1 = R_f$, alkyl, or Bn) with typical light or heavy fluorous characteristics, and yet having retained the ability to transfer reactive anionic species from water or from a solid surface into a second distinct organic or perfluorocarbon phase, have also recently emerged on the fluorous chemistry scene [11]. These initial studies were mainly focused on proving the viability of the fluorous approach in some well known PTC reactions, such as Finkelstein type ionic displacement reactions of simple fluorous alkyl halides $R_f(CH_2)_nX$ [13, 14, 37], and analogous nucleophilic substitutions on purely organic substrates [15, 16, 25]. The ease of recovery and the recycling efficiency of PT fluorous catalysts were thus demonstrated, as we just discussed. In addition, reactivities and selectivities comparable to those observed with classical PT agents could be achieved through careful design of the fluorous catalysts, even in the case of reactions run in extremely nonpolar perfluorocarbon phases [16, 37].

However, the true potential of fluorous catalysts in PTC reactions has not been fully exploited, as yet. This was particularly prevalent in the case for fluoroponytailed quaternary ammonium salts, which have not been extensively investigated as PT catalysts, in spite of the wide utility and versatility of their standard counterparts [11, 17, 38]. Following the previously discussed example of a synthetically demanding chiral fluoroponytailed ammonium salt due to Maruoka and co-workers [12, 39], a restricted number of structurally simple fluorous ammonium salts were also used by others in catalytic reactions, though not strictly speaking as PT agents [40–42]. Various synthetic methods were reported to generate fluoroponytailed ammonium salts $R_{f(4-m)}R_m^1N^+Y^-$, which could be potentially useful as

Scheme 8 Synthesis of F-TEBA

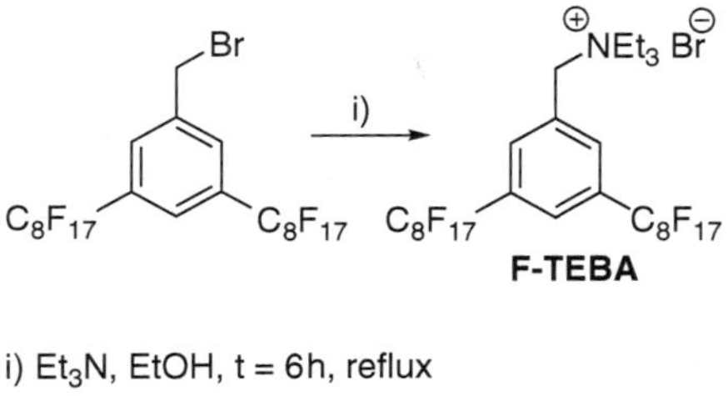

i) Et₃N, EtOH, t = 6h, reflux

PT catalysts [11], but because of the considerably lower reactivity of tertiary amines bearing R_f substituents [43], the final *N*-quaternization step often proved to be challenging. This drawback was avoided in the case of **F-TEBA** (Scheme 8), which was readily obtained in 92% yield by N-alkylation of Et₃N with the highly reactive fluoroponytailed benzyl bromide, 3,5-bis(perfluorooctyl)benzyl bromide, a valuable building-block in the synthesis of fluorous molecules [17].

F-TEBA was found to be highly soluble in polar organic solvents, such as MeOH and CH₃CN, while it was found to be insoluble in water and in lipophilic organic solvents, such as toluene and hexane. In other media, with higher affinity for fluorous substances, such as BTF or ethers, it exhibited highly temperature-dependent solubility profiles. As an example, **F-TEBA** was poorly soluble in dioxane at room temperature (1.3 mg/mL). However, by raising the T from 25 °C to 80 °C, its solubility was increased more than tenfold. Analogously, these same thermomorphic properties were previously observed in the case of quaternary phosphonium salts $(R_f)_3R^1P^+Y^-$ [13, 14, 37]. As far as nonpolar perfluorocarbon phases were concerned, **F-TEBA** has no appreciable solubilities at room temperature, but stable colloidal dispersions were formed upon refluxing and subsequent cooling. The $-NEt_3^+$ polar head of **F-TEBA**, which was instrumental for its use as a PT reagent, was likely to be responsible for this particular behavior. Indeed, related ammonium salts [37, 41] with an $-NH_3^+$ polar head attached to a 3,5-bis(perfluoroalkylated)aryl ring were found to be soluble in perfluorocarbons. Unfortunately, the reactive nature of the $-NH_3^+$ group imposed serious limitations on the applicability of such salts in many relevant PTC processes, especially those involving basic reagents.

To evaluate the efficiency of **F-TEBA**, we had focused our attention on SL-PTC, another neglected facet of fluorous PTC. Indeed, whereas a few examples of SL-PTC nucleophilic substitutions had been reported using fluoroponytailed **CE** derivatives [15, 16], fluorous onium salts had been only tested under liquid–liquid (water-organic solvent or water-perfluorocarbon) conditions [11]. Three representative reactions were thus chosen, in which a reactive anion was generated on the surface of solid K₂CO₃ suspended in an anhydrous organic solution containing the reagents and **F-TEBA**. The latter transferred the anion, as a reactive ion-pair, from the surface of the solid base into the organic phase. The epoxide ring-opening reaction, shown in Scheme 9, was selected for its versatility in the preparation of aziridines [44] and 2,6-substituted morpholines [45], both used in stereoselective synthesis as chiral auxiliaries and ligands.

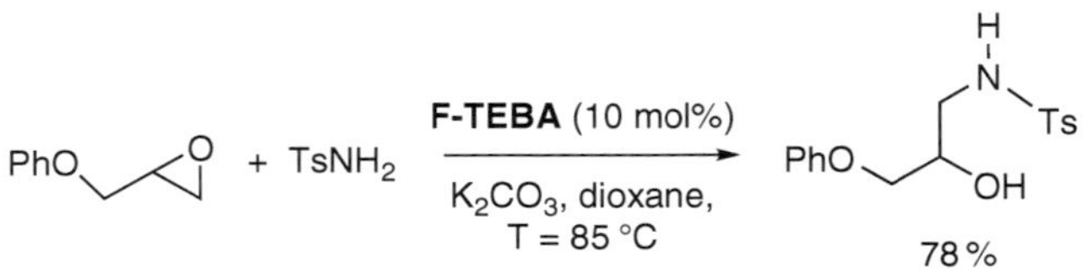

Scheme 9 Ring-opening reaction of phenyl glycidyl epoxide under SL-PTC conditions

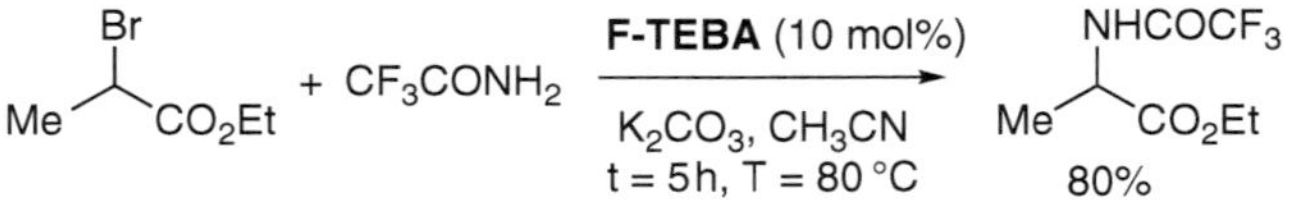

Scheme 10 N-Alkylation of trifluoroacetamide with ethyl 2-bromopropanoate under SL-PTC conditions

R^1 = Bn, Me, i-Pr, (S)-CH$_3$CH$_2$CH(CH$_3$)-

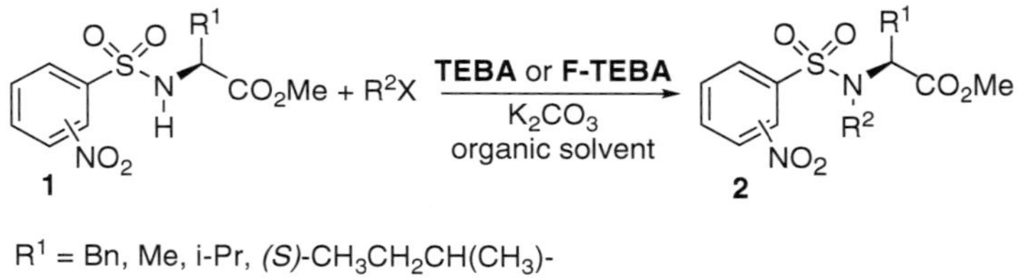

Scheme 11 N-Alkylation of N-(2(4)-nitrobenzenesulfonyl) α-amino acid esters

Furthermore, N-alkylation of trifluoroacetamide with 2-bromo esters, such as ethyl 2-bromopropanoate, provided access to a large number of natural and unnatural α-amino acids (Scheme 10) [46].

In both cases, **F-TEBA** gave results identical to those obtained with benzyltriethylammonium chloride (**TEBA**), the catalyst of choice for these reactions, thus showing that the introduction of the fluorous substituents had no adverse effects on the catalytic activity of these types of ammonium salts. More importantly, there had been considerable interest in the development of efficient procedures for the synthesis of enantiomerically pure N-alkyl amino acids and esters, due to their importance as intermediates in the synthesis of biologically active products. A functional methodology, amenable to large-scale applications, was reported by Penso and co-workers, and was based on the selective N-monoalkylation of N-(2-nitrobenzenesulfonyl) α-amino acid esters, **1**, under mild SL-PTC conditions, to give compounds of the general formula **2** (Scheme 11) [47]. The same reaction applied to N-(4-nitrobenzenesulfonyl) α-amino esters also allowed a convenient access to compounds for the synthesis of α-4-nitrobenzene α-amino acids [48], which have attracted considerable attention for the design of new oligopeptide fragments.

For example, the reaction between the L-phenylalanine derivative, **2**, where R^1 = benzyl, and a slight excess of allyl bromide, in the presence of anhydrous K_2CO_3, was chosen in order to determine suitable conditions for the use of **F-TEBA** as a **SL-PT** catalyst.

In both the case of **TEBA** (entry 5) and of **F-TEBA** (entry 8), the better results were obtained with a catalyst loading of 10 mol%, with respect to the substrate, in refluxing CH_3CN (entry 8). The reaction with **F-TEBA** was considerably slower in other solvents, such as toluene (entry 6) and BTF (entry 7). Solubility issues were not a convincing explanation for these results. Indeed, at 80 °C, **F-TEBA** readily dissolved in BTF, and the lipophilic ion-pair formed by anion exchange between **F-TEBA** and the deprotonated substrate was equally soluble in toluene. This reinforced the idea, which first emerged in the case of fluorous crown ethers [16], that the choice of the solvent in SL-PTC reactions cannot be dictated alone by its affinity for the fluorous PT reagents.

The work-up of the reaction mixture and the separation of **F-TEBA** were operationally simple. For example, fluorous solid-phase extraction, without using any fluorous solvent, provided the reaction products, in excellent yields, and without any additional purification step. The catalyst, **F-TEBA**, could also be conveniently recovered by washing the cartridge with MeOH, with no apparent degradation, and was reused. However, **F-TEBA** was partly retained on fluorous silica, and the optimal work-up method found entailed the evaporation of the clear CH_3CN phase, followed by the addition of toluene. The solid catalyst was then quantitatively recovered by filtration, and was reused: five subsequent recycles of **F-TEBA** (Table 7, entries 8–12) were conducted, with no apparent loss of activity. The resting state of the catalyst features Br^- as the counterion, and was verified by potentiometric titrations with $AgNO_3$. Indeed, multiply charged anions such as CO_3^{2-} were extracted with great difficulty by quaternary ammonium salts [49].

Table 7 N-Alkylation of N-(2-nitrobenzenesulfonyl) α-amino acid ester, **1**, with $R^2X = CH_2 = CHCH_2Br$, in CH_3CN under SL-PTC conditions

Entry	Catalyst	mol%[a]	T (°C)	t (h)	Yield[b] (%)
1	–	–	30	10	Traces[c]
2	F-TEBA	5	30	10	75
3	F-TEBA	10	30	10	86
4	TEBA	10	30	10	87
5	TEBA	10	80	1.5	98
6[d]	F-TEBA	10	80	12	98
7[e]	F-TEBA	10	80	7	97
8	F-TEBA	10	80	1.5	98
9[f]			80	1.5	99
10			80	1.5	98
11			80	1.5	98
12			80	1.5	98

[a]With respect to **2**
[b]Isolated yields
[c]Detected by ^{1}H-NMR
[d]Solvent = toluene
[e]Solvent = BTF
[f]Entries 9–12 = subsequent reaction performed with recycled **F-TEBA**

4 Conclusions

We had recently demonstrated, for the first time, that properly designed fluoroponytailed crown ethers, such as **F-CE2** and **F-CE3**, were excellent phase transfer catalysts that allowed solubilization of potassium salts in a non-coordinating, apolar perfluorocarbon solvent, leading to their reactivity in nucleophilic substitution and oxidation reactions. Recycle was shown to be facile by simply reusing the fluorous phase for each catalytic run. Moreover, **F-CE3** was found to be thermomorphic in toluene, and fully extends the PTC paradigm for catalyst recycle. At this stage, the remarkable SL-PTC activity of **F-CE3** in PFDMC can be reasonably ascribed to a poorly solvated anion effect in a hydrophobic solvent environment that provided a driving force for nucleophilic substitution at the solvent/substrate interface, and to a minor extent, the fluorous phase. Further future mechanistic investigations will lead to a better understanding of the numerous facets of PTC in a perfluorocarbon environment.

We have also recently shown that a readily accessible perfluorolakylated ammonium salt, **F-TEBA**, could be conveniently employed as a recyclable PT catalyst in a series of synthetically useful reactions that were conducted under SL-PTC conditions. Further investigations on the use of fluorous ammonium salts in PTC are underway in our laboratories.

Acknowledgments Support by CNR (G.P.) and by the US Department of Energy under Contract No DE AC02-012315CH1 (R.H.F.) is gratefully acknowledged.

References

1. Gokel G (1991) Crown ethers and cryptands. The Royal Society of Chemistry, Cambridge
2. Montanari F, Quici S, Banfi S (1996) Phase-transfer catalysis. In: Reinhoudt DN (ed) Comprehensive supramolecular chemistry, vol 10. Pergamon, Oxford
3. Yadav GD (2004) Insight into green phase transfer catalysis. Top Catal 29:145–161
4. Quici S, Manfredi A, Pozzi G (2004) Phase-transfer catalysis in environmentally benign reaction media. In: Atwood JL, Steed JW (eds) Encyclopedia of supramolecular chemistry. Dekker, New York
5. Starks CM, Liotta CL, Halpern M (1994) Phase-transfer catalysis: fundamentals, applications, and industrial perspectives. Chapman & Hall, New York
6. Sasson Y, Neumann R (eds) (1997) Handbook of phase transfer catalysis. Blackie, London
7. Desikan S, Doraiswamy LK (1995) The diffusion-reaction problem in triphase catalysis. Ind Eng Chem Res 34:3524–3537, and references cited therein
8. Albanese D, Benaglia M, Landini D et al (2002) Use of a quaternary ammonium salt supported on a liposoluble poly(ethylene glycol) matrix for laboratory and industrial synthetic applications of phase-transfer catalysis. Ind Eng Chem Res 41:4928–4935, and references cited therein
9. Horváth IT, Rábai J (1994) Facile catalyst separation without water: fluorous biphase hydroformylation of olefins. Science 266:72–75
10. Gladysz JA, Curran DP, Horváth IT (eds) (2004) Handbook of fluorous chemistry. Wiley-VCH, Weinheim

11. Pozzi G, Quici S, Fish RH (2008) Fluorous phase transfer catalysts: from onium salts to crown ethers. J Fluor Chem 129:920–929
12. Shirakawa S, Tanaka Y, Maruoka K (2004) Development of a recyclable fluorous chiral phase-transfer catalyst: application to the catalytic asymmetric synthesis of α-amino acids. Org Lett 6:1429–1431
13. Emnet C, Weber KM, Vidal JA et al (2006) Synthesis and properties of fluorous quaternary phosphonium salts that bear four ponytails: new candidates for phase transfer catalysts and ionic liquids. Adv Synth Catal 348:1625–1634
14. Consorti CS, Jurisch M, Gladysz JA (2007) Ionic transformations in extremely nonpolar fluorous media: phase transfer catalysis of halide substitution reactions. Org Lett 9:2309–2312
15. Stuart AM, Vidal JA (2007) Perfluoroalkylated 4,13-diaza-18-crown-6 ethers: synthesis, phase-transfer catalysis, and recycling studies. J Org Chem 72:3735–3740
16. Pozzi G, Quici S, Fish RH (2008) Perfluorocarbon soluble crown ethers as phase transfer catalysts. Adv Synth Catal 350:2425–2436
17. Pozzi G, Mihali V, Foschi F et al (2009) 3,5-Bis(n-perfluorooctyl)benzyltriethylammonium bromide (F-TEBA): an efficient, easily recoverable fluorous catalyst for solid-liquid PTC reactions Adv Synth Catal 351:3072–3076
18. Vincent J-M, Rabion A, Yachandra VK et al (1997) Fluorous biphasic catalysis: complexation of 1,4,7-[$C_8F_{17}(CH_2)_3$]3-1,4,7-triazacyclononane with [$C_8F_{17}(CH_2)_2CO_2)_2$] (M = Mn, Co) to provide perfluoroheptane-soluble catalysts for alkane and alkene functionalization in the presence of t-BuOOH and O_2. Angew Chem Int Ed Engl 36:2346–2349
19. Pozzi G, Cavazzini M, Quici S et al (1997) Metal complexes of a tetraazacyclotetradecane bearing highly fluorinated tails: new catalysts for the oxidation of hydrocarbons under fluorous biphasic conditions. Tetrahedron Lett 38:7605–7608
20. Pozzi G, Mercs L, Holczknecht O et al (2006) Straightforward synthesis of a fluorous tetraarylporphyrin: efficient and recyclable sensitizer for singlet oxygen generation. Adv Synth Catal 348:1611–1620
21. Dubois G, Reyé C, Corriu RJP et al (2000) Organic-inorganic hybrid materials. Preparation and properties of dibenzo-18-crown-6 ether-bridged polysilsesquioxanes. J Mater Chem 10:1091–1098
22. Klyatskaya SV, Tetyakov EV, Vasilevsky SF (2003) Synthesis and chemical properties of polyacetylenic derivatives of benzo- and dibenzo-crown ethers. ARKIVOC 13:21–34
23. Gambaretto G, Conte L, Fornasieri G et al (2003) Synthesis and characterization of a new class of polyfluorinated alkanes: tetrakis(perfluoroalkyl)alkane. J Fluor Chem 121:57–63
24. Jeffery T (1996) On the efficiency of tetraalkylammonium salts in Heck type reactions. Tetrahedron 52:10113–10130
25. Gourdet B, Singh K, Stuart AM et al (2010) Di(1 H,1 H,2 H,2 H-perfluorooctyl)-dibenzo-18-crown-6: a "light fluorous" recyclable phase transfer catalyst. J Fluor Chem 131:1133–1143
26. Pannell KH, Yee W, Lewandos GS et al (1977) Electronic substituent effects upon the selectivity of synthetic ionophores. J Am Chem Soc 99:1457–1461
27. Landini D, Maia A (2003) Phase transfer catalysis (PTC): search for alternative organic solvents, even environmentally benign. J Mol Catal A Chem 204–205:235–243, and references therein
28. Yadav GD, Lande SV (2005) Liquid-liquid-liquid phase transfer catalysis: a novel and green concept for selective reduction of substituted nitroaromatics. Adv Synth Catal 347:1235–1241, and references therein
29. Wilson MJ, Pethrick RA, Pugh D et al (1997) Nuclear magnetic resonance and ab initio theoretical studies of 18-crown-6, benzo- and dibenzo-18-crown-6 and their alkali-metal complexes. J Chem Soc Faraday Trans 93:2097–2104
30. Gladysz JA, Tesevic V (2008) Temperature-controlled catalyst recycling: new protocols based upon temperature-dependent solubilities of fluorous compounds and solid/liquid phase separations. In: Leitner W, Hölscher M (eds) Regulated systems for multiphase catalysis. Topics in Organometallic Chemistry, vol 23. Springer, Heidelberg, pp 67–89

31. Starks CM, Owens RM (1973) Phase-transfer catalysis. II. Kinetic details of cyanide displacement on 1-halooctanes. J Am Chem Soc 95:3613–3617
32. Arrad O, Sasson Y (1988) Thin-layer phase-transfer catalysis in the reaction of alkyl chlorides and solid formate salt. J Am Chem Soc 110:185–189
33. Liotta CL, Berkner J, Wright J et al (1997) Mechanisms and applications of solid-liquid phase-transfer catalysis. In: Halpern ME (ed) Phase transfer catalysis–mechanisms and synthesis. American Chemical Society, Washington DC
34. Chaouchi M, Loupy A, Marque S et al (2002) Solvent-free microwave-assisted aromatic nucleophilic substitution − synthesis of aromatic ethers. Eur J Org Chem:1278–1283 and references therein
35. Paradisi C, Quintily U, Scorrano G (1983) Anion activation in the synthesis of ethers from oxygen anions and p-chloronitrobenzene. J Org Chem 48:3022–3026
36. Starks CM, Liotta CL, Halpern M (1994) Phase-transfer-catalyzed oxidations. In: Starks CM, Liotta CL, Halpern M (eds) Phase-transfer catalysis: fundamentals, applications, and industrial perspectives. Chapman & Hall, New York
37. Mandal D, Jurisch M, Consorti CS et al (2008) Ionic transformations in extremely nonpolar fluorous media: easily recoverable phase-transfer catalysts for halide-substitution reactions. Chem Asian J 3:1772–1782
38. Mandal D, Gladysz JA (2010) Syntheses of fluorous quaternary ammonium salts and their application as phase transfer catalysts for halide substitution reactions in extremely nonpolar fluorous solvents. Tetrahedron 66:1070–1077
39. Shirakawa S, Ueda M, Tanaka Y et al (2007) Design of binaphthyl modified symmetrical chiral phase-transfer catalysts: substituent effect of 4,4',6,6'-positions of binaphthyl rings in the asymmetric alkylation of a glycine derivative. Chem Asian J 2:1276–1281
40. Maayan G, Fish RH, Neumann R (2003) Polyfluorinated quaternary ammonium salts of polyoxometalate anions: fluorous biphasic oxidation catalysis with and without fluorous solvents. Org Lett 5:3547–3550
41. Mercs L, Pozzi G, Quici S (2007) Efficient condensation of carboxylic acids with alcohols catalyzed by fluorous ammonium triflates. Tetrahedron Lett 48:3053–3056
42. Shen W, Wang L-M, Tian H (2008) Quaternary ammonium salt gemini surfactants containing perfluoroalkyl tails catalyzed one-pot Mannich reactions in aqueous media. J Fluor Chem 129:267–273
43. Slávik Z, Tárkányi G, Gömöry Á et al (2001) Convenient syntheses and characterization of fluorophilic perfluorooctyl-propyl amines and ab initio calculations of proton affinities of related model compounds. J Fluor Chem 108:7–14
44. Albanese D, Landini D, Penso M et al (1999) Synthesis of N-sulfonyl aziridines through regioselective opening of epoxides under solid-liquid PTC conditions. Tetrahedron 55:6387–6394
45. Penso M, Lupi V, Albanese D et al (2008) A straightforward synthesis of enantiopure 2,6-disubstituted morpholines by a regioselective O-protection/activation protocol. Synlett:2451–2454
46. Landini D, Penso M (1991) N-Alkylation of trifluoroacetamide with 2-bromo carboxylic esters under PTC conditions: a new procedure for the synthesis of α-amino acids. J Org Chem 56:420–423
47. Albanese D, Landini D, Lupi V et al (2000) N-Monoalkylation of α-amino acid esters under solid-liquid PTC conditions. Eur J Org Chem:1443–1449
48. Lupi V, Penso M, Foschi F et al (2009) Highly stereoselective intramolecular α-arylation of self-stabilized non-racemic enolates: synthesis of α-quaternary α-amino acid derivatives. Chem Commun:5012–5014
49. Montanari F, Landini D, Rolla F (1982) Phase-transfer catalyzed reactions. In: Vögtle F (ed) Host guest complex chemistry II. Topics in Current Chemistry, vol 101. Springer, Heidelberg

Fluorous Hydrogenation

Xi Zhao, Dongmei He, László T. Mika, and István T. Horváth

Abstract The application of fluorous phosphine-modified catalysts for the hydrogenation of olefins is reviewed.

Keywords Asymmetric hydrogenation · Biphasic catalysis · Catalysis · Fluorous · Hydrogenation · Palladium · Phosphines · Rhodium · Ruthenium

Contents

1 Introduction 234
2 Ligands for Hydrogenation in Fluorous Media 236
3 Hydrogenation in Fluorous Media 236
   3.1 Rhodium Catalyzed Hydrogenation 236
   3.2 Asymmetric Hydrogenation 243
References 244

Abbreviations

FBC	Fluorous biphasic catalysis
FBS	Fluorous biphasic system
FC-72	Mixture of perfluorohexanes
NMR	Nuclear magnetic resonance
PFMCH	Perfluoromethylcyclohexane

X. Zhao, D. He, and I.T. Horváth (✉)
Department of Biology and Chemistry, City University of Hong Kong, Kowloon, Hong Kong
e-mail: istvan.t.horvath@cityu.edu.hk

L.T. Mika
Institute of Chemistry, Eötvös University, Budapest, Hungary

TOF Turnover frequency; turnover number per unit time
TON Turnover number of moles of substrate that a mole of catalyst can convert before becoming inactivated

1 Introduction

Homogeneous catalytic hydrogenation by transition metal complexes refers to the addition of a hydrogen molecule to an unsaturated functional group of the substrate. The reaction represents one of the most environmentally benign processes in that it produces virtually no waste. Generally, the homogeneous hydrogenation of organic substrates is performed by the activation of molecular hydrogen, but the hydrogen source could be another hydrogen containing molecule, such as alcohols. The application of the latter is so-called transfer-hydrogenation. A large number of metal complexes were found to be active in hydrogen activation under mild conditions. They are usually derived from transition metals, including actinides and lanthanides having partially filled external, d or f molecule orbitals. The most active metals are Rh, Ru, and Pd and they are widely employed in organic synthesis from small to industrial scales [1]. The most important and widely studied homogeneous hydrogenation catalyst is $ClRh[P(C_6H_5)_3]_3$, usually known as Wilkinson's catalyst, which was discovered in 1965 by Geoffrey Wilkinson (Nobel prize in 1973) [2]. This complex was the first practical hydrogenation system to be used routinely. The general mechanism of homogeneous hydrogenation is a multistep process including the formation of the catalytically active organometallic species, oxidative addition, hydrogen activation, reductive elimination, isomerization, etc. It can be classified as monohydride (MH) and dihydride (MH_2) type mechanisms. The catalytic cycles, which are representative of many catalytic hydrogenations, are shown in Scheme 1. While the MH_2 mechanism (Scheme 1a) using the Wilkinson's catalyst was studied in detail by Halpern and co-workers [3, 4], the HRh(CO)[P$(C_6H_5)_3]_3$ as a monohydride containing cycle (Scheme 1b) was reported by Jardin [5].

The homogeneous hydrogenation reactions take place in conventional organic solvents, in which the transition metal complexes have good solubility under mild conditions. While this is an advantage to develop more active and selective catalysts, the product separation and catalyst recycling is a major hurdle, especially for industrial processes. Consequently, most homogeneous hydrogenations have been used for the production of high value added fine and functional chemicals including chiral molecules [6]. In order to address the product–catalyst separation issue, the liquid–liquid biphasic catalytic systems are rapidly emerging as the basis of alternative technology to recover and recycle the usually expensive chiral ligands containing catalysts. The first important application of aqueous biphasic catalysis was the TPPTS modified Rh catalyst for propylene hydroformylation [7]. An alternative biphasic system was introduced using perfluorocarbons as solvents in 1994 [8].

The fluorous biphasic system (FBS), as an alternative to aqueous biphasic catalysis, offers a facile catalyst separation and recycling protocol. Generally, the

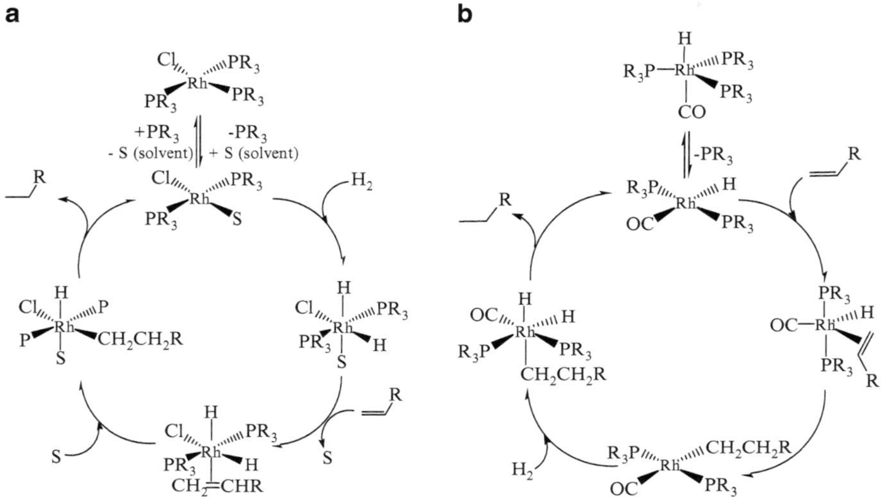

Scheme 1 Hydrogenation of terminal alkenes with (**a**) ClRh(PR$_3$)$_3$ and (**b**) HRh(CO)(PR$_3$)$_3$ catalysts (R= C$_6$H$_5$)

low miscibility of the perfluorocarbons with most organic solvents allows the separation of catalyst containing fluorous phase from the substrate and product(s) containing organic phase. One of the unique features of this system with the proper combination of the organic and fluorous solvents is that, upon warming the reaction mixture, the two phases become a single phase. Consequently, the catalysis takes place under homogeneous conditions and, after cooling the mixture, the phases are reestablished providing facile catalyst/product separation [9, 10]. This process, called "thermoregulated phase-separable catalysis", combines the advantages of homogeneous and heterogeneous catalysis. The reduction of alkenes, alkynes, and oxygen and nitrogen containing substrates leads to the formation of the corresponding products with different polarity. The efficiency of product/catalyst separation is strongly affected by the polaritiy of organic and fluorous phases, and thus the separation of the products of the reduction of ketones and N containing substrates are more favorable in FBS conditions.

The preparation of the first fluorous analog of Wilkinson's catalyst by replacing the triphenylphosphine ligand with the fluorous soluble P[CH$_2$CH$_2$(CF$_2$)$_5$CF$_3$]$_3$ was reported in 1997 [11]. The ClRh{P[CH$_2$CH$_2$(CF$_2$)$_5$CF$_3$]$_3$}$_3$ complex was used for hydrogenation of different alkenes (dodecene, cyclododecene, 4-bromostyrene, and 2-cyclohexene-1-one) (for more details see Sect. 3.1). Although the catalyst activity was lower than conventional nonfluorous analog, recovery and reuse of the catalyst was successfully demonstrated [12]. ^{31}P- and ^{1}H-NMR studies on the reaction of fluorous Wilkinson's complex with H$_2$ established the formation of a dihydrido complex, suggesting that the hydrogenation proceeded via MH$_2$ mechanism in fluorous solvents similarly to ClRh(PPh$_3$)$_3$ [12, 13].

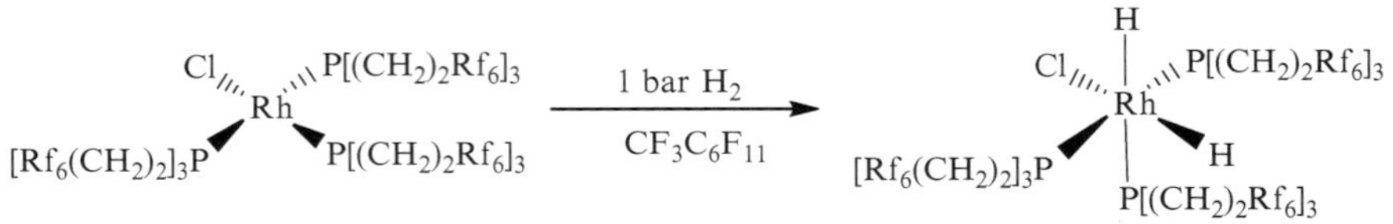

Fig. 1 Formation of fluorous dihydro Rh(III) complex in fluorous media. $Rf_6 = -(CF_2)_5CF_3$ [12]

2 Ligands for Hydrogenation in Fluorous Media

Fluorous hydrogenation catalysts can be prepared by combining fluorous soluble ligands with appropriate transition metal complexes. Most of the organic soluble ligands can be converted to fluorous soluble by introducing fluorophilic substituents to the ligand structure. The most frequently used fluorous soluble ligands are phosphorus containing ligands which could have nitrogen or sulfur in various combinations. Figure 2 summarizes the different fluorous phosphine ligands synthesized, tested, and evaluated under fluorous reaction conditions.

3 Hydrogenation in Fluorous Media

3.1 Rhodium Catalyzed Hydrogenation

The investigation of the hydrogenation of 2-cyclohexen-1-one, 1-dodecene, cyclododecene, and 4-bromostyrene in the presence of the first Wilkinson's analog (0.8–1.1 mol% of ClRh(L9)$_3$) resulted in 89–98% yield of the corresponding hydrogenated products. Good turnover numbers (TON = 87–120) were achieved in biphasic conditions (toluene/CF$_3$C$_6$F$_{11}$) under 1 bar H$_2$ at 45 °C. It was proposed that the selective hydrogenation of 2-cyclohexen-1-one to cyclohexanone goes through a different pathway than the other alkenes. The chemoselective reduction of the C,C-double bond of 2-cyclohexen-1-one and α,β-unsaturated aldehydes has been reported before with other Rh^I catalysts [12] (Table 1).

The fluorous alkylsilyl-substituted triaryl phosphines P(C$_6$H$_4$-p-SiMe$_2$-(CH$_2$)$_2$C$_n$F$_{2n+1}$)$_3$ ($n = 8$, L1), ($n = 6$, L5) were found to be active for the hydrogenation of 1-octene using [(COD)RhCl]$_2$ catalyst precursor in α,α,α-trifluoro-toluene as solvent at 1 bar of H$_2$ at 80 °C. The solubilities and the partition coefficients of the novel ligands were also reported [17]. For comparison, the p-(CH$_3$)$_3$-substituted derivative, RhCl[P(C$_6$H$_4$-p-SiMe$_3$)$_3$]$_3$ was also synthesized and characterized. A comparison of the catalytic activity of the fluorous catalysts and the nonfluorous derivatives under identical conditions in 1-octene hydrogenation revealed a decreasing activity in the order RhCl[P(C$_6$H$_4$-p-SiMe$_3$)$_3$]$_3$

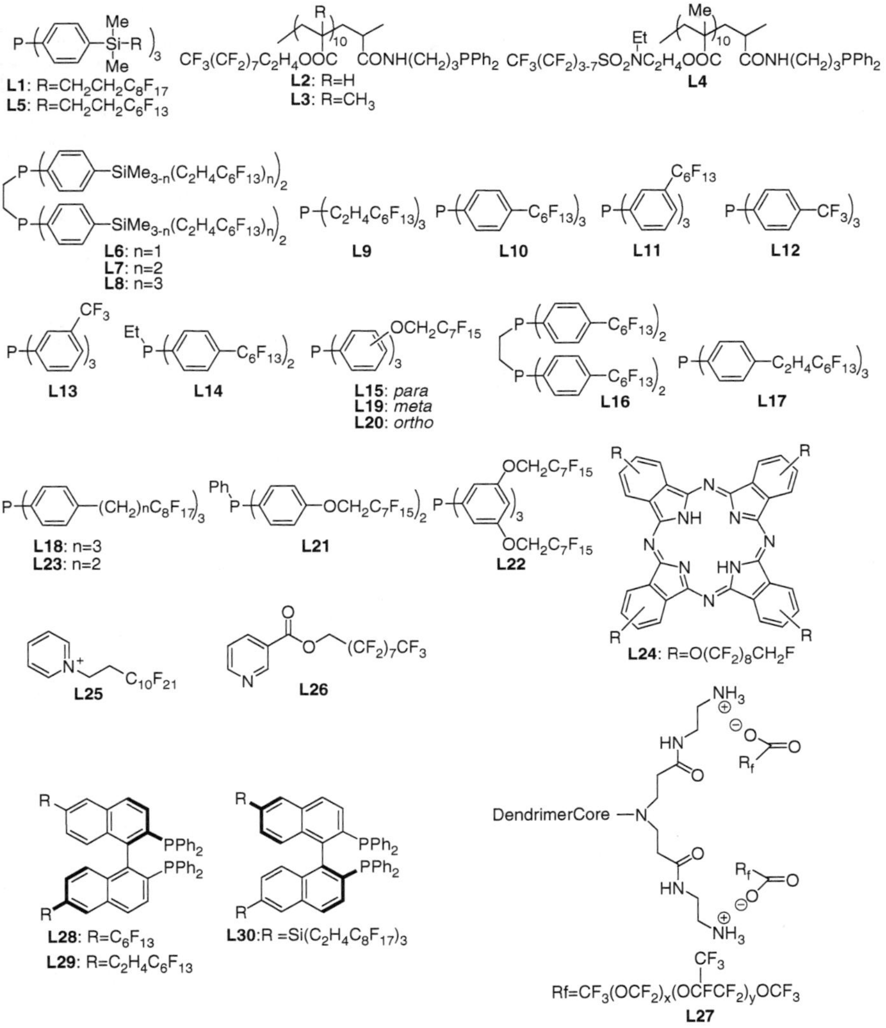

Fig. 2 Fluorous ligands for hydrogenation reactions

(TOF = 1,610 h^{-1}) > RhCl(L5)$_3$ (TOF = 1,110 h^{-1}) > RhCl(PPh$_3$)$_3$ (TOF = 960 h^{-1}) > RhCl(L1)$_3$ (TOF = 870 h^{-1}) in α,α,α-trifluoro-toluene. It also shows a very effective insulation of fluorous ponytails. The recycling efficiencies of the new catalysts (>98% for RhCl(L1)$_3$) were much better than expected on the basis of the fluorous phase affinity of the free L1 and L5 ligands. In the case of fluorous solvent PP2 and L1, the TOF drops to 177 h^{-1} under the same conditions, but at 0 °C two phases formed that allowed the separation of the product and the catalyst recycled nine times with very small Rh leaching (3 ppm) [16].

Table 1 Fluorous ligand modified rhodium catalyzed hydrogenation

Substrate	Ligand	P (MPa)	T (°C)	t (h)	Rate (mM h^{-1})	Conv. (%)	TOF (h^{-1})	Leaching (%) Rh	Leaching (%) P	References
1-Butene	L1	20	80	1.7		100	9,400			[14]
Cyclohexene	L2		25		78					[15]
	L3		25		17					[15]
	L4		25		20					[15]
2-Bicyclo[2.2.1]	L2		25		65					[15]
heptene	L3		25		14					[15]
	L4		25		56					[15]
1-Octene	L1	0.1	80			99	870			[16]
	L5	0.1	80			>95		0.3		[17]
	L5	0.1	80			99	1110			[16]
	L6	0.11	42	24			1.2	0.5		[18]
	L7	0.11	42	24			5	1		[18]
4-Octyne	L8	0.1	40	0.25		89	31	2.5	<0.08	[19]
	L8	0.1	40	0.25			12			[19]
	L8	0.1	40	0.25		59	23	2.2	0.4	[19]
1-Dodecene	L9	0.1	45	16		88	6.1			[12]
Cyclododecene	L9	0.1	45	26		94	4.6			[12]
Styrene	L9	0.1	63.5	1	79					[20]
	L10	0.1	75	1	146					[20]
	L10	0.1	63.5	1	83.9			0.32		[21]
	L10	0.1	63	1	220			0.06		[22]
	L11	0.1	63.5	1	117					[20]
	L12	0.1	63.5	1	139					[20]
	L13	0.1	63.5	1	100					[20]
	L14	0.1	63.5	1	62					[20]
	L15	0.1	63.5	1	201					[20]
	L16	0.1	63.5	1	66					[20]
	L17	9	40	1		47				[23]
4-Bromostyrene	L9	0.1	45	26		87	3.3			[12]
2-Cyclohexen-1-one	L9	0.1	45	8		98	11.7			[12]
	L18	0.1	45	2		>99				[24]
Cinnamaldehyde	L15	0.1	25	24		23				[25]
Methyl-(E)-	L15	0.1	25	20		100				[25]
cinnamate	L18	0.1	25	50		100				[25]
	L19	0.1	25	18		100		0.37	1.55	[25]
	L20	0.1	25	45		100		2.78	5.57	[25]
	L21	0.1	25	29		74				[25]
	L22	0.1	25	25		80				[25]
Cinnamaldehyde	L15	0.1	25	53		54				[25]
Methyl cinnamate + cinnamaldehyde	L15	0.1	25	31		41				[25]
Methyl α-acet amidocinnamate	L15	0.1	25	78		51				[25]

Ligands L6–L8 were applied in hydrogenation of 4-octyne under 1 bar H$_2$ at 40 °C. Higher than 99.92% catalyst retention was reported in 4-octyne hydrogenation in the presence a cationic [Rh(COD)(L8)]$^+$[BF$_4$]$^-$ in FC-75/hexane (1:3 v/v) system. In PMCHF/acetone (1:1 v/v) solvent mixture the catalyst could be recycled with 97.5%. The ligand leaching was comparable with Rh leaching in PMCHF/acetone, showing that dissociation and leaching of free ligand does not take place for this Rh diphoshine catalyst [19].

Fluorinated borane anions ($[B\{3,5-C_6H_3(C_6F_{13})_2\}_4]^-$, $[B\{4-C_6H_4(C_6F_{13})\}_4]^-$) and L6–L8 containing catalyst systems were developed including the determination of partition coefficient in PFMCH/THF and PFMCH/toluene systems. It was found that TOF = 1.2 and 5 h^{-1} for 1-octene hydrogenation in the case of $[Rh(COD)(L6)]^+[B\{3,5-C_6H_3(C_6F_{13})_2\}_4]^-$ and $[Rh(COD)(L7)]^+([B\{4-C_6H_4(C_6F_{13})\}_4]^-)$ under 1.1 bar at 42 °C, respectively [18].

L10–L16 modified Wilkinson's catalysts were tested in toluene/hexane/PP$_3$ (1:3:5) solvent mixture for hydrogenation of styrene under 1 bar of H_2 at 63.5 °C. It can be concluded that the incorporation of a fluorous ponytails decreases the reaction rate compared to the unmodified complex. The perfluoroalkyl chain reduces the electron density of the coordinated phosphorus atom. The reduction of the negative electronic effect of the fluorous ponytail increases the reaction rate, e.g., r; mmol L h^{-1}. The influence of the spacers can be ordered as follows: p-$C_6H_4OCH_2$- (r = 201) < p-C_6H_4- (r = 128) < m-C_6H_4- (r = 117) < -C_2H_4- (r = 79). In the case of spacer p-$C_6H_4OCH_2$-, a higher reaction rate could be achieved compared to PPh$_3$. The similarity in the pairs of m- and p-derivatized ligands (L10–L11 and L12–L13) suggests that the long ponytails do not exert a strong steric effect. Furthermore, the relative catalytic rates for each pair of the m- and p-derivatized aryl phosphine ligands suggest that the p-derivatized ligands are slightly more effective in these catalytic experiments and the infuence of the fluorous ponytails is purely an inductive effect. The reaction rates are compared in Fig. 3 [20].

The hydrogenation of styrene in the presence of L17 modified Wilkinson's catalyst was selected to demostrate efficient catalyst recycling, in which the

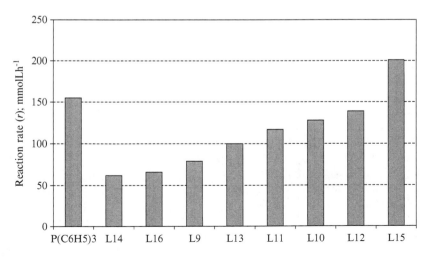

Fig. 3 Hydrogenation of styrene under 1 bar of H_2 at 63.5 °C in the presence of PPh$_3$ and various fluorous ligands modified rhodium catalysts

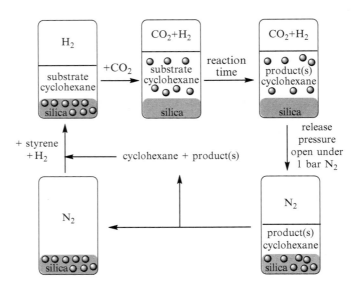

Fig. 4 CO_2 induced solubility switch using fluorous silica and $ClRh[P(p-C_6H_4CH_2CH_2C_6F_{13})]_3$ [23]

fluorous solvent was repleaced by fluorous silica phase. The fluorous catalyst was dissolved in the reaction mixture by introduction of carbon dioxide, and the recovery of the catalyst after the reaction was achieved by a simple depressurization of the system (Fig. 4). The rhodium leaching was less than 20 ppm [23]. Similarly to a silica containing system, the CO_2 induced miscibility switch of fluorous and organic solvents for catalyst recycling in the case of L27 modified Pd catalyzed hydrogenation reactions (Table 2) was reported. The catalyst containing fluorous and substrate containing organic phases were homogenized by the addition of carbon dioxide followed by depressurization after the reaction to separate fluorous insoluble product(s) and catalyst phases [30].

The fluorous triarylphosphine (L18) modified Wilkinson's analog was found to be active in the hydrogenation of 2-cyclohexene-1-one under monophasic ($CF_3C_6F_{11}$) and biphasic (toluene/$CF_3C_6F_{11}$) conditions, respectively. The formation of product was completed four times faster than in the case of L9 under 1 bar hydrogen at 45 °C. In contrast to L9, a short induction period was detected in both reaction conditions, but in the biphasic case the reaction was slower. In contrast, 1-dodecene was not as efficiently reduced mainly due to the slower reaction of internal alkenes generated during the isomerization side reactions. It should be noted that ligand L18 exhibits low solubility in fluorous solvents (e.g., ca. 10 mg/mL in $CF_3C_6F_{11}$) and virtually insoluble in nonfluorous media at ambient conditions [24].

The hydrogenation of 1-butene was performed in the presence of in situ prepared $[Rh(COD)Cl]_2$/L1 catalyst in $scCO_2$, resulting in 100% conversion. For continuous catalyst separation, a special membrane reactor was developed containing a pressure and temperature resistant silica membrane. The size of Wilkinson's catalyst

Table 2 Fluorous ligand modified palladium catalyzed hydrogenation

Substrate	Ligand	P (MPa)	T (°C)	t (h)	Conv. (%)	TOF (h^{-1})	References
Hexene	Acid-terminated perfluoropolyether	0.1	25	15		165	[26]
Cyclohexene	L24	1.5	80	6	2	2	[27]
	L25	1.5–3	25	6	99.5		[28]
	L26	1.1	47	1	74.5	166	[29]
1-Octene	L24	1.5	80	6	92	99	[27]
	L26	0.6	80	1.5	66	185	[29]
trans-2-Octene	L26	1.1	80	1.6	39.4	75	[29]
	L24	1.5	80	6	6	7	[27]
1,3-Cyclooctadiene	Acid-terminated perfluoropolyether	0.1	25	15		884	[26]
Styrene	L24	1.5	80	6	100	106	[27]
	L26	0.6	47	1	95.9	235	[29]
	L26	0.6	47	1	99.8	171	[29]
Allyl alcohol	L27		25	0.33		1093 ± 62	[30]
	L27		25	0.33		1858 ± 57	[30]
	Acid-terminated perfluoropolyether	0.1	25	15		565	[26]
Methylvinyl acrylate	Acid-terminated perfluoropolyether	0.1	25	15		10	[26]
Methyl acrylate	Acid-terminated perfluoropolyether	0.1	25	15		820	[26]

(2–4 nm) is clearly larger than the pore diameter (0.5–0.8 nm), and thus the selective permeability could be achieved. Stable operation and continuous production of n-butane was achieved at 80 °C and 200 bar H_2. After 32 h of continuous operation a TON of 1.2×10^5 was obtained. The turnover frequencies reached for the reaction carried out in carbon dioxide using the batch reactor and the membrane reactor, 4.0×10^3 h^{-1} and 9.4×10^3 h^{-1}, respectively, are significantly higher than the turnover frequency obtained in an organic (CCl_4) solvent (TOF = 100–1,000 h^{-1}) [14].

A special fluoroacrylate polymer bound rhodium based hydrogenation catalyst (($L2)_3$RhCl, ($L3)_3$RhCl, ($L4)_3$RhCl) was developed and tested for hydrogenation under ambient conditions. In comparison, rates for the reduction of 1-octene, cyclohexene, and bicyclo[2.2.1]hept-2-ene with L2 modified catalyst were 203, 78, and 65 mmol of H_2/mmol of Rh/h, respectively, on the basis of the measured Rh loading of 0.013 mmol of Rh/g of L2 [15]. It is comparable with the hydrogenation of 1-octene in p-xylene at 100 °C (190 mmol of H_2/mmol of Rh/h) with Wilkinson's catalyst [31]. The L4 and L5 containing system hydrogenated 1-octene, cyclohexene, and bicyclo[2.2.1]hept-2-ene with rates of 44, 17, 14 and 15, 20, and 56 mmol of H_2/mmol of Rh/h, respectively. The catalysts could be recycled and reused without losses in activity and selectivity, e.g., ($L2)_3$RhCl in seven cycles converted 21,700 mmol of substrate/mmol of Rh [15].

Fluorous triphenylphosphines (L15, L18, L19, L20, L21, L22) were used with [Rh(μ-Cl)(COD)]$_2$ or [Rh(COD)$_2$][PF$_6$] in hydrogenation of methyl-cinnamate, 2-cyclohexene-1-one, cinnamaldehyde, and methyl-α-acetamidocinnamate in a biphasic system D-100/ethanol under ambient conditions. Some differences could be observed in catalytic activity using the *o*-(L20), *m*-(L19), and *p*-substituted (L15) fluorous ligands in methyl-(*E*)-cinnamate reduction resulting in 100% conversion during 45, 18, and 12 h, respectively. The recycling efficiencies were high for the catalyst associated with ligand having *p*-fluorous (L15) or *m,m'*-fluorous ponytails (L22), keeping the metal loss in the range of 0.37–1.13%. It should be noted thatthe reused catalyst exhibited the same activity [25].

A new highly fluorinated insoluble perfluoroalkylated polystyrene resins was evaluated for catalyst recycling by use of supported ClRh(L10)$_3$ catalyzed hydrogenation of styrene. The reaction rate was found to be comparable with the non-supported system in the range of 220–287 mmol L^{-1} h^{-1} without significantly decreasing the activity [22]. Perfluoroalkylated micro- and mesoporous silica, along with powdered Teflon, were tested as solid supports for supported fluorous phase catalysis in the hydrogenation of styrene as a model reaction in the presence of supported ClRh(L10)$_3$ catalyst. The maximum reaction rate (83.9 mmol L^{-1} h^{-1}) was achieved by application FRP mesoporous silica. It is slightly lower than that in the case of toluene/PP3 biphase (86 mmol L^{-1} h^{-1}) at 63.5 °C [21].

The highly fluorocarbon soluble, binuclear [Ru(μ-O$_2$CMe)(CO)$_2$(L9)]$_2$ and [Ru(μ-O$_2$CMe)(CO)$_2$(L23)]$_2$ were synthesized and partition coefficients for the complexes in fluorocarbon/hydrocarbon biphases were determined. Both ruthenium complexes were found to be slightly active in the reduction of acetophenone to 1-phenylethanol (51 bar of H$_2$ at 105 °C). Turnover frequencies of 0.64 and 1.2 h^{-1} were measured for L23 and L9 containing complexes, respectively [32].

The application of dendrimer-encapsulated Pd nanoparticles for fluorous biphasic hydrogenation of alkenes, functionalized alkenes, and conjugated dienes was reported. Complexation of Pd/dendrimer composites with perfluorinated carboxylic acids renders the resulting nanocomposites preferentially soluble in fluorous media. The new catalyst system showed high activity (TOF 122–884 mol H$_2$ mol Pd^{-1}h^{-1}) and could be recycled 12 times without appreciable loss of catalytic activity [26].

A fluoroalkene-soluble L24 substituted Pd^{II}-phthalocyanine catalyst was studied for olefin (styrene, 1-octene, *trans*-2-octene, and cyclohexene) reduction with molecular hydrogen in an *n*-hexane/PFMCH biphasic system. The order of reactivity obtained for the studied olefins was found to be as follows: styrene >1-octene ≫ *trans*-2-octene > cyclohexene (Table 2) The catalyst could be recycled in nine consecutive reactions without loss of activity or metal contamination [27].

L25 modified spherical Pd^0 nanoparticles (2–8 nm) were immobilized on the nanometer-scaled platelets of montmorillonite (MMT) to prepare an active and recyclable catalyst for hydrogenation, Suzuki–Miyaura C–C coupling, and fluorous biphase catalysis [28].

The catalytic activity of the perfluoroalkylated pyridine (L26) modified Pd^{II} complex was examined for hydrogenation of styrene, 1-octene, *trans*-2-octene,

Table 3 Fluorous ligand modified asymmetric hydrogenation

Substrate	Ligand	P (MPa)	T (°C)	t (h)	Conv. (%)	e.e. (%)	References
Dimethyl itaconate	L28	2.0	25	0.25	42	95.3	[35]
	L28	5.0	22	0.25	100	73.0	[36]
	L29	2.0	25	0.25	83	95.7	[35]
	L29	5.0	22	0.25	100	74.0	[36]
	L29	15.0	100	24	100	70.6	[36]
	L29	17.0	80	0.17	82	94.5	[36]
	L30	1.0	60	6	95	95	[37]
	L30	1.0	60	1	18	92	[37]
$CH_3OCOC(=CH_2)$ NHCOCH$_3$	L30	1.0	60	6	96	90	[37]
PhC(=CH$_2$)NHAc	L30	3.0	60	8	64	44	[37]

and cyclohexene both in supercritical carbon dioxide (scCO$_2$) and in organic solvents at 80 °C. The highest activity (TON = 235 h^{-1}) was achived in the reduction of styrene in supercritical CO$_2$. It should be noted that internal olefins were hardly hydrogenated (TOF = 75 h^{-1}) under the same conditions [29].

3.2 Asymmetric Hydrogenation

Asymmetric hydrogenation is one of most important catalytic reactions using a prochiral substrate due to the potential for the enantioselective preparation of optically pure fine chemicals and pharmaceuticals. The chiral BINAP is among the most popular ligands used in catalytic asymmetric reactions and, hence, it finds widespread application in catalytic processes [33]. Recently, the development of improved strategies for the high value chiral ligand containing catalyst has gained growing attention. Although the fluorous biphasic concept (FBS) is one of the most promising approaches [8, 9] it needs to develop fluorous chiral ligands.

The preparation of chiral fluorous derivatives of BINAP was reported by Leitner in 1999 [34]. The asymmetric induction of the fluorous BINAPs L28 and L29 was investigated in the Ru catalyzed reduction of dimethylitaconate. The conversions with Ru-L28, Ru-L29, and Ru-(R)-BINAP were 42, 83, and 88%, respectively, while the enantiomeric ee values were 95.3, 95.7, and 95.4%, with the S enantiomer dominating the product in each case. It was also shown that the fluorous ponytails impose no detectable effects on the enantioselectivity but affect reaction rates, indicating that the electronic effects arising from 6,6′-fluoroalkylation of BINAP only impact on the hydrogenation activity of the Ru-BINAP catalysts [35]. L28 and L29 were also tested for asymmetric hydrogenation of dimethylitaconate in supercritical CO$_2$ resulting in the conversions of 100% with ees of 73% and 74% for Ru-L28 and Ru-29, respectively. These results are comparable with the original Ru-BINAP system (97% conversion and 95.4% ee). It should be noted that the ee could be increased up to 95.5 ± 0.2% for these ligands [36]. A trisperfluoroalkylsilyl-modified (S)-BINAP was reported and its pertinent Ru complex was applied to the asymmetric hydrogenation of functionalized olefins (Table 3). Efficient separation of the Ru catalyst by filtration and its reuse was achieved with relative low (1.6–4.9 ppm) metal leaching [37]. It should be mentioned that fluorous ponytails containing ligand ((R,S)-3-H^{2}F^6-BINAPHOS) based rhodium catalyzed enantioselective hydrogenation of polar substrates in inverted supercritical CO_2/aqueous biphasic media was also developed [38, 39].

References

1. Oro LA (2003) Hydrogenation – homogeneous. In: Horváth IT (ed) Encyclopedia of catalysis, vol 4. Wiley-Interscience, New York, pp 55–107
2. Osborn JA, Jardine FH, Young JF, Wilkinson G (1966) The preparation and properties of tris (triphenylphosphine)halogenorhodium(I) and some reactions thereof including catalytic homogeneous hydrogenation of olefins and acetylenes and their derivatives. J Chem Soc A:1711–1732
3. Halpern J, Wong CS (1973) Hydrogenation of tris(triphenylphosphine)chlororhodium(I). Chem Commun:629
4. Halpern J (1981) Mechanistic aspects of homogeneous catalytic hydrogenation and related processes. Inorg Chim Acta 50:11–19
5. Jardine FH (1982) Carbonylhydrido tris(triphenylphosphine)rhodium(I). Polyhedron 1:569–605
6. Sanfilippo D, Rylander PN (2009) Ullmann's encyclopedia of industrial chemistry
7. Cornil B, Herrmann WA (1998) Aqueous phase organometallic catalysis. Wiley-VCH, Weinheim
8. Horváth IT, Rábai J (1994) Facile catalyst separation without water. Fluorous biphase hydroformylation of olefines. Science 266:72–75
9. Horváth IT (1998) Fluorous biphase chemistry. Acc Chem Res 31:641–650
10. Gladysz JA, Curran DP, Horváth IT (2004) Handbook of fluorous chemistry. Wiley-VCH, Weinheim
11. Juliette JJJ, Horvath IT, Gladysz JA (1997) Transition metal catalysis in fluorous media: practical application of a new immobilization principle to rhodium-catalyzed hydroboration. Angew Chem Int Ed Engl 36:1610–1612
12. Rutherford D, Juliette JJJ, Rocaboy C, Horváth IT et al (1998) Transition metal catalysis in fluorous media: application of a new immobilization principle to rhodium-catalyzed hydrogenation of alkenes. Catal Today 42:381–388
13. Meakin P, Jesson JP, Tolman CA (1972) The nature of chlorotris(triphenylphosphine)rhodium in solution and its reaction with hydrogen. J Chem Soc A 94:3270–3272
14. Goetheer ELV, Verkerk AW, van den Broeke LJP, de Wolf E et al (2003) Membrane reactor for homogeneous catalysis in supercritical carbon dioxide. J Catal 219:126–133
15. Bergbreiter DE, Franchina JG, Case BL (2000) Fluoroacrylate-bound fluorous-phase soluble hydrogenation catalysts. Org Lett 2:393–395
16. Richter B, Spek AL, van Koten G, Deelman BJ (2000) Fluorous versions of Wilkinson's catalyst. Activity in fluorous hydrogenation of 1-alkenes and recycling by fluorous biphasic separation. J Am Chem Soc 122:3945–3951
17. Richter B, Deelman BJ, van Koten G (1999) Fluorous biphasic hydrogenation of 1-alkenes using novel fluorous derivatives of Wilkinson's catalyst. J Mol Catal A Chem 145:317–321
18. van den Broeke J, de Wolf E, Deelman BJ, van Koten G (2003) Enhanced hydrogenation activity and recycling of cationic rhodium diphosphine complexes through the use of highly fluorous and weakly-coordinating tetraphenylborate anions. Adv Synth Catal 345:625–634

19. de Wolf E, Spek AL, Kuipers BWM, Philipse AP et al (2002) Fluorous derivatives of Rh (COD)(dppe) BX4 (X = F, Ph): synthesis, physical studies and application in catalytic hydrogenation of 1-alkenes and 4-alkynes. Tetrahedron 58:3911–3922
20. Hope EG, Kemmitt RDW, Paige DR, Stuart AM (1999) The rhodium catalysed hydrogenation of styrene in the fluorous biphase. J Fluor Chem 99:197–200
21. Hope EG, Sherrington J, Stuart AM (2006) Supported fluorous phase catalysis on PTFE, fluoroalkylated micro- and meso-porous silica. Adv Synth Catal 348:1635–1639
22. Audic N, Dyer PW, Hope EG, Stuart AM et al (2010) Insoluble perfluoroalkylated polymers: new solid supports for supported fluorous phase catalysis. Adv Synth Catal 352:2241–2250
23. Ablan CD, Hallett JP, West KN, Jones RS et al (2003) Use and recovery of a homogeneous catalyst with carbon dioxide as a solubility switch. Chem Commun:2972–2973
24. Soós T, Bennett BL, Rutherford D, Barthel-Rosa LP et al (2001) Synthesis, reactivity, and metal complexes of fluorous triarylphosphines of the formula P(p-C$_6$H$_4$(CH$_2$)$_3$(CF$_2$)n-1CF$_3$)$_3$ (n = 6, 8, 10). Organometallics 20:3079–3086
25. Sinou D, Maillard D, Aghmiz A, i-Bulto AMM (2003) Rhodium-catalyzed hydrogenation of alkenes by rhodium/tris(fluoroalkoxy)phosphane complexes in fluorous biphasic system. Adv Synth Catal 345:603–611
26. Chechik V, Crooks RM (2000) Dendrimer-encapsulated Pd nanoparticles as fluorous phase-soluble catalysts. J Am Chem Soc 122:1243–1244
27. Yılmaz F, Özer M, Kani I, Bekaroglu Ö (2009) Catalytic activity of a thermoregulated, phase-separable Pd(II)-perfluoroalkylphthalocyanine complex in an organic/fluorous biphasic system: hydrogenation of olefins. Catal Lett 130:642–647
28. Scheuermann GM, Thomann R, Mulhaupt R (2009) Catalysts based upon organoclay with tunable polarity and dispersion behavior: new catalysts for hydrogenation, C–C coupling reactions and fluorous biphase catalysis. Catal Lett 132:355–362
29. Kani I, Sisman F (2006) Synthesis and catalytic activity of perfluoroalkylated pyridine-palladium(II) complex toward olefin hydrogenation in scCO$_2$ and conventional organic solvents. J Mol Catal A Chem 259:142–149
30. West KN, Hallett JP, Jones RS, Bush D et al (2004) CO$_2$-induced miscibility of fluorous and organic solvents for recycling homogeneous catalysts. Ind Eng Chem Res 43:4827–4832
31. Bergbreiter DE, Chandran R (1987) Polyethylene-bound rhodium(I) hydrogenation catalysts. J Am Chem Soc 109:174–179
32. Malosh TJ, Wilson SR, Shapley JR (2009) Binuclear ruthenium complexes of fluorous phosphine ligands: synthesis, properties, and biphasic catalytic activity. Crystal structure of [Ru(μ-O$_2$CMe)(CO)$_2$P(CH$_2$CH$_2$(CF$_2$)$_5$CF$_3$)$_3$]$_2$. J Organomet Chem 694:3331–3337
33. Noyori R (1994) Asymmetric catalysis in organic synthesis. Wiley, New York
34. Franco G, Leitner W (1999) Highly regio- and enantio-selective rhodium-catalysed asymmetric hydroformylation without organic solvents. Chem Commun:1663–1664
35. Birdsall DJ, Hope EG, Stuart AM, Chen W et al (2001) Synthesis of fluoroalkyl-derivatised BINAP ligands. Tetrahedron Lett 42:8551–8553
36. Hu Y, Birdsall DJ, Stuart AM, Hope EG et al (2004) Ruthenium-catalysed asymmetric hydrogenation with fluoroalkylated BINAP ligands in supercritical CO$_2$. J Mol Catal A Chem 219:57–60
37. Horn J, Bannwarth W (2007) Repetitive application of a fluorous chiral BINAP-Ru complex in the asymmetric hydrogenation of olefins. Eur J Org Chem:2058–2063
38. Burgemeister K, Francio G, Huglc H, Leitner W (2004) Enantioselective hydrogenation of polar substrates in inverted supercritical CO$_2$/aqueous biphasic media. Chem Commun:6026–6028
39. Burgemeister K, Francio G, Gego VH, Greiner L et al (2007) Inverted supercritical carbon dioxide/aqueous biphasic media for rhodium-catalyzed hydrogenation reactions. Chem Eur J 13:2798–2804

Fluorous Hydrosilylation

Monica Carreira and Maria Contel

Abstract In this review, we describe the papers and patents dealing with the fluorous biphasic system (FBS) hydrosilylation reactions reported to date. Despite the limited number of reports, the FBS hydrosilylation reaction has been extremely successful. In all cases fluorous monophosphines (either alkylic or perfluoroalkylsilyl-substituted derivatives of triphenylphosphine) have been employed as ligands to synthesize and inmobilize the metal catalysts (either rhodium(I) or gold(I) derivatives) in the fluorous solvent (including a fluorous ionic liquid). The hydrosilylation of alkenes, ketones and enones with fluorous rhodium analogs to the Wilkinson's catalyst [RhCl(PPh$_3$)$_3$], have afforded high TON/TOF and a very efficient separation and recycling of the fluorous catalyst. Modification of the fluorous content and position of the fluorous tails in the aryl groups of the phosphines have allowed for further optimization of the process and a better recovery of the catalyst with minimal leaching of rhodium and fluorous ligand to the organic phase. Moreover, the use of the so-called second generation methods which eliminate the need of fluorous solvents by exploiting the temperature-dependent solubilities of fluorous catalysts in common organic solvents (thermomorphic properties) have permitted the use and separation of fluorous alkyl-phosphine rhodium catalysts in hydrosilylation reactions in conventional organic solvents. The addition of an insoluble fluorous support such as Teflon tape allowed for an exceptionally easy and efficient recovery of fluorous rhodium catalysts ("catalyst-on-a-tape") in the hydrosilylation of ketones. In the case of the FBS gold-catalyzed hydrosilylation of aldehydes, new fluorous gold catalysts with alkylic phosphines have led to an efficient separation and recycling of the gold catalysts although the TON/TOF are lower than in the rhodium-catalyzed

---

M. Carreira and M. Contel (✉)
Department of Chemistry, Brooklyn College and The Graduate Center, The City University of New York, Brooklyn, NY 11210, USA
e-mail: mariacontel@brooklyn.cuny.edu

hydrosilylation of alkenes and ketones. A detailed study of the non-fluorous gold-catalyzed version has helped to explain how this catalytic system could be improved.

Keywords Fluorous phosphines · Gold · Hydrosilylation · Recycling · Rhodium

Contents

1 Introduction 249
2 Hydrosilylation of Alkenes 251
    2.1 FBS Rhodium-Phosphine Catalyzed Hydrosilylation of Alkenes 251
    2.2 Use of Fluorous Ionic Liquids in the FBS Hydrosilylation of Alkenes 255
3 Hydrosilylation of Ketones, Enones, and Aldehydes 257
    3.1 FBS Rhodium-Phosphine Catalyzed Hydrosilylation
        of Ketones and Enones 257
    3.2 Temperature-Controlled Fluorous Rhodium Catalyst Recycling for the
        Hydrosilylation of Ketones: Use of Teflon as Delivery/Recovery Support 259
    3.3 FBS and Temperature-Controlled Fluorous Gold Catalyst Recycling in the
        Hydrosilylation of Aldehydes: Proposal of a Unique Reaction Mechanism
        for Gold Phosphine-Catalyzed Hydrosilylation Reactions 261
4 Conclusion 268
References 269

Abbreviations

FBC	Fluorous biphasic catalysis
FBS	Fluorous biphasic system
DMF	Dimethylformamide
FC-72	Mixture of perfluorohexanes
GC	Gas chromatography
IR	Infra red
MALDI-TOF	MALDI: Matrix-assisted laser desorption/ionization; TOF: time-of-flight mass spectrometer
NMR	Nuclear magnetic resonance
PFMCH	Perfluoromethylcyclohexane
RT	Room temperature
TEM	Transmission electron microscopy
THF	Tetrahydrofuran
tht	Tetrahydrothiophene
TOF	Turnover frequency; turnover number per unit time
TON	Turnover number; number of moles of substrate that a mole of catalyst can convert before becoming inactivated
Tos	Tosyl
TXRF	Total reflection X-ray fluorescence
xantphos	4,5-Bis(diphenylphosphino)-9,9-dimethylxanthene

1 Introduction

The catalyzed hydrosilylation (or hydrosilation) reaction is the most important chemical transformation in the preparation of compounds containing silicon (either by carbon–silicon or heteroelement–silicon bond formation) [1]. There are a number of excellent reviews which have appeared in the last 15 years on the subject [2–14]. This highly flexible reaction enables a variety of functional groups to be introduced to silicones, silanes, and organic materials. The reaction is used in the hardening of silicone polymers and for making silyl derivatives with different applications (modification of surfaces like fabrics, cloths, glass, stone; pre-treatment of fillers in rubbers; paper release coatings, cosmetics, adhesives, lubricants; preparation of silsesquioxanes). Most silicon compounds still play a very important role in organic synthesis as useful intermediates to other added value products. Hydrosilylation reactions can be facilitated by transition metals and radicals but the best known examples involve platinum as catalyst. The catalytic hydrosilylation by soluble compounds was discovered by Dr. Speier at the Dow Corning Corporation in the late 1950s [15]. The catalyst used was H_2PtCl_6 although later on some other platinum derivatives were developed such as $(NBu_4)_2PtCl_6$ [16] or the Karstedt catalyst [17]. The mechanism of this reaction was first described by Chalk and Harrod [18] (Scheme 1). It involves the mechanism of H-X additions to unsaturated organic compounds, starting with the oxidative addition to zerovalent platinum compounds. Besides the insertion of alkene into a metal-hydride, the

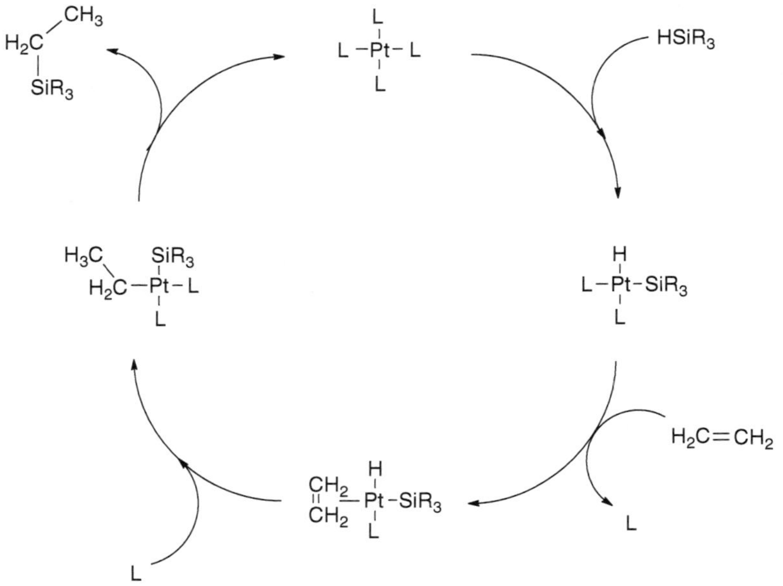

Scheme 1 Chalk–Harrod mechanism for hydrosilylation [18]

insertion of alkene in the metal–silyl bond takes place, giving rise to side-reactions which generate vinylsilanes and alkanes (1:1 mol ratio). A similar mechanism was described for rhodium(I) derivatives like Wilkinson's catalyst [RhCl(PPh$_3$)$_3$] [18]. When the substrates are ketones or aldehydes, the mechanism is the same, starting with the oxidative addition of the silane, subsequent coordination of the C=O double bond to the metal, and insertion of the ketone or aldehyde in the metal–silyl bond.

Hydrosilylation has been applied to C–C double and triple bonds [8, 9, 12, 13, 19], C=O bonds [6–8, 11, 14], and, most recently, to C=N bonds [6, 8, 10]. A variety of metal compounds can be used as catalysts. The most common and active ones for alkenes and alkynes are based on platinum and rhodium. Hydrosilylation of C=O bonds gives silyl ethers, which are subsequently hydrolyzed to their alcohols. The reaction is of higher interest in its enantioselective version in organic synthesis for preparing chiral alcohols. Thus chiral Rh, Ru, Cu, Zn, and, most recently, Ti, Fe, and Ir derivatives have been applied in the asymmetric hydrosilylation of ketones, aldehydes, and imines [6, 8, 10, 14]. As new developments, N-heterocyclic carbene groups 9 and 10 metal derivatives [20, 21] and high oxidation state metal oxo and imido complexes [22] have also been successfully used in hydrosilylation reactions.

In order to separate and recycle expensive (and sometimes toxic) metal catalysts from the products and substrates in homogeneous catalysis experiments, the biphasic solvent approach has been developed [23]. Biphasic hydrosilylation reactions with recovery of the catalyst have been described but the number of reports is scarce. The solvents employed are mainly ionic liquids with Rh [24, 25] and Pt [25–27] compounds and water with Pt compounds [28]. However, in the biphasic system version, hydrogenation and double isomerization are found as side reactions [28].

The concept of fluorous biphasic systems (FBS) and their application in catalysis was made known to the global chemical community in 1994 by Horváth and Rábai [29]. The term fluorous was introduced to emphasize the fact that one of the phases of a biphase system is richer in fluorocarbons than the other. Unlike the more classical aqueous/hydrocarbon biphasic systems, water-sensitive reactants can also be employed. Fluorocarbon solvents have unique properties which make them suitable for immobilizing catalysts in the fluorous phase. Because they are non-polar and have low intermolecular forces they are non-miscible with common organic solvents at room temperature, allowing for the formation of biphasic systems. Homogeneous catalysts can be modified to become soluble in fluorous solvents by coordination of long-chain fluoroponytailed ligands. Moreover, fluorous solvents commonly used (perfluorinated alkanes, dialkyl ethers, and trialkylamines) are inert, thermally stable, non-flammable, and practically non-toxic. Their thermal decomposition products can be toxic, but such decomposition begins only at very high temperatures [30]. In general, the higher solubility of organic substrates in fluorous solvents vs water should make the reaction kinetics of FBS more favorable than that of aqueous biphasic systems. Since Horváth and Rábai's seminal paper in 1994, which has been cited over 900 times, numerous examples of this methodology

Fluorous Hydrosilylation

for many homogenous catalytic reactions with the recovery of the catalyst have been demonstrated [31, 32]. Several of these reactions, like hydroformylation, hydrogenation, oxidation, and C–C bond formation, have been collected as chapters of the Handbook of Fluorous Chemistry published in 2004 [31] along with the properties of fluorous solvents, fluorous ponytails, and several strategies for the recovery of fluorous catalysts and reagents [31, 32]. However, to the best of our knowledge, fluorous hydrosilylation reactions have not been reviewed. We report here on all fluorous hydrosilylation reactions described to date including previous work reported by one of us [33]. The reactions include the FBS hydrosilylation of alkenes [34–36], ketones, and enones [37, 38] with rhodium compounds, and the FBS hydrosilylation of aldehydes with gold derivatives [33]. In all cases the ligands employed have been fluorous phosphines. Improved techniques for the separation/recycling of these catalysts will be discussed [39–43].

2 Hydrosilylation of Alkenes

2.1 FBS Rhodium-Phosphine Catalyzed Hydrosilylation of Alkenes

Hydrosilylation of 1-alkenes is an important reaction to produce higher alkylsilanes which are precursors for silicon-based polymers, lubricants, and water-repellent coatings. Probably Wilkinson's catalyst [RhCl(PPh$_3$)$_3$] is the best known catalyst for this type of hydrosilylation [44]. Hydrosilylation of 1-hexene using various silanes and fluorous versions of Wilkinson's catalyst in fluorous biphasic solvents systems afforded the corresponding n-hexylsilanes in high yields (1). This work developed by Deelman, van Koten and coworkers [34] was based on the use of p-silyl-substituted fluorous phosphines [36, 45] which had afforded highly active FBS rhodium hydrogenation catalysts [46, 47].

$$2\ \diagup\!\!\diagup C_4H_9 + HSiR_3 \xrightarrow[\text{PFMCH, reflux}]{0.1\ \text{mol}\%\ \text{cat.}} C_4H_9\diagup\!\!\diagdown SiR_3 + \text{hexene isomers}$$

R$_3$ = Me$_2$Ph, Me$_2$Cl or Cl$_3$
cat = 1 or 2

Equation 1 FBS Hydrosilylation of 1-hexene using fluorous Rh catalysts [34]

Fluorous catalysts **1** and **2** with these p-silyl-substituted fluorous phosphines P(C$_6$H$_4$-4-SiMe$_2$(R$_f$))$_3$ (R$_f$ = CH$_2$CH$_2$C$_6$F$_{13}$) and P(C$_6$H$_4$-4-SiMe(R$_f$)$_2$)$_3$ are depicted in Fig. 1.

The hydrosilylation was performed at reflux temperature with CF$_3$C$_6$F$_{11}$ (PFMCH: perfluoromethylcyclohexane) as the fluorous phase, while the organic phase contained the substrates (1:4 v/v). A 2:1 ratio of 1-hexane to silane was used

Fig. 1 Fluorous Rh catalysts **1** and **2** first used in the FBS hydrosilylation of 1-hexene [34]

Table 1 Catalyst recycling in the hydrosilylation of 1-hexene [34]

Catalyst precursor	Product	No. of cycles (t, min)	Leaching, % Rh	P
1	C$_6$H$_{13}$SiMe$_2$Ph	3 (15)	12 (41)	19 (58)
2	C$_6$H$_{13}$SiMe$_2$Ph	3 (15)	1.7 (5.7)	2.2 (6.6)

to prevent the formation of less fluorous [Rh(H)Cl(SiR$_3$){P(Ar$_f$)$_3$}$_2$]. Results are shown in Table 1. In all cases selective anti-Markovnikov addition and some isomerization of 1-hexene (ca. 13%) took place as had been reported for the non-fluorous Wilkinson's catalyst. In general the activities are comparable to those of Wilkinson's complex.

Catalyst recycling was performed by phase separation at ambient temperature. The organic substrates were fully miscible with the fluorous solvent at ambient temperature or slightly above. However, a biphasic system was obtained after the reaction was completed, which is due to the lower miscibility of the organic products with the fluorous phase. Full conversion of silane was achieved in all cases without an observable drop in activity (Table 1). The activity of the catalyst in subsequent cycles can be influenced not only by catalyst leaching but also by leaching of free fluorous ligand and fluorous solvent as well. Catalyst **2** was better in terms of leaching than **1** and therefore resulted in a more constant catalytic activity.

The Chalk–Harrod mechanism for hydrosilylation [18] proposes a Rh(III) hydride complex [Rh(H)(Cl)(SiR$_3$)(PAr$_3$)$_2$] as a crucial intermediate. Such a fluorous intermediate could be detected by ^{31}P and ^{1}H NMR spectroscopy by reaction of **1** with HSiCl$_3$, suggesting similar mechanisms for hydrosilylations employing Wilkinson's catalyst or fluorous catalysts **1** and **2**. This mechanism may explain the leaching of free ligand (since the mechanism involves dissociation of one of the phosphine ligands, see Scheme 1) which may also lead to increased catalyst leaching.

To enable catalyst recycling in the synthesis of fluorous alkylsilanes a reverse approach was used (Scheme 2).

The hydrosilylation reactions of *1H,1H,2H*-perfluoro-1-alkenes with [RhCl(PPh$_3$)$_3$] were performed using 2 equivalents of nonfluorous hydrosilane per equivalent of alkenes in benzene or toluene as solvents (2). When the fluorous olefin was fully converted, the mixture was extracted with FC-72 (mixture of perfluorohexanes). Results were compared to those obtained previously with the same

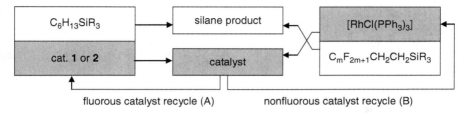

Scheme 2 Direct and reversed fluorous biphasic hydrosilylation catalyst separation [34]

approach and H$_2$PtCl$_6$ [48]. Selectivities and isolated yields were significantly better and the Wilkinson's catalyst could be recycled for three cycles without apparent loss of catalytic activity [34]. The system with Rh described in this paper was homogenous as opposed to H$_2$PtCl$_6$ [48]. No leaching of Rh or phosphine into the fluorous layer was detected.

Rf = C$_m$F$_{2m+1}$; m = 6, 8 or 10   Rf' = C$_{m-1}$F$_{2m-1}$ R$_3$ = Me$_2$Cl, MeCl$_2$, Cl$_3$ or (OMe)$_3$

Equation 2 Reversed FBS Hydrosilylation of *1H,1H,2H*-perfluoro-1-alkenes using Wilkinson's catalyst [34]

In a subsequent elegant study [35] the same laboratory prepared a library of fluorous, *1H,1H,2H,2H*-perfluoroalkylsilyl-substituted derivatives of triphenylphosphine Ph$_{3-a}$P[C$_6$H$_{5-y}$ {SiMe$_3$-b(CH$_2$CH$_2$C$_x$F$_{2x+1}$)$_b$}$_y$-pos]$_a$ [a = 1–3; b = 1–3; x = 4, 6, 8, or 10; pos = 3, 4 (y = 1) or 3, 5 (y = 2)], using parallel synthetic techniques. Each member of the library can be uniquely described by these four parameters (see structures of **3** in Fig. 2), where a is the number of silyl-substituted phenyl groups (a = 1–3), b is the number of tails attached per silicon atom (b = 1–3), x denotes the length of the perfluoroalkyl group (x = 4, 6, 8, 10), and *pos* refers to the positions at the aryl ring (pos = C3, C4, or C3C5) that are silylated. The total number of tails per phosphine is then given by ab (pos = C3, C4) or $2ab$ (pos = C3C5), and the total number of fluorinated carbon atoms is given by abx (C3, C4) and $2abx$ (C3C5). No tails longer than C$_{10}$F$_{21}$ or shorter than C$_4$F$_9$ were used [35].

Upon variation of these four parameters, a total of 108 different fluorous phosphines could be synthesized. Using factorial design, 37 phosphines were

Fig. 2 Structure of the three main groups of fluorous phosphines **3** synthesized by parallel synthesis [35, 36]

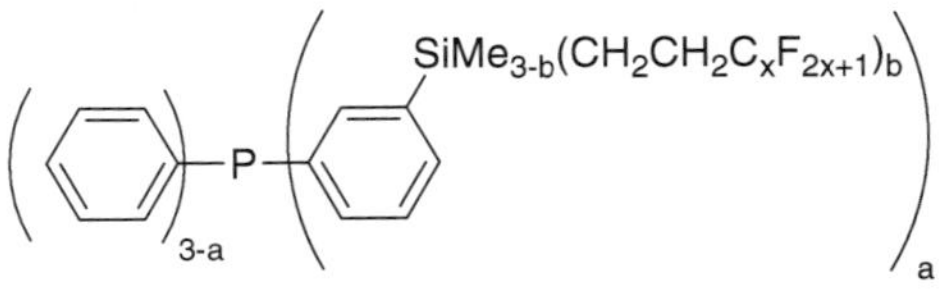

3(a, b, x, C3)

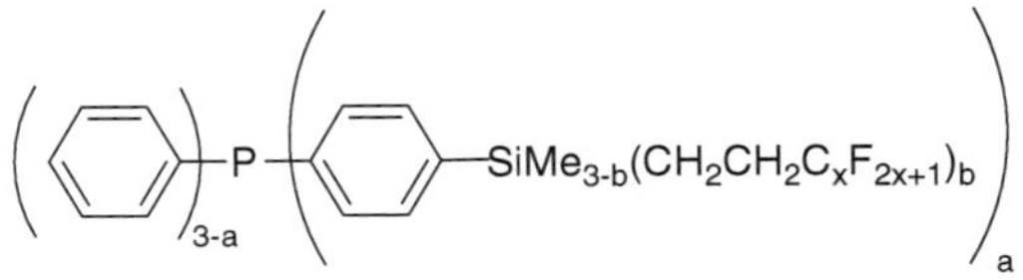

3(a, b, x, C4)

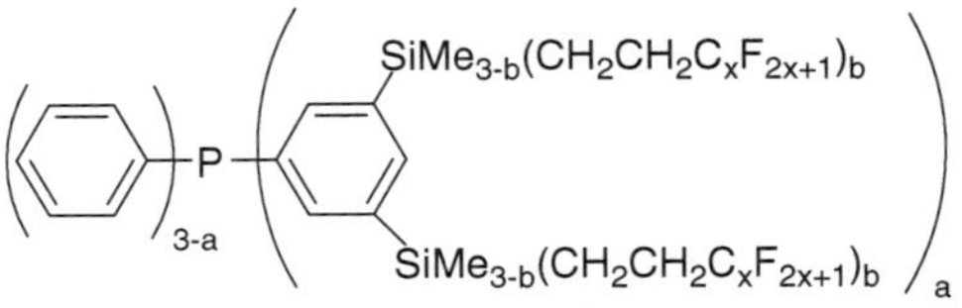

3(a, b, x, C3C5)

selected and their partition coefficients in the typical fluorous biphasic solvents system PFMCH/toluene determined. By fitting of the partition coefficient data to linear functions of the parameters a, b, and x, the partition coefficients of the remaining 71 fluorous phosphines, which were not prepared, could be predicted. Using this approach, some unexpected trends in the dependence of the partition coefficient on variations of the four parameters became clear, resulting in a better understanding of the optimum fluorous substitution pattern for obtaining the highest partition coefficient (P). In this way, the partition coefficient was increased by two orders of magnitude, i.e., from the initial value $P = 7.8$ for **3** (3, 2, 6, C4) to $P > 238$ for **3** (2, 3, 6, C3C5). Both *para*- and 3,5-substituted phosphines showed irregular behavior in the sense that elongation or increase of the number of perfluoroalkyl tails did not necessarily lead to higher partition coefficients. Particularly high values were found for phosphines containing a total of 72 fluorinated carbon atoms on the *meta* position(s) of the aryl rings. Linear relationships were found between the predicted $\log P$ of **3** (a, b, x, C4) and the experimentally determined $\log P$ values of fluorous diphosphines $[CH_2P\{C_6H_4(SiMe_{3-b}(CH_2CH_2C_6F_{13})_b)-4\}_2]_2$ and monophosphines $Ph_{3-a}P(C_6H_4(CH_2CH_2C_6F_{13})-4)_a$ [35]. The partition coefficients of phosphines which had been employed by this research group before of the type $P[C_6H_4\{SiMe_{3-b}(CH_2CH_2C_xF_{2x+1})_b\}-p]_3$ ($b = 1-3$; $x = 6, 8$) [45] and which had been applied to hydrogenation [46, 47] and the above described hydrosilylation of alkenes [34] (see Rh catalysts **1** and **2** in Fig. 1) were investigated. Surprisingly, maximum values were found in these cases only at relatively low weight percentages of fluorine ($P = 7.8$ at $b = 2$ and $x = 6$ in PFMCH/toluene at 0 °C). Therefore, not only the total number of tails and the length (x) of each tail are important, but also the specific substitution pattern/shape of the molecule.

In this subsequent work [35] one of the most fluorophilic phosphines **3** (3, 1, 8, C3C5), was applied and efficiently recycled in the FBS Rh-catalyzed hydrosilylation of 1-hexene (with catalyst **4** [RhCl(**3**)$_3$]) by HSiMe$_2$Ph using PFMCH as the fluorous phase and the substrates as the organic phase. No differences in activity and selectivity were observed compared to previously reported catalysts **1** and **2** [34]. The partition coefficient at RT was found to be 65 and this optimized value resulted in a significant decrease of catalyst leaching (<0.1% or 1 ppm of Rh) as well as leaching of free fluorous ligand (0.8%, 7 ppm) as compared to the values obtained for **1** and **2** (Table 1). The work on the library of fluorous phosphines was subsequently patented [36].

2.2 Use of Fluorous Ionic Liquids in the FBS Hydrosilylation of Alkenes

This laboratory also reported the synthesis of a fluorous room-temperature ionic liquid, 1-butyl-3-methyl-imidazolium tetrakis[p-{dimethyl(*1H,1H,2H,2H*-perfluorooctyl)silyl}phenyl]borate or [BMIm] [B{C$_6$H$_4$(SiMe$_2$CH$_2$CH$_2$C$_6$F$_{13}$)-p}$_4$] (**5**, Fig. 3) [40] by methathesis of [BMIm] and Na[B{C$_6$H$_4$(SiMe$_2$CH$_2$CH$_2$C$_6$F$_{13}$)-p}$_4$] [49].

Compound **5** is an air-stable salt which is a viscous yellow liquid at 25 °C that is transformed into a glasslike substance at subambient temperatures and it can be considered as a polar solvent [40]. Solubilities of 1-hexene (16.7 mol/mol, 9.3×10^2 g/L) and 1-octene (4.7 mol/mol, 3.5×10^2 g/L) are much higher than those reported for the closely related ionic liquids [BMIm][PF$_6$] (1-octene: 0.026 mol/mol) and [MeN-(*n*-Hex)$_3$]Tos (1-octene: 1.5 mol/mol). Because the fluorous lipophilic anion in **5** offers no possibility for hydrogen interaction it has low solubility in water and higher solubility in apolar solvents. This makes the behavior of **5** resemble that of both an ionic liquid and a fluorous solvent [40].

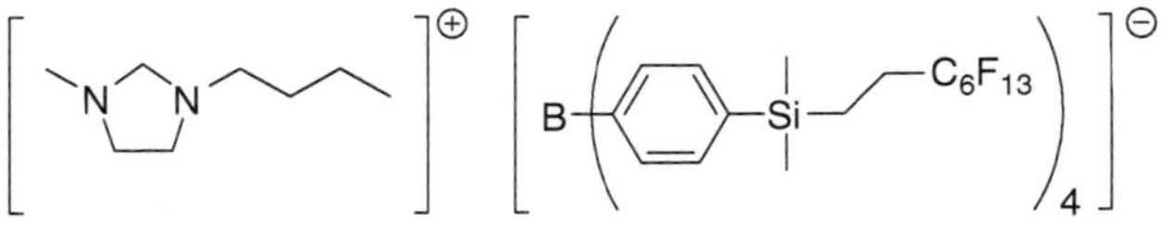

5

Fig. 3 Structure of the fluorous room-temperature ionic liquid **5** employed as a solvent for the homogeneous hydrosilylation of 1-octene catalyzed by [RhCl(P{C$_6$H$_4$(SiMe$_2$CH$_2$CH$_2$C$_6$F$_{13}$)-p}$_3$)$_3$] (**6**) [40]

Table 2 Comparison of the hydrosilylation of 1-octene using either non-fluorous or fluorous Wilkinson's catalysts in ionic liquids [40]

Catalyst	Solvent	Cycle	TOF (h^{-1})[a]	r[b]
RhCl(PPh$_3$)$_3$	Benzene	–	1.8×10^3	–
RhCl(PPh$_3$)$_3$	[BMIm][BF$_4$]	–	4.0×10^2	–
Blank	5	–	<1	–
6	5	1	4.0×10^2	–
6	5	2	3.1×10^2	0.77
6	5	3	2.8×10^2	0.91
6	5	15	1.3×10^2	<0.94>[c]
6	5 (100 °C)	–	3.2×10^2	–

[a]Average turnover frequency (TOF) defined as mol(silane)/mol(Rh)h
[b]Retention of catalyst activity = activity in cycle n/activity in cycle ($n - 1$)
[c]Average retention per cycle for cycles 4–15

The Rh-catalyzed hydrosilylation of 1-octene was studied to assess the suitability of **5** as a catalyst immobilization medium [(3), Table 2].

$$\text{PhSiH} + \text{octene} \xrightarrow[\text{ionic liquid}]{\text{RhCl(P(Ar}_f)_3)_3} \text{PhSi-octyl}$$

Ar$_f$ = C$_6$H$_4$(SiMe$_2$CH$_2$CH$_2$C$_6$F$_{13}$)-p

Equation 3 Hydrosilylation of 1-octene using either non-fluorous or fluorous Wilkinson's catalyst in ionic liquids [40]

Wilkinson's catalyst is known to form stable solutions in ionic liquids like [BMIm][BF$_4$] and [BMIm][PF$_6$]. However, in this case, Wilkinson's catalyst displayed a higher affinity for the organic phase and proved insoluble in **5**. The lightly fluorous derivative of Wilkinson's catalyst [RhCl(P{C$_6$H$_4$(SiMe$_2$CH$_2$CH$_2$C$_6$F$_{13}$)-p}$_3$)$_3$] (**6**) [36, 45] exhibited interesting solubility in **5** and concentrations of at least 1.4×10^{-2} M were attainable (Table 2).

The hydrosilylation catalyzed by **6** in fluorous ionic liquid **5** afforded the anti-Markovnikov product similar to what was reported for Wilkinson's catalysts in conventional solvents. At 84 °C (when the mixture of 1-octene and **5** becomes homogeneous) the reaction mixture was still an emulsion and at 100 °C it was homogeneous. The perfluoro tail substitution of both the ionic liquid and the catalyst does not influence the reaction rates and selectivity. Whereas the TOF observed do not make a strong case for the use of **5** in hydrosilylation catalysis, the total turnover numbers (TON) obtained after multiple cycles do. Efficient catalyst recycling was possible by phase separation at 0 °C (15 cycles with a retention of around 92% per cycle). This resulted in a TON of 4.0×10^3 mol per mol of catalyst, significantly higher than for the conventional system under monophasic conditions (of around 1,000). The drop in conversion (Table 2) is most likely due to catalyst leaching (loss of around 4% of Rh and 2% of phosphine per cycle). This is more efficient than the results of combined TOF/leaching observed with catalysts **1** and **2** (Tables 1 and 2) [34]. However, the best results obtained to date in the FBS hydrosilylation of 1-alkenes (in terms of accumulative TOF/TON, recycling and catalyst/ligand leaching) are those with catalyst **4** and fluorous phosphine **3** (3, 1, 8, C3C5) [35, 36] described in the previous section.

3 Hydrosilylation of Ketones, Enones, and Aldehydes

3.1 FBS Rhodium-Phosphine Catalyzed Hydrosilylation of Ketones and Enones

The FBS hydrosilylation of cyclic ketones and enones catalyzed by fluorous Rh catalysts was the first FBS hydrosilylation process to be reported (1999, Gladysz and Dinh [37]). The authors pre-communicated initial results [37] and more complete and detailed studies were patented in 2003 [39] and reported subsequently in 2005 [38]. The whole process is depicted in Fig. 4. The ligands employed were

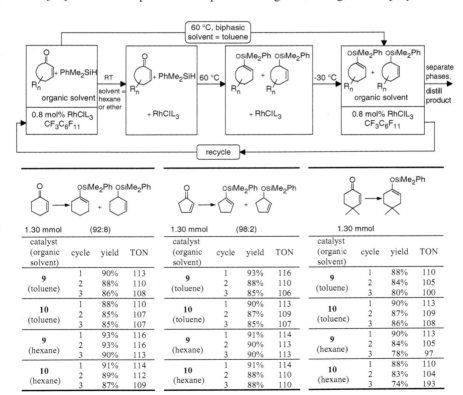

Fig. 4 Fluorous Rh-catalyzed hydrosilylation of ketones and enones: biphasic and monophasic conditions [37–39]

alkylic fluorous phosphines of the type $P(CH_2CH_2(CF_2)_{n-1}CF_3)_3$ ($n = 6$, **7**; 8, **8**) [50, 51] and the Rh derivatives were the fluorous analogs of Wilkinson's catalysts [RhCl(L)$_3$] $L = 7$, **9**; 8, **10**. Catalysts **9** and **10** had been successfully used in FBS hydroboration and hydrogenation reactions [52–54].

The reactions were run at first in toluene (Fig. 4) and reactions went to completion after 10 h under biphasic conditions at 60 °C. The fluorous solvent employed was $CF_3C_6F_{11}$ which only becomes homogeneous with toluene under further heating (but lower temperatures are preferred in synthesis). In the patent the authors also described experiments with $CF_3C_6F_{11}$/dibutyl ether at 100 °C with similar results (91% conversion, TON 114) [39]. The experiment was thus repeated with a mixture of $CF_3C_6F_{11}$/hexanes with catalyst **9** (Fig. 4) which resulted in a monophasic system at 60 °C. The product formation was completed on shorter time scales (1–2 h vs 4–10 h under biphasic conditions) and the phase separation occurred at -30 °C. Products were isolated in comparable yields and **9** could similarly be recycled. A further optimization of the process was conducted with cyclohexanone, a simple ketone that can only give a 1,2-addition product and is thus better suited for mechanistic studies. First, a study with **9** (0.2 mol%) in the addition of PhMe$_2$SiH to cyclohexanone under monophasic conditions in $CF_3C_6F_{11}$/hexanes showed that the reactions were completed in 8 h at 28 °C, with TON values of 485–481 (four cycles). Second, the rates in $CF_3C_6F_{11}$/hexanes were monitored as a function of the cycle at 40 °C vs tridecane internal standard. The first cycle showed a brief induction period. Subsequent cycles showed no induction periods, and only slightly diminutions in rates. At 40 °C the cycle took twice as long and the rates declined after 70–80% conversion [37, 38]. For comparison, an analogous set of experiments were conducted under biphasic conditions in $CF_3C_6F_{11}$-toluene [38]. The hydrosilylations became slower as compared to the monophasic experiment. However, the product yields were similar. The first cycle again showed a brief induction period and there was a similar drop in activity in the third and fourth cycles [38].

Finally, a monophasic preparative reaction was conducted (13.0 mmol cyclohexanone, 28 °C, lower catalyst loading of 0.02 mol%) and the product was obtained after 72 h in 88% yield (TON 4400). Therefore, **9** and **10** were efficient and recyclable hydrosilylation catalysts for a variety of carbonyl compounds [37–39]. However, the authors pointed out to the fact that no catalyst recovery method is without drawbacks. The liquid/liquid biphasic product/catalyst separation requires expensive fluorous solvents. A cheaper version is to use FC-72, a mixture of perfluorinated solvents. In the studies of this section $CF_3C_6F_{11}$ was selected in the interest of maximizing reproducibility. It should also be noted that small equilibrium amounts of fluorous solvents remain in organic solvents under biphasic conditions (and vice versa). Thus, there is some leaching of fluorous solvents into the organic phases under the conditions described above.

In the next section we describe alternative protocols developed by the same authors that allowed them to recycle fluorous hydrosilylation catalysts without recourse to fluorous solvents [41–43].

3.2 Temperature-Controlled Fluorous Rhodium Catalyst Recycling for the Hydrosilylation of Ketones: Use of Teflon as Delivery/Recovery Support

Up to this section, the recovery/recycling of the hydrosilylation catalyst described in this review was based on FBS techniques (although the catalytic reaction can take place in monophasic or biphasic conditions, the separation takes place in a biphase scenario). Several scientists had reported on the so-called second-generation methods [53–59] which eliminated the need of fluorous solvents by exploiting the temperature-dependent solubilities of fluorous catalysts in common organic solvents. Appropriately designed fluorous molecules are soluble only at elevated temperatures, and essentially insoluble at low temperatures. Such a thermomorphic character allows homogeneous catalysis at the high-temperature limit, and catalyst recovery by a simple liquid/solid-phase separation at the low-temperature limit. Several catalytic processes have been developed based on these second-generation protocols which have been recently reviewed by Gladysz [43, 60, 61], and others [62–64] including oxidation [65] and hydrosilylation reactions [33] described by one of us. The presence of an insoluble fluorous support can be beneficial, for example, when small quantities are involved. Initial efforts involved Teflon shavings [54, 56] and fluorous silica gel [43], and more recently insoluble perfluoroalkylated polymers [66]. An elegant protocol where CO_2 pressure was used instead of temperature to desorb a fluorous rhodium hydrogenation catalyst from fluorous silica gel is a remarkable example [67, 68].

Gladysz and co-workers described the recycling of hydrosilylation Rh fluorous catalysts **9** and **10** by liquid/solid phase separation making use of the thermomorphic properties described above [41]. The red-compounds **9** and **10** have very little or no solubility in organic solvents at room temperature but their solubility increase markedly with temperature. However, this catalyst system presents a challenge for recovery by precipitation, due to the different solubilities of the variety of rest states possible for the Rh precursor and the induction period exhibited in the first cycle [37–39].

In this work [41], the solvent of choice was dibutyl ether due to its extended liquid range (b.p. 142 °C). The system consisted of a mixture of cyclohexanone (0.53 M), PhMe$_2$SiH (1.2 equivalents), GC standard, catalysts **9** or **10** (1 mol%) and dibutyl ether which was warmed to 65 °C to achieve homogeneous conditions. After 8 h the conversion (GC analysis) was 98% and the solutions were cooled to −30 °C; supernatant solutions were then removed by syringe. The residue was washed twice with cold dibutyl ether and charged with fresh reactants. The cycle was repeated three additional times, each giving a 98% yield of hydrosilylation products [41] (Fig. 5).

The authors used Teflon tape (thickness/width 0.0075/12 mm) as a support that would facilitate the recovery of lower catalysts loadings in what was called the method of "Catalyst-on-a-tape" [41, 69] which was patented in 2006 [42]. The system was similar to that described above and homogeneous conditions could be achieved at 55 °C. Photographs of a typical sequence are shown in Fig. 6.

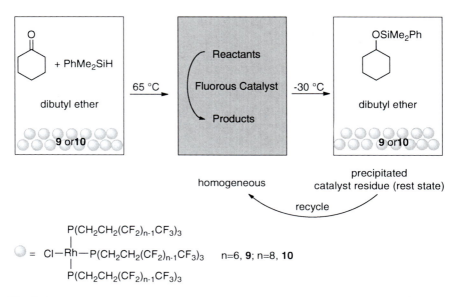

Fig. 5 Recycling of a thermomorphic fluorous rhodium hydrosilylation catalyst by liquid/solid phase separation [41]

Fig. 6 Recycling of thermomorphic fluorous Rh hydrosilylation catalyst **9** using Teflon tape. Copyright Wiley-VCH Verlag GmbH & Co. KGaA. Reproduced with permission [41]

The white tape became lightly colored during the reaction and orange-red when the sample was cooled. It is quite remarkable that the catalyst rest-state phase separates onto the tape, as opposed to giving a second solid phase. Interestingly, the Teflon stirring bar remained white.

In this case the conversion to product after 15 min was 97% and retention of activity was excellent in the second and third cycles (after the usual induction period exhibited in the first cycle). However, there was substantial loss in the fourth which was attributed to catalyst deactivation as opposed to leaching. Leaching after the three cycles was analyzed to be 11.4% of "pony-tail leaching" of ligands and 5.3% in Rh. In order to refine the process the authors precoated the catalyst **9** on the Teflon tape. When coating is uniform, this would allow low loadings to be delivered by length as opposed to mass measurements. Two 50 × 12 × 0.0075 mm strips of tape were added to a solution of **9** (0.013 g, 0.0039 mmol) in $CF_3C_6F_{11}$ (1.0 mL). The solvent was removed under an inert gas stream to give a yellowish catalyst-coated tape. This tape was applied in a three-cycle sequence and gave yield results similar to those depicted in Fig. 6. The authors were able to exclude impurities such as metal particles as the active catalytic species. Hydrosilylations of 2-octanone, acetophenone, and benzophenone to the corresponding ethers were conducted under identical conditions and similar results obtained although with a 20% activity loss in the third cycle [41].

3.3 FBS and Temperature-Controlled Fluorous Gold Catalyst Recycling in the Hydrosilylation of Aldehydes: Proposal of a Unique Reaction Mechanism for Gold Phosphine-Catalyzed Hydrosilylation Reactions

The hydrosilylation of aldehydes with fluorous gold catalysts was reported by one of us in collaboration with Prof. Istvan Horváth (Contel, Horváth and co-workers [33]). At the time of the report gold catalysis was gaining an extraordinary momentum since it became clear that gold can replace or even outperform other transition metals in the synthesis of fine chemicals. The interest of molecular catalyst designers in gold catalysis has only increased since then [70]. In order to achieve effective product separation, we modified the known [AuCl(PPh$_3$)/n PBu$_3$, $n = 6$] catalyst system [71] with fluorous phosphines which permitted the application of fluorous biphasic or thermomorphic protocols described in the previous sections.

$$[AuCl(tht)] + P\{(CH_2)_nC_8F_{17}\}_3 \xrightarrow{\text{dry CH}_2\text{Cl}_2, \text{RT}} [AuCl(P\{(CH_2)_nC_8F_{17}\}_3)]$$

$$n = 3\ (\mathbf{11}),\ 4\ (\mathbf{12}) \qquad\qquad n = 3\ (\mathbf{13}),\ 4\ (\mathbf{14})$$

Equation 4 Preparation of fluorous Au(I) hydrosilylation catalysts **13** and **14** [33]

The addition of fluorous phosphines [P{(CH$_2$)$_n$C$_8$F$_{17}$}$_3$] ($n = 3$ (**11**) and 4 (**12**)) [72] to the gold(I) complex [AuCl(tht)] (tht = tetrahydrothiophene) in CH$_2$Cl$_2$ was used for the preparation of the first fluorous gold(I) compounds (4). [AuCl(P{(CH$_2$)$_n$C$_8$F$_{17}$}$_3$)] ($n = 3$ (**13**), 4 (**14**)) are air-stable white solids that can be stored at room temperature for several months without decomposition. Compounds **13** and **14** are highly soluble in fluorous solvents at room temperature and were tested in the biphasic hydrosilylation of benzaldehyde with SiPh(Me)$_2$H (**16a**) (5) to give PhCH$_2$OSiMe$_2$Ph (**17a**).

Equation 5 FBS gold-catalyzed hydrosilylation of aldehydes [33]

The experiments were carried out under biphasic conditions using FC-72 or perfluoroheptane as the fluorous catalyst phase containing the catalyst, the upper layer being most of the substrate with **16** at the beginning or products **17** at the end of the reaction. We also noted that the reaction times were longer for the hydrosilylation in comparison to that of the reported [AuCl(PPh$_3$)]/PBu$_3$ system in DMF, acetonitrile, or THF solvents [71].

It appeared that, by increasing the phosphine **11** to gold ratio from 1:1 to 2:1, the rate of the reaction increased and the conversion to **17a** was higher (Table 3, entry 3). The performance of the catalyst system was further improved by the

Table 3 Hydrosilylation of benzaldehyde under FBC (fluorous biphasic catalysis) conditions using gold fluorous compound **13** [33]

Entry	Catalyst[a]		Conversion to **17**[b] (%)
1	–	P{Rf}$_3$ (**11**) 10 mol% Rf-(CH$_2$)$_3$CN (**18**) 25 mol%	0
2	[AuCl(P{Rf}$_3$)] (**13**) 8 mol%	– Rf-(CH$_2$)$_3$CN (**18**) 25 mol%	17
3	[AuCl(P{Rf}$_3$)] (**13**) 8 mol%	P{Rf}$_3$ (**11**) 10 mol% –	55
4	[AuCl(P{Rf}$_3$)] (**13**) 5 mol%	P{Rf}$_3$ (**11**) 10 mol% Rf-(CH$_2$)$_3$CN (**18**) 25 mol%	50
5	[AuCl(P{Rf}$_3$)] (**13**) 8 mol%	P{Rf}$_3$ (**11**) 10 mol% Rf-(CH$_2$)$_3$CN (**18**) 25 mol%	85
6	FRP[c] from 5	–	76
7	FRP[c] from 6	–	64
8	FRP[c] from 7	–	56
9	[AuCl(P{Rf}$_3$)] (**13**) 8 mol%	P{Rf}$_3$ (**11**) 10 mol% Rf-(CH$_2$)$_3$CN (**18**) 25 mol%	83
10	0.1183 g isolated from 9	–	91
11	FRP[c] from 10	–	90
12	[AuCl(P{Rf}$_3$)] (**13**) 8 mol%	–	74

[a]To a solution of **13** and additives in 1 mL of degassed FC-72 or perfluoroheptane under nitrogen, benzaldehyde (1.0 mmol) and Me$_2$PhSiH (1.2 mmol) were added. Reactions were performed at 75 °C for 18 h
[b]Conversion given by ^{1}H NMR analysis (CDCl$_3$)
[c]FRP: fluorous recovered phase

addition of the novel fluorous alkyl nitrile C$_8$F$_{17}$(CH$_2$)$_3$CN (**18**) (entries 4 and 5). Compound **18** was used as the fluorous analog of acetonitrile, which can stabilize the previous reported [AuCl(PPh$_3$)]/PBu$_3$ catalytic system. An increase of the amount of gold catalyst to 8 mol% gave a higher conversion (entry 5) as expected. In this case, the fluorous phase was separated and the recyclability of the system was demonstrated for four runs by adding benzaldehyde (**15**) and silane (**16a**) to the recovered fluorous phase (entries 6–8). The conversion decreased by about 10% in each cycle. This was probably due to the small volume of FC-72 employed (1 mL) that decreases the efficiency of the phase separation in the separatory funnel. Hydrosilylation of **15** with Si(Et)$_3$H (**16b**) was not completed under the optimal observed conditions for **16a**. It should be noted that, while electron-withdrawing groups increased, an electron-donating group decreased the catalytic activity. For example, in the case of *p*-tolualdehyde the yield was only 39%, whereas 3,4-dichlorobenzaldehyde resulted in 100% conversion under similar conditions using **13** as the catalyst (8 mol%) in the presence of 10 mol% of **11** and 25 mol% **18**.

It is also important to note that [AuCl(PPh$_3$)]/n PPh$_3$ (*n* = 0 or 6) are catalytically inactive in the hydrosilylation of **15** in DMF and the formation of a purple solution and black precipitate was reported [71]. In contrast, the [AuCl(PPh$_3$)]/n PBu$_3$ (*n* = 0 or 6) system is catalytically active and the reaction mixture remains colorless throughout the reaction indicating the stabilizing role of the much more basic PBu$_3$. Surprisingly, in the case of fluorous biphasic hydrosilylation of **15**, we observed that, although the fluorous layer became reddish purple after 1 h at 75 °C, the reaction proceeded. ^{31}P-NMR spectroscopy of the fluorous layer showed the disappearance of **13** and the appearance of a new peak at 44.7 ppm (s, br) indicating the formation of a new fluorous soluble species (**19**). This chemical shift is different from the catalyst precursor **13** and from [AuCl(P{(CH$_2$)$_3$C$_8$F$_{17}$}$_3$)$_2$] (**20**). Reaction of **13** with P{(CH$_2$)$_3$C$_8$F$_{17}$}$_3$ (**11**) in a mol ratio 1:1 at room temperature afforded **20**. In order to elucidate the nature of the novel species (**19**) a hydrosilylation of **15** with **16a** was performed up to 80% conversion and the reddish purple fluorous layer was separated and concentrated (to about 0.1 mL). Addition of diethyl ether afforded **19** as an air-stable reddish purple solid that could be stored at room temperature during several months according to MS analyses. The isolated **19** was also dissolved again in FC-72, and this reddish solution could again catalyze the hydrosilylation with reasonable yields (Table 3, entries 9–11), in spite of the fact that the isolation yield of the solid was 43%. In contrast, the original system based on PBu$_3$ was readily oxidized in air and became catalytically inactive. The chemical shift of **19** in the ^{31}P NMR spectrum suggested the presence of two fluorous phosphine ligands **11** coordinated to Au(I). The fluorous nitrile **18**, used in the hydrosilylation, was not present in the solid according to the analytical and spectroscopical data including X-ray and transmission electron microscopy (TEM) analysis. XRTF measurements indicated that this species did not contain chloride anions. The microanalysis data suggested a molecular formula of [Au(P{(CH$_2$)$_3$C$_8$F$_{17}$}$_3$)$_2$]$_2$O. While the origin of the oxygen atom is not clear, its presence is supported by one of the peaks at m/z = 4,865 observed in the MALDI-TOF spectrum of **19** in trifluorotoluene (dithranol as matrix). This peak can be assigned to a new trinuclear [Au$_3$(P{(CH$_2$)$_3$C$_8$F$_{17}$}$_3$)$_3$O$_2$] species. Complexes of Au(I) of the type [{Au(PPh$_3$)}$_3$O]BF$_4$ are well known [73] and the complex with a sulfur atom [{Au(PPh$_3$)$_2$}$_2$S] [74] has also been described. We also investigated whether **19** could form fluorous soluble colloids or nanoparticles (**nano-19**) by means of an X-ray (powdered sample of **19**) analysis. Indeed, the formation of small size gold nanoparticles was confirmed and their exact dimensions were measured by TEM. It should emphasized that the sample was very homogeneous and gold nanoparticles of mean size 2.6 nm were detected. Fluorous-soluble nanoparticles of palladium either imbedded in a fluorous dendrimer [75] or solubilized by fluorous molecules [76, 77] have been described and used in fluorous-Heck and -Suzuki couplings [76, 77]. The stability and catalytic activity of **19** under hydrosilylation conditions should be carefully investigated, since it could be in equilibria with mononuclear species [Au(P{(CH$_2$)$_3$C$_8$F$_{17}$}$_3$)$_2$]$^+$ and/or [Au(P{(CH$_2$)$_3$C$_8$F$_{17}$}$_3$)]$^+$, which could also be responsible for the catalytic activity observed. Furthermore, the gradual loss of activity (see Table 3, entry 8) could be explained by the aggregation of the bimetallic species **19** into more complicated nanostructures with limited or no solubility in fluorous environments. The MALDI-TOF spectrum in trifluorotoluene solution displays peaks that can be assigned to species of a higher nuclearity. Accordingly, the recyclability of the system was very poor under thermomorphic conditions [43] in the absence of fluorous solvents. Hydrosilylation reactions were performed with the same amounts of **13** or **14** and corresponding additives under typical hydrosilylation conditions using 1 mL of degassed DMF, THF, or CH$_3$CN (thermomorphic liquid/solid phase separation of catalyst protocol). Conversions ranged from moderate (35%) to high (92%), similar in the case of the two catalyst (**13** or **14**). The resulting reddish purple material could be separated by filtration.

In summary, novel fluorous gold(I) compounds were described and used as recoverable catalysts for the hydrosilylation of aldehydes although the catalyst loadings were high and therefore the TON and TOF values obtained were low.

Nevertheless, these results prompted us to carry out a thorough research of the non-fluorous gold-catalyzed hydrosilylation of aldehydes [78]. In previous work [71], it was noted that the initial rate of the reaction of PhCHO (**15**) with SiPh (Me)$_2$H (**16a**) (5) was hardly modified and the yield of PhCH$_2$OSiMe$_2$Ph (**17a**) was increased from 50% to almost full conversion by increasing the concentration of PBu$_3$ from 10 to 20 mol% with respect to 3 mol% of the gold catalyst [71]. These results are difficult to explain by the well accepted mechanism of hydrosilylation of aldehydes, which includes the oxidative addition of **16a** to the metal center, the coordination and insertion of the carbonyl group of **15** into the metal-silicon bond, and the reductive elimination of **17a** (Scheme 1 [18]). We studied the effects of key reaction parameters on rate and selectivity including some of the side reactions which led us to propose a plausible alternative reaction pathway [78].

The gold precursors used in the hydrosilylation of aldehydes were either [AuCl (PPh$_3$)] as reported [71], or [AuCl(tht)] (tht = tetrahydrothiophene). While these complexes are catalytically inactive even in the presence of excess PPh$_3$, the addition of PBu$_3$ results in the formation of an active species of unknown structure [71]. First, we investigated the reaction of [AuCl(tht)] with various amounts of

PBu$_3$. We performed a titration of [AuCl(tht)] with various amounts of PBu$_3$ by ^{31}P-NMR spectroscopy. The formation of [AuCl(PBu$_3$)] and [Au(PBu$_3$)$_2$]Cl could be clearly observed by the appearance of the peaks at 23.5 ppm and 33.5 ppm, respectively. Only one peak is observable above P/Au = 2, which is shifted to higher fields and becomes broader at higher ratios, indicating the possible formation of [Au(PBu$_3$)$_n$]Cl ($n > 2$) and the rapid exchange between these species (Scheme 3).

Next we investigated the performance of the PBu$_3$-modified [AuCl(tht)] catalyst in the hydrosilylation of benzaldehyde (15), propanal (21), and nonanal (22) using Me$_2$PhSiH (16a) and Et$_3$SiH (16b) in CH$_3$CN, CH$_2$Cl$_2$, or neat reaction mixture (Table 4). It is important to emphasize that in the absence of [AuCl(tht)] these hydrosilylation reactions do *not* take place. The conversions of the aldehydes with 16a were higher than with 16b. While the nature of the two solvents employed has little effect, the reaction proceeds somewhat faster in the absence of their addition.

The dependence of the reaction rate on key reaction parameters was investigated by in situ NMR measurements at room temperature. By lowering the gold concentration from 3 mol% to 1 mol%, the reaction rate decreases significantly, although not proportionally to the gold concentration. This is probably due to the changing

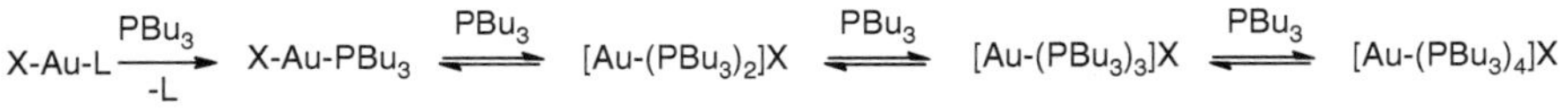

$$X\text{-Au-L} \xrightarrow[-L]{PBu_3} X\text{-Au-PBu}_3 \xrightleftharpoons{PBu_3} [Au\text{-}(PBu_3)_2]X \xrightleftharpoons{PBu_3} [Au\text{-}(PBu_3)_3]X \xrightleftharpoons{PBu_3} [Au\text{-}(PBu_3)_4]X$$

Scheme 3 Equilibrium of different gold species formed by titration of [AuCl(tht)] with PBu$_3$ [78]

Table 4 Hydrosilylation of aldehydes catalyzed with AuCl(tht) and PBu$_3$ [78]

Aldehyde	Silane	Solvent	Product[a]	Yield (%)[b]
Benzaldehyde (15)	Me$_2$PhSiH (16a)	CH$_3$CN	(Benzyloxy)dimethylphenylsilane (17a)	81
Benzaldehyde (15)	Me$_2$PhSiH (16a)	CH$_2$Cl$_2$	(Benzyloxy)dimethylphenylsilane (17a)	82
Benzaldehyde (15)	Et$_3$SiH (16b)	CH$_3$CN	(Benzyloxy)triethylsilane (17b)	78
Benzaldehyde (15)	Et$_3$SiH (16b)	CH$_2$Cl$_2$	(Benzyloxy)triethylsilane (17b)	55
Propanal (21)	Me$_2$PhSiH(16a)	CH$_3$CN	Dimethylphenylpropoxysilane (23a)	69
Propanal (21)	Me$_2$PhSiH(16a)	CH$_2$Cl$_2$	Dimethylphenylpropoxysilane (23a)	56
Propanal (21)	Et$_3$SiH (16b)	CH$_3$CN	Triethylpropoxysilane (23b)	30
Propanal (21)	Et$_3$SiH (16b)	CH$_2$Cl$_2$	Triethylpropoxysilane (23b)	5
Nonanal (22)	Me$_2$PhSiH (16a)	CH$_3$CN	Dimethylphenylsiloxynonane (24a)	82
Nonanal (22)	Me$_2$PhSiH (16a)	CH$_2$Cl$_2$	Dimethylphenylsiloxynonane (24a)	95
Nonanal (22)	Et$_3$SiH (16b)	CH$_3$CN	1-Triethylsiloxynonane (24b)	16
Nonanal (22)	Et$_3$SiH (16b)	CH$_2$Cl$_2$	1-Triethylsiloxynonane (24b)	3
Benzaldehyde (15)	Me$_2$PhSiH (16a)	–	(Benzyloxy)dimethylphenylsilane (17a)	100
Nonanal (22)	Me$_2$PhSiH (16a)	–	Dimethylphenylsiloxynonane (24a)	100

3 mol% [AuCl(tht)] and 20 mmol% PBu$_3$ in a solution of 1 mmol aldehyde and 0.9 mmol silane in 0.5 mL solvent at 70 °C for 3 h
[a]Characterized by NMR spectroscopy and GC-MS
[b]Conversion by ^{1}H-NMR

positions of the gold-phosphine equilibria shown in Scheme 3. A similar effect could be observed by changing the concentration of PBu$_3$, **15**, and **16a** [78]. It should be noted that at the Au/PBu$_3$ = 1:1 or 1:2 ratios the reaction mixture became heterogeneous, evidenced by the formation of a black precipitate, and no reaction could be observed. We therefore increased the PBu$_3$ concentration, in accordance with previous work [71] and our results. The hydrosilylation of 2 mmol of **15** with 4 mmol of **16a** in the presence of 3 mol% [AuCl(PPh$_3$)] and 20 mol% PBu$_3$, in 2 mL CH$_3$CN at 70 °C was monitored by in situ IR spectroscopy. To our surprise the colorless reaction mixture turned deep purple when the IR bands of **15** disappeared (e.g., at 100% conversion of **15**) and the formation of a black precipitate was noticed. These results have suggested that the presence of excess of PhCHO (**15**) over HSiMe$_2$Ph (**16a**) is an important factor, in addition to the excess of PBu$_3$ over gold [71]. The reaction was also performed without added solvent. When 3.22 mmol **16a** was added to a solution of 3 mol% AuCl(PPh$_3$) and 20 mol% PBu$_3$, in 20 mmol **15**, the hydrosilylation was completed at 90 °C within 25 min and the solution remained colorless. The addition of **16a** can be continued (e.g., 6.44 mmol and 3.86 mmol of **16a**) without the appearance of the purple color and the formation of a black precipitate until **15** remains in slight excess. These experiments suggest that benzaldehyde (**15**) and PBu$_3$ themselves or together play an important role in stabilizing the gold catalyst and/or forming the catalytically active species. It is also evident that the reducing power of HSiMe$_2$Ph (**16a**) is high enough to destabilize the gold(I) catalyst giving rise to gold clusters or particles. We confirmed by different experiments that small amounts of water and oxygen can lead to side reactions and the formation of species of the type PhMe$_2$SiOSiMe$_2$Ph (with H$_2$O catalyzed by Au/PBu$_3$) and the phosphonium salts [C(OH)(H)PhPBu$_3$]Cl from the oxidation of benzaldehyde **15** to benzoic acid and subsequent interaction with PBu$_3$.

Concerning the mechanism involved, our data suggest that both PBu$_3$ and the aldehydes play a crucial role in stabilizing the catalytically active gold species. In contrast, the excess of hydrosilanes is detrimental due to rapid reduction of the gold catalyst. Although the formation of a tri-coordinated AuHP$_2$ intermediate was proposed and later isolated for [AuCl(xantphos)]-catalyzed dehydrogenative silylation of alcohols [79, 80], the formation of a mononuclear species with PBu$_3$ seems unlikely. The increasing amount of PBu$_3$ with respect to gold should increase the level of substitution [81] and thus the rate of the oxidative addition of the hydrosilane should decrease (Scheme 1), resulting in a proportionally slower catalytic reaction. Since the opposite effect was observed, an alternative mechanism must be operational. One possibility is PBu$_3$ concentration dependent equilibrium positions between the [Au(PBu$_3$)$_n$]Cl (n = 2, 3, and 4) species that may exhibit significantly different reactivity towards the aldehydes depending on the electronic density on the gold center (Scheme 4). While [Au(PBu$_3$)$_4$]Cl does not have an open coordination site to activate the aldehyde, both [Au(PBu$_3$)$_n$]Cl (n = 2 and 3) could perform the activation of the aldehyde to form oxygen-bonded **A2** and **A3** adducts, respectively, with significantly different rates. While gold(I) has little affinity for oxygen donor ligands, tertiary phosphines have been shown to stabilize

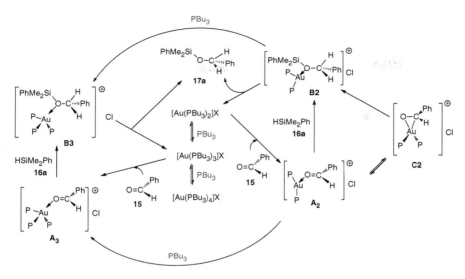

Scheme 4 Alternative mechanism for the gold-catalyzed hydrosilylation of aldehydes [78]

Au(I)–O bonds. The cation [AuL]$^+$ (L = phosphine) is isolobal with a proton and shows a great affinity for bonding to various Lewis bases [82]. The possibility for side-on coordination of the aldehyde (**C2**) cannot be ruled out, but it seems plausible for [Au(PBu$_3$)$_2$]Cl only. These intermediates could readily undergo a concerted addition of the Si$^{\delta+}$–H$^{\delta-}$-bond to the C$^{\delta+}$–O$^{\delta-}$-bond, resulting in the formation of a coordinated alkoxysilanes **B2** and **B3**, which readily eliminate alkoxysilane to regenerate the gold catalyst. The accelerating effect of the PBu$_3$ could also be the result of converting **A2** to **A3** or **B2** to **B3**.

While the operation of a novel mechanism for gold could lead to new applications in organic chemistry, the stabilizing role of one of the substrates, e.g., the aldehydes, is unusual in homogeneous transition metal catalysis and indeed surprising. A report on the hydrosilylation of aldehydes by gold indenyl phosphine complexes [83], published at the same time as our study [78], described that gold(I) indenyl phosphine derivatives are efficient catalysts (in conditions similar to ours with 3 mol% gold catalyst and 20 mol% indenyl phosphine ligand at 70 °C). They also reported that systems based on the precursor [AuCl(SMe$_2$)] and conventional alkylic phosphines PEt$_3$, P^{n}Bu$_3$, or P^{t}Bu$_3$ (mol ratio 1:6) were the most effective and that reactions could be run at 24 °C if the reaction times were longer (24 h instead of 3 h). Although the authors did not perform mechanistic studies, they pointed out the fact that the gold phosphine-catalyzed hydrosilylation of aldehydes is enabled by a strongly donating phosphine but that there is no correlation between catalytic performance and the cone angle of the phosphine. In this study [83], it was reported that gold(I)-N-heterocyclic carbenes were extremely poor catalysts as opposed to similar Cu(I) and Ag(I) derivatives, suggesting a special role for the phosphine ligands in the gold-catalyzed hydrosilylation of aldehydes.

Our study helped us to understand why the FBS gold-phosphine-catalyzed hydrosilylation was not so efficient and the TON/TOF were low [33]. It seems clear that the gold phosphine-catalyzed hydrosilylation of aldehydes requires a basic phosphine as ligand. The fluorous alkylic phosphines employed in our study (**7** and **8** [33]) are more basic than other fluorous aryl phosphines [31] or PPh$_3$. However, from our mechanistic study [78] we learnt that in order to optimize reaction conditions an excess of aldehyde over silane was needed to avoid reduction as well as a large excess of basic phosphine (conditions which had not been tested in the FBS work). We also proposed a coordination of the aldehyde and subsequent interaction with the silane (although the silane does not bind directly to the gold atom). While we were able to perform an FBS gold-catalyzed hydrosilylation of aldehydes with separation and recovery of the gold-catalyst we were not able to assess unambiguously whether the isolated fluorous gold-nanoparticles **19** were the real catalytic species or their precursors (isolated nano-**19** catalyzed the FBS hydrosilylation). It seems that this process with monophosphines is a little complex and that different gold-phosphine species are in equilibria with one another. From the work on the [AuCl(xantphos)]-catalyzed dehydrogenative silylation of alcohols [79, 80] it seems that a basic, bulky fluorous diphosphine could potentially be a much better choice as a fluorous ligand and it may lead to a conventional hydrosilylation mechanism based on the oxidative addition of the silane and the formation of AuP$_2$H species. Such a fluorous diphosphine ligand could, theoretically, prevent the issues described above with monophosphines and other undesired side-reactions.

4 Conclusion

Despite the importance of the catalyzed hydrosilylation reaction to prepare compounds containing silicon, there have not been many reports on biphasic homogeneous hydrosilylation reactions with the recovery of the metal-catalyst. The use of aqueous biphasic reactions has almost been neglected, since many metal-catalyzed hydrosilylations are quite sensitive to water. In this review we have described the few papers and patents dealing with the FBS hydrosilylation reactions reported to date. Despite the limited number of reports, the FBS hydrosilylation reaction has been extremely successful. In all cases fluorous monophosphines (either alkylic or perfluoroalkylsilyl-substituted derivatives of triphenylphosphine) have been employed as ligands to synthesize and immobilize the metal catalysts (either rhodium(I) or gold(I) derivatives) in the fluorous solvent (including a fluorous ionic liquid). The hydrosilylation of alkenes, ketones, and enones with fluorous rhodium analogs to the Wilkinson's catalyst [RhCl(PPh$_3$)$_3$] has afforded high TON/TOF and a very efficient separation and recycling of the fluorous catalyst. Modification of the fluorous content and position of the fluorous tails in the aryl groups of the phosphines has allowed for further optimization of the process and a better recovery with minimal leaching of rhodium and fluorous ligand to the organic phase. Moreover, the use of the so-called second generation methods which eliminate the need for fluorous solvents by exploiting the temperature-dependent solubilities of fluorous catalysts in common organic solvents (thermomorphic properties) has permitted the use and separation of fluorous alkyl-phosphine rhodium catalysts in hydrosilylation reactions in conventional organic solvents. The addition of an insoluble fluorous support such as Teflon tape allowed for an exceptionally easy and efficient recovery of fluorous rhodium catalysts ("catalyst-on-a-tape") in the hydrosilylation of ketones. In the case of the FBS gold-catalyzed hydrosilylation of aldehydes, new fluorous gold catalysts with alkylic phosphines have led to the separation and recycling of the gold catalysts although the TON/TOF were lower than in the rhodium-hydrosilylation of alkenes and ketones. A detailed study of the non-fluorous gold-catalyzed hydrosilylation of aldehydes has helped to explain how the catalytic system could be improved. The use of diphosphines as opposed to monophosphines may be a possible option in order to avoid equilibria reactions between different gold(I)-phosphine species, reduction by the silane to gold-nanoparticles and undesired side-reactions.

In conclusion, the hydrosilylation of alkenes, ketones and aldehydes with fluorous metal catalysts has provided a way to increase the TON/TOF obtained by the non-fluorous versions (when all the cycles are considered) and quite an efficient separation and recovery of the metal catalyst by different methods, some of them without the recourse to expensive fluorous solvents. The use of fluorous metal-catalysts based on some other fluorous ligands (such as fluorous chiral phosphines, fluorous diphosphines, fluorous N-heterocyclic carbenes) could lead to further optimization in the hydrosilylation of multiple bonds and in the enantioselective version of the hydrosilylation of $C=O$ bonds to obtain chiral alcohols with subsequent recovery and recycling of the metal-catalysts.

References

1. Marciniec B (1992) Comprehensive handbook on hydrosilylation. Pergamon, Oxford, UK
2. Marciniec B (2009) Hydrosilylation: a comprehensive review on recent advances. Springer, Berlin
3. Roy AK (2008) A review of recent progress in catalyzed homogeneous hydrosilation (hydrosilylation). Adv Organomet Chem 55:1–54
4. Brook MA (2000) Silicon in organic, organometallic and polymer chemistry. Wiley, New York
5. Lewis LN (2000) From sand to silicones: an overview of the chemistry of silicones. In: Clarkson SJ, Fitgerald JJ, Owen MJ, Smith SD (eds) Silicones and silicone-modified materials. Oxford University Press and the American Chemical Society, Washington, DC, pp 11–19
6. Arena CG (2009) Recent progress in the asymmetric hydrosilylation of ketones and imines. Mini-Rev Org Chem 6:159–167
7. Morris RH (2009) Asymmetric hydrogenation, transfer hydrogenation and hydrosilylation of ketones catalyzed by iron complexes. Chem Soc Rev 38:2282–2291
8. Bao F, Kanno KI, Takahashi T (2008) Early transition metal catalyzed hydrosilylation reaction. Trends Org Chem 12:1–17

9. Munslow IJ (2008) Alkyne reductions. In: Anderson PG, Munslow IJ (eds) Modern reduction methods. Wiley-VCH Verlag GmbH & Co. KGaA, Weinheim, pp 363–385
10. Riant O (2008) Hydrosilylation of imines. In: Anderson PG, Munslow IJ (eds) Modern reduction methods. Wiley-VCH Verlag GmbH & Co. KGaA, Weinheim, pp 321–337
11. Rendler S, Oestreich M (2008) Diverse modes of silane activation for the hydrosilylation of carbonyl compounds. In: Anderson PG, Munslow IJ (eds) Modern reduction methods. Wiley-VCH Verlag GmbH & Co. KGaA, Weinheim, pp 183–207
12. Mayes PA, Perlmutter P (2008) Alkyne reduction: hydrosilylation. In: Anderson PG, Munslow IJ (eds) Modern reduction methods. Wiley-VCH Verlag GmbH & Co. KGaA, Weinheim, pp 87–105
13. Marciniec B (2002) Catalysis of hydrosilylation of carbon-carbon multiple bonds: recent progress. Silicon Chem 1:155–175
14. Malacea R, Poli R, Manoury E (2010) Asymmetric hydrosilylation, transfer hydrogenation and hydrogenation of ketones catalyzed by iridium complexes. Coord Chem Rev 254:729–752
15. Speier JL, Hook DE (1958) Organosilicon compounds. US Patent 2,823,218
16. Iovel IG, Goldberg YSh, Shymanska MV, Lukevikc E (1987) Quaternary onium hexachloroplatinates: novel hydrosilylation catalysts. Organometallics 6:1410–1413
17. Karstedt BD (1973) Platinum-siloxane complexes as hydrosilylation catalysts. US Patent 3,775,452
18. Chalk AJ, Harrod JF (1965) Homogeneous catalysis II. The mechanism of the hydrosilylation of olefins catalyzed by group VIII metals. J Am Chem Soc 87:16–21
19. Glasser PB, Tilley TD (2003) Catalytic hydrosilylation of alkenes by a ruthenium silylene complex. Evidence for a new hydrosilylation mechanism. J Am Chem Soc 125:13640–13641
20. Normand AT, Cavell KJ (2008) Donor-functionalised N-heterocyclic carbene complexes of group 9 and 10 metals in catalysis: trends and directions. Eur J Inorg Chem 2781–2800
21. Diez-Gonzalez S, Nolan SP (2008) Copper, silver and gold complexes in hydrosilylation reactions. Acc Chem Res 41:349–358
22. Du G, Abu-Omar MM (2008) Oxo and imido complexes of rhenium and molybdenum in catalytic reductions. Curr Org Chem 12:1185–1198
23. Cornils B, Herrmann WA, Horváth IT, Leitner W, Mecking S, Olivier-Bourbigou H, Vogt D (2006) Introduction. In: Cornils B, Herrmann WA, Horváth IT, Leitner W, Mecking S, Olivier-Bourbigou H, Vogt D (eds) Multiphase homogeneous catalysis. Wiley-VCH Verlag GmbH & Co. KGaA, Weinheim, pp 1–23
24. Maciejewski H, Szubert K, Marciniec B, Pernak J (2009) Hydrosilylation of functionalised olefins catalysed by rhodium siloxide complexes in ionic liquids. Green Chem 11:1045–1051
25. Maciejewski H, Wawrzynczak A, Dutkiewicz M, Fiedorow R (2006) Silicon waxes-synthesis via hydrosilylation in homo- and heterogeneous systems. J Mol Catal A Chem 257:141–148
26. Hofmann N, Bauer A, Frey T, Auer M, Stanjek V, Schulz PS, Taccardi N, Wasserscheid P (2008) Liquid-liquid biphasic, platinum-catalyzed hydrosilylation of allyl chloride with trichlorosilane using ionic liquid catalyst phase in a continuous loop reactor. Adv Synth Catal 350:2599–2609
27. Geldbach TJ, Zhao D, Castillo NC, Laurenczy G, Weyershausen B, Dyson PJ (2006) Biphasic hydrosilylation in ionic liquids: a process set for industrial implementation. J Am Chem Soc 128:9773–9780
28. Behr A, Naendrup F, Obst D (2002) Platinum-catalyzed hydrosilylation of unsaturated fatty acids. Adv Synth Catal 344:1142–1145
29. Horváth IT, Rábai J (1994) Facile catalyst separation without water: fluorous biphase hydroformylation of olefins. Science 266:72–75
30. Clayton JW Jr (1967) Fluorocarbon toxicity and biological action. Chem Rev 1:197–252
31. Gladysz JA, Curran DP, Horváth IT (2004) Handbook of fluorous chemistry. Wiley-VCH Verlag GmbH & Co. KGaA, Weinheim

32. Horváth IT (2006) Fluorous catalysis. In: Cornils B, Herrmann WA, Horváth IT, Leitner W, Mecking S, Olivier-Bourbigou H, Vogt D (eds) Multiphase homogeneous catalysis. Wiley-VCH Verlag GmbH & Co. KGaA, Weinheim, pp 339–403
33. Lantos D, Contel M, Larrea A, Szabo D, Horváth IT (2006) Fluorous-phosphine assisted recycling of gold catalysts for hydrosilylation of aldehydes. QSAR Comb Sci 25:719–722
34. de Wolf E, Speets EA, Deelman BJ, van Koten G (2001) Recycling of rhodium-based hydrosilylation catalysts; a fluorous approach. Organometallics 20:3686–3690
35. de Wolf E, Riccomagno E, de Pater JJM, Deelman BJ, van Koten G (2004) Parallel synthesis and study of fluorous biphasic partition coefficients of 1H,1H,2H,2H-perfluoroalkylsilyl derivatives of triphenylphosphine: a statistical approach. J Comb Chem 6:363–374
36. Richter B, de Wolf ACA, Van Koten G, Deelman BJ (2000) Fluorous phosphines and process for their preparation. WO 0018774 and US 6458978
37. Dinh LV, Gladysz JA (1999) Transition metal catalysis in fluorous media: extension of a new immobilization principle to biphasic and monophasic rhodium-catalyzed hydrosilylations of ketones and enones. Tetrahedron Lett 40:8995–8998
38. Dinh LV, Gladysz JA (2005) Monophasic and biphasic hydrosilylations of enones and ketones using a fluorous rhodium catalyst that is easily recycled under fluorous-organic liquid-liquid biphasic conditions. New J Chem 29:173–181
39. Gladysz JA, Wende M, Rocaboy C (2003) Composition used for hydroformylation, hydroboration, oxidation, hydrogenation, hydrosilylation or phosphine-catalyzed organic reaction contains highly fluorinated catalyst or reagent with temperature-dependent solubility in solvent used. DE 10212424
40. van der Broeke J, Winter F, Deelman BJ, van Koten G (2002) A highly fluorous room-temperature ionic liquid exhibiting fluorous biphasic behavior and its use in catalyst recycling. Org Lett 4:3851–3854
41. Dinh LV, Gladysz JA (2005) "Catalyst-on-a-tape"-Teflon: a new delivery and recovery method for homogeneous fluorous catalysts. Angew Chem Int Ed 44:4095–4097
42. Gladysz JA, Dinh LV, Curran DP (2006) Methods, processes and materials for dispensing and recovering supported fluorous reaction components. US 20060094866 and US 2008281086
43. Gladysz JA, Tesevic V (2008) Temperature-controlled catalyst recycling: new protocols based upon temperature-dependent solubilities of fluorous compounds and solid/liquid phase separation. Top Organomet Chem 23:67–89, and refs therein
44. Jardine FH (1981) Chlorotris(triphenylphosphine rhodium(I). Its chemical and catalytic reactions. In: Lippard SJ (ed) Progress in inorganic chemistry, vol 28. Wiley, New York, pp 117–183
45. Richter B, de Wolf E, van Koten G, Deelman BJ (2000) Synthesis and properties of a novel family of fluorous triphenylphosphine derivatives. J Org Chem 65:3885–3893
46. Richter B, van Koten G, Deelman BJ (1999) Fluorous biphasic hydrogenation of 1-alkenes using novel fluorous derivatives of Wilkinson's catalyst. J Mol Catal A 145:317–321
47. Richter B, Spek AL, van Koten G, Deelman BJ (2000) Fluorous versions of Wilkinson's catalysts. Activity in fluorous hydrogenation of 1-alkenes and recycling by fluorous biphasic separation. J Am Chem Soc 122:3945–3951
48. Ameduri B, Boutevin B, Nouiri M, Talbi M (1995) Synthesis and properties of fluorosilicon-containing polybutadienes by hydrosilylation of fluorinated hydrogenosilanes. Part 1. Preparation of the silylation agents. J Fluorine Chem 74:191–197, and refs therein
49. Van den Broeke J, Lutz M, Kooijman H, Spek AL, Deelman BJ, van Koten G (2001) Increasing the lipophilic character of tetraphenylborate anions through silyl substituents. Organometallics 20:2114–2117
50. Juliette JJJ, Horváth IT, Gladysz JA (1997) Transition metal catalysis in fluorous media: practical application of a new immobilization principle to rhodium-catalyzed hydroboration. Angew Chem Int Ed 36:1610–1612

51. Juliette JJJ, Rutherford D, Horváth IT, Gladysz JA (1999) Transition metal catalysis in fluorous media: practical application of a new immobilization principle to rhodium-catalyzed hydroborations of alkenes and alkynes. J Am Chem Soc 121:2696–2704
52. Rutherford D, Juliette JJJ, Rocaboy C, Horváth IT, Gladysz JA (1998) Transition metal catalysis in fluorous media: application of a new immobilization principle to rhodium-catalyzed hydrogenation of alkenes. Catal Today 42:381–388
53. Wende M, Meier R, Gladysz JA (2001) Fluorous catalysis without fluorous solvent: a friendlier catalyst recovery/recycling protocol based upon thermomorphic properties and liquid/solid phase separation. J Am Chem Soc 123:11490–11491
54. Wende M, Gladysz JA (2003) Fluorous catalysis under homogeneous conditions without fluorous solvents: a "greener" catalyst recycling protocol based upon temperature-dependent solubilities and liquid/solid phase separation. J Am Chem Soc 125:5861–5872
55. Ishihara K, Kondo S, Yamamoto H (2011) 3,5-Bis(perfluorodecyl)phenylboronic acid as an easily recyclable direct amide condensation catalyst. Synlett 1371–1374
56. Ishihara K, Hasegawa A, Yamamoto Y (2002) A fluorous super Bronsted acid catalyst: application to fluorous catalysis without fluorous solvents. Synlett 1299–1301
57. Mikami K, Mikami Y, Matsuzawa Y, Matsumoto Y, Nishidiko J, Yamamoto F, Nakajima H (2002) Lanthanide catalysts with tris(perfluorooctanesulfonyl)methide and bis(perfluorooctanesulfonyl)amide ponytails: recyclable Lewis acid catalysts in fluorous phases or as solids. Tetrahedron 58:4015–4021
58. Otera J (2004) Toward ideal (trans)esterification by use of fluorous distannoxane catalysts. Acc Chem Res 37:288–296, and refs therein
59. Maayan G, Fish R, Neumann R (2003) Polyfluorinated quaternary ammonium salts of polyoxometalate anions: fluorous biphasic oxidation catalysis with and without fluorous solvents. Org Lett 41:3547–3550
60. Gladysz JA (2009) Catalysis involving fluorous phases: fundamentals and directions for greener methodologies. In: Anastas PT, Crabtree RH (eds) Handbook of green chemistry. Wiley-VCH Verlag GmbH & Co. KGaA, Weinheim, pp 17–38
61. Gladysz JA (2008) Thermomorphic cyclopalladated compounds. In: Dupont J, Pfeffer M (eds) Palladacycles. Wiley-VCH Verlag GmbH & Co. KGaA, Weinheim, pp 341–360
62. Bergbreiter DE (2009) Thermomorphic catalysts. In: Benaglia M (ed) Recoverable and recyclable catalysts. Wiley, Chichester, UK, pp 117–154
63. Zhang W (2009) Green chemistry aspects of fluorous techniques – opportunities and challenges for small-scale organic synthesis. Green Chem 11:911–920
64. Candeias NR, Branco LC, Gois PMP, Alfonso CAM (2009) More sustainable approaches for the synthesis of N-based heterocycles. Chem Rev 109:2703–2802
65. Contel M, Villuendas PR, Fernandez-Gallardo J, Alonso PJ, Vincent JM, Fish RH (2005) Fluorocarbon soluble copper(II) carboxylate complexes with nonfluoroponytailed nitrogen ligands as precatalysys for the oxidation of alkenols and alcohols under fluorous biphasic or thermomorphic modes: structural and mechanistic aspects. Inorg Chem 44:9771–9778
66. Audic N, Dyer PW, Hope EG, Stuart AM, Suhard S (2010) Insoluble perfluoroalkylated polymers: new solid supports for supported fluorous phase catalysis. Adv Synth Catal 352:2241–2250
67. Ablan CD, Hallet JP, West KN, Jones RS, Eckert CA, Liotta CA, Jessop PG (2003) Use and recovery of a homogeneous catalyst with carbon dioxide as a solubility switch. Chem Commun 2972–2973
68. Eckert CA, Jessop PG, Liotta CL (2002) Methods for solubilizing and recovering fluorinated compounds. WO 02096550
69. King AG (2006) Research advances: caught on tape: catalyst recovery; secondary structure switch; DNA-based chiral catalysts. J Chem Ed 83:10–14
70. Hashmi ASK (2010) Homogeneous gold catalysis beyond assumptions and proposals: characterized intermediates. Angew Chem Int Ed 49:5232–5241

71. Ito H, Yajima T, Tateiwa J, Hosomi A (2000) First gold complex-catalyzed selective hydrosilylation of organic compounds. Chem Commun 981–982
72. Vlád G, Richter FU, Horváth IT (2004) Modular synthesis of fluorous trialkylphosphines. Org Lett 6:4559–4561
73. Laguna A (1999) Gold compounds of phosphorus and the heavy group V elements. In: Schmidbaur H (ed) Gold, progress in chemistry, biochemistry and technology. Wiley-VCH Verlag GmbH & Co. KGaA, Weinheim, pp 348–428
74. Canales F, Gimeno MC, Laguna A, Villacampa MD (1996) Aurophilicity at sulfur centers. Synthesis of the polyaurated species $[S(AuPR_3)n]^{(n-2)+}$ ($n = 2-6$). Inorg Chim Acta 244:95–103
75. Crooks M, Zhao M, Sun L, Chenik V, Yeung LK (2001) Dendrimer-encapsulated metal nanoparticles: synthesis, characterization, and applications to catalysis. Acc Chem Res 34:181–190, and refs therein
76. Moreno-Mañas M, Pleixat R, Villaroya S (2002) Palladium nanoparticles stabilized by polyfluorinated chains. Chem Commun 60–61
77. Moreno-Mañas M, Pleixat R, Villaroya S (2001) Fluorous phase soluble palladium nanoparticles as recoverable catalysts for Suzuki cross-coupling and Heck reactions. Organometallics 20:4524–4528
78. Lantos D, Contel M, Sanz S, Bodor A, Horváth IT (2007) Homogeneous gold-catalyzed hydrosilylation of aldehydes. J Organomet Chem 692:1799–1805
79. Ito H, Takagi K, Miyahara T, Sawamura M (2005) Gold(I)-phosphine catalyst for the highly chemoselective dehydrogenative silylation of alcohols. Org Lett 7:3001–3004
80. Ito H, Saito T, Miyahara T, Zhong C, Sawamura M (2009) Gold(I) hydride intermediate in catalysis: dehydrogenative alcohol silylation catalyzed by gold(I) complex. Organometallics 28:4829–4840
81. Gimeno MC, Laguna A (1997) Three- and four-coordinate gold(I) complexes. Chem Rev 97:511–522
82. Fackler JP Jr, van Zyl WE, Prihoda BA (1999) Gold chalcogen chemistry. In: Schmidbaur H (ed) Gold, progress in chemistry, biochemistry and technology. Wiley, Chichester, pp 795–840
83. Wile BM, McDonald R, Ferguson MJ, Stradiotto M (2007) Au(I) complexes supported by donor-functionalized indene ligands: synthesis, characterization, and catalytic behavior in aldehyde hydrosilylation. Organometallics 26:1069–1076

Fluorous Hydroformylation

Xi Zhao, Dongmei He, László T. Mika, and István T. Horváth

Abstract The application of fluorous phosphine-modified catalysts for the hydroformylation of olefins is reviewed.

Keywords Aldehydes · Asymmetric hydroformylation · Biphasic catalysis · Carbon monoxide · Fluorous · Hydroformylation · Hydrogen · Olefins · Phosphines · Rhodium

Contents

1 Introduction 276
2 Ligands for Hydroformylation in Fluorous Media 278
3 Hydroformylation in Fluorous Media 279
   3.1 Hydroformylation of Olefins 279
   3.2 Hydroformylation of Styrene and Its Derivatives 285
   3.3 Hydroformylation of Acrylates 287
   3.4 Hydroformylation in Supercritical CO_2 287
References 287

Abbreviations

FBC	Fluorous biphasic catalysis
FC-72	Mixture of perfluorohexanes
GC	Gas chromatography
IR	Infra red

X. Zhao, D. He, and I.T. Horváth (✉)
Department of Biology and Chemistry, City University of Hong Kong, Kowloon, Hong Kong
e-mail: istvan.t.horvath@cityu.edu.hk

L.T. Mika
Institute of Chemistry, Eötvös University, Budapest, Hungary

NMR   Nuclear magnetic resonance
PFMCH   Perfluoromethyl-cyclohexane
TOF   Turnover frequency

1 Introduction

The hydroformylation reaction was discovered by Otto Roelen in 1938 [1], which leads to the formation of aldehydes by the reaction of a carbon–carbon double bond with carbon monoxide and dihydrogen in the presence of a transition metal catalyst. The accidental discovery of the oxo-reaction was made during the investigation of the cobalt-catalyzed Fischer–Tropsch reaction [2]. The formation of propanal was observed from ethylene and syngas (H_2/CO) and the yield became less and less by increasing the temperature, indicating the presence of a temperature sensitive homogeneous catalyst [$HCo(CO)_4$]. The term hydroformylation relates to the formal addition of a hydrogen (–H) and a formyl group (–COH) to a C,C– double in the presence of a Co-, Rh-, Ru-, or Pt-based catalyst (Fig. 1) [3].

The simplest "unmodified" catalysts are the hydrido cobalt or rhodium tetracarbonyls. The replacement of the coordinated carbon monoxide(s) with ligand(s) such as phosphines and phosphites leads to the formation of the "ligand-modified" catalysts with the general formula of $HM(CO)_xL_y$ (Fig. 2).

Hydroformylation could take place in both conventional and environmentally benign reaction media. The unmodified cobalt and rhodium catalysts can be dissolved in hydrocarbons such as alkanes, toluene, etc., or in the crude olefin mixtures in the case of industrial processes. The application of alternative solvents requires the use of ligand-modified catalysts to ensure high solubility, which could be designed to offer facile product separation and efficient catalyst recycling.

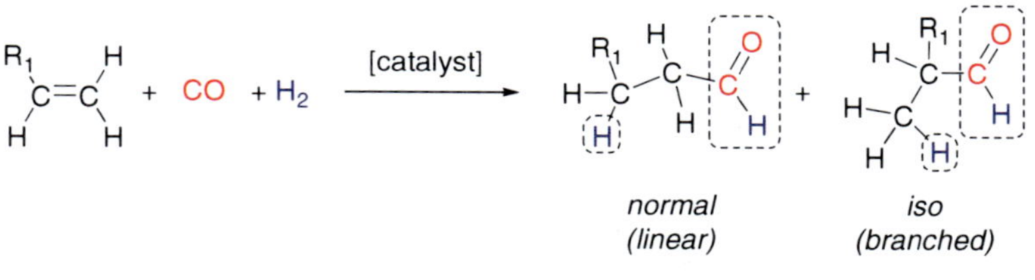

Fig. 1 Hydroformylation reaction

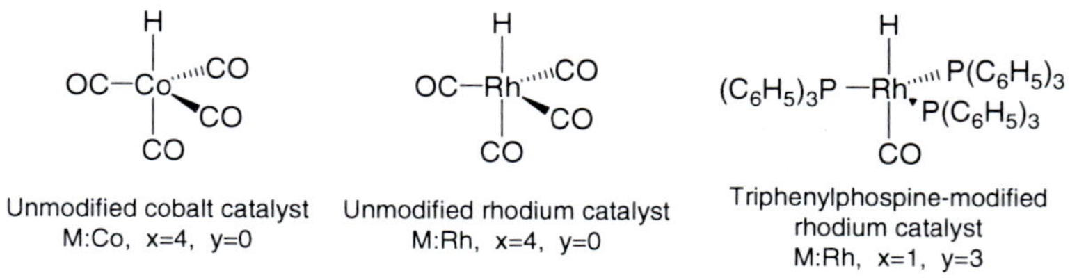

Fig. 2 Types of hydroformylation catalysts

The activity and selectivity of the catalyst can be easily affected by using different phosphorus-, nitrogen-, and sulfur-containing mono- and polydentate ligands.

One of the most challenging problems associated with the triphenylphosphine-modified rhodium catalyst system has been the separation of higher aldehydes ($C_n > 8$) from the catalyst without catalyst deactivation. The use of the concept of biphasic catalysis (Fig. 3), such as an organic/aqueous biphase system, in which the water phase contains the $P(m-C_6H_4SO_3Na)_3$-modified rhodium catalyst, offers an easy separation of the organic products from the catalyst containing aqueous phase. This process has been used commercially for the hydroformylation of propylene in the Ruhrchemie-Rhone/Poluenc process [4]. Since the catalytic reaction occurs in the aqueous phase, the potential application of this system is limited by the solubility of the olefins ($C_n > 7$) in the catalyst containing aqueous phase [5, 6].

The first fluorous hydroformylation system was designed to achieve several process requirements with a single catalyst, including (1) the possibility to use the same catalyst for the hydroformylation of lower and higher molecular weight olefins, (2) facile and effective separation of the aldehydes from the catalyst, (3) appropriate coordination power of the ligand to keep the active rhodium species stable and minimize rhodium leaching, (4) comparable activity and selectivity with the commercially used triphenylphosphine-modified (Rh/PPh$_3$) catalyst system, and (5) the advantages of single-phase catalysis with biphase product separation by running the reaction at higher temperatures, where the system forms a single phase, and separating the products from the fluorous catalyst at lower temperatures (Fig. 4) [7–9].

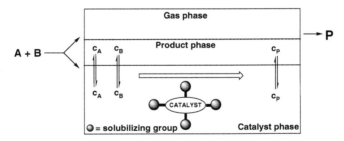

Fig. 3 Concept of biphasic catalysis

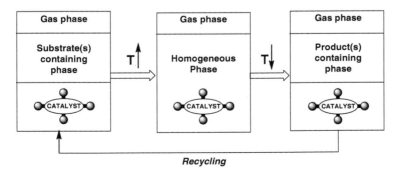

Fig. 4 Temperature dependent liquid-liquid biphasic system

2 Ligands for Hydroformylation in Fluorous Media

Fluorous hydroformylation catalysts can be prepared by combining fluorous soluble ligands with appropriate transition metal complexes. Most of the organic soluble ligands can be converted to fluorous soluble by introducing fluorous ponytails to the ligand structure. The most frequently used fluorous soluble ligands are phosphorus containing ligands which could have nitrogen or sulfur in various combinations. Figures 5 and 6 show the different fluorous phosphine and phosphite

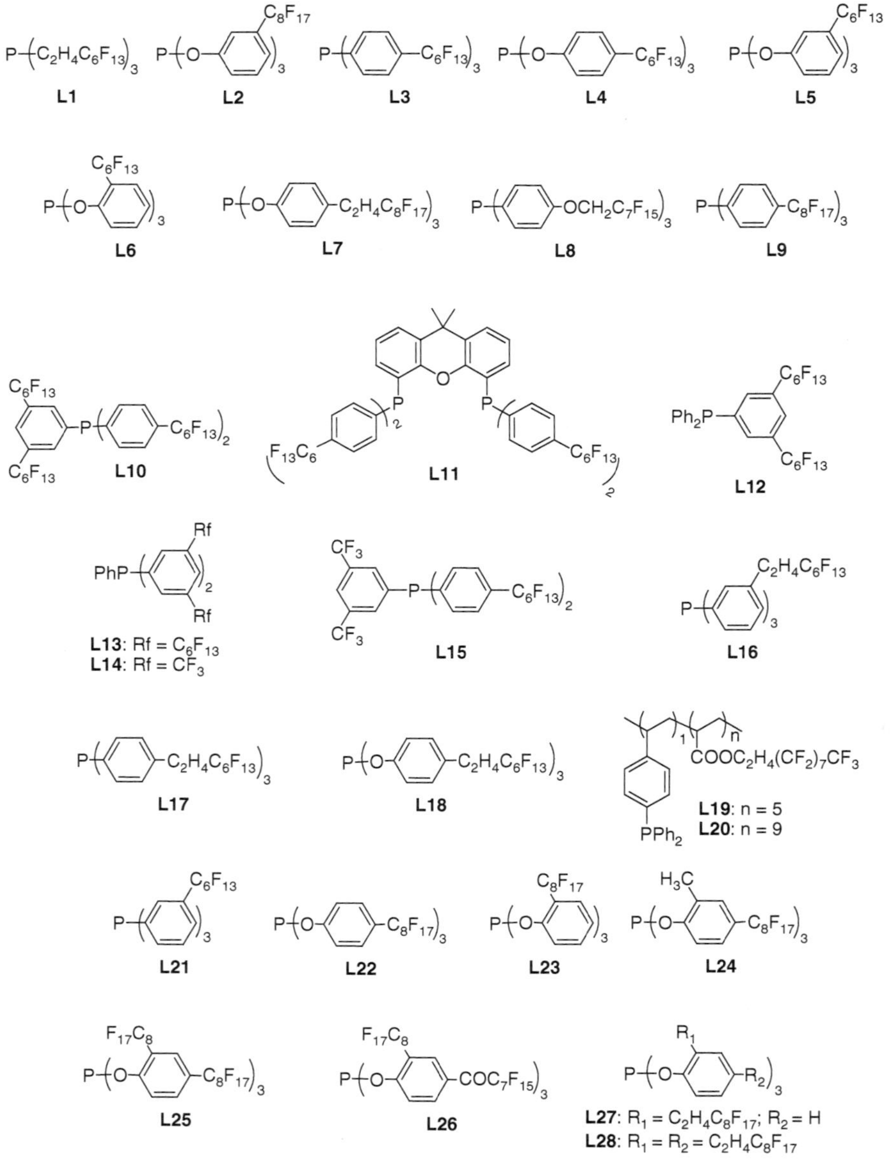

Fig. 5 Fluorous phosphines and phosphites

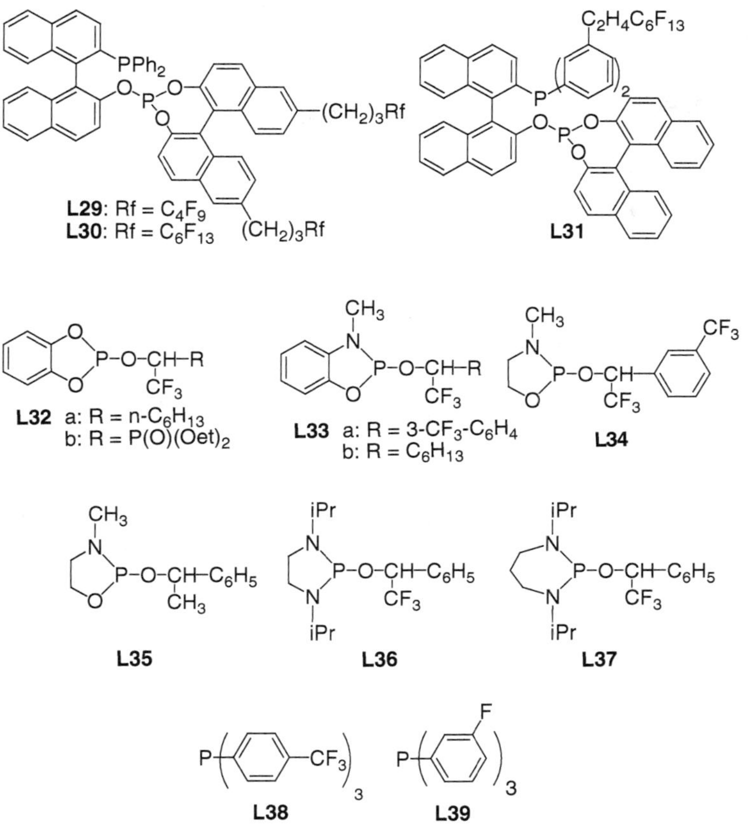

Fig. 6 Fluorous phosphines and phosphites

ligands, which have been synthesized, characterized, and tested for hydroformylation in fluorous biphasic mode and/or supercritical carbon dioxide.

3 Hydroformylation in Fluorous Media

3.1 Hydroformylation of Olefins

Fluorous catalytic systems could combine good catalyst performance, overcome the solubility limits of higher olefins, and provide facile separation of the products. For example, the hydroformylation of 1-decene was performed in the presence of the fluorous soluble $P[CH_2CH_2(CF_2)_5CF_3]_3$-modified rhodium catalyst at 100 °C and 11 bar of CO/H_2 (1:1) in 50/50 v/v% toluene/$C_6F_{11}CF_3$ solvent mixture.

Table 1 Electronic properties of $P[(CH_2)_x(CF_2)_yCF_3]_3$, $P[(CH_2)_3CH_3]$, and PPh_3[a]

Phosphines	P Mulliken Population (q)	P Lone Pair level (eV)	Protonation Energy (eV)	P–H (Å)	<P–H (°)
$P[CF_2CF_3]_3$	0.83	−11.7	−6.5	1.189	85.9
$P[CF_2CF_2CF_2CF_3]_3$	0.83	−11.7	−6.4	1.192	85.4
$P[(CH_2)_5CF_2CF_3]_3$	0.62	−10.6	−7.7	1.205	86.3
$P[CH_2CF_2CF_3]_3$	0.48	−9.9	−8.3	1.218	92.3
$P[(CH_2)_3CF_2CF_3]_3$	0.40	−9.5	−8.6	1.225	91.8
$P[(CH_2)_4CF_2CF_3]_3$	0.38	−9.3	−8.8	1.226	92.0
$P[(CH_2)_5CF_2CF_3]_3$	0.36	−9.2	−8.9	1.228	91.8
$P[CH_2CH_2CH_2CH_3]_3$	0.33	−8.7	−9.3	1.230	91.7
$P(C_6H_5)_3$	0.67	−9.0	−9.9	1.226	61.6

[a]The calculations were performed using the UniChem version of MNDO93 and employed the PM3 parameter set. Full geometry optimizations were performed

The $P[CH_2CH_2(CF_2)_5CF_3]_3$ (L1) was selected on the basis of a semiempirical calculation of the electronic properties of $P[(CH_2)_x(CF_2)_yCF_3]_3$ ($x = 0$, $y = 2, 4$ and $x = 0$–5, $y = 2$) and $P(C_6H_5)_3$ [9] (Table 1).

The attachment of the highly electron-withdrawing fluorous ponytails directly to the phosphorus atom could significantly lower the coordinating power of the fluorous phosphine. The electron-withdrawing effect can be decreased by the insertion of methylene (–CH$_2$–) groups between the fluorous ponytails and phosphorus atom. The effect of perfluoroethyl ponytails, which are sufficient models for longer perfluoroalkyl groups, is small for two ($x = 2$) and essentially negligible for three ($x = 3$) methylene units. However, the differences between the electronic properties of $P[(CH_2)_xCF_2CF_3]_3$ with more than three methylene groups ($x > 3$) and $P[(CH_2)_3CH_3]_3$ are small but finite. These results suggest that the insertion of two methylene groups ($x = 2$) is enough to lower the electron-withdrawing effect of the fluorous ponytails. It was later shown by Gladysz et al. that the electron-withdrawing effect of even five methylene units was observable according to the variation in vCO values of *trans*-[IrCl(CO)L$_2$] complexes. It appears that between eight and ten methylene groups would be needed to insulate effectively the lone pair of the phosphorus atom from the perfluoroalkyl moiety [10, 11].

It was shown by high-pressure NMR that the solution structure of the fluorous soluble HRh(CO){P[CH$_2$CH$_2$(CF$_2$)$_5$CF$_3$]$_3$}$_3$ in c-C$_6$F$_{11}$CF$_3$ is similar to that of HRh(CO)(PPh$_3$)$_3$ in toluene and HRh(CO)[P(m-C$_6$H$_4$SO$_3$Na)$_3$]$_3$ in water [9]. It has also been established that the coordinatively unsaturated {HRh(CO)(PR$_3$)$_2$} and {HRh(CO)$_2$(PR$_3$)} acts as a catalytically active species, similarly to the organic and water containing systems [9]. Their reaction with olefins leads to competing catalytic cycles involving one or two phosphine ligands on the rhodium as depicted in Scheme 1.

Kinetic studies have shown that the reaction is first order in both rhodium and 1-decene. While the reaction is inhibited by the excess of $P[CH_2CH_2(CF_2)_5CF_3]_3$ (L1), the *n*/*i* ratio of the aldehyde increases with increasing phosphine concentration, as expected. The *n*/*i* product selectivity of Rh/L1 is closer to the selectivity of

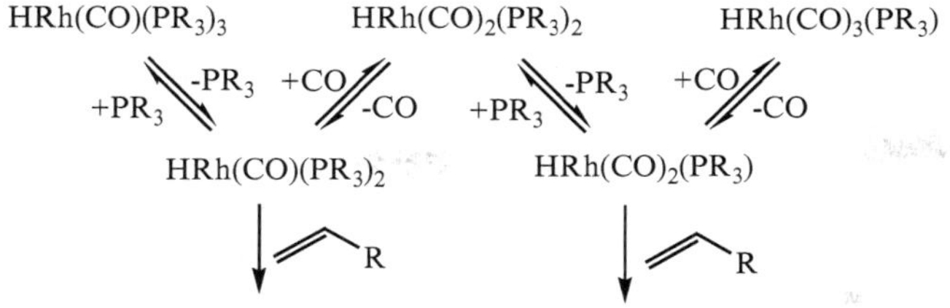

HRh(CO)(PR3)3
HRh(CO)2(PR3)2
HRh(CO)3(PR3)
-PR3
+PR3
+CO
-CO
-PR3
+PR3
+CO
-CO
HRh(CO)(PR3)2
HRh(CO)2(PR3)
R
R

Scheme 1 Ligand exchange of the catalyst species (R = $CH_2CH_2(CF_2)_5CF_3$)

C10=
Product
phase
removal
Gas Phase
Product
Phase
Fluorous Phase

Fig. 7 Semi-continuous hydroformylation of 1-decene

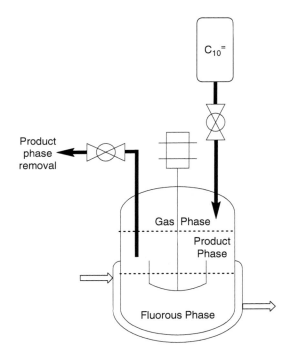

the Rh/PPh_3 catalyst than that of the Rh/P[$(CH_2)_7CH_3$]$_3$ catalyst [9]. The fluorous biphase catalyst recovery was tested in a semicontinuous hydroformylation of 1-decene with the Rh/L1 catalyst (Fig. 7). During nine consecutive reactions/separations a total turnover of more than 35,000 was achieved with a loss of 1.18 ppm of Rh/mol of undecanals (Fig. 8).

The Rh/L1 catalyst was also used for the continuous hydroformylation of ethylene using the high boiling fluorous solvent FC-70, which allows the continuous removal of propanal (Fig. 9). The long-term (60 days) stability of the Rh/L1 catalyst was better than that of the Rh/PPh_3 catalyst.

The investigation of the phase behavior of 1-octene in the biphasic hydroformylation has shown that 1-octene could be transferred from the fluorous phase into the product phase and then back again, due to the limited miscibility of the nonanal

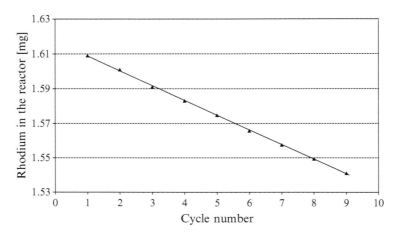

Fig. 8 Rhodium concentration during the semi-continuous hydroformylation of 1-decene

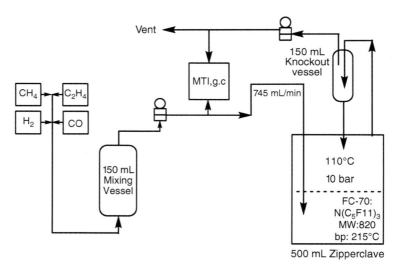

Fig. 9 Continuous hydroformylation of diluted ethylene [8]

with perfluoromethyl-cyclohexane (PFMCH). It was found that the phase behavior of terminal alkenes was dramatically affected by the temperature, by the ratio of 1-octene to PFMCH, and by the conversion of 1-octene (Table 2) [28].

Improved selectivity, catalyst retention, and product separation were reported for the hydroformylation of linear terminal alkenes (1-hexene and 1-octene) using fluorous triaryl-phosphites ligand (L4). The best results, which were comparable to those obtained with fluorous triaryl-phosphines (L3) modified systems, were obtained using rhodium (2.0 mmol L^{-1}) and P(p-O-$C_6H_4C_6F_{13}$)$_3$ (20 mmol L^{-1}) at 70 °C and 20 bar syngas (CO:H_2 = 1:1) to give a linear aldehyde selectivity of

Table 2 Fluorous Hydroformylation of Olefins

Substrate	Ligand	L/Rh	P (MPa)	T (°C)	t (min)	Conv. (%)	TOF (h^{-1})	Aldehyde selectivity (%)	n/i	Ref.
Ethylene	L1	20	1.04	100.0	–	80–95	11,851	99.3	–	[9]
	L1	14	1.04	100.0	–	80–95	11,016	98.5	–	[9]
	L1	14	1.04	110.0	–	80–95	16,744	98.2	–	[9]
1-Hexene	L2	5	4	80	9	70	7,800	94	6.5	[12]
	L3	3	2	70	60	98.4	3,600	87.8	3.8	[13, 14]
	L4	3	2	70	60	98.6	10,400	89.3	3.1	[13]
	L4	3	0.8	70	60	99.6	1,850	88.9	6.6	[13, 14]
	L5	3	2	70	60	99.9	1,300	96.0	5.2	[13]
	L6	3	2	70	60	99.9	4,200	99.1	0.9	[13]
Cyclohexene	L2	5	4	80	–	–	70	100	–	[12]
	L7	5	4	80	360	25	45	100	–	[15]
1-decene	L2	5	4	80	10	70	6,900	87	6.3	[12]
	L3	3	2	70	60	98.3	3,300	95.9	3.6	[13]
	L3	5	1.5	70	480	70	–	>90	12	[16]
	L3	10	1.5	70		–	–	90	8.8	[17]
	L3	10	2	90	30	97.9	29,800	93.3	4.4	[13]
	L3	10	2	90	30	97.9	29,800	80.2	4.4	[18]
	L4	3	2	80	30	96.6	15,600	80.5	6.3	[13]
	L4	3	2	70	60	98.2	4,400	95.8	2.1	[13]
	L4	3	2	80	30	96.6	15,600	85.0	6.3	[18]
	L7	5	4	80	60	100	3,600	95	3.0	[15]
	L8	3	20	80	60	66	–	86	76/24	[19]
	L8	5	4	80	60	98	–	97	72/28	[20]
	L9	10	2	70	90	93.1	–	90.1	6.7	[21]
	L10	10	2	70	90	98.5	–	94.8	4.7	[21]
	L11	2.2	2	80	45		–	84.5	22.9	[22]
	L12	10	2	70	90	95.9	–	93.7	4.4	[23]
	L13	10	2	70	90	98.9	–	94.7	3.9	[23]
	L14	10	2	70	90	98.3	–	94.7	5.1	[23]
	L15	10	2	70	90	97.3	–	93.2	5.4	[23]
	L16	6	22	60	1,140	92			4.6	[24]
	L16	4	21	60	1,200	97	280	97	74/23	[25]
	L17	10	20	65	1,290	99	2,080	99	81/18	[25]
	L18	10	20	65	5,280	95	2,065	95	85/10	[25]
2-Octene	L2	5	4	80	30	70	1,400	75	9/91	[12]
2-Octene	L7	5	4	80	90	100	1,200	82	5/95	[15]
4-Octene	L2	5	4	80	60	70	800	76	10/90	[12]
4-Octene	L7	5	4	80	150	100	440	77	8/92	[15]
1-Nonene	L4	3	2	70	60	99.0	1,300	85.0	6.9	[13]
2-Nonene	L4	3	2	70	60	86.1	880	74.9	0.2	[13]
1-Decene	L1	27	1.1	100	–	–	2,160	–	3.25	[9]
	L2	5	4	80	10	70	6,300	80	5.8	[12]
	L3	10	20	80	60	49	2,794	96	3.8	[26]
	L7	5	4	80	60	100-	3,800	98	3.0	[15]

(continued)

Table 2 (continued)

Substrate	Ligand	L/Rh	P (MPa)	T (°C)	t (min)	Conv. (%)	TOF (h^{-1})	Aldehyde selectivity (%)	n/i	Ref.
	L8	3	20	80	180	94	–	60	71/29	[19]
	L17	10	20	80	60	28	1,556	99	3.8	[26]
	L19	6	3	100	900	97	–	99	4.8	[27]
	L20	6	3	100	900	90	–	99	5.9	[27]
	L21	10	20	80	60	50	2,871	97	3.7	[26]
	L22	5	4	80	25	70	3,500	85	5.3	[12]
	L23	5	4	80	5	70	10,000	71	2.3	[12]
	L24	5	4	80	10	70	7,100	80	2.9	[12]
	L25	5	4	80	5	70	7,900	46	2.4	[12]
	L26	5	4	80	10	70	6,200	39	2.6	[12]
	L27	5	4	80	15	100-	10,000	85	2.0	[15]
	L28	5	4	80	12	100-	11,000	85	2.0	[15]
1-Dodecene	L7	5	4	80	60	100	2,600	94	3.0	[15]
1-Hexadecane	L3	10	20	80	60	71	2,963	95	3.6	[26]
	L17	10	20	80	60	22	957	99	3.6	[26]
	L19	6	3	100	900	78	–	98	4.8	[27]
	L20	6	3	100	900	59	–	99	5.0	[27]
	L21	10	20	80	60	73	2,998	93	3.6	[26]

80.9% and an initial productivity of 8.8 mol L^{-1} h^{-1}. It was also found that the metal and ligand leaching into the product phase was 0.05% and 3.3%, respectively [13].

Good linear aldehyde selectivity (64–80%) and excellent reaction rates (13,500–29,800 h^{-1}) were obtained by the use of P(p-O-C$_6$H$_4$C$_6$F$_{13}$)$_3$ (L4) or P(p-C$_7$F$_{15}$CH$_2$O-C$_6$H$_4$)$_3$ (L8) modified rhodium catalysts in perfluoro-1,3-dimethyl-cyclohexane for the hydroformylation of 1-octene with 0.23–0.33% rhodium loss (Table 2) [18]. For P(p-C$_7$F$_{15}$CH$_2$O-C$_6$H$_4$)$_3$ (L8), the TOF was 380 h^{-1} and the normal to branched ratio was 74/26 at 80 °C and 40 bar CO:H$_2$ = 1. The rhodium leaching was 0.78% after the first run [20]. The application of longer ponytails containing ligands (L9) or the introduction of longer fluorous groups (L10) resulted in >98% conversion and >94% selectivity at 20 bar syngas and 70 °C. The metal loss was 729 ppm in the case of L9 [21].

Fluorophilic phosphines incorporating at least one aromatic ring with two directly attached perfluoroalkyl groups (L15) were used as modifying ligands for the rhodium-catalyzed hydroformylation of 1-octene in perfluorocarbon solvents to result in good conversion (>90%) and selectivity (>97%) (Table 1) [23].

The continuous fluorous biphasic hydroformylation of 1-octene under 15 bar (CO:H$_2$ = 1:1) pressure was tested in the presence of RhH(CO)$_3$[P(p-C$_6$H$_4$C$_6$F$_{13}$)$_3$] (L3). It was observed that the system operates at higher than 15,500 turnovers at a production rate of 750 L h^{-1} [17]. This compares favorably with the commercial rhodium catalyst system for butyraldehyde production that has rates in the region of

500–700 L h^{-1}. Conversion levels were up to 70% with the linear to branched aldehyde ratios of around 12 [16].

The effects of fluorinated groups on the activity and selectivity of the biphasic hydroformylation of 1-alkenes (1-hexene, 1-octene, and 1-decene) and internal alkenes (2-octene, 4-octene, and cyclohexene) with rhodium/tris-((1 H,1 H,2 H,2 H-perfluorodecyl)phenyl)phosphites complexes (Rh/L2, L22, L23, L24, L25, L26) were investigated at 80 °C and 40 bar syngas (CO:H$_2$ = 1:1) pressure in C$_8$F$_{17}$H (Table 2). The solvent forms a homogeneous liquid phase with the olefins and/or the aldehydes under the reaction conditions, but separates at room temperature. The activities and the selectivities were found to vary markedly with the position of the perfluoroalkyl group and the presence of other electron-donating alkyl groups on the aromatic ring. Activities up to 10,000 h^{-1} and normal to branched aldehyde ratios up to 5.8 were obtained in the case of 1-decene as substrate [12, 15].

The activity and stability of the soluble fluoropolymer catalysts were evaluated in the hydroformylation of 1-hexene with olefin/Rh 200,000. At 100 °C and 50 bar syngas with L19 as the supporting ligand, the catalyst afforded a turnover of ca. 140,000 with a 98% selectivity to aldehyde (n/i = 4.4; 2% isomerization) for 58 h reaction time. It should be noted that this system was found to be active for higher molecular weight (hexadecene) and functionalized (n-butyl-acrylate) substrates as well [27].

Finally, the hydroformylation of 1-decene with P[CH$_2$CH$_2$(CF$_2$)$_5$CF$_3$] modified cobalt fluorous catalyst system showed after 5 h the formation of 37% undecanals with normal/iso ratio of 2.0, 36.5% undecanols with normal/iso ratio of 2.0, and 10% isomerization to internal decenes [8].

3.2 Hydroformylation of Styrene and Its Derivatives

Asymmetric hydroformylation of prochiral olefins to aldehydes is an important reaction due to the potential for the enantioselective synthesis of optically pure fine chemicals and pharmaceuticals (Table 3).

Table 3 Fluorous Hydroformylation of Styrene

Ligand		L/Rh	P (MPa)	T (°C)	t (h)	Conv. (%)	Aldehyde selectivity (%)	n/i	Ref.
L8		3	20	80	3	31	87	28/62	[19]
L19		6	3	80	15	85	>99	1/6.2	[27]
L20		6	3	80	15	80	>99	1/5.4	[27]
L29		4	4.1	60	12	86	100	90/10	[29]
L30		4	4.1	50–60	18	88	100	94/6	[29]
L31		4	10	40	46	42.0		95.7/4.3	[29]
4-Chlorostyrene	L31	2	150	40	15	89.0		91.9/8.1	[29]
4-Isobutylstyrene	L31	2	146	40	16	>99		95.5/4.5	[29]

Table 4 Fluorous Hydroformylation of Acrylates

Substrate	Ligand	L/Rh	P (MPa)	T (°C)	t (h)	Conv. (%)	TOF (h^{-1})	Aldehyde selectivity (%)	n/i	Ref.
Methyl acrylate	L19	8	15	80	1	38.2	2,196	98.4	1/121	[31]
Ethyl acrylate	L19	8	20	80	1	–	1,255	99.4	1/206	[31]
	L19	8	15	80	1	35.5	2,066	99.0	1/155	[31]
	L19	4	15	80	1	–	1,251	99.3	0	[31]
	L19	8	3	80	1	–	408	99.5	0	[31]
	L19	8	3	80	1	–	405	99.7	0	[31]
	L19	8	3	80	1	–	440	99.6	0	[31]
n-Butyl acrylate	L19	6	3	80	15	100	–	>99	<1/99	[27]
	L19	8	15	80	1	35.7	2,081	99.0	1/117	[31]
	L20	6	3	80	15	100	–	>99	<1/99	[27]
tert-Butyl acrylate	L19	8	15	80	1	35.6	2,087	99.2	1/111	[31]

The poly(fluoroacrylate-co-styryl-diphenylphosphine) ligands (L19 and L20) modified rhodium catalyst were found to be active and selective for the biphasic hydroformylation of styrene, resulting in >99% aldehyde selectivity and >80% conversion. It should be noted that these ligands form very efficient catalysts for the hydroformylation of long chain terminal olefines (1-decene and 1-hexadecene) under 30 bar CO:H$_2$ = 1:1 at 80–100 °C in hexane–toluene–perfluoromethylcyclohexane solvent mixture [27].

It should be noted that styrene was hydroformylated using a CO$_2$-philic fluorous ligand (L8) modified rhodium catalyst. The investigation of the effect of P/Rh molar ratio, partial pressure of CO/H$_2$, and total pressure of carbon dioxide was reported [19].

The novel fluorous BINAPHOS ligands L29 and L30 modified rhodium catalyst were found to achieve comparable or even higher regio- and enantioselectivity in the asymmetric hydroformylation of styrene in benzene or perfluorotoluene at 60 °C and 40 bar syngas (CO:H$_2$ = 1:1), as compared to those by the Rh-BINAPHOS system. Using supercritical CO$_2$ as the solvent with L29 and L30 at 100 bar total pressure (and including 20 bar CO:H$_2$ = 1:1) at same temperature, the hydroformylation of styrene gave 89:11–90:10 regioselectivity and 70–74% ee [29]. The BIPHENOS derivatives L31 was tested for styrene and different styrene derivatives (4-chloro and 4-isobutylstyrene) resulting in average 95% conversion and 96% regioselectivity.

The cyclic phosphite and amido(diamido)phosphite ligands (L32–L37) bearing different fluorous ponytails at the β-position with respect to the phosphorus atom were tested in asymmetric hydroformylation. While higher than 99% chemoselectivity and 80% stereoselectivity were achieved under 100 bar at 70 °C with Rh/P/styrene = 1:2:500 with L32-L36 in toluene, L37 resulted in 77% regioselectivity [30].

Table 5 Hydroformylation of Olefins in Super Critical Carbon Dioxide

Substrate	Ligand	L M^{-1}	Psyngas MPa	Ptotal (MPa)	T (°C)	T (h)	Conv. (%)	TOF (h^{-1})	Aldehyde selectivity (%)	n/i	Ref.
1-Octene	L1	10	2	20	65	21.5	99	2,080	98.4	81/18	[25]^a
	L16	10	2	20	65	18	93	1,950	99.4	79/14	[25]^a
	L18	10	2	20	65	88	95	2,065	99.0	85/10	[25]^a
Ethylene	L38	30	4.7	18	100	20	80	–	–	–	[33]^b
Propylene	L38	30	4.7	18	100	66	80	–	–	–	[33]^b
Styrene	L31	2.4	4	11.5	31	62	96.5	–	–	–	[32]^c

^{a}M: rhodium, density of CO_2 = 0.62 g mL^{-1}
^{b}M: cobalt, $Co_2(CO)_6(L38)_2$ was used as catalyst with L38 excess
^{c}M: rhodium, ee = 91.8% (R), regioselectivity = 95.6%, density of CO_2 = 0.80 g mL^{-1}

3.3 Hydroformylation of Acrylates

The combination of supercritical CO_2, Rh(CO)$_2$(acac), and fluorous phosphines L19 and L20 was found to be a highly effective catalytic system for the chemoselective hydroformylation of usually unreactive alkyl acrylates (Table 4) [27, 31].

3.4 Hydroformylation in Supercritical CO_2

The application of different fluorinated phosphine and phosphite ligands (L1, L16, and L18) modified rhodium catalyst using supercritical carbon dioxide as an environmentally benign reaction media should be noted. The rhodium based catalysts were formed in situ by the reaction of (cod)Rh(hfacac) and ligands L31, L38 [32]. The results are summarized in Table 5.

It was found that application of fluorous BINAPHOS derivatives (L29 and L30) resulted in high regio-and enantioselectivity for the hydroformylation of styrene in CO_2.

The first mechanistic investigation of the cobalt based system using L38 was studied by high-pressure NMR spectroscopy [33].

The cobalt-catalyzed hydroformylation of 1-octene using bis{tri(3-fluorophenyl) phosphine}hexacarbonyldicobalt (L39) and $Co_2(CO)_8$ as pre-catalysts in supercritical carbon dioxide (scCO$_2$) was investigated. The structure of $Co_2(CO)_6[P(3-FC_6H_4)_3]_2$ was also determined [34].

References

1. Roelen O (1938) DE 849548
2. Cornils B, Herrmann WA, Rasch M (1994) Otto Roelen, pioneer in industrial homogeneous catalysis. Angew Chem Int Ed Engl 33:2144–2163
3. Mika LT, Ungváry F (2011) Hydroformylation–homogeneous. In: Horváth IT (ed) Encyclopedia of Catalysis, 2nd Edition, 6 volumes. Wiley, Hoboken, NJ in press
4. Cornils B, Wiebus E (1995) Aqueous biphase catalysis. CHEMTECH 25:33–36

5. Waschen O, Himmler K, Cornils B (1998) Aqueous biphasic catalysis: where the reaction takes place. Catal Today 42(4):373–379
6. Horváth IT (1990) Hydroformylation of olefines with the water soluble HRh(CO)[P(m-C6H4SO3Na)3] 3 in supported aqueous phase. Is it really aqueous? Catal Lett 6:43–48
7. Horváth IT, Rábai J (1994) Facile catalyst separation without water. Fluorous biphase hydroformylation of olefines Science 266:72–75
8. Horváth IT, Rábai J (1995) US5463082
9. Horváth IT, Kiss G, Stevens PA, Bond JE et al (1998) Molecular engineering in homogeneous catalysis: one-phase catalysis coupled with biphase catalyst separation. The fluorous-soluble HRh(CO){(PCH2CH2(CF2)(5)(CF3)3)}(3) hydroformylation system. J Am Chem Soc 120:3133–3143
10. Alvey LJ, Meier R, Soós T, Bernatis P et al (2000) Syntheses and carbonyliridium complexes of unsymmetrically substituted fluorous trialkylphosphanes: precision tuning of electronic properties, including Insulation of the perfluoroalkyl groups. Eur J Inorg Chem 9:1975–1983
11. Jiao H, Stang SL, Soós T, Meier R et al (2002) How to insulate a reactive site from a perfluoroalkyl group: photoelectron spectroscopy, calorimetric, and computational studies of long-range electronic effects in fluorous phosphines P((CH2)m(CF2)7CF3)3. J Am Chem Soc 124:1516–1523
12. Mathivet T, Monflier E, Castanet Y (2002) Perfluorooctyl substituted triphenylphosphites as ligands for hydroformylation of higher olefins in fluorocarbon/hydrocarbon biphasic medium. C R Chem 5:417–424
13. Forster DF, Gudmussen D, Adams DJ, Stuart AM et al (2002) Hydroformylation in perfluorinated solvents; improved selectivity, catalyst retention and product separation. Tetrahedron 58:3901–3910
14. Hope EG, Stuart AM (1999) Fluorous biphase catalysis J Fluorine Chem 100:75–83
15. Mathivet T, Monflier E, Castanet Y, Mortreux A et al (2002) Hydroformylation of higher olefins by rhodium/tris((1 H,1 H,2 H,2 H-perfluorodecyl)phenyl)phosphites complexes in a fluorocarbon/hydrocarbon biphasic medium: effects of fluorinated groups on the activity and stability of the catalytic system. Tetrahedron 58:3877–3888
16. Perperi E, Huang YL, Angeli P, Manos G et al (2004) A continuous process concept for homogeneous catalysis in fluorous biphasic systems. Chem Eng Science 59:4983–4989
17. Perperi E, Huang Y, Angeli P, Manos G et al (2004) The design of a continuous reactor for fluorous biphasic reactions under pressure and its use in alkene hydroformylation. Dalton Trans 14:2062–2064
18. Foster DF, Adams DJ, Gudmunsen D, Stuart AM et al (2002) Hydroformylation in fluorous solvents. Chem Commun:722–723
19. Pedros MG, Masdeu-Bulto AM, Bayardonb J, Sinou D (2006) Hydroformylation of alkenes with rhodium catalyst in supercritical carbon dioxide. Catal Lett 107:205–208
20. Aghmiz A, Claver C, Masdeu-Bultó AM, Maillard D et al (2004) Hydroformylation of 1-octene with rhodium catalysts in fluorous systems. J Mol Catal A Chem 208:97–101
21. Adams DJ, Cole-Hamilton DJ, Hope EG, Pogorzelec PJ et al (2004) Hydroformylation in fluorous solvents. J Organomet Chem 689:722–723
22. Adams DJ, Cole-Hamilton DJ, Harding DAJ, Hope EG et al (2004) Towards the synthesis of perfluoroalkylated derivatives of Xantphos. Tetrahedron 60:4079–4085
23. Adams DJ, Bennett JA, Cole-Hamilton DJ, Hope EG et al (2005) Rhodium catalysed hydroformylation of alkenes using highly fluorophilic phosphines. Dalton Trans 24:3862–3867
24. Kainz S, Koch D, Baumann W (2005) Perfluoroalkyl-substituted arylphosphanes as ligands for homogeneous catalysis in supercritical carbon dioxide. Angew Chem Int Ed Engl 36:1628–1630
25. Koch D, Leitner W (1998) Rhodium-catalyzed hydroformylation in supercritical carbon dioxide. J Am Chem Soc 120:13398–13404

26. Osuna AMB, Chen WP, Hope EG, Kemmitt RDW, et al. (2000) Effects of the ponytails of arylphosphines on the hydroformylation of higher olefins in supercritical CO2. J Chem Soc Dalton Trans 4052–4055
27. Chen WP, Xu LJ, Xiao JL (2000) Fluorous soluble polymer catalysts for the fluorous biphase hydroformylation of olefins. Chem Commun:839–840
28. Huang YL, Perperi E, Manos G, Cole-Hamilton DJ (2004) Performance of octene in fluorous biphasic hydroformylation: octene distribution and reversible transfer between perfluoromethylcyclohexane and nonanal. J Mol Catal A Chem 210:17–21
29. Bonafoux D, Hua ZH, Wang BH, Ojima I (2001) Design and synthesis of new fluorinated ligands for the rhodium-catalyzed hydroformylation of alkenes in supercritical CO2 and fluorous solvents. J Fluorine Chem 112:101–108
30. Artyushin O, Odinets I, Goryunov E, Fedyanin I et al (2006) α-Fluorinated cyclic amidophosphite ligands. Their synthesis, Rh complexes and catalytic activity in the hydroformylation of styrene. J Organomet Chem 691:5547–5559
31. Chen W, Osuna AMB, Iggo JA, Xiao J (2002) Fast and unprecedented chemoselective hydroformylation of acrylates with a fluoropolymer ligand in supercritical CO2. Chem Commun 788–789
32. Franco G, Leitner W (1999) Highly regio- and enantio-selective rhodium-catalysed asymmetric hydroformylation without organic solvents. Chem Commun 1663–1664
33. Chen MJ, Klingler RJ, Rathke JW, Kramarz KW (2004) In situ high-pressure NMR studies of Co2(CO)6[P(p-CF3C6H4)3]2 in supercritical carbon dioxide: ligand substitution, hydrogenation, and hydroformylation reactions. Organometallics 23:2701–2707
34. Patcas F, Maniut C, Ionescu C, Pitter S et al (2007) Supercritical carbon dioxide as an alternative reaction medium for hydroformylation with integrated catalyst recycling. J Mol Catal B Env 70:630–636

Incorporation of Fluorous Glycosides to Cell Membrane and Saccharide Chain Elongation by Cellular Enzymes

Kenichi Hatanaka

Abstract A series of fluorous-tagged glycosides with different number of fluorine atoms are incorporated into the cells, transported to Golgi, elongated by cellular enzymes, and then released to the culture medium. Fluorine content strongly affects on the affinity for cell membrane and glycosylation. Essentially, the fluorocarbon chain in fluorous compound and the hydrocarbon chain are not miscible. However, the fluorous-tagged glycosides have affinity for cell membrane because of its amphiphilicity. The affinity of fluoro-amphiphilic compound for cell membrane is discussed using critical micelle concentration. The separation of glycosylated products by solvent extraction or fluorous solid phase extraction cartridges is also disscussed.

Keywords Biosynthesis · Cell membrane · Critical micelle concentration · Cytotoxicity · Fluoroalkyl · Fluorous solvent · Fluoro-amphiphilic · Glycosylation · Oligosaccharides · Saccharide production

Contents

1 Primer Method: Oligosaccharide Production by Using Mammalian Cells 292
2 Half-Fluorinated Dodecyl Lactoside as Saccharide Primer 293
3 Highly-Fluorinated Dodecyl Lactoside as Saccharide Primer 295
4 Effects of Fluorous Tags on Cellular Glycosylation of Saccharide Primers 297
5 Simultaneous Synthesis of Flurous-Tagged Oligosaccharides 300
6 Affinity of Fluoro-Amphiphilic Compound for Cell Membrane 303
7 Perspective 305
References 306

K. Hatanaka (✉)
Institute of Industrial Science, The University of Tokyo, Tokyo, Japan
e-mail: hatanaka@iis.u-tokyo.ac.jp

1 Primer Method: Oligosaccharide Production by Using Mammalian Cells

Many kinds of oligosaccharides play important roles in vivo, and therefore the mass production of oligosaccharides is an important theme. Adding to the chemical synthesis and the enzymatic synthesis of oligosaccharides, we developed the saccharide primer method using cells. Saccharide primers are synthetically accessible alkyl glycosides that resemble intermediates in the biosynthetic pathway and act as substrates (glycosylation acceptor). Long alkyl glycosides (saccharide primers) such as dodecyl lactoside is incorporated into cell membrane, transported to Golgi apparatus, glycosylated by glycosyltransferases, and the saccharide-chain-elongated products are released from the cell to the culture medium [1] (Fig. 1). The oligosaccharide production using the primer method is simpler than other methods because neither long-step chemical reactions nor unstable purified enzymes are necessary.

Differences of alkyl chain length of the saccharide primer causes large differences in the interaction with mammalian cells. Octyl(C_8) lactoside is hardly incorporated into cell membrane because of its lower hydrophobicity than membrane phospholipids which have double chains of C_{18}–C_{24} (ceramide chains). On the other hand, hexadecyl(C_{16}) lactoside can be incorporated into cell membranes and glycosylated in Golgi apparatus, but not released from the cell because of its higher hydrophobicity which causes the strong interaction with cell membrane. Compared to them, dodecyl(C_{12}) lactosides displays useful behavior. Dodecyl lactoside is incorporated well to cell membranes and the glycosylated products can be released from the cell (Fig. 2). Equilibrium of dodecyl lactoside in the medium and in the cell membrane may be adequate.

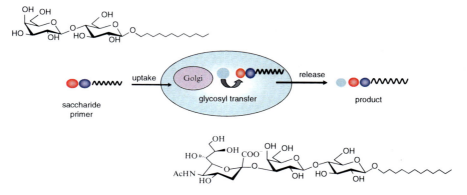

Fig. 1 Oligosaccharide production by using cells (saccharide primer method)

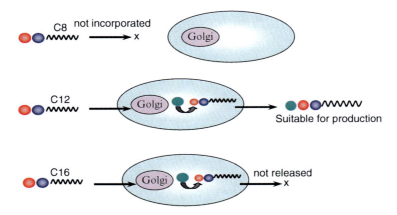

Fig. 2 Effect of alkyl chain length for saccharide production

2 Half-Fluorinated Dodecyl Lactoside as Saccharide Primer

Fluorinated alkyl glycosides can be used as saccharide primer as well as alkyl glycosides [2]. A half-fluorinated alkyl glycoside such as perfluorohexylhexyl ($C_6F_{13}C_6H_{12}$–) lactoside is one of the promising substrates. Half-fluorinated alkyl glycosides are expected to (1) show similar steric characteristic of alkyl moiety to nonfluorinated alkyl glycosides, (2) be thermally, chemically, and biologically stable, (3) form a unique hydrogen bond by using C–F linkage, (4) show low cytotoxicity, (5) behave as unique components in biological systems, (6) increase lipophilicity enhancing absorption into cell membranes, and (7) be easily purified with fluorous solvents.

Perfluorohexylhexyl ($C_6F_{13}C_6H_{12}$–) lactoside was synthesized by common glycosylation of perfluorohexylhexanol ($C_6F_{13}C_6H_{12}OH$) by lactose peracetate using BF_3OEt_2 as catalyst, followed by deacetylation. The perfluorohexylhexyl lactoside (LacH6F6: each number shows the number of methylene and the number of fluoromethylene, respectively) was added to the culture medium which contains B16 mouse melanoma cells. LacH6F6 did not show the cytotoxicity at a concentration of 50 μM. After the incubation, lipid extraction from the culture medium and the cell homogenates was carried out. HPTLC (high-performance thin-layer chromatography) of each fraction shows that each fraction contains LacH6F6 and the new product which seems to be a glycosylated product from LacH6F6. The new product was analyzed by MALDI–TOFMS and the number of m/z 1,034.86 was assigned to sialylated LacH6F6. The structure of the sialylated product was confirmed by α2–3 sialydase to be α2–3-sialylated LacH6F6 with the same oligosaccharide of GM3 which is predominantly expressed on B16 melanoma cells. Interestingly, both culture medium and cell fractions contain LacH6F6 primer, while most of the product (sialylated LacH6F6) existed in culture medium fraction. On the other hand, most of endogenous GM3 existed in cell fraction, indicating that the cells could retain a certain amount of GM3 that is inherent to B16 cells.

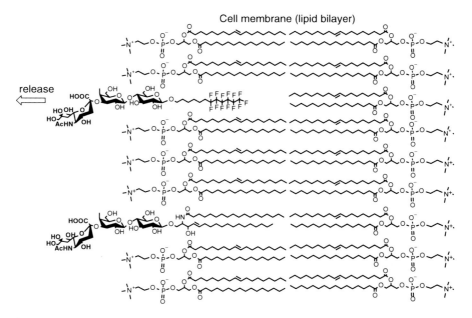

Fig. 3 Affinity of fluorous glycoside and endogenous GM3 for cell membrane

However, unnatural α2–3-sialylated LacH6F6 with single chain ($C_6F_{13}C_6H_{12}-$) hardly remained the cell surface, being released to the culture medium (Fig. 3).

Polar hydroxyl groups in the saccharide moiety make the compound difficult to penetrate into the cell membrane. Increasing the hydrophobicity of the aglycon, that is a long alkyl chain, enhances membrane permeability and cellular uptake. As mentioned above, C_{12} chain has appropriate length for saccharide production, while C_{16} chain makes the compound remain in the cell membrane because of its higher hydrophobicity. LacH6F6 primer, of which the fluorinated alkyl group ($C_6F_{13}C_6H_{12}-$) is more hydrophobic than the dodecyl ($C_{12}H_{25}-$) group because of the higher hydrophobicity of a fluorocarbon chain compared with a hydrocarbon chain, is also an appropriate primer for saccharide production, indicating that the $C_6F_{13}C_6H_{12}-$ group is less hydrophobic than the hexadecyl ($C_{16}H_{33}-$) group. Thus, half-fluorinated dodecyl lactoside diffuses through the cell membrane to Golgi apparatus where the lactoside is treated as an acceptor of sialic acid, resulting in α2–3 sialylation on the galactosyl residue of the lactose moiety to give a GM3 analog having a fluorous chain. It is noteworthy that the fluoroalkyl chain is quite stable in the cellular environment and during enzymatic reaction. The half-fluorinated dodecyl lactoside is not cytotoxic at a concentration of 50 μM and can be used for carbohydrate chain elongation by using the biosynthetic system.

The fluorinated product can be purified by extraction with fluorous solvents. The half-fluorinated dodecyl lactoside is soluble in 1,1,1,3,3,3-hexafluoro-2-propanol ($CF_3CHOHCF_3$), and partially soluble in 3,3,4,4,5,5,6,6,6-nonafluorohexanol ($C_4F_9C_2H_4OH$) or in 4,4,5,5,5-pentafluoropentanol ($C_2F_5C_3H_6OH$). However, the insoluble fluorous lactoside particles in the fluorous solvents become soluble after shaking with water. The half-fluorinated dodecyl lactoside and its sialylated product from the cellular system can be separated from other components in the culture medium by extraction with 3,3,4,4,5,5,6,6,6-nonafluorohexanol or with 4,4,5,5,5-pentafluoropentanol.

3 Highly-Fluorinated Dodecyl Lactoside as Saccharide Primer

Both fluorine content and the kind of glycosyl group in fluorous saccharide primer affect the cytotoxicity of the primer, the solubility of the product in fluorous solvent, and the yield of saccharide production [3].

As mentioned above, perfluorohexylhexyl lactoside (LacH6F6) is not cytotoxic at 50 µM. However, monosaccharide derivatives such as perfluorohexylhexyl galactoside (GalH6F6) and perfluorohexylhexyl glucoside (GlcH6F6) show remarkable cytotoxicity (Fig. 4). This phenomenon is also observed in nonfluorinated dodecyl glucoside (GluH12). On the other hand, highly-fluorinated dodecyl glycosides show different phenomena. Perfluorodecylethyl ($C_{10}F_{21}C_2H_4-$) galactoside (GalH2F10) which is insoluble in the culture medium is harmful to cells when it is present in the culture medium. Perfluorodecylethyl ($C_{10}F_{21}C_2H_4-$) glucoside (GlcH2F10) which is soluble in the culture medium is not cytotoxic but is not taken by cells. The presence of many fluorine atoms may make the penetration of the saccharide compounds into the cell membrane difficult. However, perfluorodecylethyl ($C_{10}F_{21}C_2H_4-$) lactoside (LacH2F10) of which glycosyl group is disaccharide is taken into the cell, and after the glycosylation (α2–3 sialylation), the product is released from the cell to the culture medium. Hence,

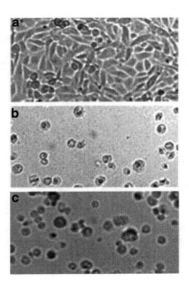

Fig. 4 Cytotoxicity of fluorus glycosides on B16 cells. (a) Control, (b) GalH6F6 (50 µM), (c) GluH6F6 (50 µM)

the balance between the hydrophilicity depending on the number of hydroxyl groups and/or saccharide steric structure and the hydrophobicity depending on the number of fluorine atoms is critical on cytotoxicity and cellular uptake. Interestingly, perfluorohexylhexyl galactoside (GalH6F6) is strongly toxic, but it can be taken by B16 cells, glycosylated, and released into the culture medium as well as nontoxic saccharide primers. This result shows that a series of processes involving cellular uptake of saccharide primer, transportation to Golgi apparatus, enzymatic glycosylation, and releasing the product can be carried out before expiring. Similarly, perfluorohexylhexyl glucoside (GlcH6F6) also penetrates into the cell membrane while it is cytotoxic.

Among highly-fluorinated dodecyl glycosides, perfluorodecylethyl ($C_{10}F_{21}C_2H_4-$) lactoside (LacH2F10) is incorporated into B16 melanoma cells resulting in sialylation of the galactose residue to give a GM3-type oligosaccharide derivative. Diffusion of fluorinated primers through the cell membrane to Golgi apparatus is necessary for saccharide chain elongation by cellular enzymes. Therefore, LacH2F10 has an appropriate structure for saccharide production.

The saccharide primer method for oligosaccharide production by using mammalian cells is not only an easy and combinatorial synthesis but is also aimed at the convenient purification of the products. A series of several kinds of chromatography is a general method for purification of the oligosaccharides. However, large amounts of organic solvents are necessary for several steps of columns. Accordingly, convenient and environmentally-friendly extraction with the solvents can be employed. The fluorous-tagged primers and their glycosylated products can be extracted with fluorous solvents to separate them from other components of the culture medium. For the extraction, inquiry into fluorous solvent that dissolves the fluorous-tagged glycosylated products is important. Fluorous-tagged primer and fluorous-tagged glycosylated product, both having several hydroxyl groups on the carbohydrate moiety and highly-fluorinated dodecyl group, are "fluoro-amphiphilic." Therefore, the fluorous solvent that can selectively dissolve the fluoro-amphiphilic compound must have both "fluoro" and hydrophilic portions. For selective dissolution of fluorous-tagged primers, appropriate fluorine content and appropriate hydroxyl content are necessary. Fluorous-tagged primer and fluorous-tagged glycosylated product have 13–16% hydroxyl groups and 24–45% fluorine. They are insoluble in solvents having relatively low content of hydroxyl groups such as perfluorohexylhexanol ($C_6F_{13}C_6H_{12}OH$) which has 4% hydroxyl groups and 59% fluorine. On the other hand, 1,1,1,3,3,3-hexafluoro-2-propanol ($CF_3CHOHCF_3$) that has 10% hydroxyl groups and 68% fluorine dissolves not only fluorous compounds but also nonfluorinated compounds, indicating that this solvent could not be used for selective extraction. The solvents having relatively high contents of hydroxyl groups may be miscible with water and therefore the fluorous compounds could not be selectively extracted from other components of the culture medium. 3,3,4,4,5,5,6,6,6-Nonafluorohexanol ($C_4F_9C_2H_4OH$) which has 6% hydroxyl groups and 65% fluorine and 4,4,5,5,5-pentafluoropentanol ($C_2F_5C_3H_6OH$) which has 9% hydroxyl groups and 53% fluorine selectively dissolved fluorous-tagged primer and fluorous-tagged glycosylated product. These kinds of fluorous (53–65% fluorine) alcohols having 6–9% hydroxyl groups are immiscible with water. Thus, the fluorous saccharide primers and the glycosylated fluorous products can be extracted from the aqueous phase into the fluorous phase. HPTLC analysis of the fluorous extract shows that only fluorous compounds, these being fluorous primer and glycosylated product, joined the fluorous phase. Seeking the appropriate fluoro-amphiphilic solvent is difficult but this method allows the simple separation of fluorous-tagged compounds from other many aqueous components in the cell culture medium.

4 Effects of Fluorous Tags on Cellular Glycosylation of Saccharide Primers

Saccharide primers of lactose (Lac) and *N*-acetyl-D-glucosamine (GlcNAc) with different length of fluorous tags can be synthesized in a few steps. Lactoside primers (LacH8F4, LacH6F6, LacH2F8, LacH3F8, LacH4F8, LacH2F10) and GlcNAc primers (GlcNAcH8F4, GlcNAcH6F6, GlcNAcH2F8, GlcNAcH3F8, GlcNAcH4F8) (Fig. 5) are synthetically accessible in a few steps. Glycosylation of the partially fluorinated alcohol with either the lactose peracetate and Lewis acid

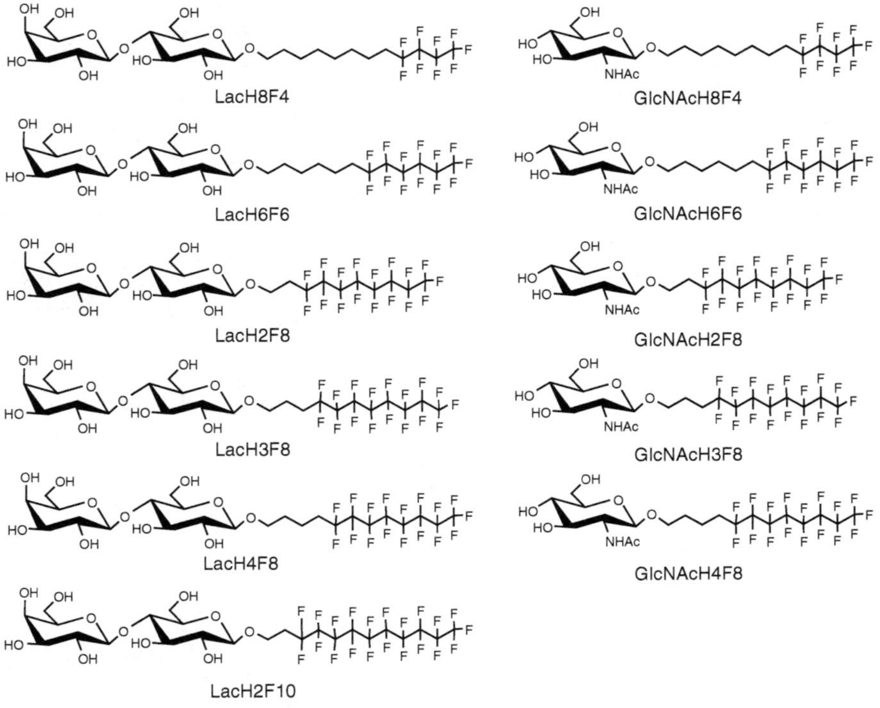

Fig. 5 Chemical structures of fluorinated alkyl lactosides and *N*-acetyl-D-glucosaminides

catalyst or GlcNAc-derived oxazoline derivative followed by deacetylation with sodium methoxide gives Lac and GlcNAc primers with different length of fluorous tags.

Saccharide primers with different lengths of fluorous tags were administered to the cell culture medium in order to investigate the effect of fluorous tags on cellular enzyme-mediated glycosylation [4]. At the concentration of 50 μM, all kinds of fluorous primers (Lac and GlcNAc) did not affect the B16 cell viability and morphology. HPTLC analyses of both fluorous Lac and GlcNAc primers indicated that (1) all the primers were taken into the cells, (2) the primer with more content of fluorine atoms was liable to remain in the cell, and (3) except for LacH4F8 and GlcNAcH4F8, glycosylated product was observed in the medium fraction. Glycosylated products were easily isolated using fluorous solid phase extraction cartridges. Mass spectra of purified products showed that sialylation occurred on fluorous Lac primers to afford GM3-like compounds and two successive reactions (galactosylation and sialylation) occurred on fluorous GlcNAc primers to give sialyl lactosamine (SA–Gal–GlcNAc) derivatives. HPTLC of the medium fraction from fluorous GlcNAc primers did not show the band corresponding to Gal–GlcNAc derivative, indicating that galactosylated fluorous GlcNAc primers were instantaneously sialylated to afford SA–Gal–GlcNAc derivatives that were subsequently released to the culture medium. The respective yield of sialylated Lac primer is 17.8% for LacH8F4, 14.1% for LacH6F6, 3.7% for LacH2F8, 1.8% for LacH3F8, and 1.2% for LacH2F10. The respective yield of sialylated galactosylated GlcNAc primer is 2.2% for GlcNAcH8F4, 1.7% for GlcNAcH6F6, 12.8% for GlcNAcH2F8, and 0.8% for GlcNAcH3F8.

Sialylation of fluorous Lac primers were just as expected, since the major glycolipid of B16 cells is GM3 which is the sialylated lactose derivative. Even if the aglycone (nonsugar compound remaining after replacement of the glycoside by a hydrogen atom) contains a lot of fluorine atoms, fluorous Lac primers are taken into the cell, transferred to Golgi, glycosylated by enzymes, transferred to cell membrane, and released to the culture medium. The lower fluorine content of fluorous Lac primer gave the higher yield of the sialylated product. Since the quantity of the cellular uptake of each fluorous Lac primer is enough, the lower yield for higher fluorinated primer is presumed to be attributable to lower transportation to Golgi or lower recognition by enzymes.

In the case of fluorous GlcNAc primers, they are first galactosylated and then sialylated. Generally, sialylation of the glycolipid occurs on β-galactosyl residue and galactosylation frequently occurs on N-acetyl-D-glucosaminyl residue, suggesting that the galactosylation of fluorous GlcNAc primer occurs prior to the sialylation. The electrospray ionization mass spectral data also supports the speculation. Fragmentation of SA–Gal–GlcNAcH6F6 (m/z 1,075.2) gives a peak of Gal–GlcNAcH6F6 (m/z 784.1) (Fig. 6). The yield of the product from fluorous GlcNAc primer was low compared with fluorous Lac primer. It is thought that the galactosylation of fluorous GlcNAc primer is slow, because galactosylation mainly occurs on glucose residue in B16 melanoma cells of which major glycolipid GM3 is the sialylated galactosyl glucose derivative. The lower fluorine content of fluorous GlcNAc primer gave the higher yield of the glycosylated product except for GlcNAcH2F8. The relatively high yield using GlcNAcH2F8 primer depends on another kind of sialylation in parallel

Incorporation of Fluorous Glycosides to Cell Membrane and Saccharide 299

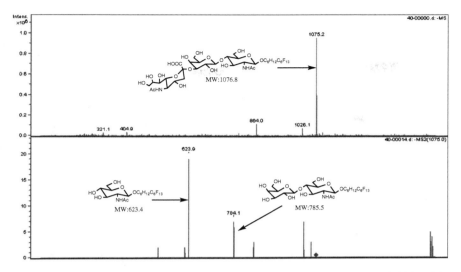

Fig. 6 Electrospray ionization mass spectra (negative mode) of SA-Gal-GlcNAcH6F6 and its fragmentation

with the normal α2–3 sialylation. The presence of two bands in the HPTLC result from GlcNAcH2F8 primer indicates two kinds of sialylated products. Mass spectra of purified products showed that both compounds have the same molecular weight of 1,120. It is expected that one is normal α2–3 sialylated GalGlcNAcH2F8 and the other one is α2–6 sialylated GalGlcNAcH2F8.

Surprisingly, neither LacH4F8 nor GlcNAcH4F8 gives any products. Significant amounts of those primers were taken up to the cell as confirmed by HPTLC analyses. However, the bands corresponding to the glycosylated primers in culture medium fraction were hardly observed in HPTLC plate, indicating that the glycosylation on LacH4F8 and GlcNAcH4F8 primers did not occur or only trace amounts of the products were produced. Transportation of those primers from cell membrane to Golgi apparatus or recognition by glycosyltransferases may not be successful. Since LacH2F10, that has more fluorine atoms than LacH4F8, gave the sialylated product, the simple increase of hydrophobicity may not be the reason. Six kinds of fluorous lactoside primers (LacH8F4, LacH6F6, LacH2F8, LacH3F8, LacH4F8, and LacH2F10) were administered to the culture medium with African green monkey Vero cells. Among the six kinds of fluorous primers, LacH4F8 primer gave no glycosylated products from Vero cells as well as B16 cells, while the other five kinds of primers gave sialylated products also from Vero cells. Moreover, LacH4F8 primer was incorporated into Vero cells similarly to B16. It is indicated that the surprising phenomenon is nonspecific for the kind of cell but specific for the unique hydrophobic fluorous chain, that is, the $C_4F_9C_8H_{16}$– group.

HPTLC results of the cell fraction indicated that the bands assignable to more fluorinated primers LacH2F8, LacH3F8, and LacH2F10 are more intense than less fluorinated primers LacH6F6 and LacH8F4, suggesting that the presence of more fluorine atoms enhances the incorporation of the primer into the cells. It is interesting that the fluorous glycoside can be incorporated into the cell membrane that consists of phospholipids containing a hydrocarbon chain, indicating that the essentially immiscible fluorous and aliphatic compounds can be miscible when they are the amphiphilic compounds having hydrophilic groups. Incorporated fluorous-tagged primers are transported to Golgi, elongated by cellular glycosyltransferases using cellular sugar nucleotide (donor substrate such as CMP–sialic acid or UDP–galactose), and then released from the cell to the culture medium without any other modification than glycosylation. The fluorous tag did not affect the cellular specific glycosylation, giving the same oligosaccharide structure as biosynthesized glycolipid, that is, GM3, the major ganglioside produced by B16 cells.

Both glycosylated product from the fluorous primer and the unreacted fluorous primer can be easily separated from the rest of the components of the culture medium by gradient elution with methanol using a fluorous solid phase extraction cartridge. For example, sialylated LacH2F8 is eluted from the cartridge with 40–50% methanol as an eluent, and the unreacted LacH2F8 primer is eluted with 60–90% methanol.

5 Simultaneous Synthesis of Flurous-Tagged Oligosaccharides

As mentioned above, a series of fluorous-tagged saccharide primers with different numbers of fluorine atoms are incorporated into the cells, transported to Golgi, elongated by cellular enzymes, and then released to the culture medium. The glycosylation product can be easily separated using a fluorous solid phase extraction cartridge. The elution pattern from the fluorous cartridge depends on the content of fluorine atoms, indicating that several kinds of glycosides with different kinds of fluorous tags can be separated using a fluorous solid phase extraction cartridge. Therefore, the products from two kinds of primers with different fluorous tags and different types of saccharide moiety can also be easily separated according to the fluorine content.

The mixture of LacH2F8 and GlcNAcH6F6 primers was administered to the culture medium containing B16 cells [5]. Each primer was not cytotoxic at the concentration of 50 μM as mentioned. Moreover, a mixture of 50 μM each of the primers was not Cytotoxic in spite of the 100 μM of total concentration. After incubation for 2 days, HPTLC analysis of the lipid extract from the medium fraction showed that the glycosylated products were detected as a single band (Fig. 7). When these primers incubated separately, the glycosylated products of SA–Gal–GlcH2F8 and SA–Gal–GlcNAcH6F6 obtained showed almost the same R_f value on HPTLC plate. Therefore, the HPTLC plate obtained from the primer mixture shows that the product obtained is SA–Gal–GlcH2F8, SA–Gal–GlcNAcH6F6, or a mixture of them. Thereupon the product from the primer mixture was separated from other components in the culture medium and analyzed by electrospray ionization mass spectroscopy. The negative mode

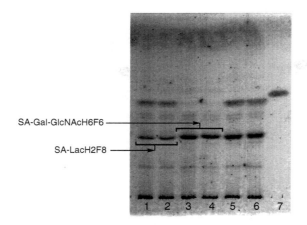

Fig. 7 HPTLC results of culture medium fractions from LacH2F8 (lanes 1 & 2), GlcNAcH6F6 (lanes 3 & 4), the mixture of LacH2F8 and GlcNAcH6F6 (lanes 5 & 6), and the chromatogram of standard LacH2F8.5 [5]

mass spectral results gave two major peaks at m/z 1,078.1 and m/z 1,075.2 (Fig. 8), indicating that the product obtained contains both SA–Gal–GlcH2F8 (SA–LacH2F8) and SA–Gal–GlcNAcH6F6 derived from the mixture of LacH2F8 and GlcNAcH6F6 primers.

When the primer mixture was incubated with B16 cells, the cellular enzymes recognized each primer separately by the substrate specificity. Galactosyltransferase recognized only GlcNAcH6F6 primer for saccharide elongation using endogenous UDP–galactose, affording Gal–GlcNAcH6F6 that was immediately converted to SA–Gal–GlcNAcH6F6 by sialylation with sialyltransferase using endogenous CMP–sialic acid. At the same time, the sialyltransferase recognized another acceptor substrate that is LacH2F8 synthesizing SA–LacH2F8. Moreover, it is considered that the same sialyltransferase catalyzes the biosynthetic reaction producing the endogenous GM3 that is the major glycolipid in B16 cells. Hence, simultaneous glycosylations occurred on both natural and artificial substrates according to the biosynthetic pathway. Although the sialyltransferase specifically acted on β-galactose residue, there was no great preference between Gal–Glc and Gal–GlcNAc confirmed by the similar quantity of each product obtained from the primer mixture as that obtained from separate incubation. Both primers have different numbers of fluorine atoms but both were likewise elongated. Significantly, the cellular enzymatic syntheses of these artificial products scarcely prevented the synthesis of endogenous GM3.

On the HPTLC plate, both SA–Gal–GlcH2F8 and SA–Gal–GlcNAcH6F6 products show almost the same R_f value, indicating that these products cannot be separated by normal column chromatography. However, the mixture can be separated using a FluoroFlash TLC plate with F254 indicator as shown Fig. 9. On the FluoroFlash TLC plate, more fluorinated product (SA–Gal–GlcH2F8) has

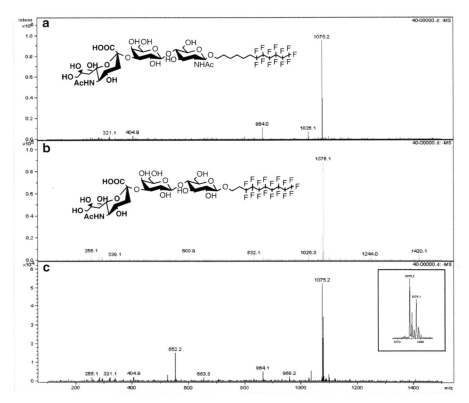

Fig. 8 Electrospray ionization mass spectra (negative mode) of SA-Gal-GlcNAcH6F6 (**a**) SA-LacH2F8, (**b**) and the product obtained from the mixture of 2 kinds of glycosides (**c**) [5]

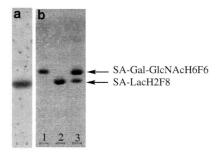

Fig. 9 Separation of two kinds of products by normal silica gel HPTLC (chloroform:methanol:0.25%KCl (5:4:1) as eluent) (a) and FluoroFlash TLC (80% methanol as eluent) (b). Lane 1: SA-Gal-GlcNAcH6F6, Lane 2: SA-LacH2F8, Lane 3: the mixture of SA-Gal-GlcNAcH6F6 and SA-LacH2F8 [5]

a smaller R_f value because of its stronger interaction with fluorous plate. It is indicated that the compounds with different fluorine content can be separated by fluorous silica gel column chromatography, even if both saccharide chain structure and hydrophobicity are quite similar.

6 Affinity of Fluoro-Amphiphilic Compound for Cell Membrane

Oligosaccharide production using fluorous glycoside primer and mammalian cells involves incorporation of the primer into cell membrane, transportation to Golgi apparatus, glycosylation by cellular enzymes using sugar nucleotides, and efficient release of products from the cell to the culture medium. The balance of hydrophobic fluorous chain and hydrophilic saccharide moiety greatly influences the oligosaccharide production. In this section, the affinity of fluoro-amphiphilic saccharide primer such as LacH6F6 for lipid bilayer in cell membrane or liposome is first discussed.

The surface tension of each fluoro-amphiphilic glycoside was measured in order to evaluate the critical micelle concentration (CMC) of each glycoside (Fig. 10) [6]. The CMC values obtained from the graph are 6.0 μM for LacH2F10, 6.9 μM for LacH3F8, 6.9 μM for LacH6F6, 12.1 μM for LacH2F8, and 57.9 μM for LacH12, indicating that the CMC values of fluoro-amphiphilic saccharide primers decrease with increasing the fluorine content. The CMC values of fluoro-amphiphilic saccharide primers were much lower than that of nonfluorinated primer (LacH12), suggesting that the fluoroalkyl chains are relatively more hydrophobic as compared

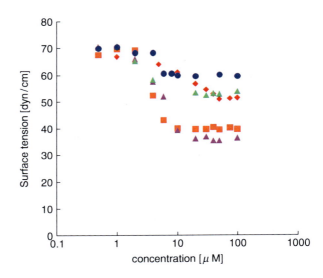

Fig. 10 Critical micelle concentration (CMC) of fluorous glycosides [6]

with alkyl chains. It has been well investigated that the fluorinated amphiphiles are more hydrophobic than nonfluorinated amphiphiles with hydrocarbon chains in terms of lowering the interfacial tension and lowering the CMC. It is estimated that the LacH2F10 primer is most hydrophobic among five kinds of primers used.

Interestingly, LacH3F8 and LacH6F6 show the same CMC value, indicating that both compounds have the same hydrophobicity because of their similar hydrophilic carbohydrate moiety. Therefore, $C_8F_{17}C_3H_6-$ and $C_6F_{13}C_6H_{12}-$ groups are equally hydrophobic. The solution of this relationship shows the hydrophobicity of the fluoromethylene group. When x and y represent the hydrophobicity of methylene ($-CH_2-$) and of fluoromethylene ($-CF_2-$), respectively, $3x + 8y = 6x + 6y$. Consequently, $3x = 2y$, indicates that the hydrophobicity of the fluoromethylene group is 1.5 times that of methylene group, supported by the reported value of 1.44 in the literature [7]. Using this equation, the hydrophobicity of each fluorous tag can be determined. The reduced chain length can be calculated as the sum of the number of methylene groups plus 1.5 times number of fluoromethylene groups. That is to say, the reduced chain length of $C_mF_{2m+1}C_nH_{2n}-$ is $n + 1.5$ m. Hydrophobicity of fluorous chains may be expressed in terms of the number of reduced chain length. For example, the number of reduced chain length of H2F10 ($C_{10}F_{21}C_2H_4-$) is 17 and that of H2F8 ($C_8F_{17}C_2H_4-$) is 14. Both LacH3F8 ($C_8F_{17}C_3H_6-$) and LacH6F6 ($C_6F_{13}C_6H_{12}-$) have the same number of reduced chain length of 15. As mentioned in the first section, hexadecyl(C_{16}) lactoside can be incorporated in the cell membrane and glycosylated in Golgi apparatus, but not released from the cell because of its higher hydrophobicity which causes the strong interaction with cell membrane. LacH2F10 whose reduced chain length is 17 is more hydrophobic than hexadecyl (C_{16}) lactoside and therefore cannot be released from the cell membrane. However, the transportation from cell membrane to Golgi apparatus and glycosylation reactions by cellular enzymes cannot be elucidated by the reduced chain length.

In order to evaluate the incorporation of fluorous glycosides, that have different number of reduced chain length, into the cell membrane, liposomal uptake of fluorous glycosides was investigated. Each fluoro-amphiphilic saccharide primer was administered to aqueous solutions of liposomes that have almost the same size as cells. The percentage of each fluoro-amphiphilic saccharide primer in liposome was quantified to be 59.9% for LacH2F10, 37.8% for LacH3F8, 42.6% for LacH6F6, 24.0% for LacH2F8, and 6.4% for LacH12, indicating that more hydrophobic primer was liable to be incorporated into liposome. The relationship between the amount of fluoro-amphiphilic saccharide primer incorporated into liposome and the reduced chain length of the primer is shown in Fig. 11, indicating the positive correlation between them. It is estimated that the minimum value of the alkyl chain length of alkyl lactoside for the liposomal uptake may be 11. This estimated value of minimum chain length for liposomal uptake can account for the facts that octyl (C_8) lactoside is hardly incorporated into cell membrane, dodecyl(C_{12}) lactoside is well incorporated into the cell membrane, and the glycosylated products can be released from the cell as mentioned in the first section.

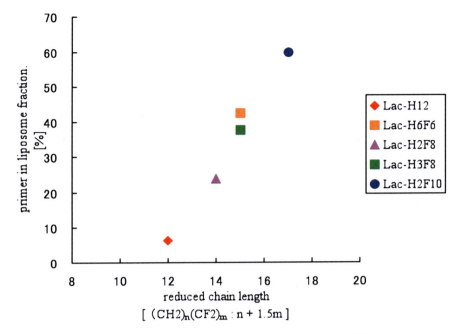

Fig. 11 Relashonship between liposomal uptake and reduced chain length [6]

7 Perspective

Fluoro-amphiphilic saccharides are of wide use because of their great number of characteristics. Fluorous compounds are not miscible with aqueous solutions nor hydrocarbon phases, and therefore fluoro-amphiphilic saccharides can be immobilized on fluorous material such as poly(tetrafluoroethylene) using fluorous–fluorous interaction and the immobilized compounds are hardly removed by washing with water or organic solvents. Moreover, the immobilized fluorous saccharides can be recognized by lectin [8]. Since the interactions between oligosaccharides and proteins are specific, the immobilized fluoro-amphiphilic saccharides are very useful for the detection of particular proteins such as viral envelope proteins. Consequently, immobilization of oligosaccharides using fluorous interaction can be one of the useful methods for preparing a glycoarray for medical use, provided that the recognition site (saccharide moiety) is free from fluorous interaction in order to prevent the steric hindrance, namely an appropriate spacer is necessary.

As mentioned in Sect. 6, the fluoro-amphiphilic saccharides are miscible with amphiphilic cell membranes in aqueous phase. Essentially, the fluorocarbon chain in fluorous compounds and the hydrocarbon chains are not miscible, indicating that the fluorous compounds in cell membranes may form aggregates. This aggregation phenomenon of fluorescent fluoro-amphiphilic saccharides was observed and the recognition by lectin became stronger [9, 10]. Such an artificial lipid-raft structure on cell membranes will be an important bio-device that can be particularly and strongly recognized. In particular, fluorocarbons are mostly inert in vivo and therefore suitable for biomaterials.

In this manuscript, in addition to the fluorous characteristics that can be used for separation and immobilization, the fluoro-amphiphilic compounds were discussed for saccharide production. The fluoro-amphiphilic compounds such as fluorous saccharide primers are new functional molecules and may be usable in the wide area of scientific investigation and bioindustry.

References

1. Kasuya MCZ, Wang LX, Lee YC, Mitsuki M, Nakajima H, Miura Y, Sato T, Hatanaka K, Yamagata S, Yamagata T (2000) Azido glycoside primer: a versatile building block for the biocombinatorial synthesis of glycosphingolipid analogues. Carbohydr Res 329:755–763
2. Kasuya MCZ, Cusi R, Ishihara O, Miyagawa A, Hashimoto K, Sato T, Hatanaka K (2004) Fluorous-tagged compound: a viable scaffold to prime oligosaccharide synthesis by cellular enzymes. Biochem Biophys Res Commun 316:599–604
3. Kasuya MCZ, Ito A, Cusi R, Sato T, Hatanaka K (2005) Cellular uptake and saccharide chain elongation of "fluoro-amphiphilic" glycosides. Chem Lett 34:856–857
4. Kasuya MCZ, Tojino M, Nakano S, Mizuno M, Hatanaka K (2009) Effect of fluorous tags on glycosylation of saccharide primers in animal cells. Bull Chem Soc Jpn 82:1409–1415
5. Kasuya MCZ, Tojino M, Mizuno M, Hatanaka K (2010) Fluorous tag method for the simultaneous synthesis of different kinds of glycolipids. J Fluor Chem 131:655–659
6. Kasuya MCZ, Nakano S, Katayama R, Hatanaka K (2011) Evaluation of the hydrophobicity of perfluoroalkyl chains in amphiphilic compounds that are incorporated into cell membrane. J Fluor Chem 132:202–206
7. Moriyama E, Lee J, Moroi Y, Abe Y, Takahashi T (2005) Micelle formation of N-(1,1-dihydroperfluorooctyl)- and N-(1,1-dihydroperfluorononyl)-N,N,N-trimethyl- ammonium chlorides. Langmuir 21:13–18
8. Ko K-S, Jaipuri FA, Pohl NL (2005) Fluorous-based carbohydrate microarrays. J Am Chem Soc 127:13162–13163
9. Webb SJ, Greenaway K, Bayati M, Trembleau L (2006) Lipid fluorination enables phase separation from fluid phospholipid bilayers. Org Biomol Chem 4:2399–2407
10. Noble GT, Flitsch SL, Liem KP, Webb SJ (2007) Assessing the cluster glycoside effect during the binding of concanavalin A to mannosylated artificial lipid rafts. Org Biomol Chem 7:5245–5254

Teflon AF Materials

Hong Zhang and Stephen G. Weber

Abstract The unique combination of chemical, thermal, and mechanical stability, high fractional free volume, low refractive index, low surface energy, and wide optical transparency has led to growing interest in Teflon Amorphous Fluoropolymers (AFs) for a wide spectrum of applications ranging from chemical separations and sensors to bioassay platforms. New opportunities arise from the incorporation of nanoscale materials in Teflon AFs. In this chapter, we highlight fractional free volume – the most important property of Teflon AFs – with the aim of clarifying the unique transport behavior through Teflon AF membranes. We then review state-of-the-art developments based on Teflon AF platforms by focusing on the chemistry behind the applications.

Keywords Electrochemical sensors · Electrowetting · Free volume · Gas sensors · Hydrophobicity · Nanocomposites · Teflon AF · Transport · Waveguides

Contents

1 Introduction 308
   1.1 Free Volume of Teflon AFs 309
2 Membrane Transport 311
   2.1 Gas Transport and Pervaporation 311
   2.2 Liquid Phase Transport 316
   2.3 Gas–Liquid Reactions 319
3 Sensors 320
   3.1 Gas Sensor for Binary Mixtures 321
   3.2 Waveguide Sensors 321
   3.3 Sorption-Based Vapor Sensors 322

H. Zhang and S.G. Weber (✉)
Department of Chemistry, University of Pittsburgh, Pittsburgh, PA 15260, USA
e-mail: sweber@pitt.edu

4	Electrochemistry	324
	4.1 Electrochemical Gas Sensors	324
	4.2 Ion Selective Electrochemical Sensors	324
5	Hydrophobic Materials	325
	5.1 Superhydrophobic Surfaces	325
	5.2 Manipulating Droplet Movement on Teflon AF Surfaces	326
6	Platform for Biological Assays	328
7	Metal-Teflon AF Nanocomposites	330
8	Summary	332
References		333

1 Introduction

Perfluoropolymers are receiving increased attention in both the scientific community and industry because they are dimensionally stable polymeric materials with the unusual properties of fluorocarbons. The most well-known property shared by perfluoropolymers is their extraordinary thermal and chemical stability due to the strong carbon–fluorine and carbon–carbon bonds [1, 2]. As an orthogonal phase to aqueous and organic media, perfluoropolymers are soluble only in perfluorinated solvents, Novec engineering liquids (hydrofluoroethers), and supercritical CO_2 under specified temperature and pressure conditions [1–4]. These intrinsic features favor the long-term chemical and morphological stability of perfluoropolymers (e.g., as coatings, membranes, tubings, fibers, seals, etc.) in hostile environments. Notably, it is of practical importance that perfluoropolymers feature low surface energy; therefore, they display much less non-specific adsorption than other polymeric materials when exposed to biological fluids [5, 6]. A wide range of perfluoropolymer-based materials have been designed and evaluated as anti-fouling coatings for biological and medical applications [7].

Among commercially available perfluoropolymers, the Teflon Amorphous Fluoropolymers (AF) family (Fig. 1) is an unusual class of amorphous glassy copolymers with a high fractional free volume (FFV) [8–10]. Teflon AFs are copolymers of tetrafluoroethylene (TFE) and 2,2-bistrifluoromethyl-4,5-difluoro-1,3 dioxole (BDD). Two types of Teflon AF, produced by DuPont, are now available on the market: Teflon AF 2400 (TFE: 13 mol%; BDD: 87 mol%) and Teflon AF 1600

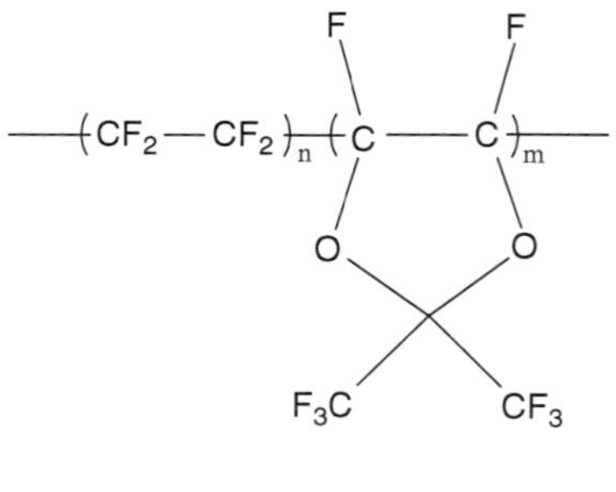

Fig. 1 Structure of Teflon AF

Table 1 Properties of Teflon AFs [11–13]

Property	Teflon AF 2400	Teflon AF 1600
Crystallinity	None	None
M_w (kDa)	300	100
T_g (°C)	240 ± 10	160 ± 10
Density (g/cm^3)	1.75	1.82
FFV (%)	33.4	30.0
Refractive index	1.29	1.31
Dielectric constant	1.90	1.93
Contact angle with water (°)	105	104
Critical surface energy (dyn/cm)	15.6	15.7
Permeability O_2 (barrer)	1140	170
Solubility		
Common organic solvents	None	None
Perfluorinated solvents	C_6F_6, $C_6H_5CF_3$, FC-72, etc.	C_6F_6, $C_6H_5CF_3$, FC-72, etc.
Tensile strength (MPa, 23°C)	26.4 ± 1.9	26.9 ± 1.5
Elongation at break (%, 23°C)	7.9 ± 2.3	17.1 ± 5.0

(TFE: 35 mol%; BDD: 65 mol%). Some important physical, chemical, and mechanical properties of Teflon AFs are listed in Table 1 [11–13].

Several further aspects deserve attention. Thin coatings and films of Teflon AFs are optically transparent over a wide range (200–2,000 nm), which makes them ideal media for the investigation of intermolecular interactions [12, 14]. Moreover, Teflon AFs exhibit a lower refractive index and dielectric constant than any other solid organic polymer known [15, 16]. Thus, Teflon AFs are recognized as eligible materials for optoelectronic devices (e.g., coatings on waveguide devices).

In this chapter, we will mainly focus our attention on a wide range of applications of Teflon AFs based on their versatile and impressive basic properties. Both the opportunities and the challenges will be addressed. In the following subsection, the most unique feature of Teflon AFs-their high free volume (size, amount, and size distribution), will be briefly introduced.

1.1 Free Volume of Teflon AFs

The high FFV of Teflon AFs is responsible for a number of their successful applications in sensors and separations. For this reason, we begin this review with a careful look at the FFV in Teflon AFs. A polymer in amorphous form usually has significantly larger free volume than the same polymer in crystalline form because of the inefficient packing of disordered chains in the amorphous regions [10]. Both Teflon AF 2400 and Teflon AF 1600 are completely amorphous. Teflon AFs possess the two criteria for permeability according to the contemporary wisdom on highly permeable polymers, namely the presence of bulky non-polar structures

Table 2 Radius of free volume elements (Å) in Teflon AFs from various methods [8, 19, 20]

Polymer	PALS R_{sp} R_3	R_4	R_c R_3	R_4	^{129}Xe-NMR R_{sp}	R_c	IGC $R(V_w)$	$R(V_b)$	$R(V_c)$
Teflon AF 2400	2.7	6.0	3.7	6.3	8.04[a] 7.83[b]	5.12[a] 5.02[b]	3.6	4.7	6.4
Teflon AF 1600	2.5	4.9		5.4	6.66	4.43	3.1	4.0	5.5

[a]Powder
[b]Film

R_{sp} represents the data obtained by assuming spherical micro voids; R_c represents the data obtained by assuming cylindrical micro voids; R_3, R_4 correspond to the radii deduced from the presence of two lifetimes. The quantitative determination of the volume of the free volume elements depends on the molecular volume of the probe solute. As the molecular volume may be determined in different ways, the measurement of the volume of the free volume elements depends on the choice of which one to use. V_w, V_b, V_c represent the volumes of the free volume elements based on the solute's van der Waals volume, molecular volume in the liquid phase at the boiling point T_b, and the critical volume respectively.

and rigid main chains to create large barriers to rotation [17, 18]. As reported by Yampolskii and his coworkers, the rotation barrier for two neighboring dioxolane rings is as high as 60 kJ/mol [10]. These structural characteristics combined with the weak van der Waals interactions between fluorocarbon chains give Teflon AFs a high free volume.

Many experimental techniques have been designed to probe the size of free volume elements in Teflon AFs: ^{129}Xe NMR spectroscopy, positron annihilation lifetime spectroscopy (PALS), and inverse gas chromatography (IGC) [8, 9, 19]. The free volume size distributions of amorphous Teflon AFs determined by various techniques are summarized in Table 2 [8, 19, 20]. Teflon AF 1600 shows smaller free volume elements than Teflon AF 2400. This is probably due to the presence of fewer dioxolane rings in Teflon AF 1600. Both ^{129}Xe NMR spectroscopy and IGC measure the average size of free volume elements, whereas PALS is able to discover both the free volume size distribution and the amount of free volume [9, 20]. Positrons in Teflon AFs have two lifetimes, indicating two populations of free volume elements [9, 20]. Computer modeling further confirms the bimodal distribution of free volume elements and their partial connectivity in Teflon AFs [20]. Figure 2 shows a qualitative impression of the free volume distribution and morphology in Teflon AFs [20]. Clearly, two types of qualitatively different free volume elements are present in Teflon AFs. The first type, with high segmental packing around it, resembles the free volume in conventional low free volume glassy polymers. The other type features a larger size. Large, preexisting free volume elements will require much less activation energy for the diffusion and dissolution/partitioning of guest molecules in the polymeric matrix. The large free volume of Teflon AFs, and of Teflon AF 2400 in particular, makes the amorphous Teflon AF materials valuable materials in the field of membrane separations. More will be discussed below on controlling the free volume in composite films.

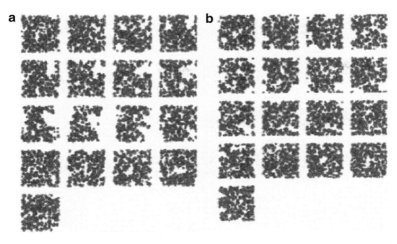

Fig. 2 Visualization of chain packing and free volume distribution from molecular modeling (InsightII/Discover software of Accelrys Inc, COMPASS force field) by assuming cubic packing of macromolecules into the blocks: (**a**) Teflon AF 2400; (**b**) Teflon AF 1600. The average length of the sides of the blocks is between 45 Å and 50 Å. The figures present a series of approximately monatomic parallel planes of the block cut perpendicular to the z axis with the distance between each two continuous planes of about 3 Å [20]

2 Membrane Transport

2.1 Gas Transport and Pervaporation

Mass transport of gaseous species across membranes depends upon the diffusion (D) and partition coefficient (S) of gases in the membrane. Penetrants (atoms or molecules in the gas phase that partition into a membrane) first dissolve into the membrane surface contacting the feed flow, and then diffuse across the membrane, finally partition into the permeate flow. The permeability coefficients (P) can be expressed as [21]

$$P = DS \tag{1}$$

The transport selectivity through membrane is defined as [21, 22]

$$S = \frac{P_A}{P_B} = \left(\frac{D_A}{D_B}\right)\left(\frac{S_A}{S_B}\right) \tag{2}$$

Glassy and rubbery polymers are known to display different behaviors in gas transport [23]. The separation of gases by selective transport through rubbery polymer membranes is primarily caused by differences in the solubility whereas the separation by glassy polymer membranes is mainly by differences in the gas diffusivity [22, 23]. Therefore, glassy polymer membranes normally separate

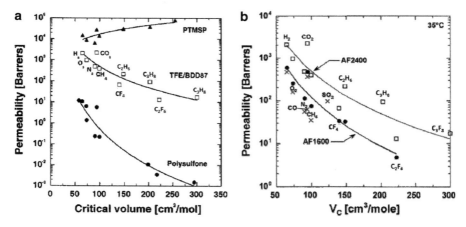

Fig. 3 Effect of critical volume on infinite dilution gas permeability at 35°C: (**a**) In PTMSP, Teflon AF 2400, and polysulfone; (**b**) In Teflon AF 2400 and Teflon AF 1600 [24, 25]

gaseous species in a size-sieving manner. Exceptions do exist for some glassy polymers with high free volume, for example, poly(1-trimethylsilyl-1-propyne) (PTMSP, FFV = 32%) [22, 24]. As shown in Fig. 3a, gas permeability through PTMSP membranes increases along with the increase of critical volume [24]. The more condensable gaseous species with higher molar volume are more permeable due to their higher solubility in PTMSP matrices [22, 24]. The gas permeability through Teflon AF 2400 is second only to PTMSP [22, 24, 25]. However, gas transport through Teflon AF 2400 membranes follows the general trend of size-sieving, which is similar (although permeabilities are larger) to the behavior of low free volume glassy polymers (e.g., polysulfone, polycarbonate, etc.) [22, 24, 25]. Notably, the decrease in permeability as critical volume increases is much less than that for polysulfone. An interesting example shown in Fig. 3 is that C_3F_8 displays an even higher transport rate than its homolog, C_2F_6, which has a smaller critical volume and is favored by diffusion. Therefore, gas transport behavior through Teflon AF 2400 membranes may be intermediate between PTMSP and classical low free volume glassy polymers depending on the solubility of the penetrant [24, 125]. Compared to Teflon AF 2400, Teflon AF 1600 has a lower free volume, and is less permeable to gases.

The gas transport rate at a certain temperature relies not only on the nature of the penetrant and polymeric membrane, but sometimes also on the feed pressure, especially when penetrants can plasticize the polymer membranes [22, 24, 25]. The dissolution of gaseous species in Teflon AFs follows the non-linear dual sorption model [24, 25]. Gases dissolve in both the free volume and the condensed polymeric matrix. The former mode obeys the Langmuir sorption isotherm while the latter follows Henry's law [24–26]. Sorption isotherms of light gases, C1–C4 hydrocarbons, and C1–C3 perfluorocarbons fit the dual sorption model very well [24, 25]. Teflon AF 2400 has a larger free volume, and displays a higher sorption capacity than Teflon AF 1600. The Henry's law constants of light gases in Teflon

AFs are much larger than in other glassy polymers. This trend is similar to that observed for gas solubility in liquids, i.e., gases are generally more soluble in solvents with lower solubility parameters [27]. Interestingly, the desorption isotherms of the hydrocarbons and perfluorocarbons display hysteresis [25]. Such hysteresis is more significant for the more soluble penetrants, which are better plasticizers [25]. The penetrant-induced changes in glassy polymers by plasticization can be ascribed to long-lived increases in free volume and therefore the Langmuir capacity parameter, which thermodynamically favors the accommodation of more penetrants [25, 28]. Therefore, the selective solubilization of the more soluble penetrants in Teflon AF membranes can be further improved by increasing the pressure of feed flow.

Because plasticization can effectively increase the polymer chain mobility, not only the gas solubility but also the diffusion coefficients will be feed-pressure dependent. The dependence of diffusion coefficients on feed pressure for poorly absorbed gases (e.g., O_2, N_2, CO_2, CH_4, CF_4) and highly absorbed gases (e.g., C_2H_6, C_3H_8, C_2F_6, C_3F_8) exhibit different behaviors [24]. The concentration dependence of diffusion coefficients is more pronounced for fluorocarbons than hydrocarbons, indicating better plasticization by the former. Because permeability coefficients are products of partition coefficients and diffusion coefficients, and certain penetrants have higher partition coefficients and diffusion coefficients at higher pressures, gas transport rates can display dramatic concentration dependent effects. Furthermore, the selectivity of gas transport can be tuned by controlling the pressure of feed flow. For example, as shown in Fig. 4, Teflon AF 2400 membranes become more selective for propane over nitrogen when the feed pressure is higher than 80 psi [22].

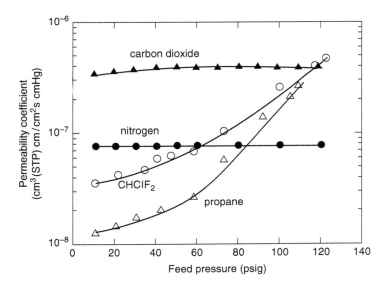

Fig. 4 The dependence of permeability coefficients through Teflon AF 2400 membranes on feed pressure at 35 °C [22]

Nanocomposite membranes based on Teflon AFs and nonporous fumed silica display distinctive transport behavior in comparison to both conventional polymeric membranes containing nonporous nano-scale fillers and pure Teflon AFs [29, 30]. Nonporous nanoparticles as dopants in polymers are known to increase the diffusion pathway and block the effective area of transport [31]. Gas solubility (measured as moles per total material volume) in a polymer/nanoparticle hybrid material ought to be lower than in the pure polymer due to the negligible gas solubility in the nonporous nanoparticles. Therefore, introducing nanoparticles into polymeric membranes should decrease permeability coefficients. Interestingly, fumed silica-doped Teflon AF 2400 membranes behave differently from this classical understanding [29, 30]. Merkel and his coworkers discovered that nonporous fumed silica nanoparticle as the dopant in Teflon AF 2400 membranes can effectively disrupt the interaction of polymeric chains and lead to a systematic increase of FFV in the membrane [29]. As shown in Fig. 5a, the longer positron lifetime, which corresponds to larger free volume elements in the composite membrane, shows a general trend of increasing along with the weight percentage of fumed silica [30]. In contrast, the shorter positron lifetime, which corresponds to small free volume elements, appears to be independent of the filler amount. Teflon AF 2400 membranes containing higher loadings of nonporous fumed silica nanoparticles become more permeable to gases, as shown in Fig. 5b [30]. Surprisingly, the dependence of the permeability on the feed pressure of highly absorbed gases in Teflon AF 2400 membranes containing large loadings of fumed silica is reversed compared to that in pure Teflon AF 2400 membranes [30]. The saturation of non-equilibrium excess free volume sorption sites at high feed pressure is responsible for the observed decrease of gas solubility in the composite membranes [30].

In view of the attractive combination of gas permeability, selectivity, and stability, Teflon AFs are potentially useful membrane materials in pervaporation,

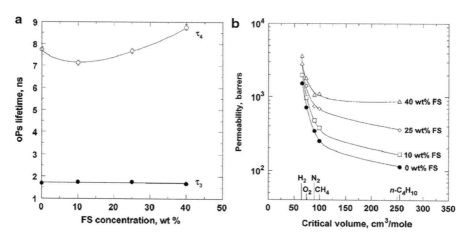

Fig. 5 (a) The dependence of positron lifetime on the dopant weight percentage; (b) Effect of critical volume on gas permeability at 25 °C. $\Delta p = 3.4$ atm for H_2, O_2, N_2, CH_4; $\Delta p = 0.33$ to 3.4 atm for n-butane [30]

a method in which components of a solution are transported selectively through a membrane into a vapor phase [13, 32, 33]. Yampolskii and co-workers systematically investigated the pervaporation and solubility of some common organic solvents (chloromethanes, acetone, alcohols, benzene, fluorobenzene, and cyclohexane) in Teflon AFs [13, 32]. Table 3 shows the solubility of some solvents in Teflon AFs. Both Teflon AFs absorb organic solvents significantly [32]. The FFV of Teflon AF 2400 is higher than Teflon AF 1600. Therefore, Teflon AF 2400 displays higher sorption capacity than Teflon AF 1600. Interestingly, the dependence of the permeability coefficient on the composition of the solution varies for different species [33]. For example, the permeability coefficient of acetone increases with the increase of acetone fraction in the liquid feed whereas the permeability coefficient of methanol decreases with the increase of methanol fraction in the liquid feed [33]. The shift of IR vibrational bands and quantum chemical calculations indicate the existence of effective hydrogen bonding of both acetone and methanol with the dioxole rings in Teflon AF 2400 (Fig. 6) [33]. However, the significant self-association of methanol molecules leads to higher activation energy of diffusion, and consequently lower diffusion coefficients compared to non-associated monomers [33]. Recently, both the dimeric and trimeric forms of methanol have

Table 3 Sorption of organic liquids in Teflon AFs at 25 °C [32]

Solvent	Sorption in Teflon AFs (g/100 g polymer)	
	Teflon AF 2400	Teflon AF 1600
CH_2Cl_2	5.15	4.36
$CHCl_3$	9.11	8.14
CCl_4	16.94	11.90
CH_3OH	1.84	0.54
C_2H_5OH	2.81	1.12
$(CH_3)_2CO$	2.43	2.38
C_6H_6	4.65	4.25
C_6H_{12}	6.75	4.21

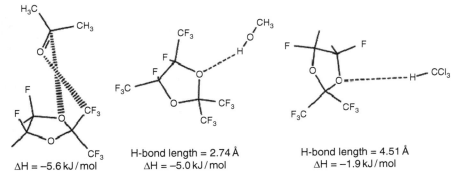

Fig. 6 The interaction of acetone, methanol and chloroform with Teflon AF 2400 from computational modeling [33]

been visualized in methanol vapor transport through Teflon AF 2400 by quantitative analysis of penetrant pressure over time [34]. The experimentally determined diffusion coefficient of methanol dimer is about one tenth of the value for methanol monomer while the diffusion coefficient of the methanol trimer is about one hundredth of that of the methanol monomer [34]. The decrease in diffusion coefficients for self-associated species is much greater than expected based on estimation from Stokes–Einstein equation. Interestingly, the molecular radius of the methanol monomer based on its critical volume is about 6.6 Å, which is similar to the feature radius of the large free volume elements in Teflon AF 2400 (Table 2). Therefore, the diffusion of the larger, self-associated species will be much slower due to a sieving effect. High concentrations of methanol shift the self-association to the dimer and trimer states which have significantly lower permeabilities than the monomer, and result in a decreased average transport rate [34].

2.2 Liquid Phase Transport

Fluorous liquids are selective solvents for fluorous tagged and F-substituted compounds, and are valuable platforms for the synthetic community [12, 35, 36]. With an eye on separation, extraction, and synthetic applications, Weber's lab initiated the investigation of Teflon AF 2400 as a dimensionally stable fluorous solvent for liquid phase transport about 10 years ago [37–40]. A thorough study on the thermodynamic cycle of vapor benzene, benzene in chloroform solution, and Teflon AF 2400 membrane reveals dramatically different behaviors of benzene permeating a Teflon AF 2400 membrane in contact with the liquid phase and benzene permeating a Teflon AF 2400 membrane in the gas phase [37]. As shown in Fig. 7, the diffusion coefficient of the solute benzene in a Teflon AF

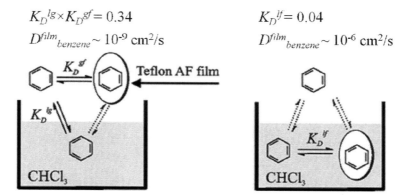

Fig. 7 The effect of chloroform sorption by Teflon AF 2400 is shown by the thermodynamic cycles of the partitioning of benzene vapor, benzene in chloroform, and benzene in Teflon AF 2400 membranes in the gas and solution phases [37]

2400 membrane in equilibrium with a chloroform solution of benzene is about three orders of magnitude larger than the value obtained from the gas transport study [37]. Meanwhile, the partition coefficient of benzene from chloroform to a Teflon AF 2400 membrane in equilibrium with chloroform is about one tenth of the value deduced from the thermodynamic cycle for partitioning from the gas phase [37]. The assessment of both diffusion and partition coefficients, combined with the fact that Teflon AF 2400 absorbs a significant amount of chloroform, indicates that chloroform fills the free volume and leads to the decrease of the benzene partition coefficient [37, 38]. At the same time, the high chloroform concentration in Teflon AF 2400 can plasticize the membrane and makes the diffusion of benzene much easier in the membrane where polymer chains have relatively high mobility [37, 38]. As a result, organic solutes with similar size display much higher permeability through Teflon AF 2400 membranes in liquid phase transport compared to gas permeation [37, 38].

Although Teflon AF 2400 membranes are highly permeable in liquid phase transport, solute transport through Teflon AF 2400 membranes displays a similar size-sieving characteristic for a homologous series of compounds as observed in the gas permeation [38]. Figure 8 shows the dependence of permeability coefficients on the molecular weight of solutes. The regression of log(Permeability) of the neutral aromatic solutes (red dots) vs the logarithm of molecular weight yields a linear relationship [38]. The permeability coefficient of benzene is 74 times that of anthracene, which is about twice the molar volume of benzene. The size sieving effect is much less significant in the gas phase. For example, in Fig. 5 (0% nanoparticles) the rate of change of permeability over critical volume (slope) becomes less steep when the critical volume increases due to the plasticization effect of organic vapors with large critical volume [25]. Even though increasing the size of a penetrant makes its diffusion less favorable, increased polymeric chain mobility due to plasticization is beneficial for diffusion through the polymer. In Fig. 8, polar hydrocarbons have lower permeability coefficients (below the line) due to the unfavorable partitioning to Teflon AF membranes. Transport of solutes that are able to dimerize shows concentration dependent permeability (e.g., benzoic acid and pentafluorobenzoic acid) due to the lower transport rate of the dimer compared to the monomer [34, 38]. Notably, fluorinated solutes show higher permeability than their hydrocarbon analogs, even though the diffusion of fluorinated analogs of hydrocarbons is less favorable because of their larger molar volume [38]. In fact, the effect is quite significant. For example, the permeability of octafluorotoluene is about 10^3 larger than would be predicted from the linear correlation based on hydrocarbons. Therefore, Teflon AFs are potentially useful materials in liquid phase membrane separation and solid phase extraction.

The solvent in contact with Teflon AF plays an important role in solute transport [37, 38, 40]. The high concentration of chloroform absorbed in Teflon AF 2400 makes the membrane less fluorous than it might otherwise be [37–40]. One can imagine that such membranes will be less selective for fluorinated solutes. In principle, a fluorous additive may plasticize Teflon AFs as well as improve the fluorous nature of the material compared to pure Teflon AFs. To maintain the

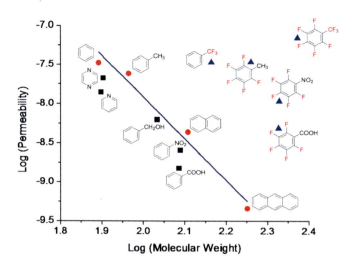

Fig. 8 Dependence of permeability coefficients on the molecular weight of solutes [38, 40]. *Red circles*: aromatic hydrocarbons (and used to establish the regression line shown); *black squares*, aromatic hydrocarbons with polar functional groups; *blue triangles*, solutes with C–F substituted for C–H

fluorophilicity of Teflon AF 2400 membranes in contact with organic liquids, Krytox 157FSH (a perfluoropolyether oil, MW 7,000–7,500) and FC-70 (perfluorotripentylamine) were separately combined with Teflon AF 2400, and relevant properties were determined [38–40] (Fig. 9).

Krytox 157FSH can effectively plasticize Teflon AF 2400 [37, 38]. Also 50 wt% Krytox 157FSH as dopant in Teflon AF 2400 membrane lowers the glass transition temperature to −40 °C, compared to 240 °C for pure Teflon AF 2400 [37, 38]. However, solute permeabilities through 50 wt% Krytox 157FSH doped Teflon AF 2400 membranes are significantly reduced compared to pure Teflon AF 2400 membranes [38]. This is attributed to the viscous environment in the membrane where the free volume is filled by Krytox oil, even though the polymeric chain mobility is enhanced [38].

FC-70 is a perfluorinated liquid, and is much less viscous than Krytox 157FSH. Teflon AF 2400 membranes containing more than 20 wt% FC-70 absorb much less solvent in contact with the membrane and show improved transport selectivity for fluorinated solutes [40]. Solute permeabilities through >30 wt% FC-70 doped Teflon AF membranes are higher than through pure Teflon AF 2400 membranes [40]. Free volume measurements and dynamic mechanical analysis of FC-70-doped Teflon AF 2400 membranes show that they are antiplasticized at <12 wt% of dopant and plasticized at >12 wt% [39, 40]. The antiplasticization of polymers has been attributed to strong polymer/dopant interactions [41]. However, fluorocarbons feature weak van der Waals interactions [42]. Interestingly, the analysis of the Teflon AF 2400 concentration (weight/volume) in doped membranes reveals a different mechanism for antiplasticization. Less than 12 wt% FC-70

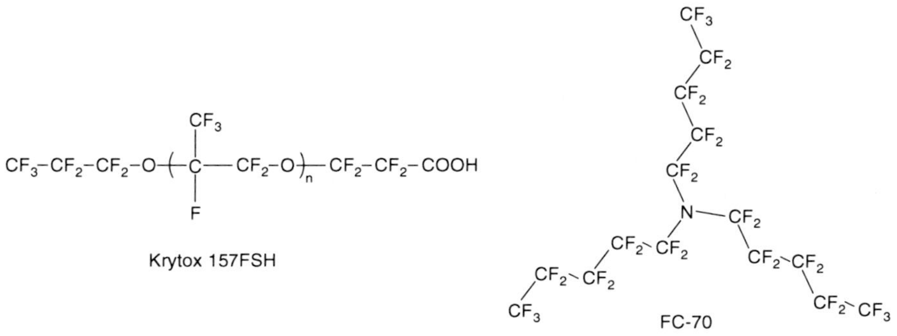

Fig. 9 Structures of Krytox 157FSH and FC-70

antiplasticizes the polymer by filling the existing free volume. This can be deduced from the observation that in this concentration range of dopant, the w/v concentration of Teflon AF 2400 in the films does not change. At concentrations of dopant greater than 12 wt%, FC-70 functions as a plasticizer and dilutes the polymer [40]. Transport behavior for a series of fluorous and organic solutes further confirms the antiplasticization and plasticization effects [39, 40]. Solute diffusion coefficients decrease dramatically when the membrane is antiplasticized, and increase significantly when the membrane is plasticized [40]. Moreover, solute partition coefficients from chloroform to the membrane containing high FC-70 content approach that from chloroform to pure FC-70 [40]. Both sorption behavior of chloroform and the solute partitioning pattern indicate the "FC-70 like" nature of doped Teflon AF 2400 membranes [40]. FC-70-doped Teflon AF 2400 membranes show a general increasing selectivity for transport of 8 F- and 5 F-substituted solute pairs as FC-70 content increases.

2.3 Gas–Liquid Reactions

A gas–liquid biphasic reaction describes the use of a gas phase reagent in a liquid phase reaction. The combined processes of mass transfer and reaction kinetics co-determine the rate of chemical reactions. The mass transport rate of the gaseous reagent in the liquid phase controls the concentration of gas as a function of position in the bulk reaction. To enhance the mass transfer between gas and liquid reagents and achieve good homogeneity of gas dispersion, mechanical agitation is the most commonly used method when a gas reagent is bubbled into a liquid phase [43, 44]. Recently, flow chemistry emerged as a new and innovative strategy, bringing opportunities for exploiting gas–liquid reactions. Instead of being bubbled into bulk liquid, the gas reagent can be transported through the wall of a highly permeable tube and react with the liquid reagent flowing through the lumen. The path length for diffusion can be modulated by controlling the diameter of the

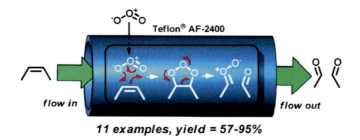

Fig. 10 Flow reactor configuration for ozonolysis [44]

tubular reactor. Compared to traditional bubbler systems, the flow reactor provides a reliable approach for efficient, well-controlled, and reproducible mass transfer of gases in the biphasic reactions.

Teflon AF 2400 is an excellent candidate material to deliver gas species in a controlled manner. While Dasgupta's group developed concentric reactors for analytical purposes as long ago as 1998, Ley and his coworkers at the University of Cambridge published the first two examples of using Teflon AF 2400 tubing to deliver gas reactants to the liquid flow stream for synthesis [14, 44, 45]. As shown in Fig. 10, the liquid flow steam containing an alkene flows inside the Teflon AF 2400 tubing while ozone can penetrate through the tubing and react with the alkene to yield a ketone or aldehyde [44]. The reaction can be optimized by modulating the flow speed of liquid phase, the pressure of ozone, and the dimensions of the Teflon tubing [44]. Further improvement of the gas–liquid reactor involves a concentric arrangement of a Teflon AF 2400 tube inside a PTFE tube [45]. Therefore, the liquid and gas flows can be independently introduced to the reactor with better control of the flow rate and pressure [45]. This approach provides significant advantages in terms of the controllability, reliability, cost efficiency, convenience, and laboratory safety especially when dealing with toxic or environmentally harmful gases.

3 Sensors

The fabrication of thin Teflon AF coatings on the surface of an existing sensor has been widely adopted to prolong the lifetime of the sensor in contact with corrosive solutions and biological systems [46–48]. Moreover, such coatings can function effectively to exclude ionic, large molecular weight, and toxic interferences by limiting their partitioning into the coating phase while remain highly permeable to target gaseous species [49]. This strategy allows real-time sensing of oxygen in proliferating cell cultures [50]. More importantly, because of certain specific properties of Teflon AFs (e.g., permselectivity, low refractive index, and high capacity for vapor adsorption), the Teflon AF can be more than a protective coating. In fact, it can be the key component in various sensors.

3.1 Gas Sensor for Binary Mixtures

As shown previously in Fig. 5b, the infinite dilution gas permeability through Teflon AFs display size permselectivity. The selectivity is not as dramatic as when membrane materials are functionalized for specific interactions [51]. For example, the transport selectivities for the H_2/CH_4 pair (based on infinite dilution permeability coefficients at 35 °C) through Teflon AFs are less than ten. However, Teflon AFs with moderate permselectivity are ideal membrane materials in sensing the composition of binary mixtures [52, 53]. At constant feed pressure of the binary mixture, the permeate flow rate (pressure) is directly related to the composition of the gas mixture. In the case of a feed flow containing H_2 and CH_4, a higher permeate pressure indicates a higher molar fraction of H_2 in the mixture because H_2 is more permeant [52, 53]. Such low cost yet sensitive gas sensors based on pressure transducers are extremely valuable in industrial applications.

3.2 Waveguide Sensors

In spectrometric sensors, the optical path length needs to match the requirements of the specific application, typically scaling inversely with the concentration and molar absorptivity of the analyte [14, 54]. Liquid-core waveguides with a long optical path length provide a convenient method for the quantitative determination of trace species [54]. Without a doubt, the selection of a suitable material for waveguide tubing with a lower reflective index than the liquid core becomes the key step. Teflon AFs have a lower refractive index than water and are transparent as mentioned above [12, 14]. Therefore, liquids can be chosen for the core with virtually no concern about the refractive index. The first example of using Teflon AF tubing in a liquid-core waveguide was reported by Waterbury and coworkers in 1997 to determine trace concentrations of Fe(II) [54]. By using a 4.47 m length of Teflon AF 2400 tubing to make the waveguide, the detection limit for Fe(II) reached 0.2 nM with a linear range between 0.5 and 10 nM [54]. One year later, Dasgupta and coworkers reported a highly sensitive gas sensor based on a liquid core waveguide by taking advantage of both the high permeability and low refractive index of Teflon AF 2400 [14]. Figure 11 shows the basic setup of the gas sensor. Gases of interest, which are introduced into the glass jacket, can penetrate into the liquid core through the Teflon AF 2400 and undergo selective chromogenic reactions for absorbance measurement [14]. This strategy provides a powerful and flexible approach for gas sensing. Sensors for H_2S, NO_2, CO_2, O_3, and chlorinated hydrocarbons have been developed successfully and validated [14, 55–58].

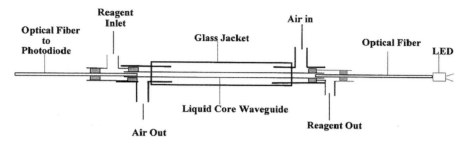

Fig. 11 The configuration of a liquid core waveguide-based gas sensor [14]

Recently, the application of waveguide detection on Teflon AF platforms has been extended to microchip-based analysis [59].

3.3 Sorption-Based Vapor Sensors

Sorption of vapor molecules in a thin polymer film can be quantitatively determined by measuring the change of refractive index [58, 60–64]. This can be achieved by coating the polymer film on a thin metal layer for surface plasmon resonance (SPR) measurement (Fig. 12) [63]. The related technique has been described as "metal-film-enhanced leaky-mode spectroscopy" [63].

Traditionally used polymer coatings (e.g., polyimides) exhibit a moderate level of moisture sensitivity, which limits their utility [64]. As an excellent humidity barrier, Teflon AFs have a high sorption capacity because of their high free volume, so they are ideal materials for this purpose. Figure 13 shows the sensor signals from toluene vapor using a 2-μm sensing layer of Teflon AF 1600 on 50 nm of silver [63]. The saturation values indicate sorption equilibrium. The detection limits for a series of aromatic vapors (benzene, toluene, o-, p-, and m-xylenes) are at ppm levels [63]. By assuming the diffusion into the polymer coating follows Fick's law, the diffusion coefficients of vapor species can be deduced by fitting the dynamic sensor response with the theoretical diffusion-induced refractive index equation [63]. This allows the differentiation of vapor species.

Another prototype sensor relies on the change of capacitance upon the sorption of gaseous species in thin films. High sensitivity (≤1 ppmv) measurements can be achieved when a target vapor is absorbed because all vapors have a higher dielectric constant than air. Teflon AFs are good candidates for the membrane materials in such sensors due to their high gas sorption capacity and excellent stability. Kebabian and Freedman demonstrated the first example of a Teflon AF 2400-based capacitive sensor for atmospheric CO_2 [65]. Figure 14 shows the design of the capacitive sensor with a parallel capacitor array. The thin Teflon AF 2400 insulators have a short path length for diffusion so that sorption equilibrium is

Fig. 12 The configuration of the gas sensor based on surface plasmon leaky mode [63]

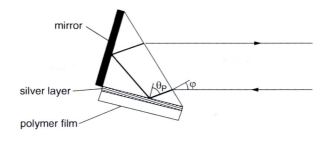

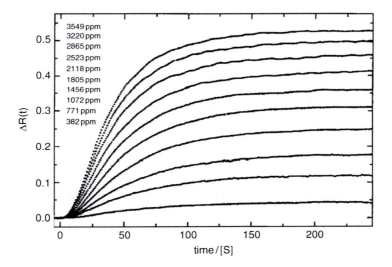

Fig. 13 The sensor signals for toluene vapor at different concentrations [63]

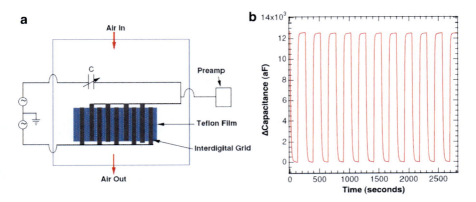

Fig. 14 (a) The configuration of the capacitive gas sensor; (b) sensor signals upon exposure to cycles of pure CO_2 and CO_2-free air [65]

reached quickly giving a fast response [65]. This CO_2 sensor displays excellent reproducibility and a wide linear response: ppmv to 100% CO_2 [65].

4 Electrochemistry

4.1 Electrochemical Gas Sensors

Teflon AFs, highly permeable and poor solvating materials, when coated on electrodes, have been found to exhibit superior performance to improve the selectivity of electrochemical sensors for non-polar neutral gases [66, 67]. Meyerhoff and coworkers have been devoted to the real time measurement of NO in biological samples. They discovered that integrating an outer gas permeable membrane (porous polytetrafluoroethylene, PTFE) with a platinized platinum electrode can decrease significantly the ionic interferences (e.g., nitrite and ascorbate) [66]. However, NH_3 might be a potential interfering species because around 1% of the total ammonia ($NH_3 + NH_4^+$) is in the neutral form at physiological pH. The oxidation of NH_3, yielding N_2, N_2O, or NO, is potential dependent, and will interfere with the signal from the NO oxidation [66]. By implanting Teflon AF in the pores of the porous PTFE, the flux of NH_3 to the internal working electrode decreased significantly, while the permeability of NO remains high [66]. The partitioning of ionic species through the membrane is further decreased compared to the porous PTFE alone, resulting in better selectivity. Starting from the nitric oxide gas sensor, the same group further developed a sensor for S-nitrosothiols by constructing a thin hydrogel layer containing an immobilized organoselenium catalyst which is capable of decomposing S-nitrosothiols and generating NO [67]. Teflon AF has also been found valuable in fuel cell technology. An example of improving the fuel cell performance has been reported by replacing PTFE membranes with Teflon AF 1600 to achieve higher oxygen accessibility [68].

4.2 Ion Selective Electrochemical Sensors

Fluorous media are known to solvate and extract non-fluorous solutes through noncovalent interactions with fluorinated receptors in the fluorous phase [12, 36, 69–71]. With the proper molecular receptor in the fluorous phase, some highly polar molecules can be extracted. For example, aminopyridines can be extracted quantitatively from $CHCl_3$ into FC-72 containing Krytox 157FSH (Fig. 9) [12, 69]. A heavy-fluorous tagged ionophore, serving as the ion receptor, can interact with target ions with high selectivity in a poorly solvating medium [6, 72]. The poor solvation of counterions further widens the response range of the ion-selective electrode. Thus, in 2009, Buhlmann and coworkers first reported a Teflon

AF 2400-based ion-selective electrode containing a linear perfluorooligoether (14.3 ether groups per molecule) as the plasticizer, sodium tetrakis [3,5-bis (perfluorohexyl)phenyl]borate as the ionic site, and bis[(perfluorooctyl)propyl]-2,2,2-trifluoroethylamine as the ionophore [72]. Teflon AF 2400-based electrodes exhibit high potentiometric selectivities, Nernstian responses to H^+ over a wide pH range (2–9) and improved mechanical properties compared to fluorous supported liquid membrane electrode (porous Teflon filters filled with perfluoroperhydrophenanthrene) [72]. In-depth study by Buhlmann's group discovered the existence of –COOH functional groups in Teflon AF 2400 [72]. The use of high ionophore concentrations diminishes the side effects caused by the –COOH groups in the electrode-matrix [72]. The –COOH group in Teflon AF 2400 (one per 854 monomer units), generated in the manufacturing process, brings both possibilities and troubles, and therefore needs special attention in the design of sensors [72].

5 Hydrophobic Materials

5.1 Superhydrophobic Surfaces

Hydrophobic materials are particularly useful for creating self-cleaning and water/ice repellent superhydrophobic surfaces. Many examples of superhydrophobic biological surfaces in nature (e.g., lotus leaves, bird feathers, etc.) indicate the two requirements to achieve superhydrophobicity. The surface material must have a low surface energy and the surface should have high surface roughness [73]. There are two common approaches to create artificial superhydrophobic surfaces. One is to prepare a smooth coating of a thin-layer hydrophobic material and then roughen the surface (e.g., by plasma etching), as shown in Fig. 15a [74]. Another well-known strategy is to fabricate a rough surface in the first step (e.g., by etching,

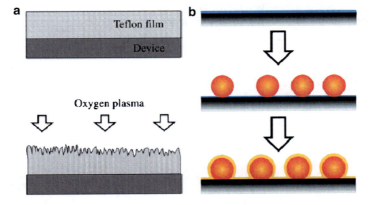

Fig. 15 Common strategies to create superhydrophobic surfaces [74, 75]

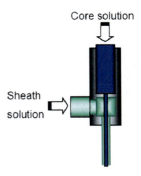

Fig. 16 The configuration of coaxial electro-spinning [73]

lithography, seeded growth, nano-imprint, etc.) and then coat the rough surface with hydrophobic material (Fig. 15b) [75].

Teflon AFs are hydrophobic. As reported by DuPont, the water contact angles of Teflon AF 2400 and Teflon AF 1600 are 105° and 104°, respectively, corresponding to their low surface energies (15.7 dynes/cm and 15.6 dynes/cm, respectively). Thus, they are attractive candidate materials for superhydrophobic surface design. The water contact angle for rough Teflon AF surfaces could be tuned between 120° to above 170° based on different substrates (e.g., Al, ZnO, ITO glasses) and fabrication strategies [74, 76, 77]. Another approach to superhydrophobic materials is to base them on fine fibers of hydrophobic material, but this requires that fibers can be created. Recently, successful electro-spinning of Teflon AFs was reported [73, 78, 79]. This is a significant step, because Teflon AFs are difficult to electrospin owing to their low dielectric constants and the low dielectric constants of solvents that they can dissolve in [79]. Only solvents with high dielectric constant can be easily polarized in response to an electric field, which in turn permits electrospinning of nanofibers. Two approaches have been developed to produce Teflon AF fiber membranes. Direct electro-spinning of Teflon AF can be achieved by adding Novec engineering fluids (hydrofluoroethers) to Teflon AF solution in perfluorocarbons to increase the dielectric constant of the solution [79]. Core-shell electro-spinning can prepare nanofibers with smaller diameters in a better controlled manner with fewer limitations compared to the direct electro-spinning by solvent manipulation [73, 78]. As shown in Fig. 16, a solution with high dielectric constant and a Teflon AF solution are fed into the sheath-flow device. As an electric potential is applied, the core solution is electrospun with the coating of Teflon AF. Membranes of Teflon AF from electrospun fibers display superhydrophobicity and are potential candidates for oxygen separator membranes in lithium-air batteries [78, 79].

5.2 *Manipulating Droplet Movement on Teflon AF Surfaces*

The manipulation of small volumes of fluid is central to future technological advances in biological assays and microfluidics. The most common and widely used method is micro-channel flow. An advanced method – sessile droplet displacement on surfaces – was introduced to avoid frequent problems in microchannel flow (e.g., clogging, and sample loss/contamination due to surface bound adsorption) [80]. Recently, innovative technologies have been focused on the control of droplet movement on hydrophobic and superhydrophobic surfaces by mechanical agitation and electrowetting [80–84]. Teflon AFs are suitable materials for fabricating such platforms considering their unique features: hydrophobic nature, weak van der Waals interactions, good insulators, and their ease of patterning. Furthermore, in applications involving surface binding, nonspecific adsorption will be significantly reduced when a fluorous surface is adopted.

In short, the manipulation of small volumes of fluid deals with the creation of a surface-energy gradient. This can be achieved via patterned surface (known as a "hard-coded track") with programmable agitation or straightforward programmable surface-energy gradient [81]. An interesting approach to create a hard-coded track is to build a rough superhydrophobic surface with a contact angle gradient [81]. Teflon AFs can be patterned with high resolution with lithography and molding due to their high mechanical strength. Shastry and coworkers reported microfabricated pillar array columns coated with Teflon AF 1600 as a platform for directing droplets [81]. As shown in Fig. 17, more air pockets are trapped under the droplet on the left side. Therefore, the left side of the droplet exhibits higher surface energy. With the assistance of mechanical vibration, the droplet can overcome contact angle hysteresis and will be propelled down the gradient [81]. The path of the droplet can be designed based on theoretical calculations. This process leads the droplet to a low-energy state and cannot be reversed.

Electrowetting, a voltage-induced surface-energy modulation, is a reversible and more flexible strategy to control fluidic movement. Rapid manipulation of droplets can be accomplished by controlling two sets of electrodes contacting liquid droplets. Figure 18 shows the configuration of a prototype electrowetting microfluidic channel [82]. Teflon AF coatings serve as the insulator and also provide hydrophobic surfaces with no significant nonspecific adsorption. The dimensions of the channel and bottom electrodes are commonly designed to satisfy the criteria that an analyte droplet overlaps two adjacent bottom electrodes. Once one side of a droplet overlaps the energized electrode, the surface energy will decrease dramatically, leading the bulk flow towards the energized electrode.

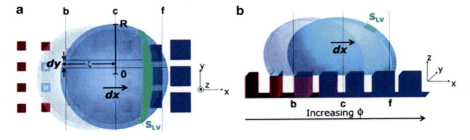

Fig. 17 A droplet moves down a contact angle gradient: (a) *top view*; (b) *side view* [81]

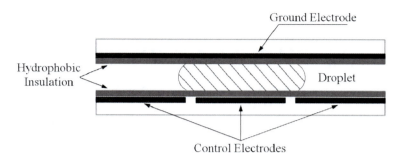

Fig. 18 The configuration of an electrowetting-based microfluidic channel [82]

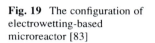

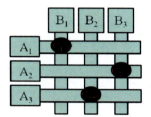

Rapid flow of droplets can be achieved by building a potential gradient on the electrode array.

The above protocol has been applied successfully for sample purification in proteomic analysis in conjunction with mass spectrometry [84]. The sample deposition, the removal of impurities by droplet-based extraction, as well as the introduction of MALDI matrices can be integrated in a single electrowetting-based microfluidic device for fast sample preparation [84]. The voltage-triggered fluid flow system also has impact on the synthetic community. Teflon AF based 2D open microfluidic chips serve as droplet reactors, a new generation of microreactor [83]. As shown in Fig. 19, a droplet containing reagents can flow on the 2D chip in one direction manipulated by electrowetting, and merge with another droplet coming from a perpendicular direction [83]. This type of microreactor exhibits unique advantages and offers opportunities for the development and optimization of synthetic methodologies as well as combinatorial chemistry. Electrowetting offers an exciting application in micro-total-analysis systems and will become extremely powerful when integrated with programmed control.

6 Platform for Biological Assays

Fluorous surfaces have become valuable in assays [85–87]. Teflon AFs, non-stick, and protective coatings have been incorporated in the fabrication of microfluidic assays to restrain contamination [88]. Recently, the plug-based microfluidic system

Fig. 20 The configuration of a plug-based microfluidic device [91]

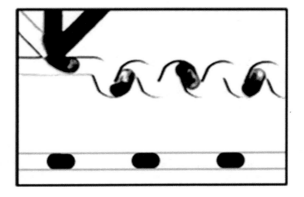

has attracted keen attention for its capability to achieve high temporal resolution in vitro/vivo measurements [89]. Compared to the traditional air-liquid segmented flow systems [90], flows of immiscible liquids are used to form plugs within the microchannels in a plug-based microfluidic system to decrease inter droplet mixing. Fluidic droplets with volumes down to nanoliters can be generated by manipulating the interfacial chemistry of the liquids and the channels [89]. For example, Ismagilov and coworkers designed a plug-based microfluidic device to handle whole blood and plasma [91]. The hydrophobic Teflon AF 1600 coating on the microfluidic channel prevents the spread of hydrophilic blood/plasma and helps permit a flow-based coagulation time assay that would seem to be impossible without the Teflon AF coating to minimize the sticking of fibrin inside the channel [91]. As shown in Fig. 20, the flow of fluorous liquid breaks up the blood/plasma flow and generates the motion of individual plugs with immiscible phases on the leading and trailing edges of the plug. As a tool that enables high throughput screening and analysis, plug-based microfluidics have raised a lot of interest in many subjects including catalyst screening and the optimization of crystallization conditions [92, 93].

Teflon AFs have been investigated as a substrate for creating vascular grafts. At this point, autologous veins are still the first choice for patients due to a variety of unsolved problems in the current synthetic vascular grafts, for example, thrombosis and restenosis which lead to restricted blood flow [94, 95]. Researchers are seeking suitable biomaterials, and modulate the surface properties of those candidate materials to approach the requirements of a functioning vascular graft [95, 96]. Endothelial cells line the interior surface of blood vessels, and serve anticoagulant, antiproliferative, and vasodilator functions [96]. Therefore, endothelialization of polymeric materials is an important step towards artificial vascular grafts [96]. Beside some widely investigated hydrocarbon polymers (e.g., polyesters, and polyurethanes), perfluoropolymers have been recognized as potential candidate materials for artificial vascular grafts due to their low friction, excellent chemical stability, and biocompatibility [95]. Spatz and coworkers accomplished well-controlled immobilization of endothelial cells on Teflon AF 1600 by decorating

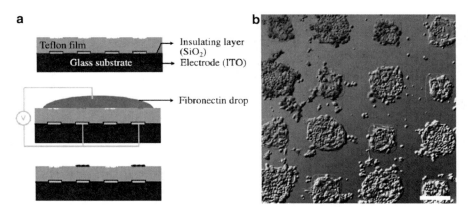

Fig. 21 (a) Figures from top to bottom: the configuration of the electrowetting device, patterning of a fibronectin array on a Teflon AF substrate by an energized electrode, and cell adhesion; (b) A microarray of HeLa cells on a Teflon AF surface (scale bar = 200 μm) [98]

the Teflon surface with thiolated gold nanoparticles [95]. Cells have been made to adhere on Teflon AF surfaces by pre-adsorbed ligands (fibronectin, and biotinylated fibronectin together with an RGD-streptavidin mutant) [97]. The endothelial cells can be patterned by creating a surface with patterned ligands. The pre-adsorption of dual-function ligands shows significantly enhanced cell adhesion to Teflon AF surfaces over surfaces with one ligand alone [97]. Chen and coworkers reported the investigation of cell adhesion on Teflon AF surfaces by cell microarrays [98]. They took advantage of the electrowetting technology as described in Sect. 5.2 to create arrays of a ligand (fibronectin) on a Teflon AF surface in order to direct cell adhesion [98, 99]. Figure 21 shows cartoons of the strategy on the left and a patterned cell array on the right [98]. The microarray technology enables high-throughput evaluation of ligand-cell recognition and optimization of cell adhesion towards artificial vascular graft.

7 Metal-Teflon AF Nanocomposites

Metal-polymer nanocomposite materials have novel properties from the metallic nanoparticles, the polymeric matrix, and the large interface. One of the most promising properties of metal-polymer nanocomposite materials is their tunable optical properties [100–103]. When a nanocomposite material is exposed to light, charge oscillations in the metallic nanoparticles lead to evanescent waves propagating along the interface, resulting in strong absorption at a particular wavelength known as the SPR wavelength [100, 102]. Teflon AFs are particularly suitable matrices in optical applications due to their wide optical transparency, and advantages of flexible molding and processing. It is known that the position, width, and intensity of plasmon resonances strongly depend on the microstructure of

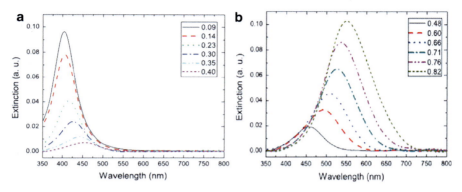

Fig. 22 Absorbance spectra of Ag-(Au)$_x$/Teflon AF nanocomposites: (**a**) $x < 0.45$; (**b**) $x > 0.45$ [100]

composite materials (e.g., the size and shape of metal nanoparticles, and the dispersion of nanoparticles in the polymeric matrix) and the chemical composition [100, 101]. During the past decade, a lot of effort has been exerted to prepare metal (Ag, Au, Ni, Pd, and Ag–Au alloy)-Teflon AF nanocomposites [100–105]. There are three strategies by which metal-Teflon AF nanocomposites with satisfactory homogeneity can be produced. The most widely used method is vapor phase co-deposition, in which the deposition rates of metals and Teflon AF are controlled to achieve different loadings of nanoparticles in Teflon AF matrix [100]. Figure 22 shows the shift of the plasmon peak position in Teflon AF films containing Ag–Au alloy nanoparticles [100]. The increase of the Au molar ratio in alloy nanoparticles leads to a gradual red shift of the plasmon peak [100]. The other two methods to prepare metal-Teflon AF composites take advantage of solubility properties [106, 107]. Metal-Teflon AF nanocomposites were obtained from both solution phase (hydrofluorocarbon fluid) and supercritical carbon dioxide phase by co-solvating Teflon AF and a metal complex with a fluorous ligand (e.g., nickel(II) hexafluoroacetylacetonate hydrate, palladium(II) hexafluoroacetylacetonate, and silver (I) 6,6,7,7,8,8,8-heptafluoro-2,2-dimethyl-3,5-octanedionate), which is reduced to metal nanoparticles [106, 107].

Recently, metal-Teflon AF nanocomposites have been found to be extremely powerful substrates for surface enhanced Raman spectroscopy (SERS) [108]. In 2010, Biswas and coworkers reported the fabrication of an ultra-sensitive SERS substrate starting from an Ag-Teflon AF composite material [108]. As shown in Fig. 23a, the team prepared Ag-Teflon AF nanocomposite by vapor phase deposition in the first step [108]. Then one side of the nanocomposite film was sputter-coated with a thin Ag layer whereas the opposite side was treated with an O_2 plasma to remove Teflon AF [108]. The resulting uniform arrangement of hot spots (Ag nanoparticles) on thin Ag film serves as the sensing phase [108]. It is important to fabricate the SERS sensing platform from an Ag-Teflon AF nanocomposite because the loadings of Ag nanoparticles can be well-controlled to be as high as

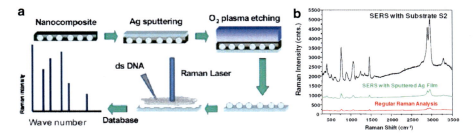

Fig. 23 (a) The preparation of SERS substrate; (b) Raman intensity of dsDNA using different sensing platforms [108]

possible without exceeding the percolation threshold for optimal signal enhancement [108]. SERS substrates prepared from Ag-Teflon AF nanocomposites display excellent reproducibility for a single sample system and significant signal enhancement (Fig. 23b) [108].

8 Summary

Teflon AF is a distinctive family of AFs. A unique combination of superior properties (e.g., high mechanical strength, wide optical transparency, low surface energy, tunable surface hydrophobicity, high permeability, etc.) of Teflon AFs has led to a wide range of innovative applications. Teflon AFs are promising matrices for transport-based separations. Tremendous progress has been made over the last two decades to understand the transport behavior of solutes/penetrants through Teflon AFs. Deep understanding of diffusion and solubility/partitioning in Teflon AFs has been developed to design better separation systems and sensors based on transport and sorption. As weakly solvating media, Teflon AFs are ideal "solvents" for the selective molecular recognition which are prototypes for selective extractions and ion-selective electrodes. The FFV of Teflon AFs is an important property in transport of molecules through these materials. The hydrophobic nature of Teflon AFs and its tunability combined with the low-biofouling characteristic make Teflon AFs promising coating materials in bioanalytical applications such as self-cleaning surfaces, microfluidics, and arrays. Recently, Teflon AF-based nanocomposite materials have been created with new properties that are important for applications in chemical separations and optical devices. Teflon AFs are very promising materials offering many advantages in practical applications due to their unique features and versatility. We expect to see more novel applications of these interesting polymers in the future.

Acknowledgements The authors are grateful for financial support from the NSF (CHE 0957038) and NIH (P50 GM067082).

References

1. Yampolskii Y, Pinnau I, Freeman BD (eds) (2006) Materials science of membranes for gas and vapor separation. Wiley, Chichester
2. Arcella V, Ghielmi A, Tommasi G (2003) High performance perfluoropolymer films and membranes. Ann NY Acad Sci 984(Advanced Membrane Technology):226
3. Rindfleisch F, DiNoia TP, McHugh MA (1996) Solubility of polymers and copolymers in supercritical CO_2. J Phys Chem 100(38):15581
4. Tuminello WH, Dee GT, McHugh MA (1995) Dissolving perfluoropolymers in supercritical carbon dioxide. Macromolecules 28(5):1506
5. Yampolskii Y, Belov N, Tokarev A, Bondarenko G (2006) Sorption and transport of F- and H-containing organic vapors in amorphous perfluorinated polymers. Desalination 199(1–3):469
6. Boswell PG, Bühlmann P (2005) Fluorous bulk membranes for potentiometric sensors with wide selectivity ranges: observation of exceptionally strong ion pair formation. J Am Chem Soc 127(25):8958
7. Genzer J, Efimenko K (2006) Recent developments in superhydrophobic surfaces and their relevance to marine fouling: a review. Biofouling 22(5):339
8. Golemme G, Nagy JB, Fonseca A, Algieri C, Yampolskii Y (2003) ^{129}Xe-NMR study of free volume in amorphous perfluorinated polymers: comparison with other methods. Polymer 44(17):5039
9. Shantarovich VP, Kevdina IB, Yampolskii YP, Alentiev AY (2000) Positron annihilation lifetime study of high and low free volume glassy polymers: effects of free volume sizes on the permeability and permselectivity. Macromolecules 33(20):7453
10. Yampolskii Y (2009) Amorphous perfluorinated membrane materials: structure, properties and application. Russ J Gen Chem 79(3):657
11. Scheirs J (1997) Modern fluoropolymers: high performance polymers for diverse applications. Wiley, Chichester
12. O'Neal KL, Zhang H, Yang Y, Hong L, Lu D, Weber SG (2010) Fluorous media for extraction and transport. J Chromatogr A 1217(16):2287
13. Polyakov AM, Starannikova LE, Yampolskii YP (2004) Amorphous Teflons AF as organophilic pervaporation materials. Separation of mixtures of chloromethanes. J Membrane Sci 238(1–2):21
14. Dasgupta PK, Zhang G, Poruthoor SK, Caldwell S, Dong S, Liu S-Y (1998) High-sensitivity gas sensors based on gas-permeable liquid core waveguides and long-path absorbance detection. Anal Chem 70(22):4661
15. Gangal SV (2002) Perfluorinated polymers, tetrafluoroethylene–perfluorodioxole copolymers. Wiley, Chichester
16. Yang MK, French RH, Tokarsky EW (2008) Optical properties of Teflon AF amorphous fluoropolymers. J Micro Nanolithogr MEMS MOEMS 7(3):033010/1
17. Finkelshtein ES, Makovetskii KL, Gringolts ML, Rogan YV, Golenko TG, Starannikova LE, Yampolskii YP, Shantarovich VP, Suzuki T (2006) Addition-type polynorbornenes with $Si(CH_3)_3$ side groups: synthesis, gas permeability, and free volume. Macromolecules 39(20):7022
18. Yampolskii YP, Finkelshtein ES, Makovetskii KL, Bondar VI, Shantarovich VP (1996) Effects of cis-trans-configurations of the main chains of poly(trimethylsilyl norbornene) on its transport and sorption properties as well as free volume. J Appl Polym Sci 62(2):349
19. Yampolskii YP, Soloviev SA, Gringolts ML (2004) Thermodynamics of sorption in and free volume of poly(5,6-bis(trimethylsilyl)norbornene). Polymer 45(20):6945
20. Hofmann D, Entrialgo-Castano M, Lerbret A, Heuchel M, Yampolskii Y (2003) Molecular modeling investigation of free volume distributions in stiff chain polymers with conventional and ultrahigh free volume: comparison between molecular modeling and positron lifetime studies. Macromolecules 36(22):8528

21. Alexander Stern S (1994) Polymers for gas separations: the next decade. J Membrane Sci 94(1):1
22. Pinnau I, Toy LG (1996) Gas and vapor transport properties of amorphous perfluorinated copolymer membranes based on 2,2-bistrifluoromethyl-4,5-difluoro-1,3-dioxole/tetrafluoroethylene. J Membrane Sci 109(1):125
23. Freeman BD, Pinnau I (1999) Polymer membranes for gas and vapor separation. ACS Symposium Series, vol 733. American Chemical Society, Washington DC, p 1
24. Merkel TC, Bondar V, Nagai K, Freeman BD, Yampolskii YP (1999) Gas sorption, diffusion, and permeation in poly(2,2-bis(trifluoromethyl)-4,5-difluoro-1,3-dioxole-co-tetrafluoroethylene). Macromolecules 32(25):8427
25. Alentiev AY, Shantarovich VP, Merkel TC, Bondar VI, Freeman BD, Yampolskii YP (2002) Gas and vapor sorption, permeation, and diffusion in glassy amorphous Teflon AF1600. Macromolecules 35(25):9513
26. Bondar VI, Freeman BD, Yampolskii YP (1999) Sorption of gases and vapors in an amorphous glassy perfluorodioxole copolymer. Macromolecules 32(19):6163
27. Hildebrand JH, Prausnitz JM, Scott RL (1970) Regular and related solutions. The solubility of gases, liquids, and solids. Van Nostrand Reinhold, New York
28. Fleming GK, Koros WJ (1986) Dilation of polymers by sorption of carbon dioxide at elevated pressures. 1. Silicone rubber and unconditioned polycarbonate. Macromolecules 19(8):2285
29. Ferrari MC, Galizia M, De Angelis MG, Sarti GC (2010) Gas and vapor transport in mixed matrix membranes based on amorphous Teflon AF1600 and AF2400 and fumed silica. Ind Eng Chem Res 49(23):11920
30. Merkel TC, He Z, Pinnau I, Freeman BD, Meakin P, Hill AJ (2003) Sorption and transport in poly(2,2-bis(trifluoromethyl)-4,5-difluoro-1,3-dioxole-co-tetrafluoroethylene) containing nanoscale fumed silica. Macromolecules 36(22):8406
31. Merkel TC, Freeman BD, Spontak RJ, He Z, Pinnau I, Meakin P, Hill AJ (2002) Ultrapermeable, reverse-selective nanocomposite membranes. Science 296(5567):519
32. Polyakov AM, Starannikova LE, Yampolskii YP (2003) Amorphous Teflons AF as organophilic pervaporation materials. Transport of individual components. J Membrane Sci 216(1–2):241
33. Polyakov A, Bondarenko G, Tokarev A, Yampolskii Y (2006) Intermolecular interactions in target organophilic pervaporation through the films of amorphous Teflon AF2400. J Membrane Sci 277(1–2):108
34. Jansen JC, Friess K, Drioli E (2010) Organic vapour transport in glassy perfluoropolymer membranes: a simple semi-quantitative approach to analyze clustering phenomena by time lag measurements. J Membrane Sci 367(1–2):141
35. Horváth IT (1998) Fluorous biphase chemistry. Accounts Chem Res 31(10):641
36. Vincent J-M (2008) Noncovalent associations in fluorous fluids. J Fluor Chem 129(10):903
37. Zhao H, Ismail K, Weber SG (2004) How fluorous is poly(2,2-bis(trifluoromethyl)-4,5-difluoro-1,3-dioxole-co-tetrafluoroethylene) (Teflon AF)? J Am Chem Soc 126(41):13184
38. Zhao H, Zhang J, Wu N, Zhang X, Crowley K, Weber SG (2005) Transport of organic solutes through amorphous Teflon AF films. J Am Chem Soc 127(43):15112
39. Zhang H, Hong L, Weber SG (2009) Dependence of the physical properties and transport behavior of perfluorotripentylamine-doped Teflon AF films on composition. PMSE Preprints 100:358
40. Zhang H, Hussam A, Weber SG (2010) Properties and transport behavior of perfluorotripentylamine (FC-70)-doped amorphous Teflon AF 2400 films. J Am Chem Soc 132(50):17867
41. Don T-M, Bell JP, Narkis M (1996) Antiplasticization behavior of polycaprolactone/polycarbonate-modified epoxies. Polym Eng Sci 36:2601
42. Goss K-U, Bronner G (2006) What is so special about the sorption behavior of highly fluorinated compounds? J Phys Chem A 110(30):9518
43. van Krevelen DW, Hoftijzer PJ (1948) Kinetics of gas-liquid reactions part I. General theory. Recl Trav Chim Pays Bas 67:563
44. O'Brien M, Baxendale IR, Ley SV (2010) Flow ozonolysis using a semipermeable Teflon AF-2400 membrane to effect gas-liquid contact. Org Lett 12(7):1596

45. Polyzos A, O'Brien M, Petersen TP, Baxendale IR, Ley SV (2011) The continuous-flow synthesis of carboxylic acids using CO_2 in a tube-in-tube gas permeable membrane reactor. Angew Chem Int Ed Engl 50(5):1190
46. Voraberger HS, Trettnak W, Ribitsch V (2003) Optochemical hydrogen peroxide sensor based on oxygen detection. Sens Actuators B Chem B90(1–3):324
47. Kurauchi Y, Ogata T, Egashira N, Ohga K (1996) Fiber-optic sensor with a dye-modified chitosan/poly(vinyl alcohol) cladding for the determination of organic acids. Anal Sci 12(1):55
48. Hughes LD, DeVol TA (2006) Characterization of a Teflon coated semiconductor detector flow cell for monitoring of pertechnetate in groundwater. J Radioanal Nucl Ch 267(2):287
49. Waich K, Mayr T, Klimant I (2007) Microsensors for detection of ammonia at ppb-concentration levels. Meas Sci Technol 18(10):3195
50. Thomas PC, Halter M, Tona A, Raghavan SR, Plant AL, Forry SP (2009) A noninvasive thin film sensor for monitoring oxygen tension during in vitro cell culture. Anal Chem 81(22):9239
51. Chen B, Xiang S, Qian G (2010) Metal-organic frameworks with functional pores for recognition of small molecules. Acc Chem Res 43(8):1115
52. Rego R, Caetano N, Mendes A (2004) Development of a new gas sensor for binary mixtures based on the permselectivity of polymeric membranes. Application to carbon dioxide/methane and carbon dioxide/helium mixtures. Anal Chim Acta 511(2):215
53. Rego R, Caetano N, Mendes A (2005) Hydrogen/methane and hydrogen/nitrogen sensor based on the permselectivity of polymeric membranes. Sens Actuators B Chem B111–B112:150
54. Waterbury RD, Yao W, Byrne RH (1997) Long pathlength absorbance spectroscopy: trace analysis of Fe(II) using a 4.5 m liquid core waveguide. Anal Chim Acta 357(1–2):99
55. Dasgupta PK, Zhang G, Li J, Boring CB, Jambunathan S, Al-Horr R (1999) Luminescence detection with a liquid core waveguide. Anal Chem 71(7):1400
56. Wang Z, Wang Y, Cai W-J, Liu S-Y (2002) A long pathlength spectrophotometric pCO_2 sensor using a gas-permeable liquid-core waveguide. Talanta 57(1):69
57. Lu Z, Dai M, Xu K, Chen J, Liao Y (2008) A high precision, fast response, and low power consumption in situ optical fiber chemical pCO2 sensor. Talanta 76(2):353
58. Murphy B, McLoughlin P (2003) Determination of Chlorinated hydrocarbon species in aqueous solution using Teflon coated ATR Waveguide/FTIR spectroscopy. Int J Environ Anal Chem 83(7–8):653
59. Du W-B, Fang Q, He Q-H, Fang Z-L (2005) High-throughput nanoliter sample introduction microfluidic chip-based flow injection analysis system with gravity-driven flows. Anal Chem 77(5):1330
60. Osterfeld M, Franke H, Brandenburg A (1994) Compensation of temperature drift in leaky-mode spectra for sensor applications. Appl Phys A Solids Surfaces A58(3):215
61. Podgorsek RP, Franke H, Caron S, Galarneau P (1998) Dynamic and polychromatic SPR-leaky mode spectroscopy with Teflon AF films on silver for chemo-sensing. In: Applications of photonic technology 3. SPIE-International Society for Optical Engineering, vol 3491, p 777
62. Podgorsek RP, Sterkenburgh T, Wolters J, Ehrenreich T, Nischwitz S, Franke H (1997) Optical gas sensing by evaluating ATR leaky mode spectra. Sens Actuators B Chem 39(1–3):349
63. Podgorsek RP, Franke H (2002) Selective optical detection of aromatic vapors. Appl Opt 41(4):601
64. Osterfeld M, Franke H, Feger C (1993) Optical gas detection using metal film enhanced leaky mode spectroscopy. Appl Phys Lett 62(19):2310
65. Kebabian PL, Freedman A (2006) Fluoropolymer-based capacitive carbon dioxide sensor. Meas Sci Technol 17(4):703
66. Cha W, Meyerhoff ME (2006) Enhancing the selectivity of amperometric nitric oxide sensor over ammonia and nitrite by modifying gas-permeable membrane with Teflon AF. Chem Analityczna 51(6):949

67. Cha W, Meyerhoff ME (2006) S-Nitrosothiol detection via amperometric nitric oxide sensor with surface modified hydrogel layer containing immobilized organoselenium catalyst. Langmuir 22(25):10830
68. Bidault F, Kucernak A (2011) Cathode development for alkaline fuel cells based on a porous silver membrane. J Power Sources 196(11):4950
69. O'Neal KL, Geib S, Weber SG (2007) Extraction of pyridines into fluorous solvents based on hydrogen bond complex formation with carboxylic acid receptors. Anal Chem 79(8):3117
70. O'Neal KL, Weber SG (2008) Molecular and ionic hydrogen bond formation in fluorous solvents. J Phys Chem B 113(1):149
71. O'Neal KL, Weber SG (2009) Extraction and metalation of porphyrins in fluorous liquids with carboxylic acids and metal salts. J Phys Chem B 113(21):7449
72. Lai C-Z, Koseoglu SS, Lugert EC, Boswell PG, Rabai J, Lodge TP, Bühlmann P (2009) Fluorous polymeric membranes for ionophore-based ion-selective potentiometry: how inert is Teflon AF? J Am Chem Soc 131(4):1598
73. Han D, Steckl AJ (2009) Superhydrophobic and oleophobic fibers by coaxial electrospinning. Langmuir 25(16):9454
74. Shiu J-Y, Whang W-T, Chen P (2008) Superhydrophobic coatings for microdevices. J Adhes Sci Technol 22(15):1883
75. Ling XY, Phang IY, Vancso GJ, Huskens J, Reinhoudt DN (2009) Stable and transparent superhydrophobic nanoparticle films. Langmuir 25(5):3260
76. Thieme M, Blank C, Pereira de Oliveira A, Worch H, Frenzel R, Hohne S, Simon F, Pryce Lewis HG, White AJ (2009) Superhydrophobic Aluminum Surfaces: Preparation Routes, Properties and Artificial Weathering Impact, in Contact Angle, Wettability and Adhesion, Vol. 6. Leiden , Koninklijke Brill NV: 251–267
77. Wu J, Xia J, Lei W, B-p W (2010) A one-step method to fabricate lotus leaves-like ZnO film. Mater Lett 65(3):477
78. Muthiah P, Hsu S-H, Sigmund W (2010) Coaxially electrospun PVDF-Teflon AF and Teflon AF-PVDF core-sheath nanofiber mats with superhydrophobic properties. Langmuir 26(15):12483
79. Scheffler R, Bell NS, Sigmund W (2010) Electrospun Teflon AF fibers for superhydrophobic membranes. J Mater Res 25(8):1595
80. Shi LT, Jiang CG, Ma GJ, Wu CW (2010) Electric field assisted manipulation of microdroplets on a superhydrophobic surface. Biomicrofluidics 4(4):041101
81. Shastry A, Case MJ, Böhringer KF (2006) Directing droplets using microstructured surfaces. Langmuir 22(14):6161
82. Pollack MG, Fair RB, Shenderov AD (2000) Electrowetting-based actuation of liquid droplets for microfluidic applications. Appl Phys Lett 77(11):1725
83. Dubois P, Marchand G, Fouillet Y, Berthier J, Douki T, Hassine F, Gmouh S, Vaultier M (2006) Ionic liquid droplet as e-microreactor. Anal Chem 78(14):4909
84. Wheeler AR, Moon H, Bird CA, Ogorzalek Loo RR, Kim C-J, Loo JA, Garrell RL (2005) Digital microfluidics with in-line sample purification for proteomics analyses with MALDI-MS. Anal Chem 77(2):534
85. Ko K-S, Jaipuri FA, Pohl NL (2005) Fluorous-based carbohydrate microarrays. J Am Chem Soc 127(38):13162
86. Collet BYM, Nagashima T, Yu MS, Pohl NLB (2009) Fluorous-based peptide microarrays for protease screening. J Fluor Chem 130(11):1042
87. Vegas AJ, Bradner JE, Tang W, McPherson OM, Greenberg EF, Koehler AN, Schreiber SL (2007) Fluorous-based small-molecule microarrays for the discovery of histone deacetylase inhibitors. Angew Chem Int Ed Engl 46(42):7960
88. Yang H, Luk VN, Abelgawad M, Barbulovic-Nad I, Wheeler AR (2009) A world-to-chip interface for digital microfluidics. Anal Chem 81(3):1061

89. Chen D, Du W, Liu Y, Liu W, Kuznetsov A, Mendez FE, Philipson LH, Ismagilov RF (2008) The chemistrode: a droplet-based microfluidic device for stimulation and recording with high temporal, spatial, and chemical resolution. Proc Natl Acad Sci USA 105(44):16843
90. Skeggs LT, Hochstrasser H (1964) Clin Chem 10:918
91. Song H, Li H-W, Munson MS, Ha TGV, Ismagilov RF (2006) On-chip titration of an anticoagulant argatroban and determination of the clotting time within whole blood or plasma using a plug-based microfluidic system. Anal Chem 78(14):4839
92. Kreutz JE, Shukhaev A, Du W, Druskin S, Daugulis O, Ismagilov RF (2010) Evolution of catalysts directed by genetic algorithms in a plug-based microfluidic device tested with oxidation of methane by oxygen. J Am Chem Soc 132(9):3128
93. Li L, Mustafi D, Fu Q, Tereshko V, Chen DL, Tice JD, Ismagilov RF (2006) Nanoliter microfluidic hybrid method for simultaneous screening and optimization validated with crystallization of membrane proteins. Proc Natl Acad Sci USA 103(51):19243
94. Miller DC, Webster TJ, Haberstroh KM (2004) Technological advances in nanoscale biomaterials: the future of synthetic vascular graft design. Expert Rev Med Devices 1(2):259
95. Kruss S, Wolfram T, Martin R, Neubauer S, Kessler H, Spatz JP (2010) Stimulation of cell adhesion at nanostructured Teflon interfaces. Adv Mater 22(48):5499
96. Ballermann BJ (1998) Adding endothelium to artificial vascular grafts. News Physiol Sci 13(3):154
97. Anamelechi CC, Clermont EC, Novak MT, Reichert WM (2009) Dynamic seeding of perfusing human umbilical vein endothelial cells (HUVECs) onto dual-function cell adhesion ligands: Arg-Gly-Asp (RGD)-streptavidin and biotinylated fibronectin. Langmuir 25(10):5725
98. Shiu J, Kuo CW, Whang W, Chen P (2010) Addressable cell microarrays via switchable superhydrophobic surfaces. J Adhes Sci Technol 24:1023
99. Shiu J-Y, Chen P (2007) Addressable protein patterning via switchable superhydrophobic microarrays. Adv Funct Mater 17(15):2680
100. Beyene HT, Chakravadhanula VSK, Hanisch C, Elbahri M, Strunskus T, Zaporojtchenko V, Kienle L, Faupel F (2010) Preparation and plasmonic properties of polymer-based composites containing Ag-Au alloy nanoparticles produced by vapor phase co-deposition. J Mater Res 45(21):5865
101. Takele H, Greve H, Pochstein C, Zaporojtchenko V, Faupel F (2006) Plasmonic properties of Ag nanoclusters in various polymer matrices. Nanotechnology 17(14):3499
102. Biswas A, Eilers H, Hidden F Jr, Aktas OC, Kiran CVS (2006) Large broadband visible to infrared plasmonic absorption from Ag nanoparticles with a fractal structure embedded in a Teflon AF matrix. Appl Phys Lett 88(1):013103/1
103. Biswas A, Aktas OC, Kanzow J, Saeed U, Strunskus T, Zaporojtchenko V, Faupel F (2004) Polymer-metal optical nanocomposites with tunable particle plasmon resonance prepared by vapor phase co-deposition. Mater Lett 58(9):1530
104. Biswas A, Bayer IS, Marken B, Pounds TD, Norton MG (2007) Networks of ultra-fine Ag nanocrystals in a Teflon AF matrix by vapor phase e-beam-assisted deposition. Nanotechnology 18(30):305602/1
105. Biswas A, Marton Z, Kanzow J, Kruse J, Zaporojtchenko V, Faupel F, Strunskus T (2003) Controlled generation of Ni nanoparticles in the capping layers of Teflon AF by vapor-phase tandem evaporation. Nano Lett 3(1):69
106. Evanoff DD Jr, Zimmerman P, Chumanov G (2005) Synthesis of metal-Teflon AF nanocomposites by solution-phase methods. Adv Mater 17(15):1905
107. Hasell T, Yoda S, Howdle SM, Brown PD (2005) Microstructural characterisation of silver/polymer nanocomposites prepared using supercritical carbon dioxide. J Phys Conf Ser 26:276
108. Biswas A, Bayer IS, Biris AS (2010) Nanocomposite route to ultra-sensitive surface enhanced Raman scattering substrates. In: Functional materials and nanostructures for chemical and biochemical sensing. Materials research society symposium proceedings, vol 1253. Warrendale, p K10

Ecotoxicology of Organofluorous Compounds

Margaret B. Murphy, Eva I.H. Loi, Karen Y. Kwok, and Paul K.S. Lam

Abstract Organofluorous compounds have been developed for myriad purposes in a variety of fields, including manufacturing, industry, agriculture, and medicine. The widespread use and application of these compounds has led to increasing concern about their potential ecological toxicity, particularly because of the stability of the C–F bond, which can result in chemical persistence in the environment. This chapter reviews the chemical properties and ecotoxicology of four groups of organofluorous compounds: fluorinated refrigerants and propellants, per- and polyfluorinated compounds (PFCs), fluorinated pesticides, and fluoroquinolone antibiotics. These groups vary in their environmental fate and partitioning, but each raises concern in terms of ecological risk on both the regional and global scale, particularly those compounds with long environmental half-lives. Further research on the occurrence and toxicities of many of these compounds is needed for a more comprehensive understanding of their ecological effects.

Keywords Chlorofluorocarbon · Fluoroquinolone · Organofluorine · Perfluorinated compound · Pesticide · Toxicity

Contents

1 Introduction	340
2 Fluorinated Refrigerants and Propellants	341
2.1 Chlorofluorocarbons and Halons	341
2.2 Ecological Effects	342

M.B. Murphy (✉), E.I.H. Loi, K.Y. Kwok, and P.K.S. Lam
Department of Biology and Chemistry and State Key Laboratory in Marine Pollution, City University of Hong Kong, Kowloon Tong, Kowloon, Hong Kong SAR
e-mail: mbmurphy@cityu.edu.hk

2.3 Hydrochlorofluorocarbons and Other Replacement Compounds 342
2.4 Global Climate Change 343
3 Poly- and Perfluorinated Compounds 344
3.1 Environmental Occurrence 344
3.2 Toxicity 344
3.3 Bioaccumulation 347
4 Organofluorous Pesticides 348
4.1 Compounds of Interest 348
4.2 Environmental Properties 352
4.3 Toxicity 352
5 Fluoroquinolone Antibiotics 353
5.1 Fluoroquinolones 354
5.2 Environmental Properties 354
5.3 Toxicity 357
6 Uncertainties and Needs 357
References 358

1 Introduction

The unique properties of the C–F bond have led to the development and application of numerous organofluorous compounds for various industrial, agricultural, manufacturing, and medical purposes. These properties result from the electronegativity of fluorine, which is highest among all known elements, and cause the C–F bond to be thermally, metabolically, oxidatively, and reactively stable [1–3]. The small size of the fluorine atom allows it to be used as a substitute for hydroxyl groups in organic compounds to introduce greater structural stability without affecting the steric characteristics of the original compound, an important consideration in the synthesis of biologically active chemicals such as pesticides [1] and therapeutic drugs [4, 5] that may require receptor-binding or other biological interactions to elicit their effects. However, the stability of the C–F bond can also result in the synthesis or production of compounds with long environmental half-lives that may possess the three characteristics that cause concern among ecotoxicologists: persistence, a high likelihood of bioaccumulation by living organisms, and toxicity. This chapter will focus on the ecotoxicological profiles of four groups of organofluorous compounds – fluorinated refrigerants and propellants, per- and polyfluorinated compounds (PFCs), organofluorous pesticides (OFPs) and fluoroquinolone antibiotics – due to their widespread use, frequent environmental occurrence and persistence, and evidence for their ecological toxicity. There is some overlap among these groups – for example, some PFCs may be used as "inert" ingredients in pesticide formulations – but each group also presents distinct potential for ecological risk, and thus they will be considered separately.

2 Fluorinated Refrigerants and Propellants

2.1 Chlorofluorocarbons and Halons

The organofluorous compounds which have had perhaps the most global effect on ecosystems and organisms are the chlorofluorocarbons (CFCs) and halons, used as refrigerants in refrigerators and air conditioning units, in portable fire extinguishers, as propellants in aerosol containers, and as components of insulation and other polymers [6]. Following work in the 1970s and 1980s that showed that these compounds undergo atmospheric photodegradation that results in the breakdown of stratospheric ozone and subsequent thinning of the ozone layer, the Montreal Protocol was put into force by the United Nations in 1989 to reduce and eventually ban the use of CFCs and halons [6, 7]. According to the text of the Protocol, CFCs, halons, and other fully halogenated CFCs were phased out as of 2010. Another group, the hydrobromofluorocarbons, was phased out by 1996.

The CFCs comprise several dozen chemicals ranging from one to three carbons in chain length with one to seven fluorine substituents, while three halons containing two to four fluorine atoms are regulated by the Protocol (Table 1) [6]. CFCs and halons are stable volatile compounds with long atmospheric half-lives that participate in cyclic stratospheric reactions, prolonging their effects on the

Table 1 Chemical formulas, ozone-depleting potential, and global warming potential of fluorinated refrigerants and propellants listed under the Montreal Protocol

Compound	Formula	Ozone-depleting potential[a]	Global warming potential (GWP)[b]
Chlorofluorocarbons (15 compounds)	$C_aF_bCl_c$ $a = 1–3$ $b = 1–7$ $c = 1–7$	0.6–1	4,750–14,420
Halons	CF_2BrCl CF_3Br $C_2F_4Br_2$	3 10 6	7,140 1,890 1,640
Hydrochlorofluorocarbons (40 compounds[c])	$C_aH_bF_cCl_d$ $a = 1–3$ $b = 1–6$ $c = 1–6$ $d = 1–6$	0.001–0.052	77–2,310
Hydrobromofluorocarbons (34 compounds[c])	$C_aH_bF_cBr_d$ $a = 1–3$ $b = 1–6$ $c = 1–6$ $d = 1–6$	0.02–1.9	1,300–7,300

[a] Defined as the ratio of the impact on ozone of a chemical compared to the impact of CFC-11 (CCl_3F) on a mass basis; data from [6]
[b] The warming effect of a greenhouse gas relative to that of CO_2 for a 100-year period; data from [8]
[c] This number does not include isomers

ozone layer. Cleavage of the CFC or halon C–Cl bond by ultraviolet radiation produces chloride ions, which react with ozone to form chlorine oxide and diatomic oxygen. Chlorine oxide dimers (ClOOCl) form, which can then be broken down again by sunlight into chloride ions, which then further react with ozone, and so on. Fluoride ions will likewise react with ozone to form fluorine oxide (FO) and diatomic oxygen, but they can also form hydrogen fluoride (HF) through reaction with atmospheric methane (CH_4); the stability of HF prevents fluoride ions from continuing to react with ozone as do chloride ions, and thus the ozone-depleting potential of fluorine is approximately 1/1,000 that of chlorine [7].

2.2 Ecological Effects

CFCs and related compounds are highly volatile, and therefore do not represent an acute toxicity hazard with the exception of occupational exposure. Rather, their toxic effects are indirect: depletion of the ozone layer results in higher levels of ultraviolet (UV) radiation at the Earth's surface, and excessive exposure to UV radiation, particularly UV-B, results in damage to cellular components, including DNA and proteins. Cellular repair systems can counteract some of this damage, but long-term UV exposure can lead to more serious health effects such as cancer in animals. Plants exposed to high levels of UV-B may show impaired growth [9]. Sensitivity to UV radiation differs across species and taxonomic groups, but younger life stages are generally more susceptible to UV damage [10].

The phase-out of CFC and halon use under the Montreal Protocol is an example of successful regulation of environmental contaminants. CFC concentrations in the atmosphere have stabilized or declined since the Protocol entered into force, particularly for those CFCs with the largest ozone-depleting potential, such as CCl_3F [7]. These relatively rapid changes over the span of 20 years have largely mitigated the effects of ozone thinning on ecosystems according to a recent review [11], but effects in humans in terms of increased incidence of melanoma and other skin cancers are an ongoing problem [12].

2.3 Hydrochlorofluorocarbons and Other Replacement Compounds

The strong and effective efforts to remove CFCs and halons from the global market have resulted in a need for alternate refrigerant and propellant compounds. The hydrochlorofluorocarbons (HCFCs) were initially adopted for this purpose, but they raise similar concerns as CFCs due to their chlorination, although the hydrogen atoms in the molecule are predicted to facilitate degradative reactions with hydroxyl groups in the atmosphere that will prevent HCFCs from reaching the stratosphere where they may affect the ozone layer [13, 14]. Nevertheless, HCFCs may still have some ozone-depleting potential, and therefore they are also scheduled to be phased out under the Montreal Protocol by 2030 [6]. Hydrofluorocarbons (HFCs) are another group of compounds that have been used as replacements for CFCs, halons, and HCFCs because of their low potential for ozone depletion [15]. HCFCs and HFCs all demonstrate low ecological and human toxicity [13, 15]. More recently, hydrofluoropolyethers (HFPEs) have been proposed as replacements for HCFCs because they also are not expected to affect stratospheric ozone levels [16].

It is important to note that the Protocol contains a 10-year delay/exemption for developing countries, and thus CFCs and halons may continue to be produced and released in some parts of the world; with the exemption, HCFCs will continue to be produced until approximately mid-century based on current targets [6]. Breakdown products of CFCs and related compounds should also be considered; one of the most ubiquitous organofluorous compounds in the environment is trifluoroacetic acid (TFA), which is a degradation product of CFCs, HCFCs, and HFCs that occurs in surface waters, precipitation, and other aqueous matrices due to its high water solubility. A recent risk assessment reported that this compound does not bioaccumulate and presents a low overall environmental risk [17], but the persistence and reactivity of CFCs and other ozone-depleting compounds may result in the formation of other chemicals that may be of concern.

2.4 Global Climate Change

Increasing awareness of the effects and scale of global climate change has prompted a reassessment of the environmental effects of CFCs, halons, HCFCs, and related compounds in recent years. These compounds, particularly the CFCs, have a high global warming potential (GWP), meaning that they may contribute to increasing atmospheric temperatures in addition to their effects on the ozone layer and levels of UV radiation [14]. HFCs and HFPEs have lower GWP than CFCs, but are still considered greenhouse gases [15, 16]. Atmospheric chemistry is complex, and understanding the effects of multiple large-scale environmental changes such as increases in temperature or oceanic acidity is not straightforward. For example, increased atmospheric CO_2 is predicted to result in increased plant growth rates, but this growth does not eliminate UV damage in some species of plants [9]. Similarly, predicting effects on the population level is difficult, as species and even populations of species differ in their tolerance of UV exposure, changes in temperature, and other stressors [9, 11, 18]. CFCs, halons, and HCFCs will persist in the atmosphere for decades after their production and release have ceased, and predicting their effects will be an important area of research.

3 Poly- and Perfluorinated Compounds

3.1 Environmental Occurrence

PFCs are used in a variety of industrial and commercial applications, such as surface protectors in textiles, furnishings and food packaging, and as surfactants in pesticides [19]. PFCs encompass a large number of compounds, including perfluorocarboxylates, perfluorosulfonates, perfluorosulfonamides, saturated and unsaturated fluorotelomer acids, and fluorotelomer alcohols (FTOHs), among others (Table 2). These compounds were first reported in biological samples 10 years ago [20], at which time the focus was on the compounds perfluorooctanesulfonate (PFOS) and perfluorooctanoate (PFOA); since this time, PFOS, PFOA, and many other PFCs have been widely detected in surface waters [21, 22], sediments [23, 24], and biota, including human blood, and studies have shown that concentrations of these chemicals in the environment have increased in recent years (reviewed by [25]). Concerns about ubiquity and toxicity led to the phase out of PFOS production in the US and EU by some manufacturers in 2002, and use of PFOS and its precursor, perfluorooctyl sulfonyl fluoride, was restricted under the Stockholm Convention on Persistent Organic Pollutants in May 2009 [26].

Although PFOS use has been phased out or restricted in some parts of the world and concentrations of this chemical have declined in human blood samples collected recently in the US [27], levels of PFOS in humans from other global regions have not shown a similar decline [28]. The phase-out of PFOA use will not be completed until 2015 [29], and these compounds may still be produced in other parts of the world. The continued use of PFOS, PFOA, and newer PFCs introduced as substitutes, as well as PFC precursor compounds, raises questions about the potential impacts of these chemicals on natural systems, especially as many of them are expected to be persistent in the environment.

3.2 Toxicity

Among the PFCs detected in the environment to date, PFOS and PFOA have also been the most studied in terms of their potential toxic effects, and have been shown to have various effects on development, reproduction, and immune function in laboratory animals (reviewed by [30–32]). One of the characteristics that distinguish PFCs from other organic pollutants is their low lipophilicity; unlike organochlorine pesticides and other "classic" contaminants, PFCs do not partition into fat tissues in animals [33]. Experiments with PFOS and PFOA showed that these compounds bind to proteins in blood and liver of mammals, birds, and fish [34]. PFOS and PFOA have also been shown to bind to peroxisome proliferator-activated receptors (PPARs) [35], a family of nuclear transcription factors that plays numerous roles in development and homeostasis; PPAR-binding has been suggested as a

Table 2 Chemical groups and representative chemicals for poly- and perfluorinated compounds

Poly- and perfluorinated compound group	Representative chemical	Chemical name	Structure	CAS number
Perfluorosulfonates	Perfluorooctanesulfonate (PFOS)	1,1,2,2,3,3,4,4,5,5,6,6,7,7,8,8,8-Heptadecafluoro-1-octanesulfonic acid		307-35-7
Perfluorocarboxylates	Perfluorooctanoate (PFOA)	2,2,3,3,4,4,5,5,6,6,7,7,8,8,8-Pentadecafluorooctanoic acid		335-67-1
Perfluorosulfonamides	Perfluorooctanesulfonamide (PFOSA)	1,1,2,2,3,3,4,4,5,5,6,6,7,7,8,8,8-Heptadecafluoro-1-octanesulfonamide		754-91-6
Fluorotelomer alcohols	8:2 Fluorotelomer alcohol (8:2 FTOH)	1-Decanol, 3,3,4,4,5,5,6,6,7,7,8,8,9,9,10,10,10-heptadecafluoro-		678-39-7

(continued)

Table 2 (continued)

Poly- and perfluorinated compound group	Representative chemical	Chemical name	Structure	CAS number
Telomer acids	3:3 Fluorotelomer unsaturated carboxylic acid (3:3 FTUCA)	4,4,5,5,6,6,6-Heptafluoro-2-hexenoic acid		356-03-6

major toxic mechanism of PFOS and PFOA (e.g., [36, 37]), although induction of PPAR activity by PFCs has been found to differ among species [38].

Field studies on PFC effects are currently lacking. A recent study in Japan reported an association between PFC concentrations in livers of common cormorants (*Phalacrocorax carbo*) and gene expression related to oxidative stress and detoxification [39], while an earlier study in freshwater fish in Belgium found that liver PFOS concentrations were negatively associated with serum protein content [40]. Similar results were found in a study of PFCs in Baikal seals (*Pusa sibirica*), in which PFC concentrations were associated with PPAR induction and detoxification [41].

Although many PFCs have been detected in the environment, toxicity studies on PFCs other than PFOS and PFOA are limited. Minor effects were observed in pregnant mice exposed to perfluorodecanoate (PFDA) [42] and perfluorobutyrate (PFBA) [43], and PFDA was also reported to bind to plasma and tissue proteins of rats in a manner similar to PFOA [44]. Exposure of male rats to PFDA resulted in decreased serum testosterone activities and changes in testicular structure [45]. Several PFCs (C6–C9) were found to be relatively nontoxic to cell lines and bacteria [46]. A new class of PFCs, perfluorinated phosphonic acids (PFPAs), was recently identified in wastewater in Canada, and these chemicals were found to have similar uptake and half-lives as PFOS and PFOA when tested in rats [47, 48]; the toxicities of these compounds are not known. Another group, polyfluoroalkyl phosphate surfactants (PAPS), which are primarily used in food packaging, have been found to migrate into food from paper packaging [49]. Although the toxicity of PAPS is also unknown, male rats exposed to PAPS showed increased blood levels of FTOHs and PFCAs, likely as a result of dephosphorylation of PAPs to 8:2 FTOH and subsequent biotransformation to PFCAs [50].

Toxic effects of PFC precursors have also been observed. *In vitro* studies showed that *N*-ethyl perfluorooctanesulfonamide (*N*-EtFOSA) was converted to PFOS in rainbow trout (*Oncorhynchus mykiss*) liver microsomes [51], while several perfluorosulfonamides were found to affect mitochondrial permeability in rats [52]. FTOHs, which degrade both in the environment and as a result of metabolism to form other PFCs [53–55], were found to have estrogenic effects in a mammalian cell line [56] and in fish liver cells [57]. The toxicities of saturated and unsaturated fluorotelomer acids, which are formed in the environment through the oxidation of FTOHs, to water fleas (*Daphnia magna*), mosquito larvae (*Chironomus tentans*), and duckweed (*Lemna gibba*), were found to be up to four orders of magnitude greater than those of other PFCs of the same chain length, although the tested concentrations were still well below environmental concentrations [58]. These studies demonstrate that transformation of PFC precursor compounds via hydrolysis, photolysis, microbial degradation, or metabolism may act as an ongoing source of PFOS, PFOA, and other PFCs in the environment, despite regulatory efforts to control their production and release.

3.3 Bioaccumulation

One of the major questions regarding PFCs is whether they bioaccumulate in individual organisms and biomagnify through food chains or food webs, resulting in higher concentrations in top predators. As summarized in a recent review [33],

PFC partitioning in the environment is largely related to the chain length and functional group of individual compounds. Longer-chain (C8 or greater) PFCs have a greater tendency to bioaccumulate than PFCs containing seven or less carbon atoms, and perfluorinated sulfonates are more likely to bioaccumulate than perfluorinated carboxylates, even when their chain lengths are the same. These properties have led some PFC manufacturers to replace PFOS/PFOA with perfluorobutanesulfonate (PFBS)/PFBA or other short-chain compounds in products. However, long-chain compounds are still used for various applications, as evidenced by their detection in biological samples (reviewed by [25]), and organisms differ in their tendency to bioaccumulate PFCs. Food web studies of PFC bioaccumulation and biomagnification are limited to a few studies in the Arctic [59, 60], North America [61], and southern China [60], indicating the need for more data on the environmental behavior of PFCs in a greater diversity of ecosystems and organisms. Toxicity data are also lacking for many long- and short-chain PFCs.

4 Organofluorous Pesticides

Organofluorous compounds are applied as herbicides, fungicides, insecticides/miticides/acaricides, and rodenticides, though most OFPs are herbicides [1]. A full survey of all OFPs is beyond the scope of this chapter; instead, three top-selling pesticides, as well as two other OFPs that are environmentally relevant, will be discussed.

4.1 Compounds of Interest

In 2006, two of the top 20 best-selling pesticides in Europe were organofluorous compounds: the insecticide fipronil and the fungicide epoxiconazole [62]; in the US in 2007, the dinitroaniline pre-emergent herbicide trifluralin ranked in the top 20 and top 10 among agricultural and domestic-use pesticides, respectively [62] (Table 3). Fipronil is a phenylpyrazole that binds to and interferes with the normal function of γ-aminobutyric acid (GABA)-gated chloride channels in the neural systems of insects, leading to hyperexcitability, paralysis, and death [63]. Epoxiconazole is a triazole that acts by inhibiting sterol synthesis [1]. Trifluralin acts by inhibiting mitosis in the root cells of plants, thereby preventing plant growth.

Another group of relatively new pesticides is the pyrethroids, derived from a naturally occurring insecticide in chrysanthemums. This group was developed in response to concerns regarding older classes of pesticides, the organophosphates and carbamates, which include compounds such as chlorpyrifos and carbofuran, respectively. These older pesticides are persistent in the environment and highly

Table 3 Chemical structures and physical properties of the organofluorous pesticide compounds of interest

Compound	Chemical name	CAS number	Structure	Log K_{ow}	Water solubility (mg/L at 25 °C)	Henry's Law constant (atm m³/mol at 25 °C)
Epoxiconazole	cis-(+/−)-1-[[3-(2-Chlorophenyl)-2-(4-fluorophenyl)oxiranyl]methyl]-1H-1,2,4-triazole	133855-98-8		3.30	8.42 (20 °C)	4.65×10^{-9}
Fipronil	(5-Amino-1-[2,6-dichloro-4-(trifluromethyl)phenyl]-4-((trifluromethyl)sulfinyl]-1H-pyrazole)-3-carbonitrile	120068-37-3		4.00	1.90 (20 °C)	8.42×10^{-10}

(continued)

Table 3 (continued)

Compound	Chemical name	CAS number	Structure	Log K_{ow}	Water solubility (mg/L at 25 °C)	Henry's Law constant (atm m³/mol at 25 °C)
Bifenthrin	2-Methyl-3-phenylphenyl) methyl (1S,3S)-3-[(Z)-2-chloro-3,3,3-trifluoroprop-1-enyl]-2,2-dimethylcyclopropane-1-carboxylate	82657-04-3		6.60	0.100	1.00×10^{-6}
λ-Cyhalothrin	(1S,3S)-3-[(1Z)-2-Chloro-3,3,3-trifluoro-1-propenyl]-2,2-dimethylcyclopropanecarboxylic acid (R)-cyano (3-phenoxyphenyl)methyl ester	76703-62-3		4.96	0.002 (20 °C)	3.19×10^{-2} (20 °C)

| Trifluralin | 2,6-Dinitro-N,N-dipropyl-4-(trifluoromethyl)aniline | 1582-09-8 | | 5.34 | 0.184 | 1.03×10^{-4} (20 °C) |

toxic to non-target organisms such as fish and birds. In addition, long-term use of these pesticides has resulted in the development of resistance in some crop pests. The pyrethroids have relatively short environmental half-lives and were designed to have higher selectivity for insects [64, 65]. Two fluorinated pyrethroids, the insecticides bifenthrin and λ-cyhalothrin, will be presented here. The five compounds discussed in this section are used both agriculturally and domestically; fipronil is also for used for veterinary applications.

4.2 Environmental Properties

The environmental fates of the organofluorous pesticides of interest vary according to their structures and properties. Fipronil is not very water soluble and degrades relatively rapidly in the environment into several metabolites via photolysis, hydrolysis, reduction, and oxidation, but can persist in anaerobic soils [63]. A study in trout (*O. mykiss*) reported rapid uptake followed by rapid elimination via metabolism, though a more persistent metabolite, fipronil sulfone, was formed [64]. In contrast, epoxiconazole is stable in water and under sunlight, and persists in soils, but its bioaccumulation tendencies are not well understood based on available data [65]. Trifluralin partitions into organic matrices such as sediment and biota and is not readily biodegraded in sediments and soils, and thus bioaccumulation of this compound may occur [66, 67]. Trifluralin degrades quickly via photolysis, but it is used in large enough quantities that it has been detected in the Arctic environment, indicating that long-range atmospheric transport occurs [68]; as such, trifluralin is currently under consideration for inclusion in the United Nations Protocol on Persistent Organic Pollutants (POPs) under the Convention on Long Range Transboundary Air Pollution [67]. The pyrethroids bifenthrin and λ-cyhalothrin have low water solubilities and log K_{ow} values similar to those of persistent organochlorine compounds, and they bind readily to soils and sediments. They are resistant to hydrolysis and photolysis and may persist, particularly in aquatic environments [65, 69, 70].

4.3 Toxicity

In comparison to compounds such as PFCs, limited toxicity data are available for OFPs. Recent studies reported that epoxiconazole exposure resulted in endocrine disruption in pregnant rats and their offspring [71, 72], as well as in quail (*Coturnix coturnix japonica*) [73]. Epoxiconazole is also acutely toxic to fish algae, aquatic invertebrates, and aquatic plants [65]. Fipronil is acutely toxic to fish and shrimp and some species of birds [63]. Pyrethroids are known to have neurotoxic effects in mammals consistent with their insecticidal mechanism of action (reviewed by [74]), and similar results have been reported in fish and invertebrates. Bifenthrin and λ-cyhalothrin are acutely toxic to several fish species at concentrations in the nanogram per liter range [69, 70]. Exposure of fathead minnows (*Pimephales promelas*) to commercial formulations of bifenthrin or fipronil resulted in altered swimming behavior, and effects were stronger than in fish exposed to the active ingredient alone, indicating that "inert" ingredients in the formulations contributed to their toxicity [75]. Altered swimming behavior was also observed in zebrafish (*Danio rerio*) embryos exposed to bifenthrin [76]. Trifluralin shows low toxicity in mammals and birds, but is acutely toxic to fish and invertebrates [66], and also has genotoxic effects at sublethal concentrations in fish [77].

One group of organisms that has been the focus of intensive research interest in recent years is honeybees (*Apis mellifera*), due to the global occurrence of a condition called colony collapse disorder which results in large-scale bee mortality and has negative consequences for crops that rely on bees for pollination [78]. Both fipronil and λ-cyhalothrin are known to be acutely toxic to honeybees [63, 70], but fipronil has also been found to affect behavior and olfaction at sublethal concentrations [79]. A pesticide analysis of 200–300 honeybee wax and hive pollen samples detected more than 180 compounds and metabolites, with bifenthrin detected in 13% of the wax samples and fipronil detected infrequently in both wax and pollen samples [78]. A three-year study of honeybees reported that bifenthrin exposure reduced the number of eggs laid by female bees and increased the maturation period of young bees [80]. It is also important to note that the degradation products of insecticides may retain their biological activity, e.g., as for fipronil [63]. The factors that cause colony collapse disorder remain unknown, but pesticides are currently thought to play at least a partial role.

5 Fluoroquinolone Antibiotics

Antibiotics, like pesticides, are designed and synthesized to be biologically active. The increased and widespread use of antibiotics for medical and veterinary purposes has attracted research attention in recent years due to concerns about both their direct toxicity and the secondary effects of elevated antibiotic levels in the environment, particularly as related to the development of antibiotic resistance among bacterial populations, strains, and species [81–83]. Direct toxicity is of concern due to the conserved nature of the pathways targeted by antibiotics; although bacterial systems are the primary target of antibiotic administration, the enzymes and receptors found in bacteria are also present across a wide range of organisms. Greater incidence of bacterial resistance to older antibiotic groups such as the β-lactams (penicillin and related compounds; [83]) has led to the development of newer antibiotic classes; of these, the fluoroquinolones are relevant here.

5.1 Fluoroquinolones

Fluroquinolone antibiotics have been in use since the 1970s–1980s. Their antibacterial activity is based on inhibition of the synthesis of DNA gyrase and topoisomerase, enzymes that are crucial for bacterial growth and reproduction. The quinolones as a class were developed in the 1960s by modifying the structure of nalidixic acid; addition of fluorinated moieties resulted in improved bactericidal activity and increased half-lives in the body, reducing the need for multiple daily doses. For these reasons, fluoroquinolones are widely administered around the world, and their use is projected to increase [84, 85].

5.2 Environmental Properties

Both photolysis and hydrolysis are important breakdown pathways for fluoroquinolones, with some studies reporting half-lives of a few hours to less than 1 day for these compounds. They are largely resistant to biodegradation because of their antibiotic properties, though fungal degradation may occur [85]. Based on this information, fluoroquinolones would be categorized as non-persistent in the environment. However, as is the case with many other pharmaceuticals, fluoroquinolones fall into a middle area termed "pseudo-persistence" – they are chemically non-persistent, but are used in large volumes and at high frequencies such that they occur commonly and ubiquitously in the environment; moreover, fluoroquinolones are largely unchanged after human excretion, meaning that the active compound is present in the environment [81, 85]. Four compounds, ciprofloxacin, enrofloxacin, levofloxacin, and norfloxacin, are often detected in environmental samples (Table 4).

The major source of fluoroquinolones in the environment is wastewater treatment plants (WWTPs), which reflect human usage in the area. Veterinary use, particularly in large-scale livestock facilities, may be an important source in some locations (reviewed by [86]). The degree and type of treatment used at particular WWTPs is a strong determinant of the types of pharmaceuticals that are released into the environment; WWTPs employing preliminary or primary treatment, which involves only screening or removal of solids, are often ineffective at removing antibiotics and other pharmaceuticals from wastewater when compared to those using secondary (biological) or tertiary (disinfection) treatment [87]. Fluoroquinolones have been widely detected in WWTP effluents around the world, and studies to date show that they are not readily removed by secondary treatment, likely because they are toxic to the bacteria used in the treatment process (reviewed by [87]). Sorption to solids appears to be one removal mechanism for these compounds; the use of biosolids as fertilizer may therefore represent a route for antibiotic transfer into agricultural soils, which could negatively affect important soil microbial communities [86].

Ecotoxicology of Organofluorous Compounds

Table 4 Chemical structures and properties of four fluoroquinolone antibiotics that commonly occur in the environment

Compound	Chemical name	CAS number	Structure	Log K_{ow}	Water solubility (mg/L)	Henry's Law constant (atm m^3/mol)
Ciprofloxacin	1-Cyclopropyl-6-fluoro-1,4-dihydro-4-oxo-7-(1-piperazinyl)-3-quinolinecarboxilyc acid	85721-33-1		0.28	3.00×10^4	5.09×10^{-19}
Levofloxacin	(−)-(S)-9-Fluoro-2,3-dihydro-3-methyl-10-(4-methyl-1-piperazinyl)-7-oxo-7H-pyrido[1,2,3-de]-1,4-benzoxazine-6-carboxylic acid	100986-85-4		−2.0	6.76×10^5	1.16×10^{-21}

(continued)

Table 4 (continued)

Compound	Chemical name	CAS number	Structure	Log K_{ow}	Water solubility (mg/L)	Henry's Law constant (atm m^3/mol)
Norfloxacin	1-Ethyl-6-fluoro-1,4-dihydro-4-oxo-7-(1-piperazinly)-3-quinolinecarboxylic acid	70458-96-7		−1.03	1.78×10^5	8.70×10^{-19}
Enrofloxacin	1-Cyclopropyl-6-fluoro-1,4-dihydro-4-oxo-7-(4-ethyl-1-piperazinyl)-3-quinoline carboxylic acid	93106-60-6		0.70	3.40×10^3	1.90×10^{-13}

5.3 Toxicity

The inherent biological activity of antibiotics and other pharmaceuticals is the primary concern associated with their release into the environment, but there is a lack of comprehensive toxicity data for most pharmaceuticals, including fluoroquinolones. A computational assessment of the acute toxicity of a large group of pharmaceuticals found little risk associated with their use, but the authors state that current methods are not suitable for antibiotics whose toxic mechanisms in animals are not well understood [88]. Fluoroquinolones may present a toxicity hazard to plants by inhibiting DNA gyrase as they do in bacteria, an effect that is particularly important because it can interfere with chloroplast replication, reducing or eliminating photosynthesis; this group of antibiotics has been reported to be the most phytotoxic [89]. However, pharmaceuticals may have toxic effects unrelated to their mechanisms of action, making it difficult to predict their impact in the environment [90].

6 Uncertainties and Needs

Scientific knowledge of the ecotoxicological properties of organofluorous compounds has increased over the last 30 years. However, there is still a need for more information on the environmental fate, transport, transformation, and toxicity of these compounds, particularly in light of their potential environmental persistence. As the case of CFCs and global climate change makes clear, emerging environmental problems may require reevaluation of so-called "legacy" contaminants that are no longer produced but still occur in the environment. Larger, more reliable datasets for "legacy" contaminants will facilitate risk assessment efforts when new environmental issues arise.

Another major uncertainty or challenge in ecotoxicology is understanding mixture effects. Organisms in the environment are exposed to hundreds if not thousands of chemicals at once, often at very low levels. Therefore, when considering the toxicity of PFCs, pesticides, or antibiotics, it is important to assess how these compounds may interact with one another and how these interactions may affect their toxicities [90]. Laboratory studies have typically focused on single compounds, a design which produces clear data for that particular chemical, but does not provide useful information for environmental risk assessments that are undertaken in complex ecological conditions. For example, it is perhaps not surprising that honeybees are affected by exposure to fluorinated insecticides, as these compounds are designed to be lethal to organisms very closely related to bees. However, PFOS exposure also affects bees [91], indicating the importance of considering the totality of chemical exposure of organisms in the environment. Fortunately, studies testing the effects of single- or multi-class chemical mixtures are becoming more common, and studies are also incorporating other stressors, such as increased temperature, UV radiation, or acidity. Data from these studies can be integrated with traditional toxicity values to assess more completely the ecotoxicity of organofluorous and other compounds.

Lastly, it is a maxim of analytical chemistry that "you find what you're looking for." As international regulations restrict the use of some organofluorous compounds, e.g., CFCs and PFOS, it is necessary to determine the environmental occurrence and properties, including toxicity, of compounds taking their place in products. One limitation on these types of determinations is the lack of analytical standards available for many compounds, even for relatively well-studied compounds such as the PFCs [92]. Furthermore, a recent series of studies has quantified the contribution of organic fluorine to total fluorine found in various biological samples based on a general extraction method, and the authors report that only a small percentage – sometimes less than 1% – of the total fluorine fraction can be extracted, with an even smaller percentage consisting of known PFCs (e.g., [60, 93, 94]). These results indicate that there are large quantities of organic fluorine compounds present in organisms or in the environment that have yet to be identified, and raise many interesting questions about the structures, properties, and sources of these unknown compounds. Development of new extraction and analytical methods will contribute to the current understanding of organofluorous compounds in the environment, and guide future ecotoxicological assessments.

Acknowledgments The authors would like to thank Ping Lam for her help in the preparation of the tables for this chapter.

References

1. Jeschke P (2004) The unique role of fluorine in the design of active ingredients for modern crop protection. ChemBioChem 5(5):571–589
2. O'Hagan D (2008) Understanding organofluorine chemistry. An introduction to the C-F bond. Chem Soc Rev 37(2):308–319
3. Murphy CD, Clark BR, Amadio J (2009) Metabolism of fluoroorganic compounds in microorganisms: impacts for the environment and the production of fine chemicals. Appl Microbiol Biotechnol 84(4):617–629
4. Biffinger JC, Kim HW, DiMagno SG (2004) The polar hydrophobicity of fluorinated compounds. ChemBioChem 5(5):622–627
5. Böhm HJ, Banner D, Bendels S, Kansy M, Kuhn B, Müller K, Obst-Sander U, Stahl M (2004) Fluorine in medicinal chemistry. ChemBioChem 5(5):637–643
6. United Nations Environment Program (UNEP) (2006) Handbook for the international treaties for the protection of the ozone layer. http://ozone.unep.org/Publications/MP_Handbook/. Accessed July 3 2011
7. Rowland FS (2006) Stratospheric ozone depletion. Philos Trans R Soc Lond B Biol Sci 361 (1469):769–790
8. Forster P, Ramaswamy V, Artaxo P, Berntsen T, Betts R, Fahey DW, Haywood J, Lean J, Lowe DC, Myhre G, Nganga J, Prinn R, Raga G, Schulz M, Van Dorland R (2007) Changes in atmospheric constituents and in radiative forcing. In: Solomon S, Qin D, Manning M, Chen Z, Marquis M, Averyt KB, Tignor M, Miller HL (eds) Climate change 2007: the physical science

basis. Cambridge University Press, Cambridge, Contribution of Working Group I to the Fourth Assessment Report of the Intergovernmental Panel on Climate Change
9. Caldwell MM, Bornman JF, Ballaré CL, Flint SD, Kulandaivelu G (2007) Terrestrial ecosystems, increased solar ultraviolet radiation, and interactions with other climate change factors. Photochem Photobiol Sci 6(3):252–266
10. Bancroft BA, Baker NJ, Blaustein AR (2007) Effects of UVB radiation on marine and freshwater organisms: a synthesis through meta-analysis. Ecol Lett 10(4):332–345
11. Ballaré CL, Caldwell MM, Flint SD, Robinson SA, Bornman JF (2011) Effects of solar ultraviolet radiation on terrestrial ecosystems. Patterns, mechanisms, and interactions with climate change. Photochem Photobiol Sci 10(2):226–241
12. Norval M, Lucas RM, Cullen AP, de Gruijl FR, Longstreth J, Takizawa Y, van der Leun JC (2011) The human health effects of ozone depletion and interactions with climate change. Photochem Photobiol Sci 10(2):199–225
13. Dekant W (1996) Toxicology of chlorofluorocarbon replacements. Environ Health Perspect 104(Suppl 1):75–83
14. Wallington TJ, Schneider WF, Worsnop DR, Nielsen OJ, Sehested J, Debruyn WJ, Shorter JA (1994) The environmental impact of CFC replacements—HFCs and HCFCs. Environ Sci Technol 28(7):320A–325A
15. Tsai WT (2005) An overview of environmental hazards and exposure risk of hydrofluorocarbons (HFCs). Chemosphere 61(11):1539–1547
16. Tsai WT (2007) Environmental risk assessment of hydrofluoroethers (HFEs). J Hazard Mater 119(1–3):69–78
17. Hanson ML, Solomon KR (2004) Haloacetic acids in the aquatic environment. Part II: ecological risk assessment. Environ Pollut 130(3):385–401
18. Häder DP, Helbling EW, Williamson CE, Worrest RC (2011) Effects of UV radiation on aquatic ecosystems and interactions with climate change. Photochem Photobiol Sci 10 (2):242–260
19. Lewandowski G, Meissner E, Milchert E (2006) Special applications of fluorinated organic compounds. J Hazard Mater 136(3):385–391
20. Giesy JP, Kannan K (2001) Global distribution of perfluorooctane sulfonate in wildlife. Environ Sci Technol 35:1339–1342
21. Wei S, Chen LQ, Taniyasu S, So MK, Murphy MB, Yamashita N, Yeung LWY, Lam PKS (2007) Distribution of perfluorinated compounds in surface seawaters between Asia and Antarctica. Mar Pollut Bull 54:1813–1818
22. Ahrens L, Barber JL, Xie Z, Ebinghaus R (2009) Longitudinal and latitudinal distribution of perfluoroalkyl compounds in the surface water of the Atlantic Ocean. Environ Sci Technol 43:3122–3127
23. Higgins CP, Field JA, Criddle CS, Luthy RG (2005) Quantitative determination of perfluorochemicals in sediments and domestic sludge. Environ Sci Technol 39:3946–3956
24. Nakata H, Kannan K, Nasu T, Cho HS, Sinclair E, Takemura A (2006) Perfluorinated contaminants in sediments and aquatic organisms collected from shallow water and tidal flat areas of the Ariake Sea, Japan: environmental fate of perfluorooctane sulfonate in aquatic ecosystems. Environ Sci Technol 40:4916–4921
25. Houde M, De Silva AO, Muir DC, Letcher RJ (2011) Monitoring of perfluorinated compounds in aquatic biota: an updated review. Environ Sci Technol. dx.doi.org/10.1021/es104326w
26. Wang T, Wang Y, Liao C, Cai Y, Jiang G (2009) Perspectives on the inclusion of perfluorooctane sulfonate into the Stockholm convention on persistent organic pollutants. Environ Sci Technol 43:5171–5175
27. Olsen GW, Mair DC, Church TR, Ellefson ME, Reagen WK, Boyd TM, Herron RM, Medhdizadehkashi Z, Nobiletti JB, Rios JA, Butenhoff JL, Zobel LR (2008) Decline in perfluorooctanesulfonate and other polyfluoroalkyl chemicals in American Red Cross adult blood donors, 2000-2006. Environ Sci Technol 42:4989–4995
28. Renner R (2008) PFOS phaseout pays off. Environ Sci Technol 42(13):4618

29. DuPont (2009) Information on PFOA. http://www2.dupont.com/PFOA/en_US/. Accessed July 5 2011
30. Lau C, Butenhoff JL, Rogers JM (2004) The developmental toxicity of perfluoroalkyl acids and their derivatives. Toxicol Appl Pharm 198:231–241
31. Younglai EV, Wu YJ, Foster WG (2007) Reproductive toxicology of environmental toxicants: emerging issues and concerns. Curr Pharm Des 13(29):3005–3019
32. DeWitt JC, Shnyra A, Badr MZ, Loveless SE, Hoban D, Frame SR, Cunard R, Anderson SE, Meade BJ, Peden-Adams MM, Luebke RW, Luster MI (2009) Immunotoxicity of perfluorooctanoic acid and perfluorooctane sulfonate and the role of peroxisome proliferator-activated receptor alpha. Crit Rev Toxicol 39(1):76–94
33. Conder JM, Hoke RA, De Wolf W, Russell MH, Buck RC (2008) Are PFCAs bioaccumulative? A critical review and comparison with regulatory criteria and persistent lipophilic compounds. Environ Sci Technol 42(4):995–1003
34. Jones PD, Hu W, De Coen W, Newsted JL, Giesy JP (2003) Binding of perfluorinated fatty acids to serum proteins. Environ Toxicol Chem 22:2639–2649
35. Abbott BD, Wolf CJ, Schmid JE, Das KP, Zehr RD, Helfant L, Nakayama S, Lindstrom AB, Strynar MJ, Lau C (2007) Perfluorooctanoic acid-induced developmental toxicity in the mouse is dependent on expression of peroxisome proliferator-activated receptor-alpha. Toxicol Sci 98:571–581
36. Asakawa A, Toyoshima M, Harada KH, Fujimiya M, Inoue K, Koizumi A (2008) The ubiquitous environmental pollutant perfluorooctanoic acid inhibits feeding behavior via peroxisome proliferator-activated receptor-alpha. Int J Mol Med 21:439–445
37. Rosen MB, Lee JS, Ren H, Vallanat B, Liu J, Waalkes MP, Abbott BD, Lau C, Corton JC (2008) Toxicogenomic dissection of the perfluorooctanoic acid transcript profile in mouse liver: evidence for the involvement of nuclear receptors PPARα and CAR 1. Toxicol Sci 103:46–56
38. Ren H, Vallanat B, Nelson DM, Yeung LW, Guruge KS, Lam PK, Lehman-McKeeman LD, Corton JC (2009) Evidence for the involvement of xenobiotic-responsive nuclear receptors in transcriptional effects upon perfluoroalkyl acid exposure in diverse species. Reprod Toxicol 27 (3–4):266–277
39. Nakayama K, Iwata H, Tao L, Kannan K, Imoto M, Kim EY, Tashiro K, Tanabe S (2008) Potential effects of perfluorinated compounds in common cormorants from Lake Biwa, Japan: an implication from the hepatic gene expression profiles by microarray. Environ Toxicol Chem 27(11):2378–2386
40. Hoff PT, Van Campenhout K, Van de Vijver K, Covaci A, Bervoets L, Moens L, Huyskens G, Goemans G, Belpaire C, Blust R, De Coen W (2005) Perfluorooctane sulfonic acid and organohalogen pollutants in liver of three freshwater fish species in Flanders (Belgium): relationships with biochemical and organismal effects. Environ Pollut 137:324–333
41. Ishibashi H, Iwata H, Kim EY, Tao L, Kannan K, Tanabe S, Batoev VB, Petrov EA (2008) Contamination and effects of perfluorochemicals in Baikal seal (*Pusa sibirica*). 2. Molecular characterization, expression level, and transcriptional activation of peroxisome proliferator-activated receptor alpha. Environ Sci Technol 42(7):2302–2308
42. Harris MW, Birnbaum LS (1989) Developmental toxicity of perfluorodecanoic acid in C57BL/ 6 N mice. Fund Appl Toxicol 12:442–448
43. Das KP, Grey BE, Zehr RD, Wood CR, Butenhoff JL, Chang S-C, Ehresman DJ, Tan Y-M, Lau C (2008) Effects of perfluorobutyrate exposure during pregnancy in the mouse. Toxicol Sci 105:173–181
44. Vanden Heuvel JP, Kuslikis BI, Peterson RE (1992) Covalent binding of perfluorinated fatty acids to proteins in the plasma, liver and testes of rats. Chem Biol Interact 82:317–328
45. Shi Z, Zhang H, Liu Y, Xu M, Dai J (2007) Alterations in gene expression and testosterone synthesis in the testes of male rats exposed to perfluorododecanoic acid. Toxicol Sci 98:206–215

46. Mulkiewicz E, Jastorff B, Składanowski AC, Kleszczyński K, Stepnowski P (2007) Evaluation of the acute toxicity of perfluorinated carboxylic acids using eukaryotic cell lines, bacteria and enzymatic assays. Environ Toxicol Pharm 23:279–285
47. D'eon JC, Crozier PW, Furdui VI, Reiner EJ, Libelo EL, Mabury SA (2009) Perfluorinated phosphonic acids in Canadian surface waters and wastewater treatment plant effluent: discovery of a new class of perfluorinated acids. Environ Toxicol Chem 28(10):2101–2107
48. D'eon JC, Mabury SA (2010) Uptake and elimination of perfluorinated phosphonic acids in the rat. Environ Toxicol Chem 29(6):1319–1329
49. Begley TH, Hsu W, Noonan G, Diachenko G (2008) Migration of fluorochemical paper additives from food-contact paper into foods and food stimulants. Food Addit Contam 25(3):384–390
50. D'eon JC, Mabury SA (2007) Production of perfluorinated carboxylic acids (PFCAs) from the biotransformation of polyfluoroalkyl phosphate surfactants (PAPS): exploring routes of human contamination. Environ Sci Technol 41:4799–4805
51. Tomy GT, Tittlemier SA, Palace VP, Budakowski WR, Braekevelt E, Brinkworth L, Friesen K (2004) Biotransformation of N-ethyl perfluorooctanesulfonamide by rainbow trout (*Onchorhynchus mykiss*) liver microsomes. Environ Sci Technol 38:758–762
52. O'Brien TM, Wallace KB (2004) Mitochondrial permeability transition as the critical target of N-acetyl perfluorooctane sulfonamide toxicity *in vitro*. Toxicol Sci 82:333–340
53. Martin JW, Mabury SA, O'Brien PJ (2005) Metabolic products and pathways of fluorotelomer alcohols in isolated rat hepatocytes. Chem Biol Interact 155:165–180
54. Wang N, Szostek B, Buck RC, Folsom PW, Sulecki LM, Capka V, Berti WR, Gannon JT (2005) Fluorotelomer alcohol biodegradation—direct evidence that perfluorinated carbon chains breakdown. Environ Sci Technol 39:7516–7528
55. Nabb DL, Szostek B, Himmelstein MW, Mawn MP, Gargas ML, Sweeney LM, Stadler JC, Buck RC, Fasano WJ (2007) *In vitro* metabolism of 8-2 fluorotelomer alcohol: interspecies comparisons and metabolic pathway refinement. Toxicol Sci 100:333–344
56. Maras M, Vanparys C, Muylle F, Robbens J, Berger U, Barber JL, Blust R, De Coen W (2006) Estrogen-like properties of fluorotelomer alcohols as revealed by MCF-7 breast cancer cell proliferation. Environ Health Perspect 114:100–105
57. Liu C, Dua Y, Zhou B (2007) Evaluation of estrogenic activities and mechanism of action of perfluorinated chemicals determined by vitellogenin induction in primary cultured tilapia hepatocytes. Aquat Toxicol 85:267–277
58. Phillips (Macdonald) MM, Dinglasan-Panlilio MA, Mabury SA, Solomon KR, Sibley PK (2007) Fluorotelomer acids are more toxic than perfluorinated acids. Environ Sci Technol 41:7159–7163
59. Martin JW, Smithwick MM, Braune BM, Hoekstra PF, Muir DC, Mabury SA (2004) Identification of long-chain perfluorinated acids in biota from the Canadian Arctic. Environ Sci Technol 38(2):373–380
60. Tomy GT, Budakowski W, Halldorson T, Helm PA, Stern GA, Friesen K, Pepper K, Tittlemier SA, Fisk AT (2004) Fluorinated organic compounds in an eastern Arctic marine food web. Environ Sci Technol 38(24):6475–6481
61. Kannan K, Tao L, Sinclair E, Pastva SD, Jude DJ, Giesy JP (2005) Perfluorinated compounds in aquatic organisms at various trophic levels in a Great Lakes food chain. Arch Environ Contam Toxicol 48(4):559–566
62. Loi EI, Yeung LW, Taniyasu S, Lam PK, Kannan K, Yamashita N (2011) Trophic magnification of poly- and perfluorinated compounds in a subtropical food web. Environ Sci Technol 45(13):5506–5513
63. Jeschke P (2010) The unique role of halogen substituents in the design of modern agrochemicals. Pest Manag Sci 66(1):10–27
64. Grube A, Donaldson D, Kiely T, Wu L (2011) Pesticides industry sales and usage: 2006 and 2007 market estimates. US EPA. http://www.epa.gov/opp00001/pestsales/07pestsales/market_estimates2007.pdf. Accessed July 4 2011

65. Gunasekara AS, Troung T (2007) Environmental fate of fipronil. California Environmental Protection Agency. http://www.cdpr.ca.gov/docs/emon/pubs/fatememo/fipronilrev.pdf. Accessed June 28 2011
66. Sudakin DL (2006) Pyrethroid insecticides: advances and challenges in biomonitoring. Clin Toxicol (Phila) 44(1):31–37
67. Spurlock F, Lee M (2008) Synthetic pyrethroid use patterns, properties, and environmental effects. In: Gan J, Spurlock F, Hendley P, Weston DP (eds) Synthetic pyrethroids: occurrence and behavior in aquatic environments, ACS Symposium Series, vol 991. American Chemical Society, Washington, DC
68. US Environmental Protection Agency (US EPA) (1996) Reregistration eligibility decision for trifluralin. http://www.epa.gov/oppsrrd1/reregistration/REDs/0179.pdf
69. UN Economic Commission for Europe (UNECE) (2007) Trifluralin risk profile. In: UNECE convention on long-range transboundary air pollution. Geneva, Switzerland, pp 29
70. Hoferkamp L, Hermanson MH, Muir DC (2010) Current use pesticides in Arctic media; 2000-2007. Sci Total Environ 408(15):2985–2994
71. Fecko A (1999) Environmental fate of bifenthrin. California Department of Pesticide Regulation. http://www.cdpr.ca.gov/docs/emon/pubs/fatememo/bifentn.pdf. Accessed July 4 2011
72. He LM, Troiano J, Wang A, Goh K (2008) Environmental chemistry, ecotoxicity, and fate of lambda-cyhalothrin. Rev Environ Contam Toxicol 195:71–91
73. Taxvig C, Hass U, Axelstad M, Dalgaard M, Boberg J, Andeasen HR, Vinggaard AM (2007) Endocrine-disrupting activities in vivo of the fungicides tebuconazole and epoxiconazole. Toxicol Sci 100(2):464–473
74. Taxvig C, Vinggaard AM, Hass U, Axelstad M, Metzdorff S, Nellemann C (2008) Endocrine-disrupting properties in vivo of widely used azole fungicides. Int J Androl 31(2):170–177
75. Grote K, Niemann L, Selzsam B, Haider W, Gericke C, Herzler M, Chahoud I (2008) Epoxiconazole causes changes in testicular histology and sperm production in the Japanese quail (*Coturnix coturnix japonica*). Environ Toxicol Chem 27(11):2368–2374
76. Shafer TJ, Meyer DA, Crofton KM (2005) Developmental neurotoxicity of pyrethroid insecticides: critical review and future research needs. Environ Health Perspect 113(2): 123–136
77. Beggel S, Werner I, Connon RE, Geist JP (2010) Sublethal toxicity of commercial insecticide formulations and their active ingredients to larval fathead minnow (*Pimephales promelas*). Sci Total Environ 408(16):3169–3175
78. Jin M, Zhang X, Wang L, Huang C, Zhang Y, Zhao M (2009) Developmental toxicity of bifenthrin in embryo-larval stages of zebrafish. Aquat Toxicol 95(4):347–354
79. Könen S, Cavaş T (2008) Genotoxicity testing of the herbicide trifluralin and its commercial formulation Treflan using the piscine micronucleus test. Environ Mol Mutagen 49:434–438
80. Mullin CA, Frazier M, Frazier JL, Ashcraft S, Simonds R, Vanengelsdorp D, Pettis JS (2010) High levels of miticides and agrochemicals in North American apiaries: implications for honey bee health. PLoS One 5(3):e9754
81. Aliouane Y, El Hassani AK, Gary V, Armengaud C, Lambin M, Gauthier M (2009) Subchronic exposure of honeybees to sublethal doses of pesticides: effects on behavior. Environ Toxicol Chem 28:113–122
82. Dai PL, Wang Q, Sun JH, Liu F, Wang X, Wu YY, Zhou T (2010) Effects of sublethal concentrations of bifenthrin and deltamethrin on fecundity, growth, and development of the honeybee *Apis mellifera ligustica*. Environ Toxicol Chem 29:644–649
83. Daughton CG, Ternes TA (1999) Pharmaceuticals and personal care products in the environment: agents of subtle change? Environ Health Perspect 107(Suppl 6):907–938
84. Richardson BJ, Lam PK, Martin M (2005) Emerging chemicals of concern: pharmaceuticals and personal care products (PPCPs) in Asia, with particular reference to Southern China. Mar Pollut Bull 50:913–920
85. Hawkey PM (2008) Molecular epidemiology of clinically significant antibiotic resistance genes. Br J Pharmacol 153(Suppl 1):S406–S413

86. Emmerson AM, Jones AM (2003) The quinolones: decades of development and use. J Antimicrob Chemother 51(Suppl 1):13–20
87. Sukul P, Spiteller M (2007) Fluoroquinolone antibiotics in the environment. Rev Environ Contam Toxicol 191:131–162
88. Sarmah AK, Meyer MT, Boxall AB (2006) A global perspective on the use, sales, exposure pathways, occurrence, fate and effects of veterinary antibiotics (VAs) in the environment. Chemosphere 65(5):725–759
89. Le-Minh N, Khan SJ, Drewes JE, Stuetz RM (2010) Fate of antibiotics during municipal water recycling treatment processes. Water Res 44(15):4295–4323
90. Sanderson H, Johnson DJ, Wilson CJ, Brain RA, Solomon KR (2003) Probabilistic hazard assessment of environmentally occurring pharmaceuticals toxicity to fish, daphnids and algae by ECOSAR screening. Toxicol Lett 144(3):383–395
91. Brain RA, Hanson ML, Solomon KR, Brooks BW (2008) Aquatic plants exposed to pharmaceuticals: effects and risks. Rev Environ Contam Toxicol 192:67–1189
92. Jones OA, Voulvoulis N, Lester JN (2004) Potential ecological and human health risks associated with the presence of pharmaceutically active compounds in the aquatic environment. Crit Rev Toxicol 34(4):335–350
93. Mommaerts V, Hagenaars A, Meyer J, De Coen W, Swevers L, Mosallanejad H, Smagghe G (2011) Impact of a perfluorinated organic compound PFOS on the terrestrial pollinator Bombus terrestris (Insecta, Hymenoptera). Ecotoxicology 20(2):447–456
94. Lehmler HJ (2005) Synthesis of environmentally relevant fluorinated surfactants—a review. Chemosphere 58(11):1471–1496
95. Yeung LW, Miyake Y, Taniyasu S, Wang Y, Yu H, So MK, Jiang G, Wu Y, Li J, Giesy JP, Yamashita N, Lam PK (2008) Perfluorinated compounds and total and extractable organic fluorine in human blood samples from China. Environ Sci Technol 1 42(21):8140–8145
96. Yeung LW, Miyake Y, Li P, Taniyasu S, Kannan K, Guruge KS, Lam PK, Yamashita N (2009) Comparison of total fluorine, extractable organic fluorine and perfluorinated compounds in the blood of wild and pefluorooctanoate (PFOA)-exposed rats: evidence for the presence of other organofluorine compounds. Anal Chim Acta 635(1):108–114

Biology of Fluoro-Organic Compounds

Xiao-Jian Zhang, Ting-Bong Lai, and Richard Yuen-Chong Kong

Abstract Investigations on diverse aspects of fluoro-organic compounds have rapidly increased during the past decades. Because natural sources of fluoro-organic compounds are extremely rare, the industrial synthesis of fluorinated organic compounds and production of fluorinated natural product derivatives have greatly expanded in recent years because of their increasing importance in the agrochemical and pharmaceutical industries. Due to structural complexity or instability, synthetic modification is often not possible, and various biofluorination strategies have been developed in recent years for applications in the anti-cancer, anti-viral and anti-infection fields. Despite the industrial importance of fluorinated compounds, there have been serious concerns worldwide over the levels and synthetic routes of certain fluorinated organic compounds, in particular perfluorinated chemicals (PFCs). PFCs are emerging and recalcitrant pollutants which are widely distributed in the environment and have been detected in humans and wildlife globally. PFCs have been demonstrated to be potentially carcinogenic, adversely affect the neuroendocrine and immune systems, and produce neurotoxicity, heptatotoxicity and endocrine disrupting effects in vertebrate animals. Here, we provide an overview of recent advances in our understanding of the biology of various fluoro-organic compounds and perspectives for new enzymes and metabolic pathways for bioremediation of these chemicals.

Keywords Biodegradation · Defluorogenase · Environmental toxicity · Fluorinase · Human health · Perfluorinated compounds · Polyfluorinated compounds

---

X.-J. Zhang, T.-B. Lai, and R.Y.-C. Kong (✉)
Department of Biology and Chemistry, State Key Lab of Marine Pollution, City University of Hong Kong, 83 Tat Chee Avenue, Kowloon Tong, Hong Kong
e-mail: bhrkong@cityu.edu.hk

Contents

1 Introduction 366
   1.1 Natural Sources of Fluorinated Compounds 366
   1.2 Biofluorination and Fluorinase 366
   1.3 (Bio) Synthesis and Pharmaceutical Applications of Fluorinated Compounds 368
   1.4 Perfluorinated Compounds 368
2 Biodegradation of Organofluorinated Compounds 370
   2.1 Fluoroaliphatics 371
   2.2 Fluoroaromatics 374
   2.3 Biodegradation of Polyfluorinated Compounds 381
   2.4 Perspectives for the Biodegradation of Perfluorinated Compounds 387
3 Defluorination Pathways and Defluorogenases 391
   3.1 Enzymatic Metabolic Pathways 391
   3.2 Defluorinases 394
   3.3 Perspectives for New Enzymes and Metabolic Pathways 396
4 Summary and Perspectives 397
References 398

1 Introduction

1.1 Natural Sources of Fluorinated Compounds

Fluorine exists naturally in the Earth's crust and is the most abundant halogen and the 13th most abundant element. Compared with other halogens, fluorine shows very low levels in surface water and exists mainly in an insoluble form (CaF_2) in nature, and thus has very little effects on the environment and biota. More than 4,000 natural products that contain chlorine, bromine, and even iodine have been reported in living organism, whereas only about a dozen fluorinated natural products have been isolated to date [1]. Fluoroacetate was the first natural organofluorinated compound to be identified in 1943 as a metabolite from *Dichapetalum cymosum* [2]. The low bioavailability of natural fluorinated compounds and fluorine's very low concentration in surface water may be due to its largely insoluble form (CaF_2). The fluoride ion has a high heat of hydration in aqueous solution, which thus limits its participation in displacement reactions. Fluorine cannot be transformed into organic substrates by haloperoxidases (which is a family of peroxidase enzymes that mediate the oxidation of halides by hydrogen peroxide) [3].

1.2 Biofluorination and Fluorinase

Fluorine substitution is widely used in pharmaceutical and agricultural applications because of the effects of fluorine on membrane permeability, metabolic stability, and receptor-binding properties [4, 5]. Because fluorinated products are extremely rare in nature, a number of methods have been developed for synthesis of fluorinated compounds [6, 7]. However, the greatest progress has been in the generation of nonselective fluorinated products, which often cause toxicity and are difficult to handle. Selective incorporation of fluorine is challenging; therefore, development of biologically-based methods for fluorochemical production is needed.

Some enzymatic systems have been reported to utilize fluoride ions. For example, pyruvate kinase is known to catalyze the generation of fluorophosphate from ATP fluoride [8], and more recently, mutant glycosyl transferases were reported to fluorinate 2,4-dinitrophenyl-activated sugars to form α-fluoroglycosides [9, 10]. However, these reactions are adventitious or the intermediates are unstable. In 2002, the first fluorinase was reported in *Streptomyces cattleya* (O'Hagan et al. 2002), which uses *S*-adenosyl-L-methionine (SAM) and a fluoride ion as substrates to catalyze the formation of 5-fluoro-5-deoxyadenosine (5-FDA) and L-methionine (L-Met), which is the first step in the biosynthetic pathway of the fluorometabolites, fluoroacetate and 4-fluorothreonine (Fig. 1; Hagan et al. 2002). As the only native fluorination enzyme that has been identified so far, fluorinase was used to explore the syntheses of diverse fluorinated derivatives. For example, an engineered organofluorine biosynthetic metabolite that is a potent anticancer agent, fluorosalinosporamide, was produced by introducing a fluorinase gene (*flA*) into *Salinispora tropica* using recombinant DNA technology [12]. This study showed that selective fluorination of drugs and drug candidates could be expanded by inserting the *flA* gene into a variety of microorganisms to initiate the biosynthesis of novel organofluorine compounds.

In 2009, a chemo-enzymatic approach for selective fluorination was established whereby fluorine substitutions were used to produce a set of organic molecules including some prodrugs via a two-step regio- or stereo-selective procedure. The initial reaction is catalyzed by cytochrome P450 monooxygenases to insert oxygen selectively into non-reactive C–H bonds with deoxofluorination. The generated hydroxyl group was substituted by a nucleophilic fluorinating reagent, leading to selective fluorine substitution [13].

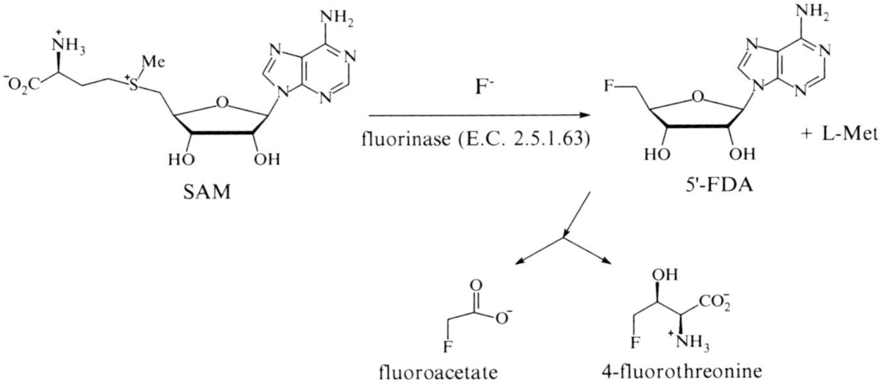

Fig. 1 The fluorinase enzyme of *S. cattleya* is the first committed step in the biosynthetic pathway to produce fluoroacetate and 4-fluorothreonine [11]

1.3 (Bio) Synthesis and Pharmaceutical Applications of Fluorinated Compounds

Burgeoning after the 1970s, the industrial synthesis of fluorinated organic compounds expanded because of their applications in pharmaceutical, agricultural, and other industrial areas. In medical applications, fluorine substitution often increases the hydrophobicity, metabolic stability, bioactivity, and bioavailability of molecules, thus improving their therapeutic indices. Medicinal production has focused on fluorinated drugs and drug candidates based on natural product analogs. While fluorinated natural products are very rare, the production of fluorinated natural product derivatives is increasingly common. Due to structural complexity or instability, synthetic modification is often not possible, and alternative strategies have been sought. In the past 20 years, synthetic methodologies in organic fluorine chemistry have focused on the biosynthesis of fluorinated analogs of natural products. Precursor-directed biosynthesis and mutasynthesis are two of the main industrial approaches for biosynthesis of fluorinated natural products. For example, fluorinated diazepinomicin analogs with modest anti-bacterial activity against *Staphylococcus aureus* have been generated through precursor-directed biosynthesis by supplementing *Micromonospora* cultures with various indole-related derivatives [14]. Using the mutasynthesis approach, auxotrophic strains of bacteria (which are unable to produce specific amino acids) have been successfully exploited to produce a number of fluorinated natural products [15]. For example, several new calcium-dependent antibiotics were produced by feeding 5-fluorotryptophan to a *Streptomyces coelicolor* tryptophan-auxotrophic strain [16].

Derivatives of anti-cancer drugs and other compounds such as the anti-inflammatory drugs fluorouracil and fluorocorticoids have been successfully biosynthesized. Other recent efforts have led to the development of fluorinated natural product derivatives, such as fluorine-substituted nucleosides, alkaloids, macrolides, steroids, amino acids, and prostaglandins, for applications in the anti-cancer, anti-viral, and anti-infection fields [15, 17]. Almost 20% of all pharmaceutical drugs on the market contain at least one fluorine atom, including the two best selling compounds, Lipitor (Atorvastatin; Fig. 2a), an inhibitor of cholesterol biosynthesis, and Advair Discus (a mixture of fluticasone (Fig. 2b) and salmeterol), a steroidal anti-inflammatory [18].

1.4 Perfluorinated Compounds

In industrial applications, fluorinated compounds, especially perfluorinated compounds perfluorooctanesulfonate (PFOS) and perfluorooctanoic acid (PFOA), play important roles in material science, including fluoropolymers, liquid crystals, and fire extinguishing products, due to their thermal and oxidative stability [19]. The phase-partitioning behavior of perfluoroalkanes makes them a prominent class of surfactants widely used in fire-fighting applications, herbicide and insecticide

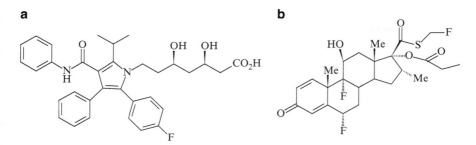

Fig. 2 Structures of the fluorine containing market leading pharmaceuticals. (**a**) Lipitor (Atorvastatin, (3*R*,5*R*)-7-[2-(4-fluorophenyl)-3-phenyl-4-(phenylcarbamoyl)-5-(propan-2-yl)-1*H*-pyrrol-1-yl]-3,5-dihydroxyheptanoic acid). (**b**) Advair Discus (a combination of fluticasone [*S*-(fluoromethyl) (6 *S*,8 *S*,9*R*,10 *S*,11 *S*,13 *S*,14 *S*,16*R*,17*R*)-6,9-difluoro-11,17-dihydroxy-10,13,16-trimethyl-3-oxo-6,7,8,11,12,14,15,16-octahydrocyclopenta [*a*] phenan-threne-17-carbothioate] and salmeterol-2-(hydroxymethyl)-4-{1-hydroxy-2-[6-(4-phenylbutoxy) hexylamino] ethyl} phenol) (O'Hagan 2010).

formulations, cosmetics, greases and lubricants, paints, polishes, and adhesives. In addition, poly/perfluorine derivatives are applied as oxygen carriers in blood substitutes [20]. Although production of many perfluorinated compounds such as PFOA and PFOS has ended in the USA and EU, these compounds are still produced in China and other developing countries.

1.4.1 Environmental Fate and Toxicity

Thousands of tons of fluorinated organic compounds have been emitted into the environment [19]. In recent years, concerns over the levels and synthetic routes of fluorinated organic compounds, especially perfluorinated compounds, have increased. Perfluorinated compounds show thermal, chemical, and biological stability, lipophilicity, worldwide distribution and accumulation in the atmosphere [21], river water [22], wildlife [22, 23], and in humans [24], which may lead to serious problems. The detection of organofluorines in wildlife and humans has been increasingly reported since 1968 [25, 26]. In 2003–2004, >99% of individuals sampled in one study in the US showed detectable PFOA in their serum [27]. In 2009, PFOS was included in Annex B of the Stockholm Convention on Persistent Organic Pollutants.

1.4.2 Fluorinated Compounds and Human Health

While fluorine is regarded as an essential element and is beneficial to human health at low concentrations, the environmental distribution of fluorinated organic compounds is dangerous to humans due to their bioaccumulation and potential impacts on metabolism. During the last two decades, concerns about the toxicity of fluorinated organic compounds, especially perfluorinated compounds, have increased.

Most toxicological studies on PFCs have been conducted on rats or monkeys. In animal research, common PFCs such as PFOA and PFOS have been demonstrated to be potentially carcinogenic, to affect the neuroendocrine and immune systems, to cause neurotoxicity and hepatotoxicity, and to reduce serum cholesterol and triglycerides [28–30]. Effects on gestational and developmental toxicity were also confirmed [31]. In vitro studies on human cells also demonstrated the toxicity of PFCs on DNA integrity, intracellular organelles, and hormones ([32]; Vanden Heuvel et al. 2006; [33]). In population studies, some PFCs were reported to act as hormone disruptors and thus to affect human fecundity [34]. Human fetal birth weight was also reported to be impaired by background exposure to PFOA [35]. Additionally, exposure to PFCs causes altered hepatic function, immune function, thyroid function, and cholesterol metabolism, and has carcinogenic potential in humans [36].

2 Biodegradation of Organofluorinated Compounds

Biodegradation is the chemical dissolution of materials by bacteria or through other biological means. Over the years, scientists and engineers have developed a number of bioremediation and biotransformation methods to degrade, transform, or accumulate a huge range of man-made contaminants. A great variety of microbes such as *Burkholderia*, *Rhodococcus*, *Pseudomonads*, *Aspergillus*, and *Beauveria* have shown an extraordinary capability to degrade artificial pollutants such as hydrocarbons (e.g., oil), polychlorinated biphenyls (PCBs), polyaromatic hydrocarbons (PAHs), heterocyclic compounds (such as pyridine or quinoline), and pharmaceutical substances [37]. Biodegradation by microorganisms is perhaps one of the most effective methods to remove organic pollutants from the environment and has attracted considerable interest in bioremediation of organofluorinated compounds.

Although fluoroorganic compounds are well known for their inertness and contain the strong C–F bond, some organisms such as bacteria, fungi, algae, and even vertebrates can still biotransform and biodegrade fluoroorganic compounds because of the steric size similarity between fluorine, hydrogen, and hydroxyl groups. To date, little is known about the bacterial metabolism of fluoroorganic compounds, even though several reports have been published on the degradation of monofluorinated compounds. In 1954, the first report on biological defluorination described fluoride elimination of *p*-fluoroaniline using a horseradish peroxidase. Fluoroaliphatics such as fluoroacetate can be degraded with monofluoroacetate dehalogenase (*Pseudomonas indoloxidans, Pseudomonas cepacia, Moraxella* sp., *Burkholderia* sp., etc.) and biodegradation of trifluoroacetic acid has also been reported [38, 39]. Fluoroaromatic compounds can be biodegraded aerobically and anaerobically. However, the biodegradation pathways of perfluorinated chemicals are still not known.

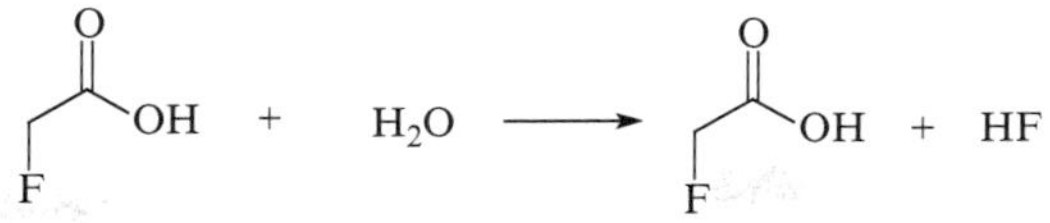

Biology of Fluoro-Organic Compounds

Fig. 3 Hydrolytic defluorination of fluoroacetate [40]

$$\text{FCH}_2\text{COOH} + \text{H}_2\text{O} \longrightarrow \text{HOCH}_2\text{COOH} + \text{HF}$$

2.1 Fluoroaliphatics

2.1.1 Fluoroacetate

Fluoroacetate is one of the most highly toxic compounds for mammals [40]. The dissociation energy of its C–F bond is among the highest found in natural products [41]. The presence of fluoroacetate in the environment and biota results from its industrial use as a vertebrate pest control agent as well as from metabolites of other compounds such as fluoroacetamide, which is used to control rodents, the anticancer drugs 5-fluorouracil and fluoroethyl nitrosourea, and the industrial chemical fluoroethanol [42].

Microbial defluorination of fluoroacetate was first reported in 1961 [43], followed by reports of the first enzymatic release of fluoride ion from fluoroacetate in both vertebrates and bacteria [44]. A wide variety of microorganisms such as *Moraxella*, *Pseudomonas*, and *Burkholderia* were isolated and shown to be capable of defluorinating fluoroacetate [39, 45]. Fluoroacetate dehalogenases have been characterized in *Pseudomonas* strains as well as other bacteria for decades (Fig. 3) [46–48]. Microbial degradation of fluoroacetate is now well understood at the mechanistic level. Two possible mechanisms were delineated from the enzyme reaction [49]. The ester intermediate pathway has been examined for fluoroacetate dehalogenases and other enzymes such as rat liver microsomal epoxide hydrolase [45, 50–52]. The carboxylate group of the aspartate residue at the active site acts as a nucleophile and first attacks the α-carbon atom of fluoroacetate to displace the fluorine atom, leading to the release of a fluoride ion. An ester intermediate is formed, which is subsequently hydrolyzed by a water molecule activated by a histidine residue, thereby regenerating the carboxylate group of the aspartate molecule [53].

2.1.2 Fluoropyruvate

Fluoropyruvate is often used in the laboratory as an inhibitor to inactivate pyruvate carboxylase, lactate dehydrogenase, and the pyruvate dehydrogenase complex [54]. In recent years, there has been increasing focus on the use of 3-halopyruvate as an anti-cancer agent because it acts as an irreversible inhibitor of metabolic enzyme(s) associated with glycolysis. For example, it has been demonstrated that 3-bromo pyruvate shows high in vivo toxicity on tumors but has no adverse effect on healthy tissue [55]. In 1978, a pyruvate dehydrogenase component of *Escherichia coli* that catalyzes the conversion of 3-fluoropyruvate to acetate and fluoride ions was reported [56]. Fluoride is eliminated by β-elimination, which is the classical

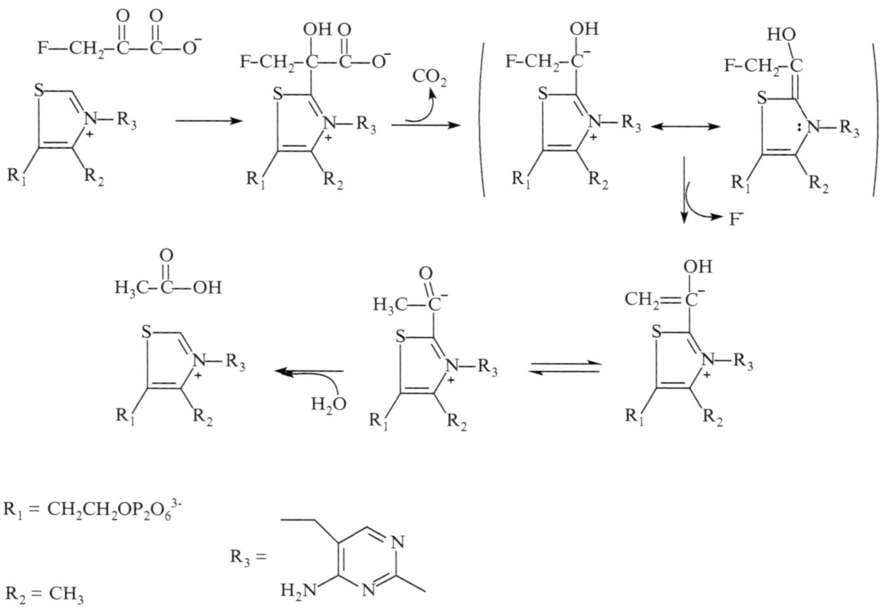

Fig. 4 Proposed enzymatic defluorination of 3-fluoropyruvate [38]

mechanism for dehydrogenases (Fig. 4). Recently, ^{19}F NMR spectroscopy studies demonstrated the conversion of fluoropyruvate to fluoroacetate by *D. cymosum*, where fluoroacetate is mineralized followed by the release of fluoride [57].

2.1.3 Maleylacetate

Fluorinated maleylacetates have been investigated as substrates of maleylacetate reductase for a number of years [58–61]. A maleylacetate reductase enzyme was first isolated in 1995 from *Pseudomonas* sp. strain B13 that catalyzes the halo-elimination of 2-fluoromaleylacetate as well as other halomaleylacetates (Fig. 5). This enzyme consumes two moles of NADH per mole of maleylacetate that contains a fluorine substituent in the 2-position, while only one mole of NADH is required for halide elimination in substrates without a fluorine substituent in the 2-position [58].

2.1.4 Fluorinated Cycloalkyl *N*-Phenylcarbamates

Fluorine substitution of a hydrocarbon position in fluorinated cycloalkyl *N*-phenylcarbamates occurs in hydroxylation reactions by *Beauveria bassiana*, a soil-borne filamentous fungus. The hydroxylation of 4-*cis*-fluorocycloalkyl

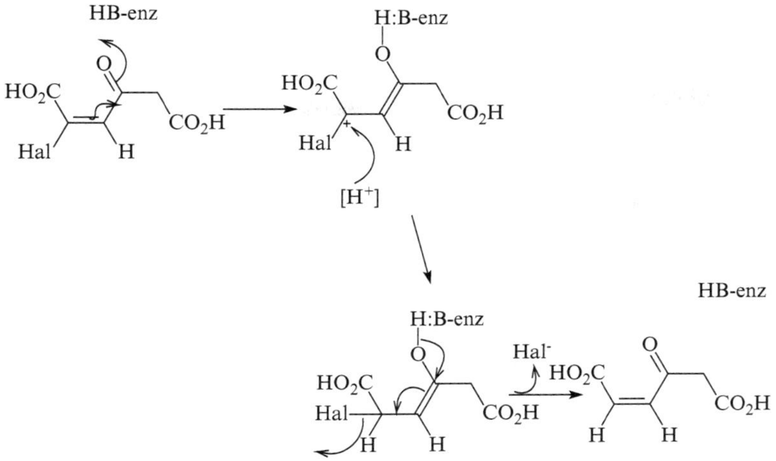

Biology of Fluoro-Organic Compounds

Fig. 5 Proposed mechanism for the elimination of halogen substituents from the 2-position of maleylacetate [58]

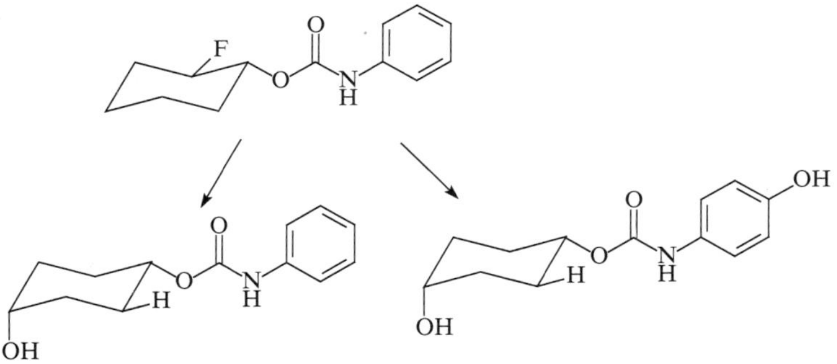

Fig. 6 Defluorination of trans-2-fluorocycloalkyl N-phenylcarbamate by *Beauveria bassiana* [63]

N-phenylcarbamates probably produces terminal fluorohydrins, which are not stable and thus are subsequently dehydrofluorinated to give the corresponding ketones [62]. Recently, the defluorination of *trans*-2-fluorocycloalkyl N-phenylcarbamate by *B. bassiana* was also reported, in which fluorine elimination could occur either via hydroxylation of the six member ring at C-4 or p-hydroxylation of the aromatic ring (Fig. 6).

2.1.5 Fluorinated Carbohydrates

Fluorinated carbohydrates have a broad range of pharmaceutical and biomedical applications ranging from metabolic and biochemical studies to disease diagnoses. Replacement of a hydroxyl group with a fluorine atom in carbohydrates can affect their metabolic and biochemical behavior, including enzyme-carbohydrate interactions, lectin-carbohydrate affinities, antibody-carbohydrate binding [64, 65], and application in positron emission tomography for cancer diagnosis [66]. Therefore, fluorinated compounds are important reagents in metabolic studies and for disease diagnoses. The microbial catalytic defluorination of fluoromonosaccharides has been reported [67, 68]. Expression of a 65.5 kDa membrane protein is induced by 4-deoxy-4-fluoro-D-glucose (4-FG) or glucose and is associated with the active D-glucose transporter system in *Pseudomonas putida* [68]. *P. putida* defluorination of fluoro-D-glucose is stereospecific. In addition, 4-FG is converted to 2,3-dideoxy-D-glycero-pentonic acid with fluoride elimination while 3-deoxy-3-fluoro-D-glucose (3-FG) is metabolized without defluorination. Electron donors such as L-malate are required in these defluorination metabolic pathways [69].

2.2 Fluoroaromatics

Fluoroaromatics are widely used in industry as intermediates or end-products in the synthesis of pharmaceuticals, insecticides, plastics, and molecules related to liquid crystal technology [15, 70]. The broad applications of fluoroaromatics have led to their accumulation in the environment. Their widespread occurrence and potential toxicity have led to increasing interest in biodegradation and treatment processes for fluoroaromatics.

2.2.1 Fluorobenzoates

As model compounds of other fluoro-substituted aromatic compounds, fluorobenzoates have been widely used to study bacterial metabolism of fluorinated aromatics. For example, bacteria such as *Pseudomonas* [71, 72], *Xanthobacter* [73], and *Sphingomonas* [74] have been reported to exhibit fluorobenzoate degradation. In addition, the metabolism of 2-, 3-, and 4-fluorobenzoic acid has been well studied [71, 74, 75]. Using $^{18}O_2$, *Pseudomonas* sp. was shown to form catechol from 2-fluorobenzoic acid by incorporation of two oxygen atoms from a single dioxygen molecule. This defluorination proceeds through a cyclic peroxide intermediate. In the major pathway, 1,2-dioxygenation of 2-fluorobenzoic acid leads to an unstable fluorohydrin, which is then defluorinated to catechol. Muconate is finally formed, which subsequently goes in the TCA cycle to produce energy (Fig. 7, pathway a). The minor pathway, 1,6-dioxygenation, also takes place, leading to the formation of 3-fluorocatechol and then 2-fluoro-*cis-cis*-muconate (Fig. 7, pathway b) [75]. 3-Fluorobenzoate is degraded by 1,2-dioxygenation to yield fluorocatechol, which is metabolized to 2-fluorobenzoic acid in the minor pathway (Fig. 7, pathway c) [74, 75]. The predominant pathway of 3-fluorobenzoate includes a 1,6-dioxygenation reaction to yield fluoromuconic acids. Defluorination then occurs to yield muconate [74] (Fig. 7, pathway d).

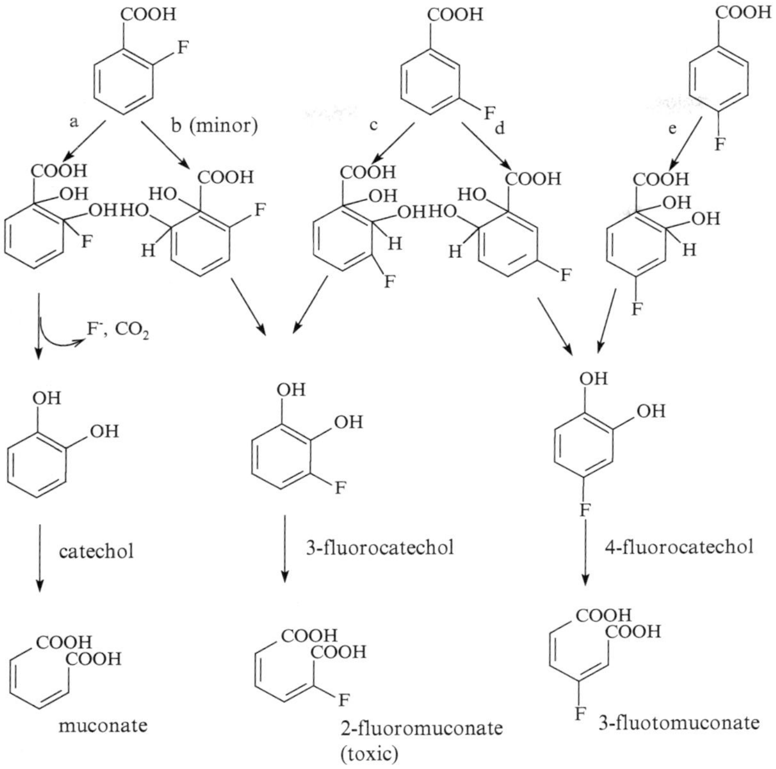

Biology of Fluoro-Organic Compounds

Fig. 7 Metabolism of 2-, 3-, and 4-fluorobenzoic acid [38]

4-Fluorobenzoate is degraded by *Pseudomonas* sp. in similar pathways to 3-fluorobenzoate (Fig. 7, pathway e) [75, 76].

The anaerobic degradation of monofluorobenzoates under various electron-accepting conditions including denitrifying, sulfate-reducing, iron-reducing, and methanogenic conditions has also been studied [77–80]. After long-term incubation, 2-fluoro- and 4-fluorobenzoates are degraded by *Pseudomonas* with fluoride elimination [79]. Recently, dehalogenated 3-fluorobenzoate was investigated in *Syntrophus aciditrophicus* culture. Two hydrogen atoms are added to 3-fluorobenzoate to form a 3-fluorocyclohexadiene metabolite, leading to stoichiometric accumulation of benzoate and fluorine [80].

2.2.2 Fluorophenols

Fluorophenolic compounds are widely used in agricultural industries as herbicides, insecticides, and fungicides [81]. Fluorophenols are transferred to fluorocatechols and fluoromuconates via microbial degradation [82]. The fluorophenol

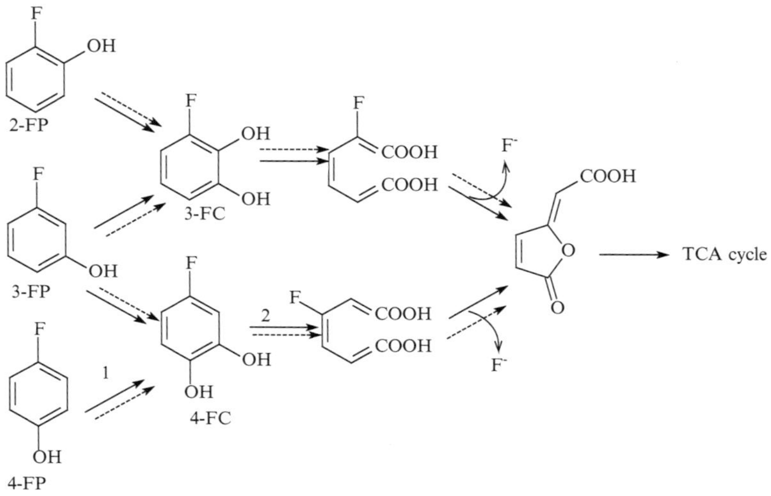

Fig. 8 Pathways for defluorination of fluorinated phenols by *P. benzenivorans*. *Dashed arrows* show the known monofluorophenols pathway for comparison. *1*: phenol hydroxylase, *2*: catechol 1,2-dioxygenase [83]

metabolites of *Exophiala jeanselmei*, a yeast-like fungus, which are converted by the phenol hydroxylase and catechol 1,2-dioxygenase enzymes, have been characterized by ^{19}F NMR spectroscopy. The conversion of fluorophenols (i.e., 3-fluoro-, 4-fluoro-, and 3,4-difluorophenol) via catechol 1,2-dioxygenase involves two common steps [81, 83]: (1) the introduction of *ortho*-hydroxyl groups and (2) ring cleavage by catechol dioxygenase. The resulting muconates and accumulation of stoichiometric amounts of fluoride anions have been detected (Fig. 8).

2.2.3 Fluorotoluene

3-Fluorotoluene was reported to be accumulated and co-metabolized by *Cladosporium sphaerospermum*, a fungi culture grown on toluene [84]. ^{19}F NMR was used to determine the catabolic pathway. A methyl group is first oxidized by the toluene monooxygenase enzyme followed by ring hydroxylation to form fluoroprotocatechuate. The remaining steps include decarboxylation of the fluoroprotocatechuate followed by *ortho*-cleavage (Fig. 9).

2.2.4 Fluorobiphenyls

Fluorobiphenyls can be co-metabolized via the classical aromatic degradation pathways by fungi and bacteria [85–87]. Recently, the degradation pathway of 4,4-difluorobiphenyl was proposed. The hydrolase BphD catalyzes the transformation from 3-fluoro-2-hydroxy-6-oxo-6-(4-fluorophenyl)-hexa-2,4-dienoate to 3-fluoro-2-hydroxypenta-2,4-dienoate. Then, (Z)-3-fluoro-2-oxo-pent-3-enoate is

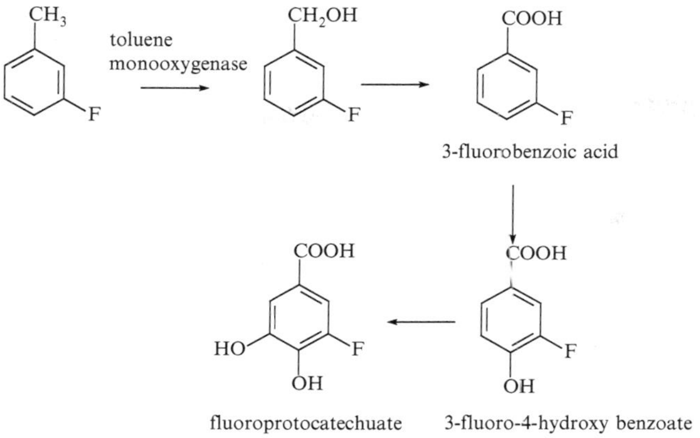

Biology of Fluoro-Organic Compounds 377

Fig. 9 Proposed fungal catabolism of fluorotoluene [38]

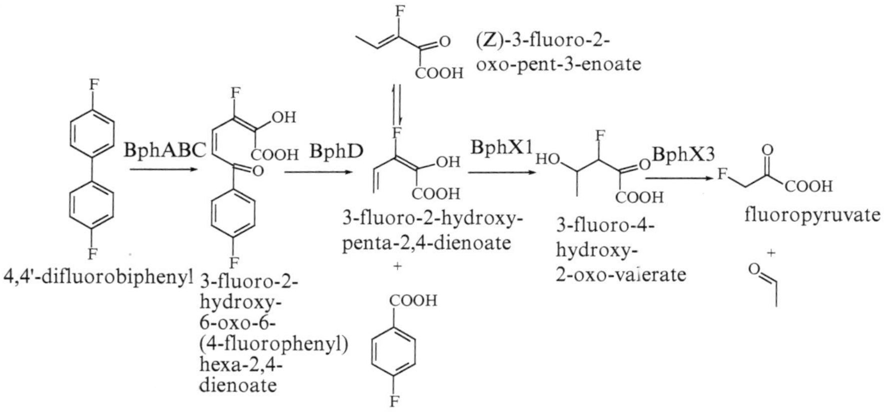

Fig. 10 Proposed catabolism of 4,4-difluorobiphenyl along the upper and lower pathways. *BphA* biphenyl 2,3-dioxygenase; *BphB* dehydrogenase; *BphC* 2,3-dihydroxybiphenyl 1,2-dioxygenase; *BphD* 2-hydroxy-6-oxo-6-phenylhexa-2,4-dienoate hydrolase; *BphX1* 2-hydroxypenta-2,4-dienoate hydratase; *BphX3* 4-hydroxy-2-oxovalerate hydrolase [88]

formed and further catabolized, eventually yielding acetaldehyde and fluoropyruvate (Fig. 10) [88].

2.2.5 Fluorophenylacetic Acid

The defluorination of *p*-fluorophenylacetic acid by *Pseudomonas* has been studied [76]. First, the aromatic ring is cleaved between C-1 and C-2. Then, C-2 is further modified by two alternative pathways. Hydrolyzation occurs to give 3-hydroxy-3-fluoroadipic acid. Fluorine elimination occurs and yields β-ketoadipic acid (Fig. 11, pathway a). Alternatively, after lactonization and formation of 4-carboxymethyl-

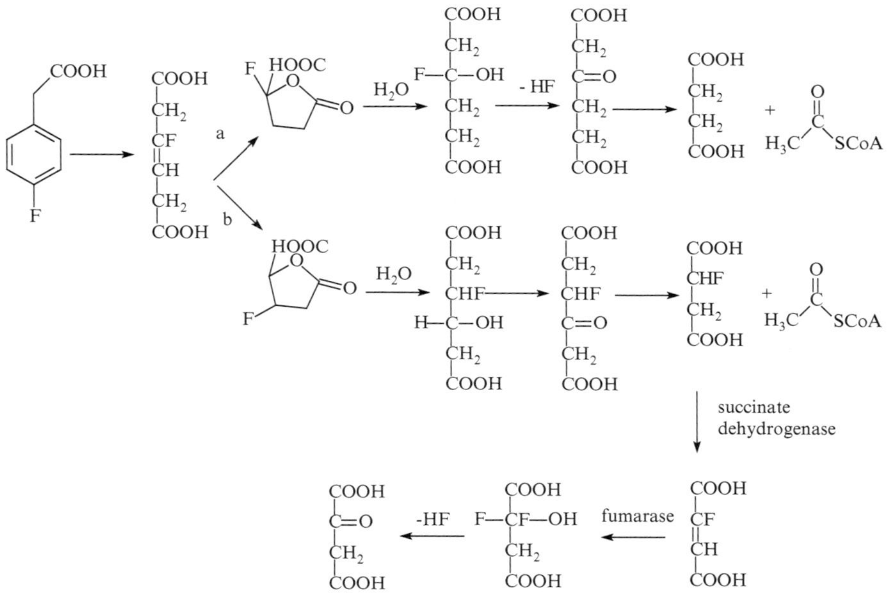

Fig. 11 Degradative pathways for 3-fluoro-3-hexenedioic acid [76]

3-fluoro-butanolide, hydrolyzation and cleavage of C–C bonds yield acetate and monofluorosuccinic acid (Fig. 11, pathway b). The latter compound is converted to oxaloacetate and hydrogen fluoride.

2.2.6 Fluorobenzene

A microbial consortium containing *Sphingobacterium*, *Flavobacterium*, and *β-Proteobacterium* was shown by Carvalho et al. in 2002 [89] to be capable of defluorinating fluorobenzene. In addition, a bacterial strain from the *Labrys portucalensis* group that uses fluorobenzene as a sole carbon and energy source has been purified [90]. The degradation of fluorobenzene via *ortho* cleavage of 4-fluorocatechol and catechol by *Rhizobiales* strain F11 has been investigated by Carvalho et al. in 2006 [91]. It was found that the initial attack on fluorobenzene by a dioxygenase enzyme could lead to two different pathways. In one pathway, a dihydrodiol dehydrogenase enzyme (step 1) transforms 4-fluoro-*cis*-benzene-1, 2-dihydrodiol to 4-fluorocatechol. In the second pathway (step 10), 1-fluoro-*cis*-benzene-1,2-dihydrodiol is converted to catechol (Fig. 12).

2.2.7 Fluoroquinolones

Fluoroquinolones are some of the most widely used antimicrobial agents for treating both Gram-negative and Gram-positive infections. Their widespread

Biology of Fluoro-Organic Compounds

Fig. 12 Proposed pathway for fluorobenzene metabolism. The enzyme activities are denoted as follows: *1*: fluorobenzene dioxygenase; *2*: fluorobenzene dihydrodiol dehydrogenase; *3*: fluoro-catechol 1,2-dioxygenase; *4*: fluoromuconate cycloisomerase; *5* and *6*: possible side reactions to *cis*-dienelactone by fluoromuconate cycloisomerase (activity 5) or by slow spontaneous conversion (activity 6); *7*: *trans*-dienelactone hydrolase; *8*: maleylacetate reductase; *9*: fluorobenzene dioxygenase; *10*: nonenzymatic defluorination; *11*: catechol-1,2-dioxygenase; *12*: muconate cycloisomerase; *13*: muconolactone isomerase; *14*: 3-oxoadipate enol-lactone hydrolase [91]

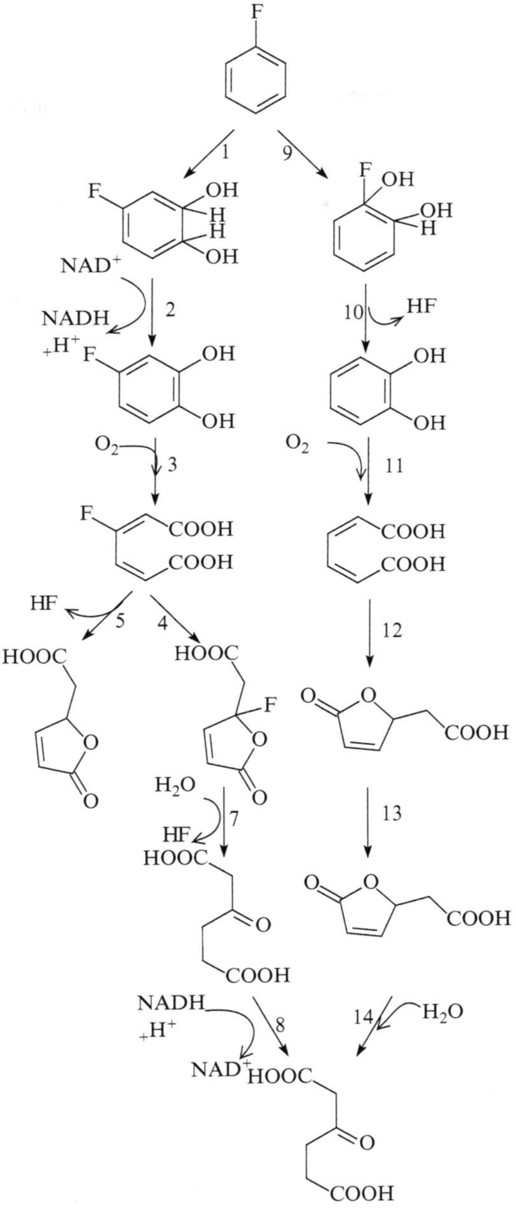

presence has been detected at multiple locations around the world [92]. Other reports have suggested their potential toxicity to plants and aquatic organisms [93, 94]. Many clinically relevant bacterial species including *S. aureus* and *Pseudomonas aeruginosa* are capable of developing resistance to quinolones [95].

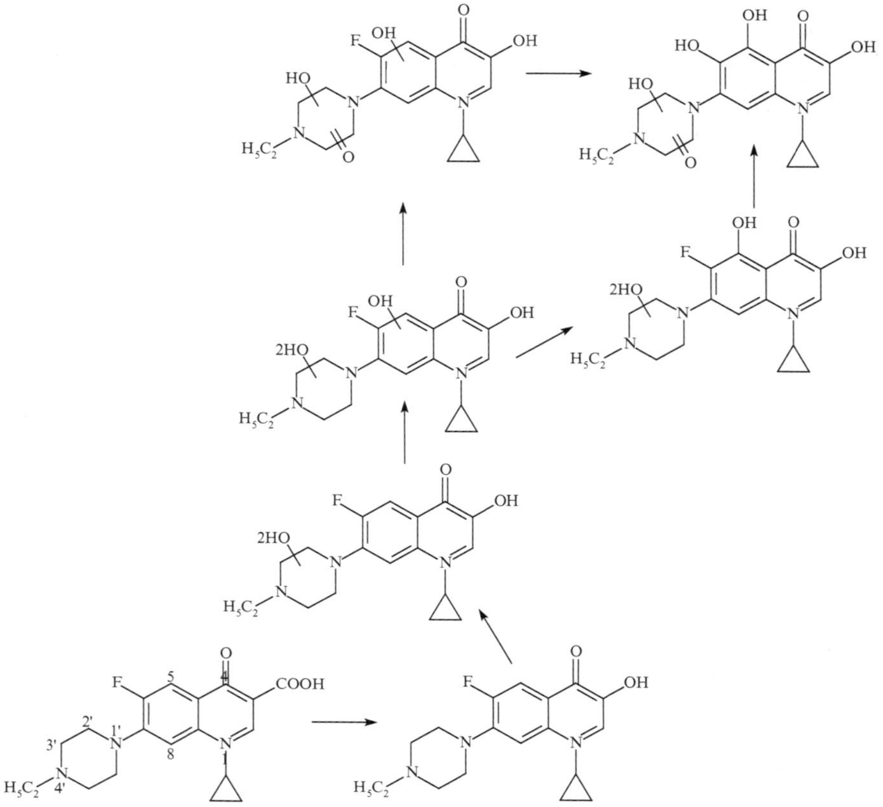

Fig. 13 Degradation of enrofloxacin by basidiomycetous fungi [97]

Degradation of the fluoroquinolone, enrofloxacin, was observed in *Gloeophyllum striatum*, a brown rot fungus where a hydroxyl radical attacks fluorine at the C-6 position to form 6-hydroxyen-rofloxacin which is further hydroxylated to 5,6- and 6,8-dihydrox-yenrofloxacin [96]. The metabolism of enrofloxacin by seven basidiomycetous fungi from agricultural sites was recently reported by Wetzstein et al. [97]. Oxidative decarboxylation of enrofloxacin first occurs, then defluorination takes place in multiply hydroxylation and acetylation steps (Fig 13) [97].

2.2.8 Fluorinated Anilines

Microsomal NADPH-dependent reaction pathways for biodehalogenation of fluorinated anilines have been investigated [98]. Three possible pathways for dehalogenation of fluorinated anilines, such as 2-fluoro-4-hydroxyaniline and pentafluoroaniline, in the presence of xanthine glutathione and NADPH were

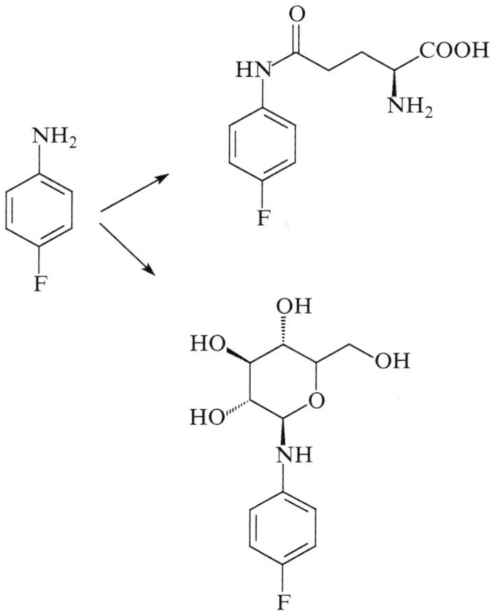

Biology of Fluoro-Organic Compounds

Fig. 14 Metabolism of 4-fluoroaniline in *Eisenia veneta* [100]

proposed. A study of the metabolism of 3,4-difluoroaniline with *Pseudomonas fluorescens* 26-K showed the formation of 3-fluoro-4-hydroxyaniline and the release of a fluoride ion [99]. Recently, biotransformation of 4-fluoroaniline was observed in the earthworm *Eisenia veneta*. The catabolic products were analyzed using 19-F NMR, but no fluoride ion was detected (Fig. 14) [100, 101].

2.3 Biodegradation of Polyfluorinated Compounds

The degradation of polyfluorinated compounds, such as fluorotelomer alcohols (FTOHs), fluorotelomer ethoxylates, and polyfluoroalkyl phosphates, in atmospheric and aqueous systems has been established and has been reported to be a source of perfluorinated carboxylic acids (PFCAs). However, published information on the biodegradation of PFCAs is very limited. The aerobic and anaerobic biodegradability of three fluorinated surfactants have been described [102]. However, no release of fluoride has been found.

2.3.1 Fluorotelomer Alcohols

FTOH is the generic name of fluorinated compounds that contain even-numbered fluorocarbon chains and an ethanol moiety [103]. FTOHs are used in fire-fighting foams, grease-resistant food packaging, leather protectants, and stain-resistant carpeting and textiles. In addition, FTOHs are used industrially to generate acrylate

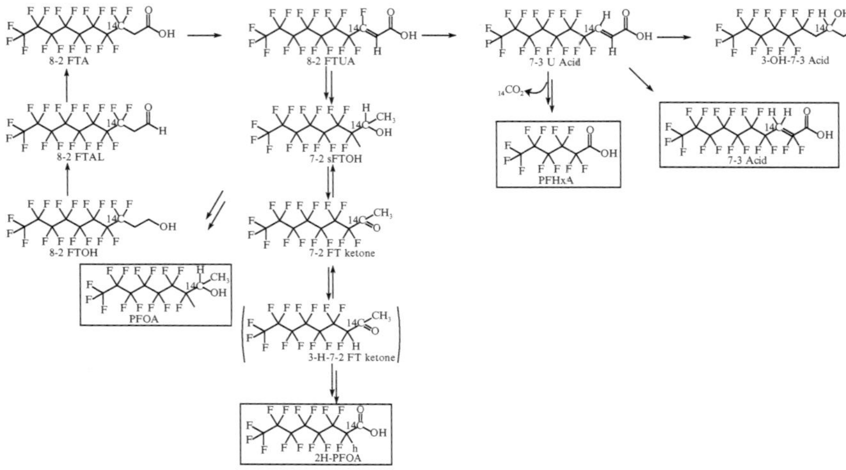

Fig. 15 Proposed biodegradation pathways of 8–2 FTOH [109]

polymers and as intermediates in the production of fluorinated surfactants. Consequently, FTOHs are widely detected in air. Furthermore, estrogen-like properties have been reported for these compounds [104].

8–2 FTOH degradation was first reported in detail in reactions catalyzed by a mixed microbial consortium [105–109]. Based on ^{14}C analysis, 8–2 FTOH biodegradation in aerobic soils was proposed (Fig. 15). 8–2 FTOH is converted rapidly to 8–2 fluorotelomer aldehyde (FTAL) by an alcohol dehydrogenase and to 8–2 fluorotelomer acid (8–2 FTA) by an aldehyde dehydrogenase. The conversion of 8–2 FTA to 8–2 fluorotelomer unsaturated acid (8–2 FTUA) in soils is so rapid that no 8–2 FTA above the limit of quantification was observed.

Recently, the first study to investigate aerobic biodegradation of 6–2 FTOH [F(CF$_2$)$_6$CH$_2$CH$_2$OH] was described by Liu et al. [110]. Based on this investigation and previous studies on the mechanism of 8–2 FTOH biodegradation [107–109, 111, 112], several pathways for 6–2 FTOH degradation have been proposed. 6–2 FTOH is first converted to 6–2 FTAL through oxidation by alcohol dehydrogenase or cytochrome P450, and then to 6–2 FTA by aldehyde dehydrogenase. Using the 2,4-dinitrophenylhydrazine (DNPH) derivatization method previously described for the detection of 8–2 FTAL from 8–2 FTOH degradation in soil and mammals [108, 109], 6–2 FTAL was not detected in the soil extracts. Hydrogen fluoride (HF) is removed from 6–2 FTA to form 6–2 FTUA either because α-oxidation is not operable or because rapid HF elimination to 6–2 FTUA supersedes 6–2 FTA α-hydroxylation, which is necessary for α-oxidation (Martin et al. 2005). 6–2 FTUA degradation proceeds by two pathways (Fig. 16).

Biodegradation of a novel fluorotelomer alcohol, 1H,1H,2H,2H,8H,8H-perfluorododecanol (degradable telomer fluoroalcohol, DTFA), was investigated in a mixed bacterial culture obtained from activated sludge and the pathway was also proposed (Fig. 17) [103], First, through the catalytic activity of alcohol dehydrogenase and

Biology of Fluoro-Organic Compounds 383

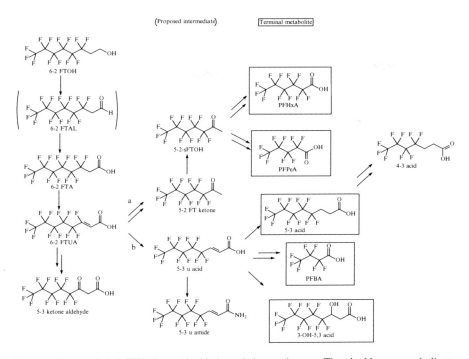

Fig. 16 Proposed 6–2 FTOH aerobic biodegradation pathways. The *double arrows* indicate multiple transformation steps [110]

aldehyde dehydrogenase, DTFA is oxidized to 2*H*,2*H*,8*H*,8*H*-perfluorododecanoic acid (2*H*,2*H*,8*H*,8*H*-PFDoA) which is then defluorinated to 2*H*,8*H*,8*H*-2-perfluorododecenoic acid (2*H*,8*H*,8*H*-2-PFUDoA). Double bonds are formed between the internal –CH2– and –CF2– groups in 2*H*,8*H*,8*H*-2-PFUDoA which is then further degraded via two different β-oxidation pathways. In pathway a, through the removal of –CF2–, 2*H*,8*H*-2,8-PFUDoA is transformed into three different long-chain carboxylic acids which are further degraded into perfluorobutanoic acid (PFBA) with dicarboxylic acids containing different fluorocarbon lengths (C4–C6 compounds), whereas in pathway b, 2*H*,8*H*-2,8-PFUDoA is transformed into perfluoropentanoic acid (PFPeA) with three different fluorinated dicarboxylic acids (Fig. 17).

2.3.2 Fluorotelomer Ethoxylates

Fluorotelomer ethoxylates [F–(CF$_2$–CF$_2$–)$_x$–(CH$_2$–CH$_2$–O)$_y$–H] are an important class of non-ionic fluorinated surfactants and are regarded as a potential source of per- and polyfluorinated organic pollutants. Aerobic biotransformation of FTEOs was recently demonstrated by Frömel and Knepper [113]. Distinct from the biodegradation of FTOHs, ω-oxidation occurs and is responsible for the transformation of FTEOs to FTEO carboxylates (FTEOCs). After oxidation of the terminal

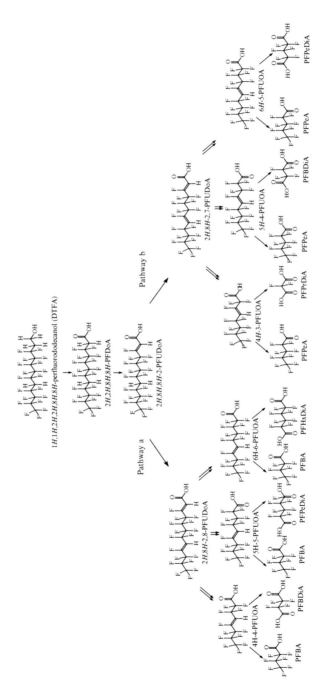

Fig. 17 Proposed biodegradation pathway for DTFA. The *double arrows* indicate multiple transformation steps [103]

hydroxyl group to a carboxylic acid, the carbon chain is subsequently shortened whereby the short-chained FTEOCs are not further degraded. In this chapter, no PFCA formation attributable to FTEO degradation was observed.

2.3.3 Fluorotelomer-Based Urethanes

Fluorotelomer-based urethanes are urethane polymers consisting of a series of fluorotelomer side-chains attached to a hydrocarbon backbone and are commercially used as stains and soil repellents for textiles. Russell et al. [114] tested the potential for microbial activities in four different soil samples to degrade a fluorotelomer-based urethane polymer under aerobic conditions over a 24-month period and demonstrated that fluorotelomer side-chains were released and transformed to perfluorocarboxylic acids including PFOA.

2.3.4 Polyfluoroalkyl Phosphates

Polyfluoroalkyl phosphates (PAPs) are used as commercial surfactants for oil repelling applications, and have been shown to be degraded to PFCAs in a rat model and waste water treatment plant system [115]. The pathway in Fig. 18

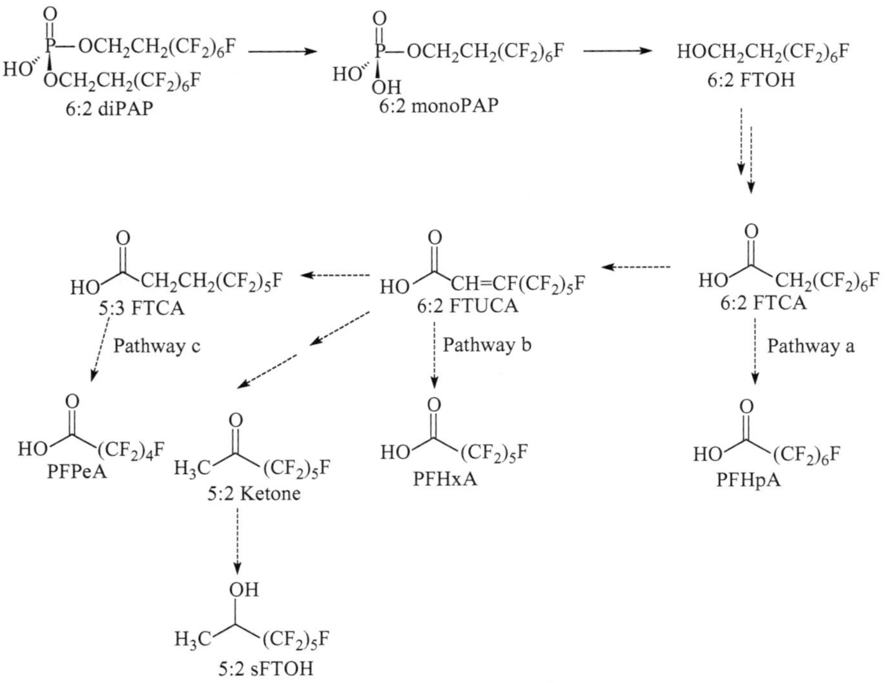

Fig. 18 Proposed degradation pathway of 6:2 diPAP and 6:2 monoPAP [115]

describes the aerobic degradation routes of 4:2, 6:2, 8:2, and 10:2 monosubstituted PAPs (monoPAPs) and 6:2 disubstituted PAP (diPAP) by a microbial mixture collected from sewage of a wastewater treatment plant. In the microbial system, 6:2 FTOH was initially oxidized into a series of acid metabolites. The intermediate metabolite, 6:2 saturated fluorotelomer carboxylic acid (6:2 FTCA), is converted via β-oxidation to 6:2 unsaturated fluorotelomer carboxylic acid (6:2 FTUCA) and perfluorohexanoic acid (PFHxA) (Fig. 18, pathway b), while 5:3 FTCA is transformed to perfluoropentanoic acid (PFPeA) (Fig. 18, pathway c). However, the production of PFPeA may also be attributed to other precursors. For example, 6:2 FTUCA may degrade into 5:2 fluorotelomer ketone ($F(CF_2)_5CH(OH)CH_3$) which is further reduced to 5:2 secondary fluorotelomer alcohol (sFTOH, $F(CF_2)_5CH(OH)CH_3$), and subsequently transformed to PFPeA (Fig. 18, pathway d). 6:2 FTCA and PFHpA production have been observed, supporting the possibility of oxidation of the α-carbon in FTCA to form odd-chain PFCAs (Fig. 18, pathway a).

2.3.5 ω-(Bis(trifluoromethyl)amino)alkane-1-Sulfonates

Biodegradation of ω-(bis(trifluoromethyl)amino)alkane-1-sulfonates was detected in a fixed-bed bioreactor. Its incomplete mineralization revealed that degradation mostly takes place via desulfonation, oxidation, and further β-oxidation [116]. The C–F and C–N bonds in the bis(tri-fluoromethyl)amino (BTFMA) group cannot be accessed by microbes for biodegradation; therefore no defluorination was observed (Fig. 19).

2.3.6 *N*-Ethyl Perfluorooctane Sulfonamide Ethanol

N-Ethyl perfluorooctane sulfonamide ethanol (*N*-EtFOSE) is present in protective paper coatings. Although the only producer in the USA, 3M, has stopped production since 2002, *N*-EtFOSE can still be detected in the North American atmosphere [117]. Aerobic biotransformation of *N*-EtFOSE in activated sludge has been studied [118]. Fast oxidation of *N*-EtFOSE forms *N*-ethyl perfluorooctane sulfonamido acetic acid (*N*-EtFOSAA) through an aldehyde intermediate. *N*-Ethyl perfluorooctane sulfonamide (*N*-EtFOSA) undergoes direct dealkylation to perfluorooctane sulfonamide (FOSA), while perfluorooctane sulfonamido acetic acid (FOSAA) production proceeds at a slower rate. The extremely stable compound PFOS was observed as the final product (Fig. 20) [118].

2.3.7 10-(Trifluoromethoxy) Decane-1-Sulfonate

10-(Trifluoromethoxy) decane-1-sulfonate is a fluorinated surfactant that has been globally distributed, thus leading to increasing concern on its environmental fate and toxicity. Biomineralization of 10-(trifluoromethoxy) decane-1-sulfonate was reported by Peschka et al. in 2008. Two proposed pathways, major and minor, have been

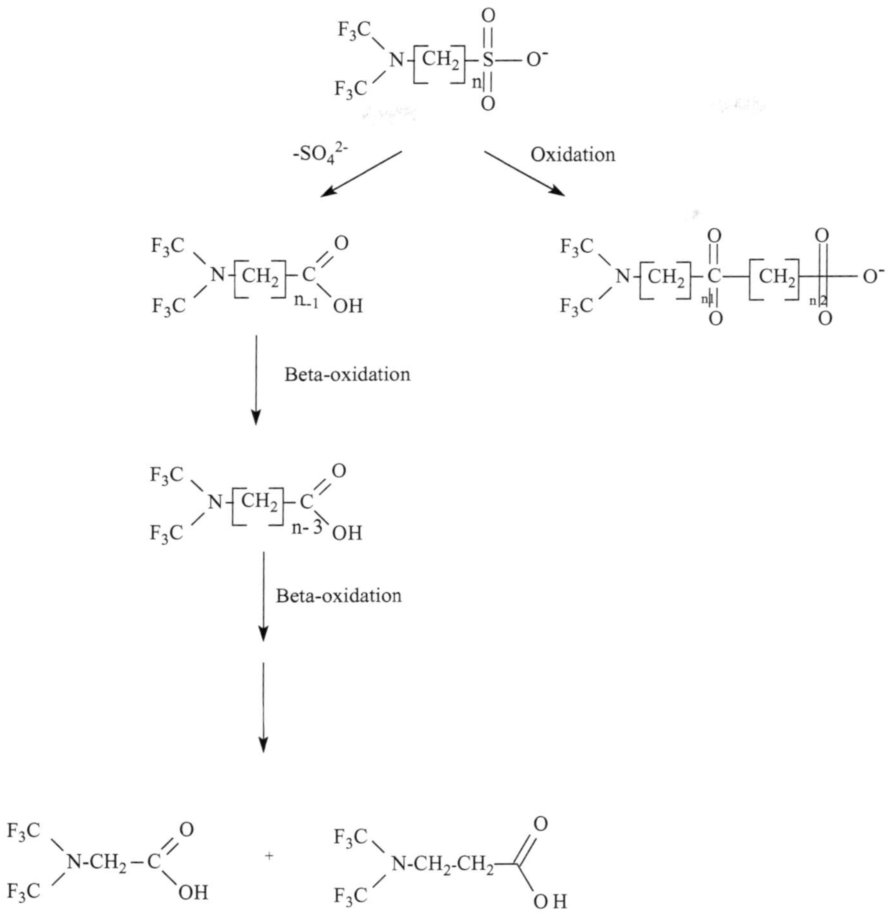

Biology of Fluoro-Organic Compounds

Fig. 19 Biotransformation pathways of ω-(bis(trifluoromethyl)amino)alkane-1-sulfonates [116]

described (Fig. 21). In the major degradation pathway, the carbon chain of the fluorinated alkylsulfonate derivative is shortened by β-oxidation after desulfonation and oxidation. The formed trifluoromethanol is unstable and mineralizes immediately (Fig. 21, pathway a). In the minor degradation pathway, insertion of oxygen occurs, and then, the molecule is subsequently cleaved and degraded (Fig. 21, pathway b).

2.4 Perspectives for the Biodegradation of Perfluorinated Compounds

As previously mentioned, biodegradation and biotransformation of several polyfluorinated compounds such as FTOHs, FTEOs, and PAPs have been reported.

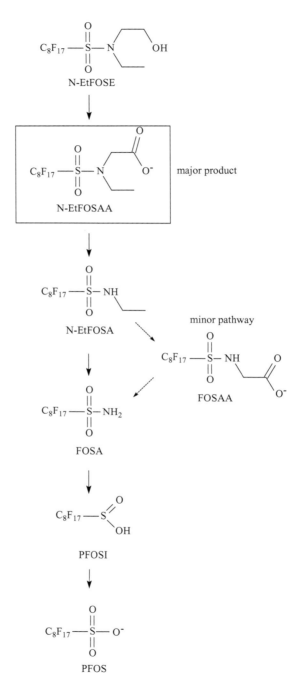

Fig. 20 Proposed transformation pathway of N-EtFOSE in activated sludge [118]

Biology of Fluoro-Organic Compounds

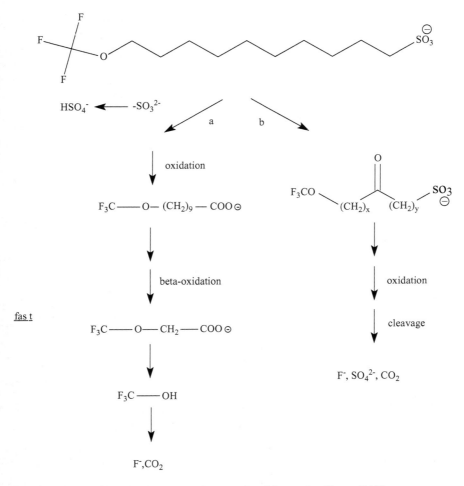

Fig. 21 Degradation pathways of 10-(trifluoromethoxy)decane-1-sulfonate [119]

But to date there are still no reliable reports on the biodegradation or biotransformation of perfluoroalkyl compounds such as PFOS and PFOA. To date, the studies examining biodegradation and transformation of PFCs is very limited. PFCAs, including PFOA, are common transformation products from fluorotelomer chemicals [105, 106, 109, 120, 121]. No evidence about the biodegradation and biotransformation of PFCAs has been found. A recent study about the biodegradability of PFOA using five different microbial communities incubated for up to 259 days showed that PFOA is still microbiologically inert and thus is environmentally persistent [122]. Because of the high stability of the strong C–F covalent bonds, the rigidity of the perfluoroalkyl chain, and the lack of reactive substituent, PFOS is highly recalcitrant to biodegradation or chemical degradation under ambient conditions. Only two reports about the chemical degradation of PFOS have been published [123, 124]. No studies on biodegradation or biotransformation have been reported. Recently, the first report of reductive dehalogenation of PFOS catalyzed by vitamin B12 was published, in which PFOA was reduced and dehalogenated by Ti(III)-citrate [125]. These results suggested the potential for reductive dehalogenation of PFCs.

2.4.1 Thermodynamics of Organofluorine Biodegradation

Thermodynamics can be used to evaluate whether organisms can obtain energy for growth by catalyzing certain reactions. This approach has been applied to the study of reductive biodechlorination of chlorinated compounds such as 3-chlorobenzoate and chloromethanes. The amount of energy available from reductive dechlorination was reported to be between 100 and 180 kJ/mol [126], which is enough to support microorganism growth by using halogenated compounds as electron acceptors. The Gibbs free energy values of fluorinated compounds showed that the amount of energy obtained from defluorination is similar to the amount available from dechlorination and could support microorganism growth. Therefore, organisms may be able to obtain energy by catalyzing certain defluorination reactions for growth.

Although the thermodynamic properties of perfluoroalkylated compounds are not available, the Gibbs free energy values for the reductive removal of one fluorine atom from fluoropropane molecules (Table 1) showed that the energy yields from the hydrogenolysis of perfluorinated compounds and from less fluorinated compounds are similar. These results reveal the thermodynamic basis for reductive biodefluorination of perfluoroalkylated compounds, especially under anaerobic conditions.

Table 1 Gibbs free energy values for reductive dehalogenation (hydrogenolysis) of selected fluorinated aromatic and aliphatic compounds and their chlorinated analogs [127]

Substrate	Product	$\Delta G°$ (kJ/mol)
$CF_3CF_2CH_3$	$CHF_2CF_2CH_3$	−100.0
$CF_3CF_2CH_3$	CF_3CHFCH_3	−117.4
CF_3CHFCH_3	CHF_2CHFCH_3	−88.2
$CHF_2CF_2CH_3$	CHF_2CHFCH_3	−105.5
CF_3CHFCH_3	CF_3CH2CH_3	−163.2
CF_3CH2CH_3	CHF_2CH2CH_3	−78.8
CHF_2CHFCH_3	$CH_2FCHFCH_3$	−96.0
CHF_2CHFCH_3	$CHF_2CH_2CH_3$	−153.9
$CHF_2CH_2CH_3$	$CFH_2CH_2CH_3$	−88.4
$CH_2FCHFCH_3$	$CFH_2CH_2CH_3$	−146.3
$CFH_2CH_2CH_3$	$CH_3CH_2CH_3$	−140.0

All calculations used the following standard conditions: T=298.15K, pH=7, methanes and H_2 in the gas phase at 1 atm, and benzoates and halides in the aqueous phase at 1 M

2.4.2 Perspectives for the Biodegradation of Perfluorinated Compounds

Thermodynamically, perfluorinated compounds should be potentially biodegradable, especially under anaerobic conditions [127]. To date, the reductive biodefluorination of perfluorinated compounds has not yet been observed. Moreover, the rate at which microorganisms can evolve the capability to grow on this potential source of energy and the function of the enzymatic machinery that catalyzes this reaction are largely unknown. However, co-metabolic degradation of several polyfluorinated compounds under aerobic conditions and without thermodynamic facilitation has been studied in detail [109, 110, 115]. This information formed the basis for technology that has been applied in the field for the degradation of other polyhalogenated compounds such as trichloroethene. The search for co-metabolic degradation of poly- and perfluorinated compounds, and studies to understand its mechanisms better will continue.

3 Defluorination Pathways and Defluorogenases

3.1 Enzymatic Metabolic Pathways

3.1.1 Aerobic Metabolism

Limited reports are available about the biodegradation of fluorinated organic compounds, and therefore little is known about the enzyme-catalyzed defluorination pathway. Under aerobic conditions, fluorinated organic compounds are usually degraded via the electron donor or co-metabolic pathways. It has been reported that 4-fluorophenol can be utilized as the sole source of carbon and energy for *Arthrobacter* sp. strain IF1, and that two gene clusters are involved [128]. Cluster A harbors *fpd*A1DE and includes an FADH2-dependent monooxygenase, a putative maleylacetate reductase, and a hydrolase gene. In Cluster B, *fpd*A2 encodes a 4-FP monooxygenase, *fpd*B encodes a flavin reductase, and *fpd*C encodes a putative hydroxyquinol dioxygenase (Fig. 22). The proposed catabolic pathway is shown in Fig. 23.

To date, the well-known enzymes involved in fluoro-degradation are normally responsible for the catabolism of non-fluorinated compounds. Evidence suggests that enzymes are specifically employed for the catabolism of these substrates. Enzymes for degrading aromatic compounds such as monooxygenases, cleavage dioxygenases, and maleylacetate reductase have exhibited biodefluorination activity. As shown in Fig. 24, a number of enzymes that do not show specific activity for fluoroaromatic compounds are involved in the catabolism of 3-fluorobenzoic acid [86].

Because of the similar steric sizes of hydrogen and fluorine, substitution of hydrogen for fluorine is considered to have an important role in defluorination. Many oxygenases and (de)hydroxylases make up a group of defluorinating enzymes

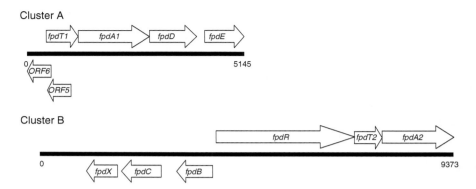

Fig. 22 Organization of the open reading frames (ORFs) in the *fpd* gene regions of *Arthrobacter* sp. strain IF1. *Open arrows* indicate the size and direction of each ORF [128]

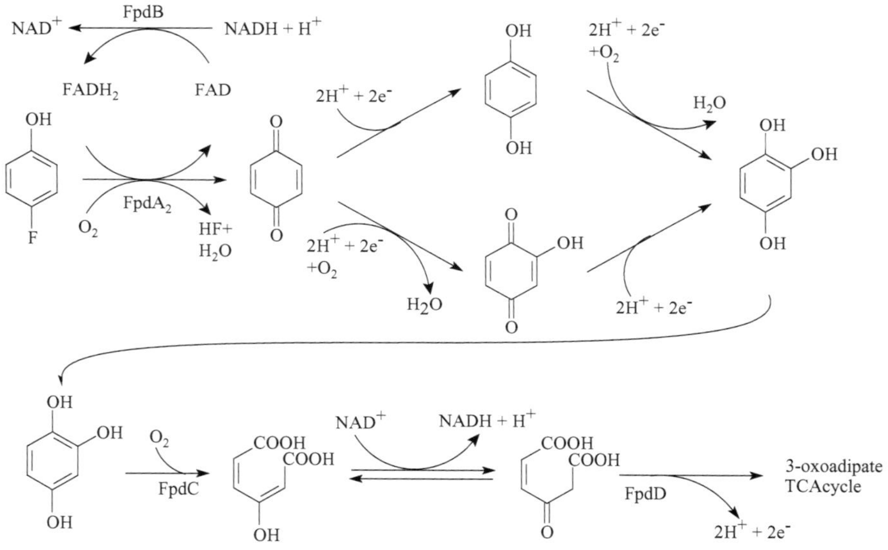

Fig. 23 Proposed catabolic pathway for 4-FP degradation [128]

that act on both fluorinated aromatics and fluoroaliphatics. In 1978, a pyruvate dehydrogenase component of *E. coli* was found to catalyze the conversion of 3-fluoropyruvate to acetate and fluoride ions and to eliminate the fluorine [56]. Later, the proposed mechanism of fluorine elimination by dehydrogenases was proposed. *p*-Hydroxybenzoate hydroxylase, a NADPH-dependent flavin-containing monooxygenase from *P. fluorescens* and *Candida parapsilosis*, was reported to

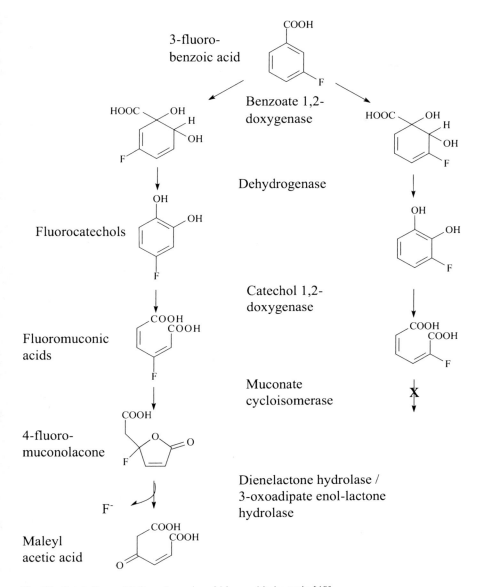

Fig. 24 Catabolism of 3-fluorobenzoic acid in aerobic bacteria [40]

degrade several fluorine-substituted *p*-hydroxybenzoates such as fluorohydroxybenzoate [129, 130]. Fluorobiphenyl metabolism is catalyzed by a series of dioxygenases dehydrogenases and hydrolases to yield fluoropyruvate and 4-fluorobenzoate [86, 87].

3.1.2 Anaerobic Metabolism

Less is known about the degradation of fluoroaromatic compounds under anaerobic conditions. Defluorination of 2-hydroxybenzoate and 3-fluorobenzoate was observed in *S. aciditrophicus* cultures under anaerobic conditions (Mouttaki et al. 2009). Recently, co-metabolism in a bacterial culture was found to catalyze 4-fluorobiphenyl to a carboxylic acid derivative [131]. Both methanogenic and sulfate-reducing defluorination generated trifluoroacetic acid (TFA) via a co-metabolism pathway [132, 133]. Denitrifying bacteria have also been reported to mineralize 2- and 4-fluorobenzoate [79].

3.2 Defluorinases

Among all the dehalogenations, defluorination is most difficult because the C–F bond is one of the most stable bonds in nature. Partly because of limited studies on defluorinases, most man-made organofluorine compounds are degraded via co-metabolism pathways. Enzymes with alterable substrates play an important role, although few fluorine-specific enzymes have been identified.

3.2.1 Fluoroacetate Dehalogenase

As the most common fluorinated natural product, fluoroacetate was reported in 1965 to be degraded by *Pseudomonas* fluoroacetate dehalogenase, which catalyzes the hydrolytic cleavage of the C–F bond to yield glycolate and a fluoride ion [46]. Other fluoroacetate dehalogenases have been isolated from microorganisms such as *Moraxella*, *Delftia*, and *Burkholderia* [45, 48, 134]. Fluoroacetate dehalogenase belongs to the α/β hydrolase superfamily protein. The mechanism of C–F bond cleavage by fluoroacetate dehalogenase has been extensively investigated (Keuning et al. 1985; [45]). The three-dimensional structure of FAc-DEX FA1, a fluoroacetate dehalogenase from *Burkholderia sp.* strain FA1 [53], suggested a mechanism whereby fluoroacetate is degraded by an initial nucleophilic attack on the α-carbon atom by the carboxylate group of Asp104 which displaces the fluoride ion to form an ester intermediate. The ester intermediate is then hydrolyzed by a His271-activated water molecule, which yields glycolate and regenerates the carboxylate group of Asp104 (Fig. 25).

3.2.2 Fluoroacetate-Specific Defluorinases

Detoxification of fluoroacetate in mammals is catalyzed by fluoroacetate-specific defluorinases (FSDs) such as the glutathione-*S*-transferase isozyme GSTZ which is distinct from bacterial fluoroacetate dehalogenases (Fig 26) [135]. Two distinct

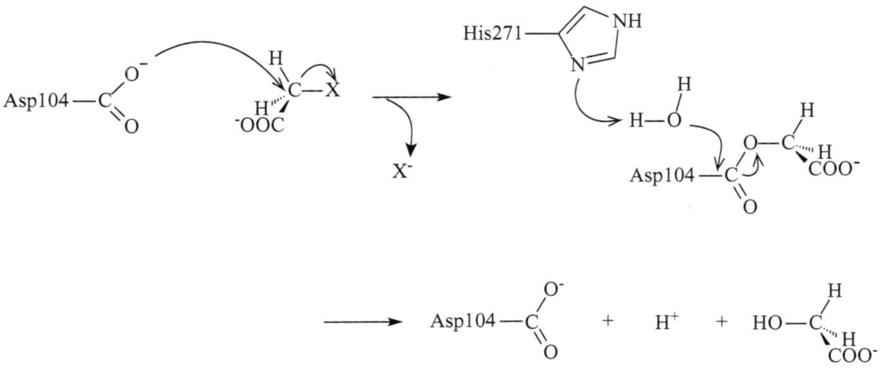

Biology of Fluoro-Organic Compounds

Fig. 25 Proposed reaction mechanism of FAc-DEX FA1 [53]

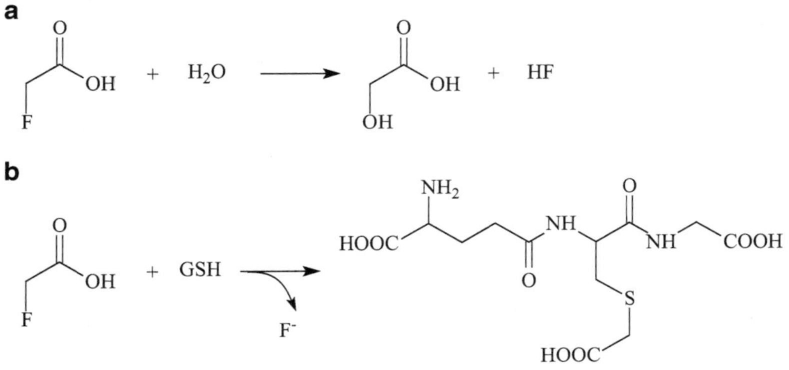

Fig. 26 Enzymatic C–F bond cleavage. (**a**) Fluoroacetate dehalogenase. (**b**) Fluoroacetate-specific defluorinase [40]

FSD activities have been identified in rat liver: one is glutathione-*S*-transferase-like and the other more predominant activity is apparently a new type of dehalogenase, which is considered to be an FSD. Interestingly, the amino acid sequence of the latter FSD is similar to the sorbitol dehydrogenase protein, which does not show defluorination activity on fluoroacetate.

3.2.3 4-Fluorobenzoate Dehalogenase

4-Fluorophenol (4-FP) monooxygenase (FpdA2) was first cloned and purified from *Arthrobacter* sp. strain IF1. In combination with FpdB, which uses NADH to reduce either flavine-adenine dinucleotide (FAD) or flavin Mononucleotide (FMN), FpdA2 transforms various halogenated phenols via *para*-substitution, leading to halide release and hydroquinone formation (Fig. 27) [128].

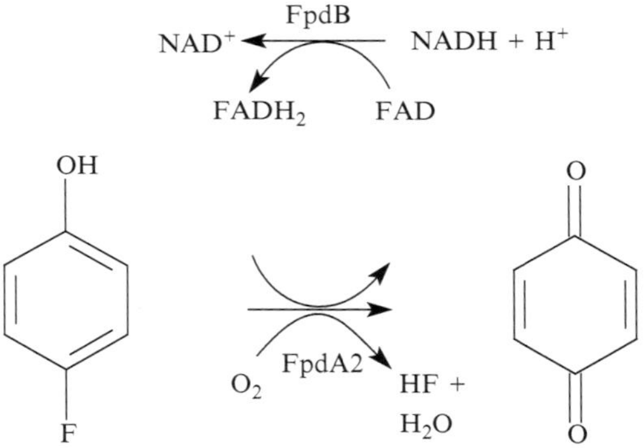

Fig. 27 4-FP monooxygenase [128]

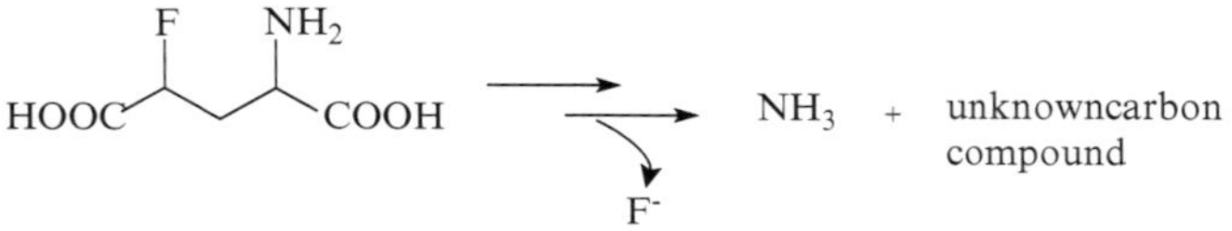

Fig. 28 4-Fluoroglutamate dehalogenase/deaminase. *Arrows* indicate that more than one reaction might be occurring. GSH=glutathione [40]

3.3 Perspectives for New Enzymes and Metabolic Pathways

As mentioned above, co-metabolism is the main degradation pathway for mono- and polyfluorinated organic compounds. However, due to very limited research, only a few co-metabolism pathways have been found in the laboratory. Further studies in this area will help to investigate the pathways and enzymes involved in defluorination. Compared with a series of dechlorinases that catalyze various kinds of reactions, only three kinds of defluorinases have been identified to date. In 2007, biodegradation of 4-fluoroglutamate was reported via an unusual pathway, yielding equimolar concentrations of fluoride ions and ammonia, indicating that an enzyme such as glutamate dehydrogenase is not responsible for the biotransformation (Fig. 28) [136]. In addition, the defluorinating/deaminating activity was found in the soluble fraction of the cell and was not related to the dechlorinating/deaminating activity, which was located in the cell membrane. These results suggest the existence of a potential new fluoroglutamate dehalogenase. Under anaerobic conditions, defluorination was detected in methanogenic, sulfate-reducing, and denitrifying bacteria, indicating that extensive defluorination occurs under anaerobic conditions [79, 132]. Reductive defluorination is a thermodynamically feasible mechanism to derive cellular energy under anaerobic conditions. However, microbes that are able to obtain energy for growth by reductive defluorination have yet to be isolated [127]. And there is much be done to elucidate the defluorination mechanisms and properties of the enzymes involved.

4 Summary and Perspectives

Brominated and chlorinated compounds have been investigated in previous research on the biodegradation of halogenated compounds. However, fluorinated chemicals have thus far received much less attention [127]. The inertness of fluorine results in persistence and leads to accumulation in the environment, making it necessary to explore microbial degradation of fluoroorganic compounds. Until recently, only a few microbes including bacteria, fungi, and algae have been found to be capable of fluoro-degradation. For most fluorinated substrates, the mechanism of fluoro-degradation is still not clear. Several monofluorinated compounds, including fluoroaliphatics [57, 136, 137], fluoroaromatics [71, 74, 81, 88], and a few other polyfluorinated compounds [105, 106, 109, 110], can be degraded. However, the mechanisms of these degradation reactions are largely unknown. No biodegradation of perfluorinated compounds has been observed [25, 122]. Perfluorinated and polyfluorinated compounds are widely used as surfactants, catalysts, and insecticides [18, 19]. These compounds are highly recalcitrant and have been detected throughout the global environment [26, 27]. Biodegradation of perfluorinated compounds is thermodynamically possible under reductive conditions, but has not been measured [127]. Despite a great increase in knowledge over the last few decades, we are still far from being able to predict the biodegradation of fluorinated organic compounds as well as the mechanism of defluorination. Although the dehalogenation of both fluorinated and chlorinated organic compounds is largely mediated by soil microflora, limited knowledge of the factors influencing these microorganisms is available. Development of systematic biological and molecular genetics studies will help in the study of soil microbial species and communities, thus facilitating the discovery of new microbes capable of defluorination.

New technologies for chemical analysis have made highly sophisticated studies practical in the laboratory. Fluorine-19 nuclear magnetic resonance spectroscopy (^{19}F NMR) and isotopic labeling techniques have helped to contribute to a deeper understanding of several key processes in the catalyzed reactions of fluorinated substances [74, 109, 110]. The rapid growth of bioinformatics has led to the development of databases that search for organic persistence information. Furthermore, scientists have created computer programs such as MultiCASE based on general quantitative structure-degradation relationships (QSDRs) to predict the degradation/persistence of organic chemicals in the environment that have not been characterized [138]. One of these popular computer programs, BIOWIN, contains a series of models collectively referred to as biodegradability probability. Based on QSDR models as well as six aerobic biodegradation models and one anaerobic model, BIOWIN can predict the biodegradability probability under aerobic and anaerobic conditions. If a metabolic pathway is available for a chemical, it is assumed to be biodegradable [139]. Similar programs, including the UM-BBD Pathway Prediction and System CATABOL program, have also been developed for determining the biodegradability probability [140, 141]. These tools provide unique approaches to studying biodifluorination.

In general, biodegradation studies need interdisciplinary collaborations between microbiology, ecology, genetics, biochemistry, and analytical chemistry to resolve complex problems. As more research attention is given to this field and more technologies are developed and applied, further mechanisms of the biodegradation of fluorine-containing organic compounds will be elucidated.

References

1. Gribble GW (2004) Natural organohalogens: a new frontier for medical agents? J Chem Educ 81:1441–1449
2. Deng H, O'Hagan D, Schaffrath C (2004) Fluorometabolite biosynthesis and the fluorinase from *Streptomyces cattleya*. Nat Prod Rep 21:773–784
3. Murphy CD, Schaffrath C, O'Hagan D (2003) Fluorinated natural products: the biosynthesis of fluoroacetate and 4-fluorothreonine in *Streptomyces cattleya*. Chemosphere 52:455–461
4. Park BK, Kitteringham NR, O'Neill PM (2001) Metabolism of fluorine-containing drugs. Annu Rev Pharmacol Toxicol 41:443–470
5. Muller K, Faeh C, Diederich F et al (2007) Fluorine in pharmaceuticals: looking beyond intuition. Science 317:1881–1886
6. Kirsch P (2004) Modern fluoroorganic chemistry: synthesis, reactivity, applications. Wiley-VCH, Weinheim
7. Shimizu M, Hiyama T (2005) Modern synthetic methods for fluorine-substituted target molecules. Angew Chem Int Ed Engl 44:214–231
8. Flavin M, Castro-Mendoza H, Ochoa S (1957) Metabolism of propionic acid in animal tissues. J Biol Chem 229:981–996
9. Nashiru O, Zechel DL, Stoll D et al (2001) β–Mannosynthase:synthesis of β-mannosides with a mutant β-mannosidase. Angew Chem Int Ed Engl 113:431–434
10. Zechel DL, Reid SP, Nashiru O et al (2001) Enzymatic synthesis of carbon-fluorine bonds. J Am Chem Soc 123:4350–4351
11. Cobb SL, Deng H, McEwan AR et al (2006) Substrate specificity in enzymatic fluorination. The fluorinase fromStreptomyces cattleya accepts 2-deoxyadenosine substrates. Org Biomol Chem 4:1458–1460
12. Eustáquio AS, O'Hagan D, Moore BS (2010) Engineering fluorometabolite production fluorinase expression in *Salinispora tropica* yields fluorosalinosporamide. J Nat Prod 73: 378–382
13. Rentmeister A, Arnold FH, Fasan R (2009) Chemo-enzymatic fluorination of unactivated organic compounds. Nat Chem Biol 5:26–28
14. Ratnayake AS, Janso JE, Feng X, Schlingmann G, Goljer I, Carter GT (2009) Evaluating indole-related derivatives as precursors in the directed biosynthesis of diazepinomicin analogues. J Nat Prod 72:496–499
15. Murphy CD, Clark BR, Amadio J (2009) Metabolism of fluoroorganic compounds in microorganisms: impacts for the environment and the production of fine chemicals. Appl Microbiol Biotechnol 84:617–629
16. Amir-Heidari B, Thirlway J, Micklefield J (2008) Auxotrophic-precursor directed biosynthesis of nonribosomal lipopeptides with modified tryptophan residues. Org Biomol Chem 6: 975–978
17. Begue JP, Bonnet-Delpon D (2006) Recent advances (1995–2005) in fluorinated pharmaceuticals based on natural products. J Fluor Chem 127:8992–1012
18. Ojima (2009) Fluorinein medicinal chemistry and chemical biology. Blackwell, Chichester

19. Prevedouros K, Cousins IT, Buck RC et al (2006) Sources, fate and transport of perfluorocarboxylates. Environ Sci Technol 40:32–44
20. Lewandowski G, Meissner E, Milchert E et al (2006) Special applications of fluorinated organic compounds. J Hazard Mater 136:385–391
21. Barber JL, Berger U, Chaemfa C et al (2007) Analysis of per- and polyfluorinated alkyl substances in air samples from Northwest Europe. J Environ Monit 9:530–541
22. Senthilkumar K, Ohi E, Sajwan K et al (2007) Perfluorinated compounds in river water, river sediment, market fish, and wildlife samples from Japan. Bull Environ Contam Toxicol 79: 427–431
23. Houde M, Martin JW, Letcher RJ et al (2006) Biological monitoring of polyfluoroalkyl substances: a review. Environ Sci Technol 40:3463–3473
24. Olsen GW, Burris JM, Ehresman DJ et al (2007) Half-life of serum elimination of perfluorooctanesulfonate, perfluorohexanesulfonate, and perfluorooctanoate in retired fluorochemical production workers. Environ Health Perspect 115:1298–1305
25. Kennedy GL, Butenhoff JL, Olsen GW et al (2004) The toxicology of perfluorooctanoate. Crit Rev Toxicol 34:351–84
26. Lau C, Butenhoff JL, Rogers JM et al (2004) The developmental toxicity of perfluoroalkyl acids and their derivatives. Toxicol Appl Pharmacol 198:231–41
27. Calafat AM, Wong LY, Kuklenyik Z et al (2007) Polyfluoroalkyl chemicals in the U.S. population: data from the National Health and Nutrition Examination Survey (NHANES) 2003–2004 and comparisons with NHANES 1999–2000. Environ Health Perspect 115: 1596–602
28. Yang Q, Xie Y, Eriksson AM et al (2001) Further evidence for the involvement of inhibition of cell proliferation and development in thymic and splenic atrophy induced by the peroxisome proliferator perfluoroctanoic acid in mice. Biochem Pharmacol 62:1133–40
29. Nakayama S, Harada K, Inoue K et al (2005) Distributions of perfluorooctanoic acid (PFOA) and perfluorooctane sulfonate (PFOS) in Japan and their toxicities. Environ Sci 12:293–313
30. Tilton SC, Orner GA, Benninghoff AD et al (2008) Genomic profiling reveals an alternate mechanism for hepatic tumor promotion by perfluorooctanoic acid in rainbow trout. Environ Health Perspect 116:1047–55
31. Asakawa A, Toyoshima M, Fujimiya M et al (2007) Perfluorooctane sulfonate influence feeding behavior and gut motility via the hypothalamus. Int J Mol Med 19:733–739
32. Yao X, Zhong L (2005) Genotoxic risk and oxidative DNA damage in HepG2 cells exposed to perfluorooctanoic acid. Mutat Res 587:38–44
33. Ishibashi H, Ishida H, Matsuoka M et al (2007) Estrogenic effects of fluorotelomer alcohols for human estrogen receptor isoforms alpha and beta in vitro. Biol Pharm Bull 30:1358–1359
34. Fei C, McLaughlin JK, Lipworth L et al (2009) Maternal levels of perfluorinated chemicals and subfecundity. Hum Reprod 24:1200–1205
35. Fei C, McLaughlin JK, Tarone RE et al (2008) Fetal growth indicators and perfluorinated chemicals:a study in the Danish National Birth Cohort. Am J Epidemiol 168:66–72
36. Frisbee SJ (2008) The C8 health project:how a class action lawsuit caninteract with public health – history of events. Robert C. Byrd Health Science Center. Available at: http://www.hsc.wvu.edu/som/cmed/c8/index.asp. Accessed 14 Feb 09
37. Díaz E (2008) Microbial biodegradation: genomics and molecular biology. Caister Academic Press, Norfolk
38. Natarajana R, Azerada R, Badet B et al (2005) Microbial cleavage of C–F bond. J Fluor Chem 126:425–436
39. Yu M, Faan Y-W, Chung WYK et al (2007) Isolation and characterization of a novel haloacid permease from *Burkholderia cepacia* MBA4. Appl Environ Microbiol 73:4874–4880
40. Murphy CD (2010) Biodegradation and biotransformation of organofluorine compounds. Biotechnol Lett 32:351–359
41. Hiyama T (2000) Organofluorine compounds: chemistry and applications. Springer, Berlin

42. Goncalves LPB, Antunes OAC, Pinto GF et al (2003) Kinetic aspects involved in the simultaneous enzymatic synthesis of (S)-3-fluoroalanine and (R)-3-fluorolactic acid. J Fluor Chem 124:219–227
43. Horiuchi N, Agric J (1961) The CF bond rupture of monofluoroacetate by soil microbes. Chem Soc Jpn 35:870–873
44. Horiuchi N (1962) The CF bond rupture in monofluoroacetate by soil microbes. II. Some properties of the bacteria and the enzyme. Jpn J Biochem Soc 34:92–98
45. Liu JQ, Kurihara T, Ichiyama S et al (1998) Reaction mechanism of fluoroacetate dehalogenase from *Moraxella* sp.B. J Biol Chem 273:30897–30902
46. Goldman P (1965) The enzymatic cleavage of the carbon-fluorine bond in fluoroacetate. J Biol Chem 240:3434–3438
47. Tonomura K, Futai F, Tanabe O et al (1965) Defluorination of monofluoroacetate by bacteria. Agric Biol Chem 29:124–128
48. Kawasaki H, Yahara H, Tonomura K (1984) Cloning and expression in *Escherichia coli* of the haloacetate dehalogenase genes from *Moraxella* plasmid pUO1. Agric Biol Chem 48: 2627–2632
49. Goldman P, Milne GWA, Keister DB et al (1968) Carbon-halogen bond cleavage. J Biol Chem 243:428–434
50. Lacourciere GM, Armstrong RN (1993) The catalytic mechanism of microsomal epoxide hydrolase involves an ester intermediate. J Am Chem Soc 115:10466–10467
51. Verschueren KHG, Seljee F, Rozeboom HJ et al (1993) Crystallographic analysis of the catalytic mechanism of haloalkane dehalogenase. Nature 363:693–698
52. Tzeng H-F, Laughlin LT, Armstrong RN et al (1998) Semifunctional site-specific mutants affecting the hydrolytic half-reaction of microsomal epoxide hydrolase. Biochemistry 37:2905–2911
53. Jitsumori K, Omi R, Kurihara T et al (2009) X-Ray crystallographic and mutational studies of fluoroacetate dehalogenase from *Burkholderia* sp. strain FA1. J Bacteriol 191:2630–2637
54. Zeczycki TN, Maurice MS, Attwood PV (2010) Inhibitors of pyruvate carboxylase. Open Enzym Inhib J 3:8–26
55. Vali M, Vossen JA, Buijs M et al (2008) Targeting of VX2 rabbit liver tumor by selective delivery of 3-bromopyruvate: a biodistribution and survival study. J Pharmacol Exp Ther 327:32–37
56. Leung LS, Frey PA (1978) Fluoropyruvate: an unusual substrate for pyruvate dehydrogenase. Biochem Biophys Res Commun 81:274–279
57. Meyer JJM, O'Hagan D (1992) Conversion of fluoropyruvate to fluoroacetate by *Dichapetalum cymosum*. Phytochemistry 31:499–501
58. Kaschabek SR, Reineke W (1995) Maleylacetate reductase of *Pseudomonas* sp strain B13: specificity of substrate conversion and halide elimination. J Bacteriol 177:320–325
59. Muller D, Schlomann M, Reineke W et al (1996) Maleylacetate reductases in chloroaromatic-degrading bacteria using the modified *ortho* pathway: comparison of catalytic properties. J Bacteriol 178:298–300
60. Bertau M (2001) Novel unusual microbial dehalogenation during enantioselective reduction of ethyl 4,4,4-trifluoro acetoacetate with baker's yeast. Tetrahedron Lett 42:1267–1268
61. Perez-Pantoja D, Donoso RA, Sanchez MA et al (2009) Genuine genetic redundancy in maleylacetate-reductase-encoding genes involved in degradation of haloaromatic compounds by *Cupriavidus necator* JMP134. Microbiology 155:3641–3651
62. Haufe G, Pietz S, Wölker D et al (2003) Synthesis of fluorinated cycloalkyl N-phenyl-carbamates and their microbial defluorination/oxygenation by *Beauveria bassiana*. J Org Chem 21:2166–2175
63. Zhan J, Gunatilaka AAL (2009) Microbial transformation by *Beauveria bassiana*. In: Rai M (ed) Advances in fungal biotechnology. I. K, International, New Delhi

64. Akiyama Y, Hiramatsu C, Fukuhara T et al (2006) Selective introduction of a fluorine atom into carbohydrates and a nucleoside by ring-opening fluorination reaction of epoxides. J Fluor Chem 127:920–923
65. Li X, Turánek J, Knötigová P et al (2009) Hydrophobic tail length, degree of fluorination and headgroup stereochemistry are determinants of the biocompatibility of (fluorinated) carbohydrate surfactants. Colloids Surf B Biointerfaces 73:65–74
66. Moller AKH, Loft A, Berthelsen AK et al (2011) ^{18}F-FDG PET/CT as a diagnostic tool in patients with extracervical carcinoma of unknown primary site: a literature review. Oncologist 16:445–451
67. D'Amore TD, Taylor NF (1982) The reaction of 4-deoxy-4-fluoro-D-glucose with an outer membrane protein of *Pseudomonas putida*. FEBS Lett 143:247–251
68. Sbrissa D, McIntosh JM, Taylor NF (1990) The metabolism of 4-deoxy-4-fluoro-D-glucose in *Pseudomonas putida*. Carbohydr Res 203:271–280
69. Tejada ML, Green JR, Taylor NF (1993) The defluorination of 4-deoxy-4-fluoro-d-glucose in the cytoplasmic membrane of *Pseudomonas putida*. Carbohydr Res 249:207–219
70. Zhang C, Zhou Q, Chen L et al (2007) Biodegradation of meta-fluorophenol by an acclimated activated sludge. J Hazard Mater 141:295–300
71. Engesser KH, Auling G, Busse J et al (1990) 3-Fluorobenzoate enriched bacterial strain FLB-300 degrades benzoate and all 3 isomeric monofluorobenzoates. Arch Microbiol 153: 193–199
72. Misiak K, Casey E, Murphy CD (2011) Factors influencing 4-fluorobenzoate degradation in biofilm cultures of *Pseudomonas knackmussii* B13. Water Res 45:3512–3520
73. Emanuelsson MAE, Osuna ME, Jorge RMF (2009) Isolation of a *Xanthobacter* sp. degrading dichloromethane and characterization of the gene involved in the degradation. Biodegradation 20:235–244
74. Boersma FGH, McRoberts WC, Cobb SL et al (2004) A F-19 NMR study of fluorobenzoate biodegradation by *Sphingomonas* sp HB-1. FEMS Microbiol Lett 237:355–361
75. Schreiber A, Hellwig M, Dorn E et al (1980) Critical reactions in fluorobenzoic acid degradation by *Pseudomonas* sp. B13. Appl Environ Microbiol 39:58–67
76. Harper DB, Blakley ER (1971) The metabolism of p-fluorobenzoic acid by a *Pseudomonas* sp. Can J Microbiol 17:1015–1023
77. Schennen U, Braun K, Knackmuss H-J (1985) Anaerobic degradation of 2-fluorobenzoate by benzoate–degrading, denitrifying bacteria. J Bacteriol 161:321–325
78. Drzyzga O, Jannsen S, Blotevogel KH (1994) Mineralization of monofluorobenzoate by a diculture under sulfate-reducing conditions. FEMS Microbiol Lett 116:215–219
79. Vargas C, Song B, Camps M et al (2000) Anaerobic degradation of fluorinated aromatic compounds. Appl Microbiol Biotechnol 53:342–347
80. Mouttaki H, Nanny MA, McInerney MJ et al (2009) Metabolism of hydroxylated and fluorinated benzoates by *Syntrophus aciditrophicus* and detection of a fluorodiene metabolite. Appl Environ Microbiol 75:998–1004
81. Shimoda K, Hamada H (2010) Bioremediation of fluorophenols by glycosylation with immobilized marine microalga *Amphidinium Crassum*. Environ Health Insights 4:87–91
82. Bondar VS, Boersma MG, Vervoort J et al (1998) ^{19}F NMR study on the biodegradation of fluorophenols by various *Rhodococcus* species. Biodegradation 9:475–486
83. Kim EJ, Jeon JR, Kim YM et al (2010) Mineralization and transformation of monofluorophenols by *Pseudonocardia benzenivorans*. Appl Microbiol Biotechnol 87:1569–1577
84. Prenafeta-Boldu FX, Luykx D, Vervoort J et al (2001) Fungal metabolism of toluene: monitoring of fluorinated analogs by F-19 nuclear magnetic resonance spectroscopy. Appl Environ Microbiol 67:1030–1034
85. Green NA, Meharg AA, Till C et al (1999) Degradation of 4-fluorobiphenyl by mycorrhizal fungi as determined by 19 F nuclear magnetic resonance spectroscopy and 14 C radiolabelling analysis. Appl Environ Microbiol 65:4021–4027

86. Murphy CD, Quirke S, Balogun O (2008) Degradation of fluorobiphenyl by *Pseudomonas pseudoalcaligenes* KF707. FEMS Microbiol Lett 286:45–49
87. Amadio J, Murphy CD (2010) Biotransformation of fluorobiphenyl by *Cunninghamella elegans*. Appl Microbiol Biotechnol 86:345–351
88. Hughes D, Clark BR, Murphy CD (2011) Biodegradation of polyfluorinated biphenyl in bacteria. Biodegradation 22:741–749
89. Carvalho MF, Alves CCT, Ferreira MIM et al (2002) Isolation and initial characterization of a bacterial consortium able to mineralize fluorobenzene. Appl Environ Microbiol 68:102–105
90. Carvalho MF, Marco PDe, Duque AF et al (2008) *Labrys portucalensis* sp. nov., a fluorobenzene-degrading bacterium isolated from an industrially contaminated sediment in northern Portugal. Int J Syst Evol Microbiol 58:692–698
91. Carvalho MF, Ferreira MIM, Moreira IS et al (2006) Degradation of fluorobenzene by *Rhizobiales* strain F11 via ortho cleavage of 4-fluorocatechol and catechol. Appl Environ Microbiol 72:7413–7417
92. Gros M, Petrovic M, Barcelo D (2007) Wastewater treatment plants as a pathway for aquatic contamination by pharmaceuticals in the Ebro river basin (Northeast Spain). Environ Toxicol Chem 26:1553–1562
93. Brain RA, Johnson DJ, Richards SM et al (2004) Effects of 25 pharmaceutical compounds to *Lemna gibba* using a seven-day static-renewal test. Environ Toxicol Chem 23:371–382
94. Robinson AA, Belden JB, Lydy MJ et al (2005) Toxicity of fluoroquinolone antibiotics to aquatic organisms. Environ Toxicol Chem 24:423–430
95. Hernández A, Sánchez MB, Martínez JL (2011) Quinolone resistance: much more than predicted. Front Microbio 2:22
96. Wetzstein HG, Schmeer N, Karl W (1997) Degradation of the fluoroquinolone enrofloxacin by the brown rot fungus *Gloeophyllum striatum*: identification of metabolites. Appl Environ Microbiol 63:4272–4281
97. Wetzstein H-G, Schneider J, Karl W et al (2006) Patterns of metabolites produced from the fluoroquinolone enrofloxacin by Basidiomycetes indigenous to agricultural sites. Appl Microbiol Biotechnol 71:90–100
98. Rietjens IM, Tyrakowska B, Veeger C et al (1990) Reaction pathways for biodehalogenation of fluorinated anilines. Eur J Biochem 194:945–954
99. Travkin VM, Solyanikova IP, Rietjens IMCM et al (2003) Degradation of 3,4-dichloro- and 3,4-difluoroaniline by *Pseudomonas fluorescens* 26-K. J Environ Sci Health B 38:121–132
100. Cobb SL, Murphy CD (2009) 19 F NMR applications in chemical biology. J Fluor Chem 130:132–143
101. Duckett CJ, Wilson ID, Douce DS et al (2007) Metabolism of 2-fluoro-4-iodoaniline in earthworm Eisenia veneta using super (19) F-NMR spectroscopy, HPLC-MS, and HPLC-ICPMS. Xenobiotica 37:1378–1393
102. Remde A, Debus R (1996) Biodegradability of fluorinated surfactants under aerobic and anaerobic conditions. Chemosphere 32:1563–1574
103. Arakaki A, Ishii Y, Tokuhisa T et al (2010) Microbial biodegradation of a novel fluorotelomer alcohol, 1H,1H,2H,2H,8H,8H-perfluorododecanol, yields short fluorinated acids. Appl Microbiol Biotechnol 88:1193–1203
104. Maras M, Vanparys C, Muylle F et al (2006) Estrogen-like properties of fluorotelomer alcohols as revealed by MCF-7 breast cancer cell proliferation. Environ Health Perspect 114:100–105
105. Wang N, Szostek B, Folsom PW et al (2005) Aerobic biotransformation of ^{14}C-labeled 8–2 telomer B alcohol by activated sludge from a domestic sewage treatment plant. Environ Sci Technol 39:531–538
106. Wang N, Szostek B, Buck RC et al (2005) Fluorotelomer alcohol biodegradation:direct evidence that perfluorinated carbon chains breakdown. Environ Sci Technol 39:7516–7528
107. Fasano WJ, Carpenter SC, Gannon SA et al (2006) Absorption, distribution, metabolism, and elimination of 8–2 fluorotelomer alcohol in the rat. Toxicol Sci 91:341–355

108. Nabb DL, Szostek B, Himmelstein MW et al (2007) In vitro metabolism of 8–2 fluorotelomer alcohol: interspecies comparisons and metabolic pathway refinement. Toxicol Sci 100:333–344
109. Wang N, Szostek B, Buck RC et al (2009) 8–2 Fluorotelomer alcohol aerobic soil biodegradation: pathways, metabolites, and metabolite yields. Chemosphere 75:1089–1096
110. Liu J, Wang N, Szostek B et al (2010) 6–2 Fluorotelomer alcohol aerobic biodegradation in soil and mixed bacterial culture. Chemosphere 78:437–444
111. Fasano WJ, Sweeney LM, Mawn MP et al (2009) Kinetics of 8–2 fluorotelomer alcohol and its metabolites, and liver glutathione status following daily oral dosing for 45 days in male and female rats. Chem Biol Interact 180:281–295
112. Martin JW, Chan K, Mabury SA et al (2009) Bioactivation of fluorotelomer alcohols in isolated rat hepatocytes. Chem Biol Interact 177:196–203
113. Frömel T, Knepper TP (2010) Fluorotelomer ethoxylates:sources of highly fluorinated environmental contaminants part I: Biotransformation. Chemosphere 80:1387–1392
114. Russell MH, Berti WR, Szostek B et al (2010) Evaluation of PFO formation from the biodegradation of a fluorotelomer-based urethane polymer product in aerobic soils. Polym Degrad Stabil 95:79–85
115. Lee H, D'eon J, Mabury SA (2010) Biodegradation of polyfluoroalkyl phosphates as a source of perfluorinated acids to the environment. Environ Sci Technol 44:3305–3310
116. Frömel T, Peschka1 M, Fichtner N et al (2008) ω-(Bis(trifluoromethyl)amino)alkane-1-sulfonates:synthesis and mass spectrometric study of the biotransformation products. Rapid Commun Mass Spectrom 22:3957–3967
117. Stock NL, Lau FK, Ellis DA et al (2004) Polyfluorinated telomer alcohols and sulfonamides in the North American troposphere. Environ Sci Technol 38:991–996
118. Rhoads KR, Janssen EML, Luthy RG et al (2008) Aerobic biotransformation and fate of N-ethyl perfluorooctane sulfonamidoethanol (N-EtFOSE) in activated sludge. Environ Sci Technol 42:2873–2878
119. Peschka M, Fichtner N, Hierse W et al (2008) Synthesis and analytical follow-up of the mineralization of a new fluorosurfactant prototype. Chemosphere 72:1534–1540
120. Dinglasan MJA, Ye Y, Edwards EA et al (2004) Fluorotelomer alcohol biodegradation yields poly- and perfluorinated acids. Environ Sci Technol 38:2857–2864
121. Liu J, Lee LS, Nies LF et al (2007) Biotransformation of 8: 2 fluorotelomer alcohol in soil and by soil bacteria isolates. Environ Sci Technol 41:8024–8030
122. Liou JS-C, Szostek B, DeRito CM et al (2010) Investigating the biodegradability of perfluorooctanoic acid. Chemosphere 80:176–183
123. Moriwaki H, Takagi Y, Tanaka M et al (2005) Sonochemical decomposition of perfluorooctane sulfonate and perfluorooctanoic acid. Environ Sci Technol 39:3388–3392
124. Hori H, Nagaoka Y, Yamamoto A et al (2006) Efficient decomposition of environmentally persistent perfluorooctanesulfonate and related fluorochemicals using zerovalent iron in subcritical water. Environ Sci Technol 40:1049–1054
125. Ochoa-herrera V, Sierra-alvarez R, Somogy A et al (2008) Reductive defluorination of perfluorooctane sulfonate. Environ Sci Technol 42:3260–3264
126. Dolfing J (2003) Thermodynamic considerations for dehalogenation. In: Häggblom MM, Bossert ID (eds) Dehalogenation:microbial processes and environmental applications. Springer, Berlin
127. Parsons JR, Sáez M, Dolfing J et al (2009) Biodegradation of perfluorinated compounds. Rev Environ Contam Toxicol 196:53–71
128. Ferreira MIM, Iida T, Hasan SA et al (2009) Analysis of two gene clusters involved in the degradation of 4-fluorophenol by *Arthrobacter* sp. Strain IF1. Appl Environ Microbiol 75: 7767–7773
129. Husain M, Entsch B, Ballou DP et al (1980) Fluoride elimination from substrates in hydroxylation reactions catalyzed by p-hydroxybenzoate hydroxylase. J Biol Chem 255:4189–4197

130. van der Bolt FJT, van den Heuvel RHH, Vervoort J et al (1997) ^{19}F NMR study on the regiospecificity of hydroxylation of tetrafluoro-4-hydroxybenzoate by wild-type and Y385F p-hydroxybenzoate hydroxylase: evidence for a consecutive oxygenolytic dehalogenation mechanism. Biochemistry 36:14192–14201
131. Selesi D, Meckenstock RU (2009) Anaerobic degradation of the aromatic hydrocarbon biphenyl by a sulfate-reducing enrichment culture. FEMS Microbiol Ecol 68:86–93
132. Visscher PT, Culbertson CW, Oremland RS (1994) Degradation of trifluoroacetate in oxic and anoxic sediments. Nature 369:729–731
133. Kim BR, Suidan MT, Wallington TJ et al (2000) Biodegradability of trifluoroacetic acid. Environ Eng Sci 17:337–342
134. Ichiyama S, Kurihara T, Miyagi M et al (2002) Catalysis-linked inactivation of fluoroacetate dehalogenase by ammonia: a novel approach to probe the active-site environment. J Biochem 131:671–677
135. Tu LQ, Wright PFA, Rix CJ et al (2006) Is fluoroacetate-specific defluorinase a glutathione S-transferase? Comp Biochem Physiol C 143:59–66
136. Donnelly C, Murphy CD (2007) Bacterial defluorination of 4-fluoroglutamic acid. Appl Microbiol Biotechnol 77:699–703
137. Peters RA (1957) Mechanism of the toxicity of the active constituent of dichapetalum cymosum and related compounds. Adv Enzymol Relat Subj Biochem 18:113–159
138. Howard PH (2009) Howard predicting the persistence of organic compounds. Hdb Env Chem 2:17–41
139. Meylan WM, Boethling RS, Aronson D et al (2007) Chemical structure-based predictive model for methanogenic anaerobic biodegradation potential. Environ Toxicol Chem 26:1785–1792
140. Dimitrov S, Pavlov T, Nedelcheva D et al (2007) A kinetic model for predicting biodegradation. SAR QSAR Environ Res 18:443–457
141. Wicker J, Fenner K, Ellis L et al (2010) Predicting biodegradation products and pathways: a hybrid knowledge- and machine learning-based approach. Bioinformatics 26:814–821
142. O'Hagan D, Schaffrath C, Cobb S et al (2002) Biosynthesis of an organofluorine molecule. Nature 416:279
143. O'Hagan D (2010) Fluorine in health care: Organofluorine containing blockbuster drugs. J Fluorine Chem 131:1071–1081
144. Vanden Heuvel JP, Thompson JT, Frame SR et al (2006) Differential activation of nuclear receptors by perfluorinated fatty acid analogs and natural fatty acids: a comparison of human, mouse, and rat peroxisome proliferator-activated receptor-alpha, -beta, and -gamma, liver X receptor-beta, and retinoid X receptor-alpha. Toxicol Sci 92:476–89
145. Martin JW, Mabury SA, O'Brien PJ (2005) Metabolic products and pathways of fluorotelomer alcohols in isolated rat hepatocytes. Chem Biol Int 155:165–180
146. Mouttaki H, Nanny MA, McInerney MJ et al (2009) Metabolism of hydroxylated and fluorinated benzoates by *Syntrophus aciditrophicus* and detection of a fluorodiene metabolite. Appl Environ Microbiol 75:998–1004
147. Keuning S, Janssen DB, Witholt B (1985) Purification and characterization of hydrolytic haloalkane dehalogenase from *Xanthobacter autotrophicus* gj10. J Bacteriol 163:635–639

Index

A

N-Acetyl-D-glucosamine (GlcNAc), 297
Acrylates, hydroformylation, 286, 287
Acylations, 72
Acyl-Strecker reaction, Schreiner thiourea, 205
Adamantyl bromide, fluorous tin hydride reduction, 81
Adsorption, 7
Aerobic metabolism, 391
Alcohols, 158, 250
    acylation, 169
    fluorous, 14, 94, 296
    oxidation, 85, 87, 162, 168
    secondary, 84, 88, 186, 197
    silylation, 266
Alcoholysis, 207
Aldehyde dehydrogenase, 382
Aldehydes, 32, 86, 183, 197, 203, 254, 276, 301, 310
    amination, 206
    α-chlorination, 207
    hydrosilylation, 247, 268, 389
    nucleophilic addition, 38
Alkenes, hydroformylation, 285
    hydrosilylation, 251
    metathesis, 161
N-Alkylation, 213
Alkyl lactosides, 297
Allylchlorodiisopropylsilane, 48
Aminations, F-Boc carbamates, 76
    reductive, fluorous thiourea, 206
Amines, 93
    fluorous, 14
    secondary, 177
Ammonium salts, 213
Anaerobic metabolism, 394
Anilines, fluorinated, 380
Antibiotics, 353
Arenes, 14

Asymmetric hydroformylation, 275
Asymmetric hydrogenation, 233
Asymmetric synthesis, 175
Atorvastatin, 368
ATP fluoride, 367
Azacrowns, separation/reuse, 216

B

Beauveria bassiana, 372
Benzamides, radiolabeled, 61
Benzene, 14
Benzo–1,4-diazepine–2,5-diones, 55, 57
Benzoic acid, 140
Benzophenone imine, 76
Benzotrifluoride, 137
Benzylic compounds, chromium-catalyzed oxidation, 140
Benzyltriethylammonium chloride (TEBA), 216, 228
Bifenthrin, 350, 352
BINAP, 158, 243
BINAPHOS, 286
Binaphthyl-thiourea, 196
Bioaccumulation, 347
Biocatalysis, 162
Biodegradation, 365, 370
Biofluorination, 366
Biological assays, platform, 328
Biosynthesis, 291
Biotin, fluorous-tagged, 63
BIOWIN, 397
Biphasic catalysis, 233, 275
3,5-Bis(perfluorooctyl)benzyl bromide, 213
Bis[(perfluorooctyl)propyl]–2,2,2-trifluoroethylamine, 325
Bis(trifluoroacetoxy)perfluoroalkyliodines, 84
ω-(Bis(trifluoromethyl)amino)alkane–1-sulfonates, 386

405

2,2-Bistrifluoromethyl–4,5-difluoro–1,3-
  dioxole (BDD), 308
N,N-Bis(trifluoromethyl)phenyl thiourea, 203
Bond energies 1, 6
Bonding motifs, 6
3-Bromo pyruvate, 371
4-Bromostyrene, hydrogenation, 236
Buchwald–Hartwig amination, 76
1-Butene, hydrogenation, 240

C
Calixarenes, 4
Capping, 105, 114, 129
Carbamate, 77
Carbanions, fluorine substitution, 28
Carbodiimides, 72
Carbohydrates, fluorinated, 373
Carbon dioxide, 8, 322
Carbon monoxide, 275
Carbon–carbon bonds, fluorous tin reagents, 89
CATABOL, 397
Catalysis, 153, 213, 233
  biphasic, 233, 275
Catalyst-on-a-tape, 247
Cell membrane, 291
  liposomal uptake, 304
Chalk–Harrod mechanism, hydrosilylation, 249
α-Chlorination, aldehydes, 207
Chlorine oxide dimers, 342
Chloro-dimethoxy-triazine (CDMT), 72
Chlorofluorocarbons (CFCs), 339, 341
Chlorpyrifos, 348
Cinchona alkaloids, catalysts, 181
  thiourea, bifunctional, So€s-Connon, 197
Ciprofloxacin, 353
Conformations, 1
Corey–Bakshi–Shibata (CBS) reduction, 82
Critical micelle concentration (CMC), 291
  glycosides, 303
Crown ethers, 213
Curran's diphenyl urea, 194
Cycloalkyl N-phenylcarbamates, 372
Cyclododecene, hydrogenation, 236
Cyclohexanol, 14
2-Cyclohexen–1-one, hydrogenation, 236
Cyhalothrin, 350
Cytotoxicity, 291

D
DABCO. See Diazabicyclooctane (DABCO)
1-Decene, hydroformylation, 279, 281

Defluorinases, 394
  fluoroacetate-specific (FSDs), 394
Defluorination, 391
Defluorogenases, 365, 391
Degradable telomer fluoroalcohol (DTFA), 382
Deoxy-fluorination, F-DFMBA, 90
3-Deoxy–3-fluoro-D-glucose (3-FG), 374
4-Deoxy–4-fluoro-D-glucose (4-FG), 374
Detagging, oxidative, 49
Diamines, 183
Diazabicyclooctane (DABCO), 169, 188, 204
Diazabicyclo[5.4.0]undec–7-ene (DBU), 186
4,13-Diaza–18-crown–6 ethers, 215
Diazepinomicin, 368
Diazodicarboxylates, 78
Dibenzo–18-crown–6 ether, 213, 217
Dibutyltin oxide, 75
Dichapetalum cymosum, 366
2,4-Dichloro–1,3,5-triazines, 96
Diels–Alder, fluorous thiourea, 203
Dienophile scavenger, 97
Diethylaminosulfur trifluoride (DAST), 90
Difluoroacetophenone, 40
2,2-Difluoro–1-aminocyclohexanecarboxylic
  acid, 48
3,4-Difluoroaniline, 381
Difluorocarbene, 25, 39
Difluoromethylation, nucleophilic, 32
Difluoromethyl ethers, 40
Difluoromethyl phenyl sulfone, 29, 32
Difluoromethyl 2-pyridyl sulfone, 36
Diisopropylcarbodiimide, 48
Dimethylaminopyridine (DMAP), 185
Dimethylphenyl silyl ether, 14
Dipeptides, 72
Diphenylpyrrolinol silyl ether, 180
Distannoxanes, 75
Diversity-oriented synthesis (DOS), 45, 50
1-Dodecene, hydrogenation, 236
Dodecyl lactoside, 293

E
Electrochemical sensors, 307
Electronic effects, 1, 15
Electrowetting, 307
Enones, hydrosilylation, 257
Enrofloxacin, 354
  degradation, 380
Environmental toxicity, 365, 369
Enzymatic metabolic pathways, 391
Ephedrine, 184
Epoxiconazole, 349

Index

Epoxides, nucleophilic ring-opening fluoroalkylation, 29
Ethoxylates, 383
Ethylene, hydroformylation, 282
N-Ethyl perfluorooctane sulfonamide (N-EtFOSA), 347
N-Ethyl perfluorooctane sulfonamide ethanol (N-EtFOSE), 386
Etter's urea, 193
Exophiala jeanselmei, 376

F

F-Boc, 118
  carbamate, 77
FC-70, 319
[^{18}F]-2-Fluoroethylazide, 62
F-Fmoc, 112, 122
  chloroformate, 113
F-Fmoc-O-succinimide ester, 113
F-HPLC, 106
Fipronil, 348
FLLE. See Fluorous liquid-liquid extraction (FLLE)
Fluorinase, 365, 366
Fluorinated sulfones, 25
Fluorinated sulfoximines, 25
Fluorine, 366
Fluoroacetate, 366, 371
  hydrolytic defluorination, 371
Fluoroacetate dehalogenase, 394
Fluoroaliphatics, 371
Fluoroalkylations, 25, 291
  negative fluorine effect, 28
Fluoroalkyl tag thiourea, 199
Fluoro-amphiphilic, 291
4-Fluoroaniline, 381
Fluoroaromatics, 374
Fluorobenzene, 378
Fluorobenzoates, 374
2-Fluorobenzoic acid, 374
Fluorobiphenyl metabolism, 393
Fluorobiphenyls, 376
Fluorocatechols, 374, 393
trans-2-Fluorocycloalkyl N-phenylcarbamate, 373
3-Fluorocyclohexadiene, 375
5-Fluoro-5-deoxyadenosine (5-FDA), 367
Fluoroepoxides, 35
4-Fluoroglutamate, 396
α-Fluoroglycosides, 367
3-Fluoro-3-hexenedioic acid, degradation, 378
2-Fluoro-4-hydroxyaniline, 380

3-Fluoro-2-hydroxypenta-2,4-dienoate, 376
Fluoromethylation, 27
Fluoro(phenylsulfonyl)-methyllithium, 29
Fluoromonosaccharides, 374
Fluoromuconic acids, 374
4-Fluorophenol (4-FP) monooxygenase (FpdA2), 395
Fluorophenols, 375
Fluorophenylacetic acid, 377
Fluorophilicities, 14
Fluorophosphate, 367
Fluoroponytails, 213
Fluoroprotocatechuate, 377
Fluoropyruvate, enzymatic defluorination, 371, 392
Fluoroquinolones, 339, 353, 378
α-Fluorosulfoximines, 35
Fluorotelomer alcohols (FTOHs), 345, 381
Fluorotelomer ethoxylates (FTEOs), 383
4-Fluorothreonine, 367
Fluorotoluene, 376
Fluorous based trichloroethoxycarbonyl (Froc), 118
Fluorous biphasic catalysis (FBC), 154, 156, 215
Fluorous biphasic system (FBS), 247, 250
Fluorous catalysis, 153, 213
Fluorous compounds, 1
Fluorous diastereomeric mixture synthesis (FDMS), 58
Fluorous end tagging, SPPS, 110
Fluorous ligand capture (FLC), 98
Fluorous liquid-liquid extraction (FLLE), 71, 106
Fluorous mixture synthesis (FMS), 56
Fluorous modified (diacetoxy) iodobenzenes (F-DAIB), 83
Fluorous peptide synthesis, 105
Fluorous phosphines, 247
Fluorous reagents, 69, 71
Fluorous scavengers, 69
Fluorous solid-phase extraction (FSPE), 46, 69, 71, 105
Fluorous solvents, 2, 291
  miscibilities, 9
Fluorous synthesis, 45
Fluorous tagged benzyloxycarbonyl (F-Cbz), 111, 117, 120
Fluorous tagged methylsulfonylmethoxycarbonyl (F-Msc), 112
Fluorous tags, 105
Fluorous target-oriented synthesis, 47

Fluticasone, 368
Free volume, 307
Friedel–Crafts alkylation, fluorous thiourea, 204
F-TEBA, 216, 227
F-Teoc, 120
F-TMSE, 117, 119

G

Gas sensors, 307
   electrochemical, 324
Global climate change, 343
Glutamate dehydrogenase, 396
Glutathione-S-transferase isozyme (GSTZ), 394
Glycosides, 303
Glycosylation, 291
Gold, 247
   catalysts, 261
Gold phosphine complexes, 4
Green chemistry, 135, 153, 175

H

Halons, 341
3-Halopyruvate, 371
Heptafluoro-2-hexenoic acid, 346
Hexafluorobenzene, 14
Hexahydrochromeno[4,3-b]pyrroles, 58
Hfa/Hfb, 121
HFE-7100, 50
HF elimination, 18
Human health, 365, 369
Hydrobromofluorocarbons, 341
Hydrochlorofluorocarbons (HCFCs), 341
Hydrocinnamaldehyde, 208
Hydrofluorocarbons (HFCs), 343
Hydrofluoropolyethers (HFPEs), 343
Hydroformylation, 136, 154, 156, 161, 234, 251, 275
   acrylates, 287
   olefins, 279
   styrene, 285
Hydrogen, 275
Hydrogenation, 233
   asymmetric, 243
   rhodium-catalyzed, 236
Hydrogenolysis, 390
Hydrophobicity, 307
Hydrosilylation, 247
   gold-catalyzed, 266
p-Hydroxybenzoate hydroxylase, 392

I

[^{125}I]-5-Iodo-2'-deoxyuridine, 60
[^{125}I]-meta-Iodobenzylguanidine, 60
Iododifluoroacetamides, Cu-mediated fluoroalkylation, 38
1-Iodooctane, 221
Iodooperfluoroalkane, 186
2-Iodooxybenzoic acid, 185
Isatoic anhydride, 95
Isocyanate, 95
Itopride, fluorous ammonia surrogate, 49

J

Jacobsen's Schiff base (thio)urea, 194

K

KDP 4606, 163
Ketimines, nucleophilic difluoromethylation, 37
Ketones, hydrosilylation, 257
Krytox 157FSH, 319, 324
Krytox 157FSL, 163

L

Lactate dehydrogenase, 371
Lactose, 297
Lawesson's reagent, 90
Levofloxacin, 353
Lewis acids, 76
Library scaffolds, 45
Ligand capture, 98
Lipitor, 368

M

MacMillan imidazolidinone, 178
Maleylacetate, 372
Metal sulfonamides, 76
Metal-Teflon AF nanocomposites, 330
N-Methyl-morpholine (NMM), 73
Methyl perfluorobutyl ether (Novec 7100), 147
Methyl vinyl ketone, cyclopentadiene, 203
Microarrays, 45, 62
Miscibilities, 1, 9
Mitsunobu reaction, 78
Mobile Order and Disorder (MOD) theory, 15
Monofluoroacetate dehalogenase, 370
Monofluoromethylation, nucleophilic, 33
Monophosphines, 247
Morita–Baylis–Hillman, fluorous thiourea, 203

Index

Muconates, 374
Mukaiyama's salt, 74
MultiCASE, 397

N

Nanocomposites, 333
Negative fluorine effect (NFE), 25, 28, 29
2-Nitrobenzenesulfonyl, 54
N-(2(4)-Nitrobenzenesulfonyl) α-amino acid esters, 228
Nitroolefins, 205
NMR, 1, 16
Norfloxacin, 354
Nucleophilic substitution, 213

O

Olefins, 275
   hydroformylation, 279
Oligonucleotide synthesis, 105, 123
Oligopeptides, 228
Oligosaccharides, 291
Onium salts, 214
Organic synthesis, 135
Organocatalysis, 162, 176, 191
   thiourea, 198
Organofluorines, 339
   biodegradation, 370
      thermodynamics, 390
Oxazaborolidine, 82
1,3-Oxazoles, 53
Oxidations, 83
   chemoselective, 209
Ozonolysis, 17, 320

P

Palladium, 233
Partition coefficients, 1, 10
Pentafluoroaniline, 380
Pentafluorobenzene, 14
Pentafluorobutane (Solkane 365mfc), 144
Peptide synthesis, 105, 108
Peptidic adamantane, 198
Peptidomimetics, 47
Perfluorinated carboxylic acids (PFCAs), 381
Perfluorinated compounds, 339, 365, 368
Perfluorinated phosphonic acids (PFPAs), 347
n-Perfluoroalkanes, 6
Perfluorobutyrate (PFBA), 347
Perfluorocarbons, 153
Perfluorodecanoate (PFDA), 347
Perfluoro–1,3-dimethylcyclohexane (PFDMC), 213, 215
Perfluorododecanoic acid, 383
Perfluorododecanol, 382
Perfluorohexanes, 3, 136
Perfluoromethylcyclohexane (PFMCH), 76
   solubility of oxygen, 7
n-Perfluorononane, 8
Perfluorononanyl-N-methylformamide (F-DMF), 148
n-Perfluorooctane, 8
Perfluorooctane sulfonamide (FOSA), 386
Perfluorooctanesulfonamide (PFOSA), 345
Perfluorooctane sulfonamido acetic acid (FOSAA), 386
Perfluorooctanesulfonate (PFOS), 344, 368
Perfluorooctanoic acid (PFOA), 344, 368
Perfluorooctyl 1,3-dimethylbutyl ether (F–626), 145
1-[4-(Perfluorooctyl)phenyl]–3-phenyl thiourea, 201, 204
4-Perfluorooctylphenyl thiourea, 201
Perfluoropolyether carboxylate, 163
Perfluorotriethylamine (PF-TEA), 147
Peroxisome proliferator-activated receptors (PPARs), 344
Pesticides, 339, 348
Petrocortyne A, 57
PFMC, solubility of oxygen, 7
Phase transfer catalysis, 214
Phenylboronic acid, 185
Phenyl glycidyl epoxide, 228
Phenyl selenide, 85
Phosphines, 78, 233, 275, 278
Phosphites, 278
Phosphonium salts, 214
Phosphoramidite monomers, 123
Plug-based microfluidic device, 329
Poly(ethylene glycol)s (PEGs), 215
Poly(fluoroacrylate-co-styryl-diphenylphosphine), 286
Polyfluorinated compounds, 365
   biodegradation, 381
Polyfluoroalkyl phosphates (PAPs), 385
Polystyrene, 167
Polytetrafluoroethylene (PTFE), 324
Propellants, 341
Proteomics, 64
Purification procedures, 153
Pyrethroids, 348, 352
Pyrolinol, 180

410   Index

Pyrrolidine-thiourea, 196
(S)-Pyrrolidine-thiourea bifunctional organocatalyst, 201
Pyruvate carboxylase, 371
Pyruvate dehydrogenase, 371

Q

Quantitative structure-degradation relationships (QSDRs), 397
Quaternary onium salts, 214
Quinolones, 186

R

Radiochemistry, 45, 59
Reaction solvent, 135
Reactivity, 1
Recycling, 202, 247
Reductions, 81
Reductive amination, fluorous thiourea, 206
Refrigerants, 341
Resorcylic acid lactones, 59
Retropeptides, fluorinated, 47
Reverse fluorous solid phase extraction (RFSPE), 72
Rhodium, 14, 233, 247, 258, 275
Ruthenium, 233
    fluorous phosphine, 3

S

Saccharide-amine thiourea, bifunctional, 198
Saccharide production, 291
Salmeterol, 368
Scavengers, 69, 91
    nucleophilic, 92
Schreiner's thiourea, 199
Selenium compounds, 85
Sensors, 320
Separation techniques, 153
Silicon-tethered diols, 55
Sodium naphthalenide, 16
Solid/liquid phase separation, 164
Solid/liquid phase transfer catalysis (SL-PTC), 216, 226
Solid phase peptide synthesis (SPPS), 108
Solubilities, 1, 7
Solution phase syntheses, 117, 130
Soós-Connon bifunctional cinchona alkaloid-thiourea, 197
Streptomyces cattleya, 367

Styrene, hydroformylation, 285
    hydrogenation, 239
Styrene oxides, alcoholysis, Schreiner thiourea, 207
Sulfonamide, 184
Sulfones, fluorinated, 25
    gem-difluoroolefination, 36
Sulfoxidation, Schreiner thiourea, 209
Sulfoxides, 86
Sulfoximines, fluorinated, 25
Superhydrophobic surfaces, 325
Swern reaction, 88
Syntrophus aciditrophicus, 375

T

Tags, 46, 105
Takemoto's bifunctional chiral thiourea, 195
Tang's chiral bifunctional pyrrolidine-thiourea, 196
Teflon, 8, 16
    AF, 307
TEMPO, 87
Tetrafluoroethylene (TFE), 16, 334
Thiazoles, 53
Thioisocyanates, 95
Thiols, 93
Thiouracils, 53
Thiourea, 183, 191, 194
    bifunctional chiral, Takemoto, 195
    fluoroalkyl tag, 199
    organocatalysts, 198
    Schreiner's, 199
Thiourea-binaphthyl, bifunctional, Wang, 196
Thiourea-cinchona alkaloid, 197
Thiourea-pyrrolidine, chiral bifunctional, Tang, 196
Thyrotropin-releasing hormone (TRH), 121
Tin hydrides, 81
Tin oxides, 75
Toluene vapor, sensors, 323
Toxicity, 339, 344, 365, 369
Transesterification, 75
Transfer hydrogenation, 234
Transport, 307
Triazines, 72
Tributyltin hydride, 81
Trifluoroacetamide, ethyl 2-bromopropanoate, 228
Trifluoroacetic acid, 343, 370, 394
10-(Trifluoromethoxy) decane–1-sulfonate, 386

Index 411

Trifluoromethylbenzene, 15
Trifluoromethylcyclohexane (TFMC), 143
Trifluoromethyltrimethylsilane (TMSCF3), 27, 41
Trifluralin, 348, 351
Trimethylsilyl cyanide (TMSCN), 205
2-(Trimethylsilyl)ethanol, 47
Triphenylphosphines, 242

U
Uracils, 53
Urea, hydrogen bonds, 193

Urethanes, fluorotelomer-based, 385

V
Vapor sensors, sorption-based, 322
Vibrational circular dichroism (VCD), 3

W
Wang's bifunctional binaphthyl-thiourea, 196
Waveguides, 307
　　sensors, 321

Printed by Publishers' Graphics LLC USA
MO20120306-195
2012